Instructor's Solutions Manual

Srdjan Divac
Harvard University
LTS

Linear Algebra
With Applications

Otto Bretscher

PRENTICE HALL, Upper Saddle River, NJ 07458

Supplement Editor: Audra Walsh
Special Projects Manager: Barbara A. Murray
Production Editor: Shea Oakley
Supplement Cover Manager: Paul Gourhan
Supplement Cover Designer: Liz Nemeth
Manufacturing Buyer: Alan Fischer

Simon & Schuster / A Viacom Company
Upper Saddle River, NJ 07458

Printed in the United States of America

10 9 8 7 6 5 4 3 2 1

ISBN 0-13-576265-0

Prentice-Hall International (UK) Limited, *London*
Prentice-Hall of Australia Pty. Limited, *Sydney*
Prentice-Hall Canada, Inc., *London*
Prentice-Hall Hispanoamericana, S.A., *Mexico*
Prentice-Hall of India Private Limited, *New Delhi*
Prentice-Hall of Japan, Inc., *Tokyo*
Simon & Schuster Asia Pte. Ltd., *Singapore*
Editora Prentice-Hall do Brazil, Ltda., *Rio de Janeiro*

Chapter 1

1.1

1. $$\left|\begin{array}{r} x+2y=1 \\ 2x+3y=1 \end{array}\right| \xrightarrow[-2\times\text{2nd equation}]{} \left|\begin{array}{r} x+2y=1 \\ -y=-1 \end{array}\right| \xrightarrow[\div(-1)]{}$$

$$\left|\begin{array}{r} x+2y=1 \\ y=1 \end{array}\right| \xrightarrow{-2\times\text{2nd equation}} \left|\begin{array}{l} x=-1 \\ y=1 \end{array}\right|,$$ so that $(x, y) = (-1, 1)$.

2. $$\left|\begin{array}{r} 4x+3y=2 \\ 7x+5y=3 \end{array}\right| \begin{array}{l} \div 4 \\ \\ \end{array} \longrightarrow \left|\begin{array}{r} x+\frac{3}{4}y=\frac{1}{2} \\ 7x+5y=3 \end{array}\right| \xrightarrow[-7\times\text{1st equation}]{}$$

$$\left|\begin{array}{r} x+\frac{3}{4}y=\frac{1}{2} \\ -\frac{1}{4}y=-\frac{1}{2} \end{array}\right| \begin{array}{l} \\ \times -4 \end{array} \longrightarrow \left|\begin{array}{r} x+\frac{3}{4}y=\frac{1}{2} \\ y=2 \end{array}\right| \xrightarrow{-\frac{3}{4}\times\text{2nd equation}} \left|\begin{array}{l} x=-1 \\ y=2 \end{array}\right|,$$ so that $(x, y) = (-1, 2)$.

3. $$\left|\begin{array}{r} 2x+4y=3 \\ 3x+6y=2 \end{array}\right| \begin{array}{l} \div 2 \\ \\ \end{array} \longrightarrow \left|\begin{array}{r} x+2y=\frac{3}{2} \\ 3x+6y=2 \end{array}\right| \xrightarrow[-3\times\text{1st equation}]{} \left|\begin{array}{r} x+2y=\frac{3}{2} \\ 0=-\frac{5}{2} \end{array}\right|$$

There is no solution.

4. $$\left|\begin{array}{r} 2x+4y=2 \\ 3x+6y=3 \end{array}\right| \begin{array}{l} \div 2 \\ \\ \end{array} \longrightarrow \left|\begin{array}{r} x+2y=1 \\ 3x+6y=3 \end{array}\right| \xrightarrow[-3\times\text{1st equation}]{} \left|\begin{array}{r} x+2y=1 \\ 0=0 \end{array}\right|$$

This system has infinitely many solutions: if we choose $y = t$, an arbitrary real number, then the equation $x+2y = 1$ gives us $x = 1-2y = 1-2t$. Therefore the general solution is $(x, y) = (1-2t, t)$, where t is an arbitrary real number.

5. $$\left|\begin{array}{r} 2x+3y=0 \\ 4x+5y=0 \end{array}\right| \begin{array}{l} \div 2 \\ \\ \end{array} \longrightarrow \left|\begin{array}{r} x+\frac{3}{2}y=0 \\ 4x+5y=0 \end{array}\right| \xrightarrow[-4\times\text{1st equation}]{}$$

$$\left|\begin{array}{r} x+\frac{3}{2}y=0 \\ -y=0 \end{array}\right| \xrightarrow[\div(-1)]{} \left|\begin{array}{r} x+\frac{3}{2}y=0 \\ y=0 \end{array}\right| \xrightarrow{-\frac{3}{2}\times\text{2nd equation}} \left|\begin{array}{l} x=0 \\ y=0 \end{array}\right|,$$ so that $(x, y) = (0, 0)$.

6. $$\left|\begin{array}{r} x+2y+3z=8 \\ x+3y+3z=10 \\ x+2y+4z=9 \end{array}\right| \begin{array}{l} \\ -\text{I} \\ -\text{I} \end{array} \longrightarrow \left|\begin{array}{r} x+2y+3z=8 \\ y=2 \\ z=1 \end{array}\right| \begin{array}{l} -2(\text{II}) \\ \\ \\ \end{array} \longrightarrow$$

$$\left|\begin{array}{r} x+3z=4 \\ y=2 \\ z=1 \end{array}\right| \begin{array}{l} -3(\text{III}) \\ \\ \\ \end{array} \longrightarrow \left|\begin{array}{l} x=1 \\ y=2 \\ z=1 \end{array}\right|,$$ so that $(x, y, z) = (1, 2, 1)$.

7. $$\left|\begin{array}{r} x+2y+3z=1 \\ x+3y+4z=3 \\ x+4y+5z=4 \end{array}\right| \begin{array}{l} \\ -\text{I} \\ -\text{I} \end{array} \longrightarrow \left|\begin{array}{r} x+2y+3z=1 \\ y+z=2 \\ 2y+2z=3 \end{array}\right| \begin{array}{l} -2(\text{II}) \\ \\ -2(\text{II}) \end{array} \longrightarrow \left|\begin{array}{r} x+z=-3 \\ y+z=2 \\ 0=-1 \end{array}\right|.$$

This system has no solution.

8. $\left| \begin{array}{r} x+2y+3z=0 \\ 4x+5y+6z=0 \\ 7x+8y+10z=0 \end{array} \right| \begin{array}{l} \\ -4(\text{I}) \\ -7(\text{I}) \end{array} \longrightarrow \left| \begin{array}{r} x+2y+3z=0 \\ -3y-6z=0 \\ -6y-11z=0 \end{array} \right| \begin{array}{l} \\ \div(-3) \\ \\ \end{array} \longrightarrow \left| \begin{array}{r} x+2y+3z=0 \\ y+2z=0 \\ -6y-11z=0 \end{array} \right| \begin{array}{l} -2(\text{II}) \\ \\ +6(\text{II}) \end{array} \longrightarrow$

$\left| \begin{array}{r} x-z=0 \\ y+2z=0 \\ z=0 \end{array} \right| \begin{array}{l} +\text{III} \\ -2(\text{III}) \\ \\ \end{array} \longrightarrow \left| \begin{array}{l} x=0 \\ y=0 \\ z=0 \end{array} \right|$, so that $(x, y, z) = (0, 0, 0)$.

9. $\left| \begin{array}{r} x+2y+3z=1 \\ 3x+2y+z=1 \\ 7x+2y-3z=1 \end{array} \right| \begin{array}{l} \\ -3(\text{I}) \\ -7(\text{I}) \end{array} \longrightarrow \left| \begin{array}{r} x+2y+3z=1 \\ -4y-8z=-2 \\ -12y-24z=-6 \end{array} \right| \begin{array}{l} \\ \div(-4) \\ \\ \end{array} \longrightarrow \left| \begin{array}{r} x+2y+3z=1 \\ y+2z=\frac{1}{2} \\ -12y-24z=-6 \end{array} \right| \begin{array}{l} -2(\text{II}) \\ \\ +12(\text{II}) \end{array} \longrightarrow$

$\left| \begin{array}{r} x-z=0 \\ y+2z=\frac{1}{2} \\ 0=0 \end{array} \right|$

This system has infinitely many solutions: if we choose $z=t$, an arbitrary real number, then we get $x=z=t$ and $y=\frac{1}{2}-2z=\frac{1}{2}-2t$. Therefore, the general solution is $(x, y, z)=\left(t, \frac{1}{2}-2t, t\right)$, where t is an arbitrary real number.

10. $\left| \begin{array}{r} x+2y+3z=1 \\ 2x+4y+7z=2 \\ 3x+7y+11z=8 \end{array} \right| \begin{array}{l} \\ -2(\text{I}) \\ -3(\text{I}) \end{array} \longrightarrow \left| \begin{array}{r} x+2y+3z=1 \\ z=0 \\ y+2z=5 \end{array} \right| \begin{array}{l} \\ (\text{II}) \leftrightarrow (\text{III}) \\ \\ \end{array} \longrightarrow \left| \begin{array}{r} x+2y+3z=1 \\ y+2z=5 \\ z=0 \end{array} \right| \begin{array}{l} -2(\text{II}) \\ \\ \\ \end{array} \longrightarrow$

$\left| \begin{array}{r} x-z=-9 \\ y+2z=5 \\ z=0 \end{array} \right| \begin{array}{l} +(\text{III}) \\ -2(\text{III}) \\ \\ \end{array} \longrightarrow \left| \begin{array}{l} x=-9 \\ y=5 \\ z=0 \end{array} \right|$, so that $(x, y, z) = (-9, 5, 0)$.

11. $\left| \begin{array}{r} x-2y=2 \\ 3x+5y=17 \end{array} \right| \begin{array}{l} \\ -3(\text{I}) \end{array} \longrightarrow \left| \begin{array}{r} x-2y=2 \\ 11y=11 \end{array} \right| \begin{array}{l} \\ \div 11 \end{array} \longrightarrow \left| \begin{array}{r} x-2y=2 \\ y=1 \end{array} \right| \begin{array}{l} +2(\text{II}) \\ \\ \end{array} \longrightarrow \left| \begin{array}{l} x=4 \\ y=1 \end{array} \right|$, so that $(x, y) = (4, 1)$.

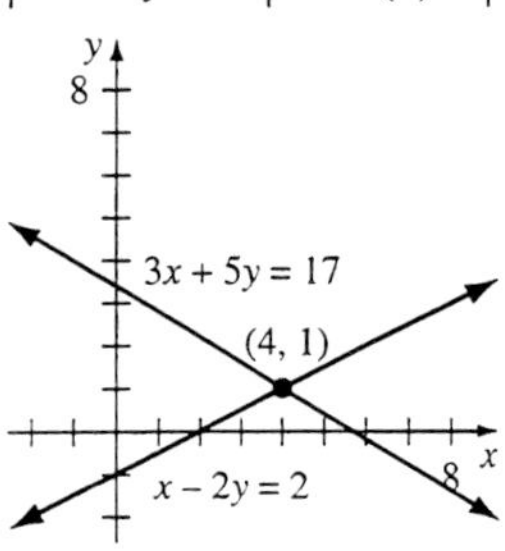

12. $\begin{vmatrix} x-2y=3 \\ 2x-4y=6 \end{vmatrix} \xrightarrow{-2(\text{I})} \begin{vmatrix} x-2y=3 \\ 0=0 \end{vmatrix}$

This system has infinitely many solutions: If we choose $y = t$, an arbitrary real number, then the equation $x - 2y = 3$ gives us $x = 3 + 2y = 3 + 2t$. Therefore the general solution is $(x, y) = (3 + 2t, t)$, where t is an arbitrary real number.

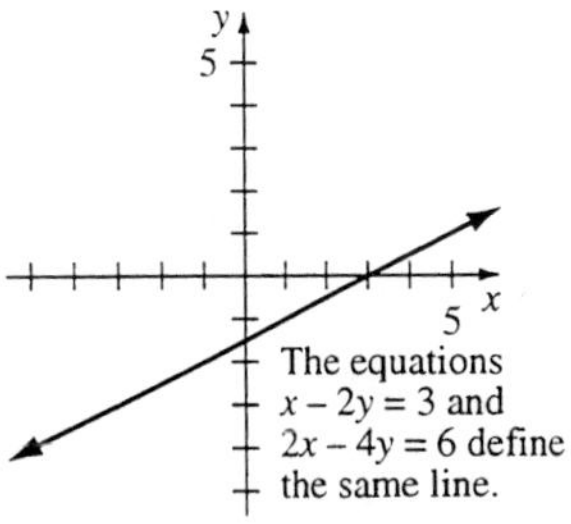

13. $\begin{vmatrix} x-2y=3 \\ 2x-4y=8 \end{vmatrix} \xrightarrow{-2(\text{I})} \begin{vmatrix} x-2y=3 \\ 0=2 \end{vmatrix}$

This system has no solution.

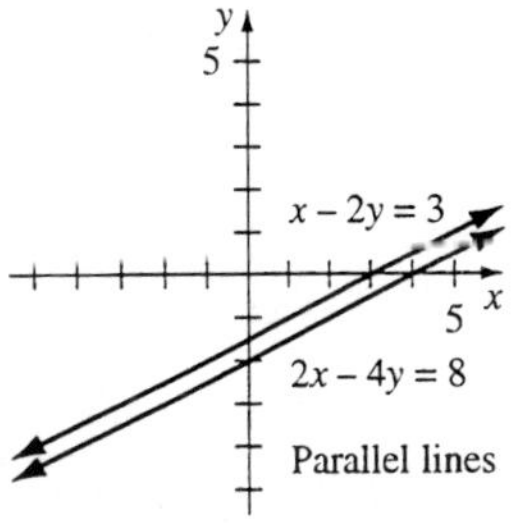

14. The system reduces to $\begin{vmatrix} x+5z=0 \\ y-z=0 \\ 0=1 \end{vmatrix}$, so that there is no solution; no point in space belongs to all three planes.

Compare with Figure 2b.

15. The system reduces to $\begin{vmatrix} x=0 \\ y=0 \\ z=0 \end{vmatrix}$ so the unique solution is $(x, y, z) = (0, 0, 0)$. The three planes intersect at the origin.

16. The system reduces to $\begin{vmatrix} x+5z=0 \\ y-z=0 \\ 0=0 \end{vmatrix}$, so the solutions are of the form $(x, y, z) = (-5t, t, t)$, where t is an arbitrary number. The three planes intersect in a line; compare with Figure 2a.

17. $\begin{vmatrix} x+2y=a \\ 3x+5y=b \end{vmatrix} \xrightarrow{-3(\text{I})} \begin{vmatrix} x+2y=a \\ -y=-3a+b \end{vmatrix} \xrightarrow{\div(-1)} \begin{vmatrix} x+2y=a \\ y=3a-b \end{vmatrix} \xrightarrow{-2(\text{II})}$

$\begin{vmatrix} x=-5a+2b \\ y=3a-b \end{vmatrix}$, so that $(x, y) = (-5a + 2b, 3a - b)$.

18. $\begin{vmatrix} x+2y+3z=a \\ x+3y+8z=b \\ x+2y+2z=c \end{vmatrix} \begin{matrix} \longrightarrow \\ -\text{I} \\ -\text{I} \end{matrix} \begin{vmatrix} x+2y+3z=a \\ y+5z=-a+b \\ -z=-a+c \end{vmatrix} \begin{matrix} -2(\text{II}) \\ \longrightarrow \\ \end{matrix} \begin{vmatrix} x-7z=3a-2b \\ y+5z=-a+b \\ -z=-a+c \end{vmatrix} \begin{matrix} \\ \longrightarrow \\ \div(-1) \end{matrix}$

$\begin{vmatrix} x-7z=3a-2b \\ y+5z=-a+b \\ z=a-c \end{vmatrix} \begin{matrix} +7(\text{III}) \\ -5(\text{III}) \\ \longrightarrow \end{matrix} \begin{vmatrix} x=10a-2b-7c \\ y=-6a+b+5c \\ z=a-c \end{vmatrix}$, so that $(x, y, z) = (10a-2b-7c, -6a+b+5c, a-c)$

19. a. Note that the demand D_1 for product 1 increases with the increase of price P_2; likewise the demand D_2 for product 2 increases with the increase of price P_1. This indicates that the two products are competing; some people will switch if one of the products gets more expensive.

b. Setting $D_1 = S_1$ and $D_2 = S_2$ we obtain the system $\begin{vmatrix} 70-2P_1+P_2=-14+3P_1 \\ 105+P_1-P_2=-7+2P_2 \end{vmatrix}$, or $\begin{vmatrix} -5P_1+P_2=-84 \\ P_1-3P_2=-112 \end{vmatrix}$, which yields the unique solution $P_1 = 26$ and $P_2 = 46$.

20. The total demand for the product of Industry A is 1000 (the consumer demand) plus $0.1b$ (the demand from Industry B). The output a must meet this demand: $a = 1000 + 0.1b$.

Setting up a similar equation for Industry B we obtain the system $\begin{vmatrix} a=1000+0.1b \\ b=780+0.2a \end{vmatrix}$ or $\begin{vmatrix} a-0.1b=1000 \\ -0.2a+b=780 \end{vmatrix}$, which yields the unique solution $a = 1100$ and $b = 1000$.

21. The total demand for the products of Industry A is 310 (the consumer demand) plus $0.3b$ (the demand from Industry B). The output a must meet this demand: $a = 310 + 0.3b$.

Setting up a similar equation for Industry B we obtain the system $\begin{vmatrix} a=310+0.3b \\ b=100+0.5a \end{vmatrix}$ or $\begin{vmatrix} a-0.3b=310 \\ -0.5a+b=100 \end{vmatrix}$, which yields the solution $a = 400$ and $b = 300$.

22. Since $x(t) = a\sin(t) + b\cos(t)$ we can compute $\frac{dx}{dt} = a\cos(t) - b\sin(t)$ and $\frac{d^2x}{dt^2} = -a\sin(t) - b\cos(t)$.

Substituting these expressions into the equation $\frac{d^2x}{dt^2} - \frac{dx}{dt} - x = \cos(t)$ and simplifying gives $(b-2a)\sin(t) + (-a-2b)\cos(t) = \cos(t)$.

Comparing the coefficients of $\sin(t)$ and $\cos(t)$ on both sides of the equation then yields the system $\begin{vmatrix} -2a+b=0 \\ -a-2b=1 \end{vmatrix}$, so that $a = -\frac{1}{5}$ and $b = -\frac{2}{5}$.

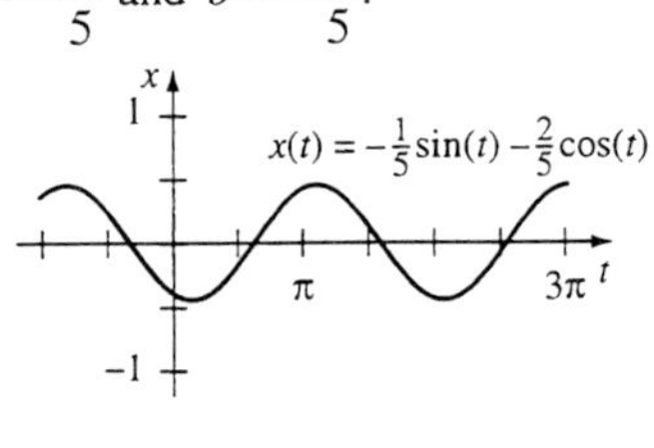

23. a. Substituting $\lambda = 5$ yields the system $\begin{vmatrix} 7x-y=5x \\ -6x+8y=5y \end{vmatrix}$ or $\begin{vmatrix} 2x-y=0 \\ -6x+3y=0 \end{vmatrix}$ or $\begin{vmatrix} 2x-y=0 \\ 0=0 \end{vmatrix}$.

There are infinitely many solutions, of the form $(x, y) = \left(\frac{t}{2}, t\right)$, where t is an arbitrary real number.

b. Proceeding as in part (a), we find $(x, y) = \left(-\frac{1}{3}t, t\right)$.

c. Proceeding as in part (a), we find only the solution (0, 0).

24. Let v be the speed of the boat relative to the water, and s be the speed of the stream; then the speed of the boat relative to the land is $v + s$ downstream and $v - s$ upstream.

Using the fact that (distance) = (speed)(time), we obtain the system $\left| \begin{array}{l} 8 = (v+s)\frac{1}{3} \\ 8 = (v-s)\frac{2}{3} \end{array} \right| \begin{array}{l} \leftarrow \text{downstream} \\ \leftarrow \text{upstream} \end{array}$

The solution is $v = 18$ and $s = 6$.

25. The system reduces to $\left| \begin{array}{r} x + z = 1 \\ y - 2z = -3 \\ 0 = k - 7 \end{array} \right|$.

a. The system has solutions if $k - 7 = 0$, or $k = 7$.

b. If $k = 7$ then the system has infinitely many solutions.

c. If $k = 7$ then we can choose $z = t$ freely and obtain the solutions $(x, y, z) = (1 - t, -3 + 2t, t)$.

26. The system reduces to $\left| \begin{array}{r} x - 3z = 1 \\ y + 2z = 1 \\ (k^2 - 4)z = k - 2 \end{array} \right|$.

This system has a unique solution if $k^2 - 4 \neq 0$, that is, if $k \neq \pm 2$.
If $k = 2$ the last equation is $0 = 0$, and there will be infinitely many solutions.
If $k = -2$ the last equation is $0 = -4$, and there will be no solutions.

27. Let x = the number of male children and y = the number of female children.
Then the statement "Emile has twice as many sisters as brothers" translates into $y = 2(x - 1)$ and "Gertrude has as many brothers as sisters" translates into $x = y - 1$.

Solving the system $\left| \begin{array}{r} -2x + y = -2 \\ x - y = -1 \end{array} \right|$ gives $x = 3$ and $y = 4$.

There are seven children in this family.

28. The thermal equilibrium condition requires that $T_1 = \frac{T_2 + 200 + 0 + 0}{4}$, $T_2 = \frac{T_1 + T_3 + 200 + 0}{4}$, and $T_3 = \frac{T_2 + 400 + 0 + 0}{4}$.

We can rewrite this system as $\left| \begin{array}{r} -4T_1 + T_2 = -200 \\ T_1 - 4T_2 + T_3 = -200 \\ T_2 - 4T_3 = -400 \end{array} \right|$.

The solution is $(T_1, T_2, T_3) = (75, 100, 125)$.

29. To assure that the graph goes through the point $(1, -1)$, we substitute $t = 1$ and $f(t) = -1$ into the equation $f(t) = a + bt + ct^2$ to give $-1 = a + b + c$.

Proceeding likewise for the two other points, we obtain the system $\begin{vmatrix} a+b+c=-1 \\ a+2b+4c=3 \\ a+3b+9c=13 \end{vmatrix}$.

The solution is $a = 1$, $b = -5$, and $c = 3$, and the polynomial is $f(t) = 1 - 5t + 3t^2$.

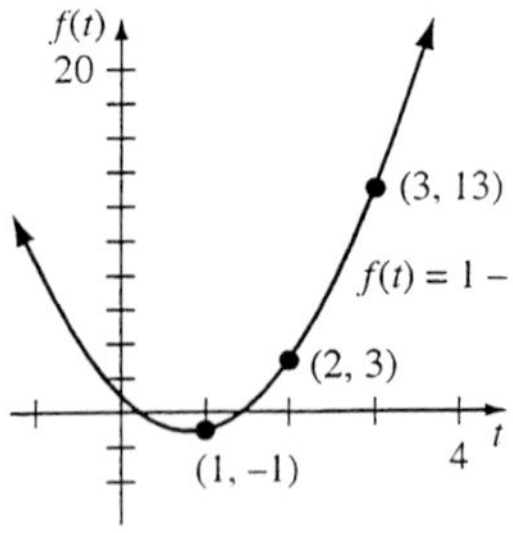

30. Proceeding as in the previous exercise, we obtain the system $\begin{vmatrix} a+b+c=p \\ a+2b+4c=q \\ a+3b+9c=r \end{vmatrix}$.

The unique solution is $\begin{vmatrix} a = 3p - 3q + r \\ b = -2.5p + 4q - 1.5r \\ c = 0.5p - q + 0.5r \end{vmatrix}$.

Only one polynomial of degree 2 goes through the three given points, namely,
$f(t) = 3p - 3q + r + (-2.5p + 4q - 1.5r)t + (0.5p - q + 0.5r)t^2$.

31. The given system reduces to $\begin{vmatrix} x - z = \frac{-5a+2b}{3} \\ y + 2z = \frac{4a-b}{3} \\ 0 = a - 2b + c \end{vmatrix}$.

This system has solutions (in fact infinitely many) if $a - 2b + c = 0$.
The points (a, b, c) with this property form a plane through the origin.

32. a. $x_1 = -3$
$x_2 = 14 + 3x_1 = 14 + 3(-3) = 5$
$x_3 = 9 - x_1 - 2x_2 = 9 + 3 - 10 = 2$
$x_4 = 33 + x_1 - 8x_2 + 5x_3 - x_4 = 33 - 3 - 40 + 10 = 0$,
so that $(x_1, x_2, x_3, x_4) = (-3, 5, 2, 0)$.

b. $x_4 = 0$
$x_3 = 2 - 2x_4 = 2$
$x_2 = 5 - 3x_3 - 7x_4 = 5 - 6 = -1$
$x_1 = -3 - 2x_2 + x_3 - 4x_4 = -3 + 2 + 2 = 1$,
so that $(x_1, x_2, x_3, x_4) = (1, -1, 2, 0)$

33. **a.**

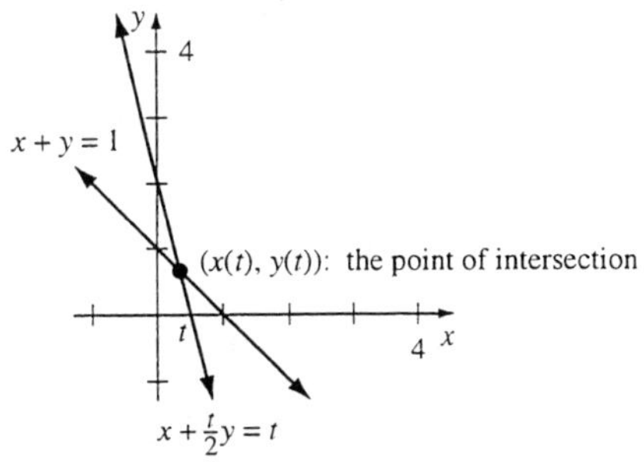

The two lines intersect unless $t = 2$ (in which case both lines have slope -1).
To draw a rough sketch of $x(t)$, note that

$\lim_{t\to\infty} x(t) = \lim_{t\to-\infty} x(t) = -1$ (the line $x + \frac{t}{2}y = t$ becomes almost horizontal)

and

$\lim_{t\to 2^-} x(t) = \infty$, $\lim_{t\to 2^+} x(t) = -\infty$.

Also note that $x(t)$ is positive if t is between 0 and 2, and negative otherwise.
Apply similar reasoning to $y(t)$.

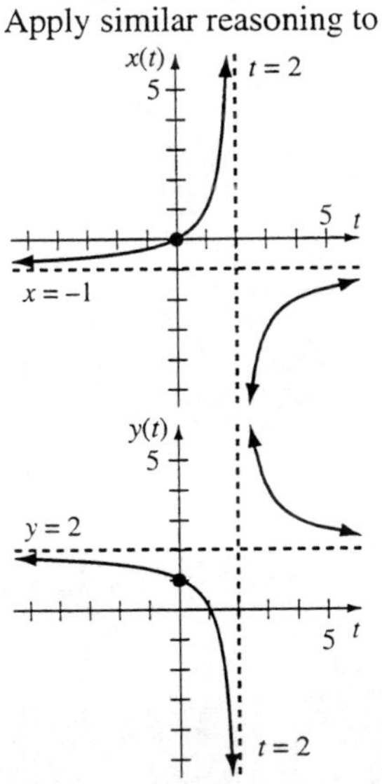

b. $x(t) = \dfrac{-t}{t-2}$, and $y(t) = \dfrac{2t-2}{t-2}$

34. Let x = number of sheep, y = number of goats, and z = number of hogs.

We are told that $\begin{vmatrix} x+y+z=100 \\ \frac{1}{2}x+\frac{4}{3}y+\frac{7}{2}z=100 \end{vmatrix}$.

This system reduces to $\begin{vmatrix} x-2.6z=40 \\ y+3.6z=60 \end{vmatrix}$.

The solutions of this system are of the form $(x, y, z) = (40 + 2.6t, 60 - 3.6t, t)$, where t is an arbitrary real number. In terms of the given word problem, it is required that x, y and z be positive integers (or zero). Therefore t (= z) must be a positive integer (or zero); in fact, t must be a multiple of 5 so that $40 + 2.6t$ and $60 - 3.6t$ are integers. Finally, $60 - 3.6t$ must be positive or zero; this leaves only the possibilities $t = 0, 5, 10$, and 15.

Solutions:

# sheep	# goats	# hogs
40	60	0
53	42	5
66	24	10
79	6	15

35. To eliminate the arbitrary constant t, we can solve the last equation for t to give $t = z - 2$, and substitute $z - 2$ for t in the first two equations, obtaining $\begin{vmatrix} x = 6 + 5(z-2) \\ y = 4 + 3(z-2) \end{vmatrix}$ or $\begin{vmatrix} x - 5z = -4 \\ y - 3z = -2 \end{vmatrix}$.

This system does the job.

36. Let b = Boris' money, m = Marina's money, and c = cost of a chocolate bar.

We are told that $\begin{vmatrix} b + \frac{1}{2}m = c \\ \frac{1}{2}b + m = 2c \end{vmatrix}$, with solution $(b, m) = (0, 2c)$.

Boris has no money.

37. Let us start by reducing the system:

$$\begin{vmatrix} x + 2y + 3z = 39 \\ x + 3y + 2z = 34 \\ 3x + 2y + z = 26 \end{vmatrix} \begin{matrix} \longrightarrow \\ -\text{I} \\ -3(\text{I}) \end{matrix} \begin{vmatrix} x + 2y + 3z = 39 \\ y - z = -5 \\ -4y - 8z = -91 \end{vmatrix}$$

Note that the last two equations are exactly those we get when we substitute $x = 39 - 2y - 3z$: either way, we end up with the system $\begin{vmatrix} y - z = -5 \\ -4y - 8z = -91 \end{vmatrix}$.

1.2

1. $\left[\begin{array}{ccc|c} 1 & 1 & -2 & 5 \\ 2 & 3 & 4 & 2 \end{array}\right] \xrightarrow{\;\;-2(\text{I})} \left[\begin{array}{ccc|c} 1 & 1 & -2 & 5 \\ 0 & 1 & 8 & -8 \end{array}\right] \xrightarrow{-\text{II}} \left[\begin{array}{ccc|c} 1 & 0 & -10 & 13 \\ 0 & 1 & 8 & -8 \end{array}\right]$

$\begin{vmatrix} x - 10z = 13 \\ y + 8z = -8 \end{vmatrix} \rightarrow \begin{vmatrix} x = 13 + 10z \\ y = -8 - 8z \end{vmatrix}$

$\begin{bmatrix} x \\ y \\ z \end{bmatrix} = \begin{bmatrix} 13 + 10t \\ -8 - 8t \\ t \end{bmatrix}$, where t is an arbitrary real number.

2. $\left[\begin{array}{ccc|c} 3 & 4 & -1 & 8 \\ 6 & 8 & -2 & 3 \end{array}\right] \xrightarrow{\div 3} \left[\begin{array}{ccc|c} 1 & \frac{4}{3} & -\frac{1}{3} & \frac{8}{3} \\ 6 & 8 & -2 & 3 \end{array}\right] \xrightarrow{\;\;-6(\text{I})} \left[\begin{array}{ccc|c} 1 & \frac{4}{3} & -\frac{1}{3} & \frac{8}{3} \\ 0 & 0 & 0 & -13 \end{array}\right]$

This system has no solutions, since the last row represents the equation $0 = -13$.

3. $x = 4 - 2y - 3z$

y and z are free variables; let $y = s$ and $z = t$.

$\begin{bmatrix} x \\ y \\ z \end{bmatrix} = \begin{bmatrix} 4 - 2s - 3t \\ s \\ t \end{bmatrix}$, where s and t are arbitrary real numbers.

4. $\left[\begin{array}{cc|c} 1 & 1 & 1 \\ 2 & -1 & 5 \\ 3 & 4 & 2 \end{array}\right] \begin{array}{l} \\ -2(\text{I}) \\ -3(\text{I}) \end{array} \longrightarrow \left[\begin{array}{cc|c} 1 & 1 & 1 \\ 0 & -3 & 3 \\ 0 & 1 & -1 \end{array}\right] \begin{array}{l} \\ \div(-3) \\ \\ \end{array} \longrightarrow \left[\begin{array}{cc|c} 1 & 1 & 1 \\ 0 & 1 & -1 \\ 0 & 1 & -1 \end{array}\right] \begin{array}{l} -\text{II} \\ \\ -\text{II} \end{array} \longrightarrow \left[\begin{array}{cc|c} 1 & 0 & 2 \\ 0 & 1 & -1 \\ 0 & 0 & 0 \end{array}\right]$, so that $\begin{array}{l} x = 2 \\ y = -1 \end{array}$.

5. $\left[\begin{array}{cccc|c} 0 & 0 & 1 & 1 & 0 \\ 0 & 1 & 1 & 0 & 0 \\ 1 & 1 & 0 & 0 & 0 \\ 1 & 0 & 0 & 1 & 0 \end{array}\right] \xrightarrow{\text{I} \leftrightarrow \text{III}} \left[\begin{array}{cccc|c} 1 & 1 & 0 & 0 & 0 \\ 0 & 1 & 1 & 0 & 0 \\ 0 & 0 & 1 & 1 & 0 \\ 1 & 0 & 0 & 1 & 0 \end{array}\right] \begin{array}{l} \\ \\ \\ -\text{I} \end{array} \longrightarrow \left[\begin{array}{cccc|c} 1 & 1 & 0 & 0 & 0 \\ 0 & 1 & 1 & 0 & 0 \\ 0 & 0 & 1 & 1 & 0 \\ 0 & -1 & 0 & 1 & 0 \end{array}\right] \begin{array}{l} -\text{II} \\ \\ \\ +\text{II} \end{array} \longrightarrow$

$\left[\begin{array}{cccc|c} 1 & 0 & -1 & 0 & 0 \\ 0 & 1 & 1 & 0 & 0 \\ 0 & 0 & 1 & 1 & 0 \\ 0 & 0 & 1 & 1 & 0 \end{array}\right] \begin{array}{l} +\text{III} \\ -\text{III} \\ \\ -\text{III} \end{array} \longrightarrow \left[\begin{array}{cccc|c} 1 & 0 & 0 & 1 & 0 \\ 0 & 1 & 0 & -1 & 0 \\ 0 & 0 & 1 & 1 & 0 \\ 0 & 0 & 0 & 0 & 0 \end{array}\right]$

$\left| \begin{array}{rrrrrrr} x_1 & & & + & x_4 & = & 0 \\ & x_2 & & - & x_4 & = & 0 \\ & & x_3 & + & x_4 & = & 0 \end{array} \right| \rightarrow \left| \begin{array}{l} x_1 = -x_4 \\ x_2 = x_4 \\ x_3 = -x_4 \end{array} \right|$

$\begin{bmatrix} x_1 \\ x_2 \\ x_3 \\ x_4 \end{bmatrix} = \begin{bmatrix} -t \\ t \\ -t \\ t \end{bmatrix}$, where t is an arbitrary real number.

6. The system is in rref already.

$\left| \begin{array}{l} x_1 = 3 + 7x_2 - x_5 \\ x_3 = 2 + 2x_5 \\ x_4 = 1 - x_5 \end{array} \right|$

Let $x_2 = t$ and $x_5 = r$.

$\begin{bmatrix} x_1 \\ x_2 \\ x_3 \\ x_4 \\ x_5 \end{bmatrix} = \begin{bmatrix} 3 + 7t - r \\ t \\ 2 + 2r \\ 1 - r \\ r \end{bmatrix}$

7. $\left[\begin{array}{ccccc|c} 1 & 2 & 0 & 2 & 3 & 0 \\ 0 & 0 & 1 & 3 & 2 & 0 \\ 0 & 0 & 1 & 4 & -1 & 0 \\ 0 & 0 & 0 & 0 & 1 & 0 \end{array}\right] \begin{array}{l} \\ \\ -\text{II} \\ \\ \end{array} \longrightarrow \left[\begin{array}{ccccc|c} 1 & 2 & 0 & 2 & 3 & 0 \\ 0 & 0 & 1 & 3 & 2 & 0 \\ 0 & 0 & 0 & 1 & -3 & 0 \\ 0 & 0 & 0 & 0 & 1 & 0 \end{array}\right] \begin{array}{l} -2(\text{III}) \\ -3(\text{III}) \\ \\ \\ \end{array} \longrightarrow$

$\left[\begin{array}{ccccc|c} 1 & 2 & 0 & 0 & 9 & 0 \\ 0 & 0 & 1 & 0 & 11 & 0 \\ 0 & 0 & 0 & 1 & -3 & 0 \\ 0 & 0 & 0 & 0 & 1 & 0 \end{array}\right] \begin{array}{l} -9(\text{IV}) \\ -11(\text{IV}) \\ +3(\text{IV}) \\ \\ \end{array} \longrightarrow \left[\begin{array}{ccccc|c} 1 & 2 & 0 & 0 & 0 & 0 \\ 0 & 0 & 1 & 0 & 0 & 0 \\ 0 & 0 & 0 & 1 & 0 & 0 \\ 0 & 0 & 0 & 0 & 1 & 0 \end{array}\right]$

$\left| \begin{array}{r} x_1 + 2x_2 = 0 \\ x_3 = 0 \\ x_4 = 0 \\ x_5 = 0 \end{array} \right| \rightarrow \left| \begin{array}{l} x_1 = -2x_2 \\ x_3 = 0 \\ x_4 = 0 \\ x_5 = 0 \end{array} \right|$

Let $x_2 = t$.

$$\begin{bmatrix} x_1 \\ x_2 \\ x_3 \\ x_4 \\ x_5 \end{bmatrix} = \begin{bmatrix} -2t \\ t \\ 0 \\ 0 \\ 0 \end{bmatrix}$$, where t is an arbitrary real number.

8. $$\left[\begin{array}{ccccc|c} 0 & 1 & 0 & 2 & 3 & 0 \\ 0 & 0 & 0 & 4 & 8 & 0 \end{array}\right] \xrightarrow[\div 4]{} \left[\begin{array}{ccccc|c} 0 & 1 & 0 & 2 & 3 & 0 \\ 0 & 0 & 0 & 1 & 2 & 0 \end{array}\right] \xrightarrow{-2(\text{II})} \left[\begin{array}{ccccc|c} 0 & 1 & 0 & 0 & -1 & 0 \\ 0 & 0 & 0 & 1 & 2 & 0 \end{array}\right]$$

$$\left| \begin{aligned} x_2 - x_5 &= 0 \\ x_4 + 2x_5 &= 0 \end{aligned} \right| \rightarrow \left| \begin{aligned} x_2 &= x_5 \\ x_4 &= -2x_5 \end{aligned} \right|$$

Let $x_1 = r$, $x_3 = s$, $x_5 = t$.

$$\begin{bmatrix} x_1 \\ x_2 \\ x_3 \\ x_4 \\ x_5 \end{bmatrix} = \begin{bmatrix} r \\ t \\ s \\ -2t \\ t \end{bmatrix}$$, where r, t, and sw are arbitrary real numbers.

9. $$\left[\begin{array}{cccccc|c} 0 & 0 & 0 & 1 & 2 & -1 & 2 \\ 1 & 2 & 0 & 0 & 1 & -1 & 0 \\ 1 & 2 & 2 & 0 & -1 & 1 & 2 \end{array}\right] \xrightarrow{\text{I} \leftrightarrow \text{II}} \left[\begin{array}{cccccc|c} 1 & 2 & 0 & 0 & 1 & -1 & 0 \\ 0 & 0 & 0 & 1 & 2 & -1 & 2 \\ 1 & 2 & 2 & 0 & -1 & 1 & 2 \end{array}\right] \begin{matrix} \longrightarrow \\ \\ -\text{I} \end{matrix}$$

$$\left[\begin{array}{cccccc|c} 1 & 2 & 0 & 0 & 1 & -1 & 0 \\ 0 & 0 & 0 & 1 & 2 & -1 & 2 \\ 0 & 0 & 2 & 0 & -2 & 2 & 2 \end{array}\right] \xrightarrow{\text{II} \leftrightarrow \text{III}} \left[\begin{array}{cccccc|c} 1 & 2 & 0 & 0 & 1 & -1 & 0 \\ 0 & 0 & 2 & 0 & -2 & 2 & 2 \\ 0 & 0 & 0 & 1 & 2 & -1 & 2 \end{array}\right] \begin{matrix} \\ \div 2 \rightarrow \\ \\ \end{matrix}$$

$$\left[\begin{array}{cccccc|c} 1 & 2 & 0 & 0 & 1 & -1 & 0 \\ 0 & 0 & 1 & 0 & -1 & 1 & 1 \\ 0 & 0 & 0 & 1 & 2 & -1 & 2 \end{array}\right]$$

$$\left| \begin{aligned} x_1 + 2x_2 + x_5 - x_6 &= 0 \\ x_3 - x_5 + x_6 &= 1 \\ x_4 + 2x_5 - x_6 &= 2 \end{aligned} \right| \rightarrow \left| \begin{aligned} x_1 &= -2x_2 - x_5 + x_6 \\ x_3 &= 1 + x_5 - x_6 \\ x_4 &= 2 - 2x_5 + x_6 \end{aligned} \right|$$

Let $x_2 = r$, $x_5 = s$, and $x_6 = t$.

$$\begin{bmatrix} x_1 \\ x_2 \\ x_3 \\ x_4 \\ x_5 \\ x_6 \end{bmatrix} = \begin{bmatrix} -2r - s + t \\ r \\ 1 + s - t \\ 2 - 2s + t \\ s \\ t \end{bmatrix}$$, where r, s, and t are arbitrary real numbers.

10. The system reduces to $\left|\begin{array}{ccccccc} x_1 & & & + & x_4 & = & 1 \\ & x_2 & & - & 3x_4 & = & 2 \\ & & x_3 & + & 2x_4 & = & -3 \end{array}\right| \to \left|\begin{array}{l} x_1 = 1 - x_4 \\ x_2 = 2 + 3x_4 \\ x_3 = -3 - 2x_4 \end{array}\right|$

Let $x_4 = t$.

$\begin{bmatrix} x_1 \\ x_2 \\ x_3 \\ x_4 \end{bmatrix} = \begin{bmatrix} 1-t \\ 2+3t \\ -3-2t \\ t \end{bmatrix}$, where t is an arbitrary real number.

11. The system reduces to $\left|\begin{array}{ccccccc} x_1 & & + & 2x_3 & & = & 0 \\ & x_2 & - & 3x_3 & & = & 4 \\ & & & & x_4 & = & -2 \end{array}\right| \to \left|\begin{array}{l} x_1 = -2x_3 \\ x_2 = 4 + 3x_3 \\ x_4 = -2 \end{array}\right|$.

Let $x_3 = t$.

$\begin{bmatrix} x_1 \\ x_2 \\ x_3 \\ x_4 \end{bmatrix} = \begin{bmatrix} -2t \\ 4+3t \\ t \\ -2 \end{bmatrix}$

12. The system reduces to $\left|\begin{array}{ccccccccc} x_1 & & & & + & 3.5x_5 & + & x_6 & = 0 \\ & x_2 & & & + & x_5 & & & = 0 \\ & & x_3 & & & & - & \frac{5}{3}x_6 & = 0 \\ & & & x_4 & + & 3x_5 & + & x_6 & = 0 \end{array}\right| \to \left|\begin{array}{l} x_1 = -3.5x_5 - x_6 \\ x_2 = -x_5 \\ x_3 = \frac{5}{3}x_6 \\ x_4 = -3x_5 - x_6 \end{array}\right|$.

Let $x_5 = r$ and $x_6 = t$.

$\begin{bmatrix} x_1 \\ x_2 \\ x_3 \\ x_4 \\ x_5 \\ x_6 \end{bmatrix} = \begin{bmatrix} -3.5r - t \\ -r \\ \frac{5}{3}t \\ -3r - t \\ r \\ t \end{bmatrix}$

13. The system reduces to $\left|\begin{array}{ccccccc} x & & & - & z & = & 0 \\ & & y & + & 2z & = & 0 \\ & & & & 0 & = & 1 \end{array}\right|$.

There are no solutions.

14. The system reduces to $\left|\begin{array}{ccccc} x & + & 2y & & = -2 \\ & & & z & = 2 \end{array}\right| \to \left|\begin{array}{l} x = -2 - 2y \\ z = 2 \end{array}\right|$.

Let $y = t$.

$\begin{bmatrix} x \\ y \\ z \end{bmatrix} = \begin{bmatrix} -2-2t \\ t \\ 2 \end{bmatrix}$

15. The system reduces to $\begin{vmatrix} x & & & = & 4 \\ & y & & = & 2 \\ & & z & = & 1 \end{vmatrix}$.

16. The system reduces to $\begin{vmatrix} x_1 + 2x_2 + 3x_3 & & + 5x_5 & = & 6 \\ & x_4 & + 2x_5 & = & 7 \end{vmatrix} \to \begin{vmatrix} x_1 = 6 - 2x_2 - 3x_3 - 5x_5 \\ x_4 = 7 - 2x_5 \end{vmatrix}$.

Let $x_2 = r$, $x_3 = s$, and $x_5 = t$.

$$\begin{bmatrix} x_1 \\ x_2 \\ x_3 \\ x_4 \\ x_5 \end{bmatrix} = \begin{bmatrix} 6 - 2r - 3s - 5t \\ r \\ s \\ 7 - 2t \\ t \end{bmatrix}$$

17. The system reduces to $\begin{vmatrix} x_1 & & & & & = & -\frac{8221}{4340} \\ & x_2 & & & & = & \frac{8591}{8680} \\ & & x_3 & & & = & \frac{4695}{434} \\ & & & x_4 & & = & -\frac{459}{434} \\ & & & & x_5 & = & \frac{699}{434} \end{vmatrix}$.

18. **a.** No, since the third column contains two leading ones.

b. Yes

c. No, since the third row contains a leading one, but the second row does not.

19. $\begin{bmatrix} 0 \\ 0 \\ 0 \\ 0 \end{bmatrix}$ and $\begin{bmatrix} 1 \\ 0 \\ 0 \\ 0 \end{bmatrix}$

20. Four, namely $\begin{bmatrix} 0 & 0 \\ 0 & 0 \end{bmatrix}, \begin{bmatrix} 1 & k \\ 0 & 0 \end{bmatrix}, \begin{bmatrix} 0 & 1 \\ 0 & 0 \end{bmatrix}, \begin{bmatrix} 1 & 0 \\ 0 & 1 \end{bmatrix}$ (k is an arbitrary constant.)

21. Four, namely $\begin{bmatrix} 0 & 0 \\ 0 & 0 \\ 0 & 0 \end{bmatrix}, \begin{bmatrix} 1 & k \\ 0 & 0 \\ 0 & 0 \end{bmatrix}, \begin{bmatrix} 0 & 1 \\ 0 & 0 \\ 0 & 0 \end{bmatrix}, \begin{bmatrix} 1 & 0 \\ 0 & 1 \\ 0 & 0 \end{bmatrix}$ (k is an arbitrary constant.)

22. Seven, namely $\begin{bmatrix} 0 & 0 & 0 \\ 0 & 0 & 0 \end{bmatrix}, \begin{bmatrix} 1 & a & b \\ 0 & 0 & 0 \end{bmatrix}, \begin{bmatrix} 0 & 1 & c \\ 0 & 0 & 0 \end{bmatrix}, \begin{bmatrix} 0 & 0 & 1 \\ 0 & 0 & 0 \end{bmatrix}, \begin{bmatrix} 1 & 0 & d \\ 0 & 1 & e \end{bmatrix}, \begin{bmatrix} 1 & f & 0 \\ 0 & 0 & 1 \end{bmatrix}, \begin{bmatrix} 0 & 1 & 0 \\ 0 & 0 & 1 \end{bmatrix}$

Here, $a, b, \ldots, f$ are arbitrary constants.

23. We need to show that the matrix has the three properties listed on page 19.
Property a holds by Step 2 of the Gauss-Jordan algorithm (page 21).
Property b holds by Step 3 of the Gauss-Jordan algorithm.
Property c holds by Steps 1 and 4 of the algorithm.

24. Yes; each elementary row operation is reversible, that is, it can be "undone." For example, the operation of row swapping can be undone by swapping the same rows again. The operation of dividing a row by a scalar can be reversed by multiplying the same row by the same scalar.

25. Yes; if A is transformed into B by a sequence of elementary row operations, then we can recover A from B by applying the inverse operations in the reversed order (compare with Exercise 24).

26. Yes, by Exercise 25, since rref(A) is obtained from A by a sequence of elementary row operations.

27. No; whatever elementary row operations you apply to $\begin{bmatrix} 1 & 2 & 3 \\ 4 & 5 & 6 \\ 7 & 8 & 9 \end{bmatrix}$, you cannot make the last column equal to zero.

28. Suppose $(c_1, c_2, \ldots, c_n)$ is a solution of the system $\left| \begin{array}{l} a_{11}x_1 + a_{12}x_2 + \cdots + a_{1n}x_n = b_1 \\ a_{21}x_1 + a_{22}x_2 + \cdots + a_{2n}x_n = b_2 \\ \quad \ldots\ldots\ldots\ldots \end{array} \right|$.

To keep the notation simple, suppose we add k times the first equation to the second; then the second equation of the new system will be $(a_{21} + ka_{11})x_1 + \cdots + (a_{2n} + ka_{1n})x_n = b_2 + kb_1$.
We have to verify that $(c_1, c_2, \ldots, c_n)$ is a solution of this new equation. Indeed,
$(a_{21} + ka_{11})c_1 + \cdots + (a_{2n} + ka_{1n})c_n = a_{21}c_1 + \cdots + a_{2n}c_n + k(a_{11}c_1 + \cdots + a_{1n}c_n) = b_2 + kb_1$.
We have shown that any solution of the "old" system is also a solution of the "new." To see that, conversely, any solution of the new system is also a solution of the old system, note that elementary row operations are reversible (compare with Exercise 24); we can obtain the old system by <u>subtracting</u> k times the first equation from the second equation of the new system.

29. Since the number of oxygen atoms remains constant, we must have $2a + b = 2c + 3d$.

Considering hydrogen and nitrogen as well, we obtain the system $\left| \begin{array}{rcl} 2a + b &=& 2c + 3d \\ 2b &=& c + d \\ a &=& c + d \end{array} \right|$ or

$\left| \begin{array}{rcl} 2a + b - 2c - 3d &=& 0 \\ 2b - c - d &=& 0 \\ a - c - d &=& 0 \end{array} \right|$, which reduces to $\left| \begin{array}{rcl} a - 2d &=& 0 \\ b - d &=& 0 \\ c - d &=& 0 \end{array} \right|$.

The solutions are $\begin{bmatrix} a \\ b \\ c \\ d \end{bmatrix} = \begin{bmatrix} 2t \\ t \\ t \\ t \end{bmatrix}$.

To get the smallest positive integers, we set $t = 1$:
$2NO_2 + H_2O \rightarrow HNO_2 + HNO_3$

30. Plugging the points into $f(t)$, we obtain the system $\left| \begin{array}{ccccccccc} a & & & & & & & = & 1 \\ a & + & b & + & c & + & d & = & 0 \\ a & - & b & + & c & - & d & = & 0 \\ a & + & 2b & + & 4c & + & 8d & = & -15 \end{array} \right|$,

with unique solution $a = 1$, $b = 2$, $c = -1$, and $d = -2$, so that $f(t) = 1 + 2t - t^2 - 2t^3$.

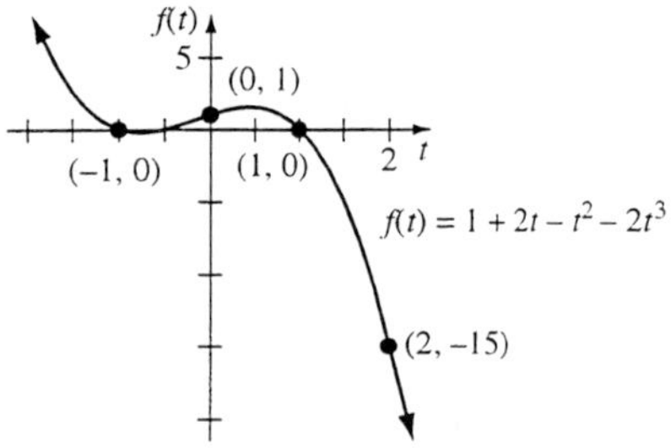

31. Let $f(t) = a + bt + ct^2 + dt^3 + et^4$. Substituting the points in, we get

$$\left| \begin{array}{ccccccccccc} a & + & b & + & c & + & d & + & e & = & 1 \\ a & + & 2b & + & 4c & + & 8d & + & 16e & = & -1 \\ a & + & 3b & + & 9c & + & 27d & + & 81e & = & -59 \\ a & - & b & + & c & - & d & + & e & = & 5 \\ a & - & 2b & + & 4c & - & 8d & + & 16e & = & -29 \end{array} \right|.$$

This system has the unique solution $a = 1$, $b = -5$, $c = 4$, $d = 3$, and $e = -2$, so that $f(t) = 1 - 5t + 4t^2 + 3t^3 - 2t^4$.

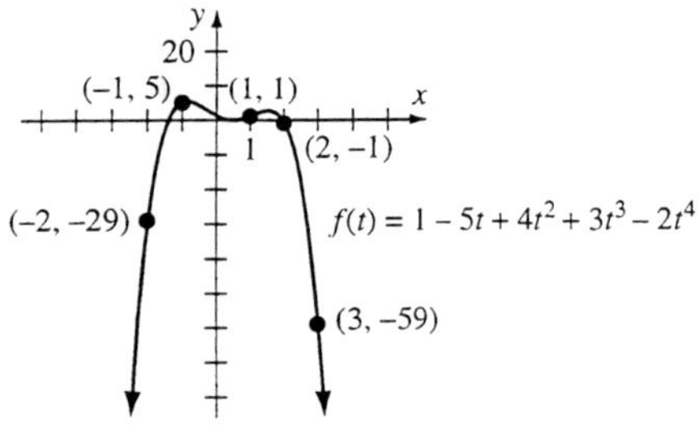

32. The requirements $f_i'(a_i) = f_{i+1}'(a_i)$ and $f_i''(a_i) = f_{i+1}''(a_i)$ ensure that at each junction two different cubics fit "into" one another in a "smooth" way, since they must have the same slope and be equally curved.
The requirement that $f_1'(a_0) = f_n'(a_n) = 0$ ensures that the track is horizontal at the beginning and at the end.
How many unknowns are there? There are n pieces to be fit, and each one a cubic of the form
$f(t) = p + qt + rt^2 + st^3$, with p, q, r, and s to be determined; therefore, there are $4n$ unknowns.
How many equations are there?

$f_i(a_i) = b_i$	for $i = 1, 2, \ldots, n$	gives n equations
$f_i(a_{i-1}) = b_{i-1}$	for $i = 1, 2, \ldots, n$	gives n equations
$f_i'(a_i) = f_{i+1}'(a_i)$	for $i = 1, 2, \ldots, n-1$	gives $n - 1$ equations
$f_i''(a_i) = f_{i+1}''(a_i)$	for $i = 1, 2, \ldots, n-1$	gives $n - 1$ equations
$f_1'(a_0) = 0$, $f_n'(a_n) = 0$		gives 2 equations

Altogether, we have $4n$ equations; convince yourself that all these equations are linear.

33. Let $f(t)=a+bt+ct^2+dt^3$, so that $f'(t)=b+2ct+3dt^2$.

Substituting the given points into $f(t)$ and $f'(t)$ we obtain the system $\begin{vmatrix} a & + & b & + & c & + & d & = & 1 \\ a & + & 2b & + & 4c & + & 8d & = & 5 \\ & & b & + & 2c & + & 3d & = & 2 \\ & & b & + & 4c & + & 12d & = & 9 \end{vmatrix}$.

This system has the unique solution $a=-5$, $b=13$, $c=-10$, and $d=3$, so that $f(t)=-5+13t-10t^2+3t^3$.

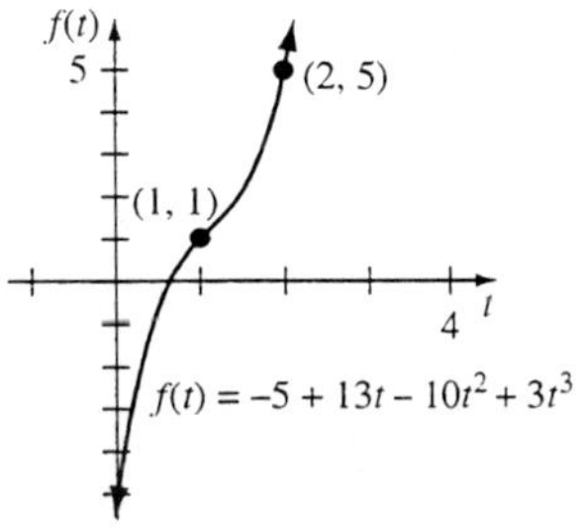

34. We want all vectors $\begin{bmatrix} x \\ y \\ z \end{bmatrix}$ in $\mathbb{R}^3$ such that $\begin{bmatrix} x \\ y \\ z \end{bmatrix} \cdot \begin{bmatrix} 1 \\ 3 \\ -1 \end{bmatrix} = x+3y-z=0$. The endpoints of these vectors form a plane.

These vectors are of the form $\begin{bmatrix} x \\ y \\ z \end{bmatrix} = \begin{bmatrix} -3r+t \\ r \\ t \end{bmatrix}$, where r and t are arbitrary real numbers.

35. We need to solve the system $\begin{vmatrix} x_1+x_2+x_3+x_4=0 \\ x_1+2x_2+3x_3+4x_4=0 \\ x_1+9x_2+9x_3+7x_4=0 \end{vmatrix}$, which reduces to $\begin{vmatrix} x_1 & & & + & 0.25x_4 & = & 0 \\ & x_2 & & - & 1.5x_4 & = & 0 \\ & & x_3 & + & 2.25x_4 & = & 0 \end{vmatrix}$.

The solutions are of the form $\begin{bmatrix} x_1 \\ x_2 \\ x_3 \\ x_4 \end{bmatrix} = \begin{bmatrix} -0.25t \\ 1.5t \\ -2.25t \\ t \end{bmatrix}$, where t is an arbitrary real number.

36. Writing the equation $\vec{b}=x_1\vec{v}_1+x_2\vec{v}_2+x_3\vec{v}_3$ in terms of its components, we obtain the system

$$\begin{vmatrix} x_1+2x_2+4x_3=-8 \\ 4x_1+5x_2+6x_3=-1 \\ 7x_1+8x_2+9x_3=9 \\ 5x_1+3x_2+x_3=15 \end{vmatrix}.$$

The system has the unique solution $x_1=2$, $x_2=3$, and $x_3=-4$.

37. Compare with the solution of Exercise 1.1.21.

The diagram tells us that $\begin{vmatrix} x_1=0.2x_2+0.3x_3+320 \\ x_2=0.1x_1+0.4x_3+90 \\ x_3=0.2x_1+0.5x_2+150 \end{vmatrix}$ or $\begin{vmatrix} x_1-0.2x_2-0.3x_3=320 \\ -0.1x_1+x_2-0.4x_3=90 \\ -0.2x_1-0.5x_2+x_3=150 \end{vmatrix}$.

This system has the unique solution $x_1=500$, $x_2=300$, and $x_3=400$.

38. a. $\vec{v}_1 = \begin{bmatrix} 0 \\ 0.1 \\ 0.2 \end{bmatrix}, \vec{v}_2 = \begin{bmatrix} 0.2 \\ 0 \\ 0.5 \end{bmatrix}, \vec{v}_3 = \begin{bmatrix} 0.3 \\ 0.4 \\ 0 \end{bmatrix}, \vec{b} = \begin{bmatrix} 320 \\ 90 \\ 150 \end{bmatrix}$

b. Recall that x_j is the output of industry I_j, and the ith component a_{ij} of $\vec{v}_j$ is the demand of Industry I_j on industry I_i *for each dollar of output* of industry I_j.

Therefore, the product $x_j a_{ij}$ (that is, the ith component of $x_j\vec{v}_j$), represents the total demand of industry I_j on Industry I_i (in dollars).

c. $x_1\vec{v}_1 + \cdots + x_n\vec{v}_n + \vec{b}$ is the vector whose ith component represents the total demand on industry I_i (consumer demand and interindustry demand combined).

d. The ith component of the equation $x_1\vec{v}_1 + \cdots + x_n\vec{v}_n + \vec{b} = \vec{x}$ expresses the requirement that the output x_i of industry I_i equal the total demand on that industry.

39. a. These components are zero because neither manufacturing nor the energy sector directly require agricultural products.

b. We have to solve the system $x_1\vec{v}_1 + x_2\vec{v}_2 + x_3\vec{v}_3 + \vec{b} = \vec{x}$ or

$$\left| \begin{array}{lllllll} 0.707x_1 & & & & & = & 13.2 \\ -0.014x_1 & + & 0.793x_2 & - & 0.017x_3 & = & 17.6 \\ -0.044x_1 & - & 0.01x_2 & + & 0.784x_3 & = & 1.8 \end{array} \right|.$$

The unique solution is approximately $x_1 = 18.67$, $x_2 = 22.60$, and $x_3 = 3.63$.

40. We want to find m_1, m_2, m_3 such that $m_1 + m_2 + m_3 = 1$ and $\frac{1}{1}\left(m_1\begin{bmatrix} 1 \\ 2 \end{bmatrix} + m_2\begin{bmatrix} 2 \\ 3 \end{bmatrix} + m_3\begin{bmatrix} 4 \\ 1 \end{bmatrix}\right) = \begin{bmatrix} 2 \\ 2 \end{bmatrix}$, that is, we have to solve the system $\left| \begin{array}{l} m_1 + m_2 + m_3 = 1 \\ m_1 + 2m_2 + 4m_3 = 2 \\ 2m_1 + 3m_2 + m_3 = 2 \end{array} \right|$.

The unique solution is $m_1 = \frac{1}{2}$, $m_2 = \frac{1}{4}$, and $m_3 = \frac{1}{4}$.

We will put $\frac{1}{2}$ kg at the point $\begin{bmatrix} 1 \\ 2 \end{bmatrix}$ and $\frac{1}{4}$ kg at each of the two other vertices.

41. We know that $m_1\vec{v}_1 + m_2\vec{v}_2 = m_1\vec{w}_1 + m_2\vec{w}_2$ or $m_1(\vec{v}_1 - \vec{w}_1) + m_2(\vec{v}_2 - \vec{w}_2) = \vec{0}$ or $\left| \begin{array}{l} -3m_1 + 2m_2 = 0 \\ -6m_1 + 4m_2 = 0 \\ -3m_1 + 2m_2 = 0 \end{array} \right|$.

We can conclude that $m_1 = \frac{2}{3}m_2$.

42. Let x_1, x_2, x_3, and x_4 be the traffic volume at the four locations indicated below.

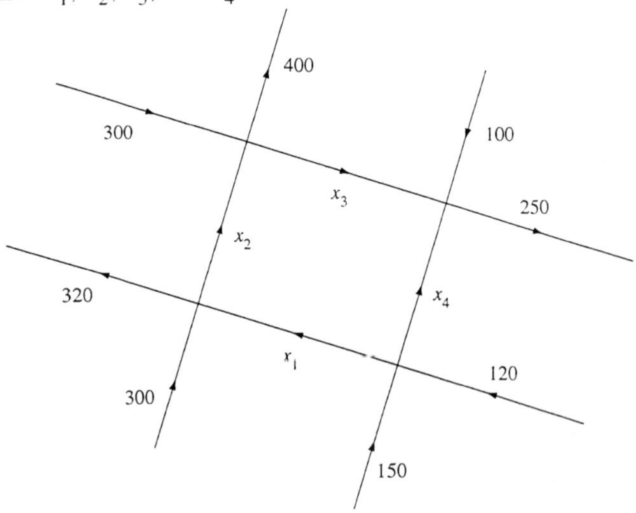

We are told that the number of cars coming into each intersection is the same as the number of cars coming out:

$$\begin{vmatrix} x_1+300=320+x_2 \\ x_2+300=400+x_3 \\ x_3+x_4+100=250 \\ 150+120=x_1+x_4 \end{vmatrix} \text{ or } \begin{vmatrix} x_1 & - & x_2 & & & & & = & 20 \\ & & x_2 & - & x_3 & & & = & 100 \\ & & & & x_3 & + & x_4 & = & 150 \\ x_1 & & & & & + & x_4 & = & 270 \end{vmatrix}.$$

The solutions are of the form $\begin{bmatrix} x_1 \\ x_2 \\ x_3 \\ x_4 \end{bmatrix} = \begin{bmatrix} 270-t \\ 250-t \\ 150-t \\ t \end{bmatrix}$.

Since the x_i must be positive integers (or zero), t must be an integer with $0 \le t \le 150$.
The lowest possible values are $x_1 = 120$, $x_2 = 100$, $x_3 = 0$, and $x_4 = 0$, while the highest possible values are $x_1 = 270$, $x_2 = 250$, $x_3 = 150$, and $x_4 = 150$.

43. Plugging the data into the function $S(t)$ we obtain the system

$$\begin{vmatrix} a+\cos\left(\frac{2\pi\cdot 47}{365}\right)b+\sin\left(\frac{2\pi\cdot 47}{365}\right)c = 11.5 \\ a+\cos\left(\frac{2\pi\cdot 74}{365}\right)b+\sin\left(\frac{2\pi\cdot 74}{365}\right)c = 12 \\ a+\cos\left(\frac{2\pi\cdot 273}{365}\right)b+\sin\left(\frac{2\pi\cdot 273}{365}\right)c = 12 \end{vmatrix}.$$

The unique solution is approximately $a = 12.17$, $b = -1.15$, and $c = 0.18$, so that

$$S(t) = 12.17 - 1.15\cos\left(\frac{2\pi t}{365}\right) + 0.18\sin\left(\frac{2\pi t}{365}\right).$$

The longest day is about 13.3 hours.

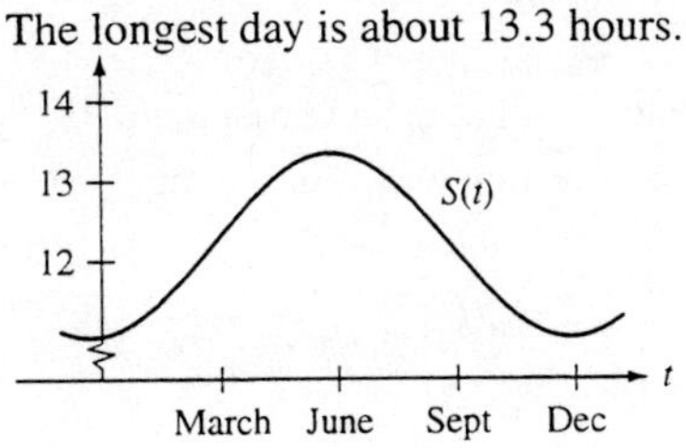

1.3

1. **a.** No solution, since the last row indicates 0 = 1.

   b. The unique solution is $x = 5$, $y = 6$

   c. Infinitely many solutions; the first variable can be chosen freely.

2. The rank is 3 since each row contains a leading one.

3. This matrix has rank 1 since its rref is $\begin{bmatrix} 1 & 1 & 1 \\ 0 & 0 & 0 \\ 0 & 0 & 0 \end{bmatrix}$.

4. This matrix has rank 2 since its rref is $\begin{bmatrix} 1 & 0 & -1 \\ 0 & 1 & 2 \\ 0 & 0 & 0 \end{bmatrix}$.

5. **a.** $x\begin{bmatrix} 1 \\ 3 \end{bmatrix} + y\begin{bmatrix} 2 \\ 1 \end{bmatrix} = \begin{bmatrix} 7 \\ 11 \end{bmatrix}$

   b. The solution of the system in part (a) is $x = 3$, $y = 2$.

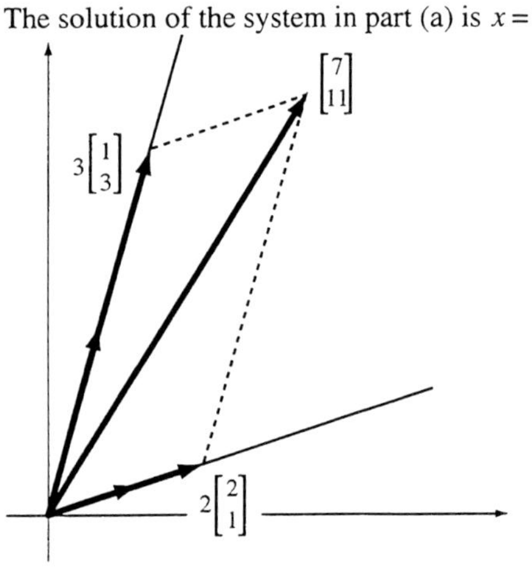

6. No solution, since any linear combination $x\vec{v}_1 + y\vec{v}_2$ of $\vec{v}_1$ and $\vec{v}_2$ will be parallel to $\vec{v}_1$ and $\vec{v}_2$.

7. A unique solution, since there is only one parallelogram with sides along $\vec{v}_1$ and $\vec{v}_2$ and one vertex at the tip of $\vec{v}_3$.

8. Infinitely many solutions. There are at least two obvious solutions: Write $\vec{v}_4$ as a linear combination of $\vec{v}_1$ and $\vec{v}_2$ alone or as a linear combination of $\vec{v}_3$ and $\vec{v}_2$ alone. Therefore, this linear system has infinitely many solutions, by Fact 1.3.1.

9. $\begin{bmatrix} 1 & 2 & 3 \\ 4 & 5 & 6 \\ 7 & 8 & 9 \end{bmatrix}\begin{bmatrix} x \\ y \\ z \end{bmatrix} = \begin{bmatrix} 1 \\ 4 \\ 9 \end{bmatrix}$

10. $\begin{bmatrix}1\\2\\3\end{bmatrix}\cdot\begin{bmatrix}1\\-2\\1\end{bmatrix}=1\cdot 1+2\cdot(-2)+3\cdot 1=0$

11. Undefined since the two vectors do not have the same number of components.

12. $[1\quad 2\quad 3\quad 4]\cdot\begin{bmatrix}5\\6\\7\\8\end{bmatrix}=1\cdot 5+2\cdot 6+3\cdot 7+4\cdot 8=70$

13. $\begin{bmatrix}1&2\\3&4\end{bmatrix}\begin{bmatrix}7\\11\end{bmatrix}=7\begin{bmatrix}1\\3\end{bmatrix}+11\begin{bmatrix}2\\4\end{bmatrix}=\begin{bmatrix}29\\65\end{bmatrix}$ or $\begin{bmatrix}1&2\\3&4\end{bmatrix}\begin{bmatrix}7\\11\end{bmatrix}=\begin{bmatrix}1\cdot 7+2\cdot 11\\3\cdot 7+4\cdot 11\end{bmatrix}=\begin{bmatrix}29\\65\end{bmatrix}$

14. $\begin{bmatrix}1&2&3\\2&3&4\end{bmatrix}\begin{bmatrix}-1\\2\\1\end{bmatrix}=-1\begin{bmatrix}1\\2\end{bmatrix}+2\begin{bmatrix}2\\3\end{bmatrix}+1\begin{bmatrix}3\\4\end{bmatrix}=\begin{bmatrix}6\\8\end{bmatrix}$ or $\begin{bmatrix}1&2&3\\2&3&4\end{bmatrix}\begin{bmatrix}-1\\2\\1\end{bmatrix}=\begin{bmatrix}1\cdot(-1)+2\cdot 2+3\cdot 1\\2\cdot(-1)+3\cdot 2+4\cdot 1\end{bmatrix}=\begin{bmatrix}6\\8\end{bmatrix}$

15. $[1\quad 2\quad 3\quad 4]\begin{bmatrix}5\\6\\7\\8\end{bmatrix}=5\cdot 1+6\cdot 2+7\cdot 3+4\cdot 8=70$ either way

16. $\begin{bmatrix}0&1\\3&2\end{bmatrix}\begin{bmatrix}2\\-3\end{bmatrix}=\begin{bmatrix}0\cdot 2+1\cdot(-3)\\3\cdot 2+2\cdot(-3)\end{bmatrix}=\begin{bmatrix}-3\\0\end{bmatrix}$

17. Undefined, since the matrix has three columns, but the vector has only two components.

18. $\begin{bmatrix}1&2\\3&4\\5&6\end{bmatrix}\begin{bmatrix}1\\2\end{bmatrix}=1\begin{bmatrix}1\\3\\5\end{bmatrix}+2\begin{bmatrix}2\\4\\6\end{bmatrix}=\begin{bmatrix}5\\11\\17\end{bmatrix}$

19. $\begin{bmatrix}1&1&-1\\-5&1&1\\1&-5&3\end{bmatrix}\begin{bmatrix}1\\2\\3\end{bmatrix}=1\begin{bmatrix}1\\-5\\1\end{bmatrix}+2\begin{bmatrix}1\\1\\-5\end{bmatrix}+3\begin{bmatrix}-1\\1\\3\end{bmatrix}=\begin{bmatrix}0\\0\\0\end{bmatrix}$

20. **a.** $\begin{bmatrix}9&8\\7&6\\6&6\end{bmatrix}$

b. $\begin{bmatrix}9&-9&18\\27&36&45\end{bmatrix}$

21. $\begin{bmatrix} 158 \\ 70 \\ 81 \\ 123 \end{bmatrix}$

22. By Fact 1.3.4, the rref is $\begin{bmatrix} 1 & 0 & 0 \\ 0 & 1 & 0 \\ 0 & 0 & 1 \end{bmatrix}$.

23. All variables are leading, that is, there is a leading one in each column of the rref: $\begin{bmatrix} 1 & 0 & 0 \\ 0 & 1 & 0 \\ 0 & 0 & 1 \\ 0 & 0 & 0 \end{bmatrix}$.

24. By Fact 1.3.4, $\text{rref}(A) = \begin{bmatrix} 1 & 0 & 0 & 0 \\ 0 & 1 & 0 & 0 \\ 0 & 0 & 1 & 0 \\ 0 & 0 & 0 & 1 \end{bmatrix}$.

Using Fact 1.3.4 again, we can conclude that the system $A\vec{x} = \vec{c}$ has a unique solution as well.

25. In this case, rref(A) has a row of zeros, so that rank(A) < 4; there will be a nonleading variable. The system $A\vec{x} = \vec{c}$ could have infinitely many solutions (for example, when $\vec{c} = \vec{0}$) or no solutions (for example, when $\vec{c} = \vec{b}$), but it cannot have a unique solution, by Fact 1.3.4.

26. From Exercise 22 we know that $\text{rref}(A) = \begin{bmatrix} 1 & 0 & 0 \\ 0 & 1 & 0 \\ 0 & 0 & 1 \\ 0 & 0 & 0 \end{bmatrix}$.

Since all variables are leading, the system $A\vec{x} = \vec{c}$ cannot have infinitely many solutions, but it could have a unique solution (for example, if $\vec{c} = \vec{b}$) or no solutions at all (compare with Example 13).

27. **a.** True, since $\text{rref}\left[A \,\vdots\, \vec{b}\right]$ contains a row $[0 \;\; 0 \;\; \ldots \;\; 0 \,\vdots\, 1]$.

b. False; as a counterexample, consider the case when A is the zero matrix and $\vec{b}$ the zero vector.

28. **a.** True; you can check that the new matrix satisfies the three requirements for an rref listed on page 19.

b. False; to get a counterexample, remove the first column in $\begin{bmatrix} 1 & 2 & 0 \\ 0 & 0 & 0 \end{bmatrix}$.

29. True; if $A\vec{x}=\vec{b}$ is inconsistent then $\text{rank}\left[A \mid \vec{b}\right]=\text{rank}(A)+1$, since there will be an extra leading one in the last column of the augmented matrix:

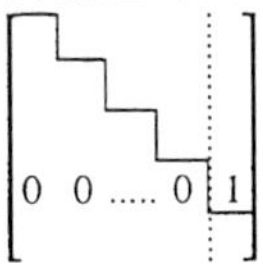

However, if $A\vec{x}=\vec{b}$ is consistent then $\text{rank}\left[A \mid \vec{b}\right]=\text{rank}(A)$, since we do not have an extra leading one in the last column.

30. There must be a leading one in each column: $\text{rref}(A)=\begin{bmatrix}1&0&0\\0&1&0\\0&0&1\\0&0&0\\0&0&0\end{bmatrix}$

31. By Fact 1.3.4, $\text{rref}(A)=\begin{bmatrix}1&0&0&0\\0&1&0&0\\0&0&1&0\\0&0&0&1\end{bmatrix}$.

32. True, by Fact 1.3.4

33. The ith component of $A\vec{x}$ is $[0\quad 0\quad \dots\quad 1\quad \dots\quad 0]\begin{bmatrix}x_1\\x_2\\\dots\\x_i\\\dots\\x_n\end{bmatrix}=x_i$. (The 1 is in the ith position.)

Therefore, $A\vec{x}=\vec{x}$.

34. **a.** $A\vec{e}_1=\begin{bmatrix}a\\d\\g\end{bmatrix}$, $A\vec{e}_2=\begin{bmatrix}b\\e\\h\end{bmatrix}$, and $A\vec{e}_3=\begin{bmatrix}c\\f\\k\end{bmatrix}$.

b. $B\vec{e}_1=\begin{bmatrix}\vec{v}_1 & \vec{v}_2 & \vec{v}_3\end{bmatrix}\begin{bmatrix}1\\0\\0\end{bmatrix}=1\vec{v}_1+0\vec{v}_2+0\vec{v}_3=\vec{v}_1$

Likewise, $B\vec{e}_2=\vec{v}_2$ and $B\vec{e}_3=\vec{v}_3$.

35. Write $A = \begin{bmatrix} \vec{v}_1 & \vec{v}_2 & \dots & \vec{v}_i & \dots & \vec{v}_n \end{bmatrix}$. Then

$$A\vec{e}_i = \begin{bmatrix} \vec{v}_1 & \vec{v}_2 & \dots & \vec{v}_i & \dots & \vec{v}_n \end{bmatrix} \begin{bmatrix} 0 \\ 0 \\ \dots \\ 1 \\ \dots \\ 0 \end{bmatrix} = 0\vec{v}_1 + 0\vec{v}_2 + \dots + 1\vec{v}_i + \dots + 0\vec{v}_n = \vec{v}_i = i\text{th column of } A.$$

36. By Exercise 35, the ith column of A is $A\vec{e}_i$, for $i = 1, 2, 3$. Therefore, $A = \begin{bmatrix} 1 & 4 & 7 \\ 2 & 5 & 8 \\ 3 & 6 & 9 \end{bmatrix}$.

37. We have to solve the system $\begin{vmatrix} x_1 + 2x_2 & = 2 \\ x_3 & = 1 \end{vmatrix}$ or $\begin{vmatrix} x_1 = 2 - 2x_2 \\ x_3 = 1 \end{vmatrix}$.

Let $x_2 = t$. Then the solutions are of the form $\begin{bmatrix} x_1 \\ x_2 \\ x_3 \end{bmatrix} = \begin{bmatrix} 2-2t \\ t \\ 1 \end{bmatrix}$, where t is an arbitrary real number.

38. We will illustrate our reasoning with an example. We generate the "random" 3×3 matrix

$$A = \begin{bmatrix} 0.141 & 0.592 & 0.653 \\ 0.589 & 0.793 & 0.238 \\ 0.462 & 0.643 & 0.383 \end{bmatrix}.$$

Since the entries of this matrix are chosen from a large pool of numbers (in our case 1000, from 0.000 to 0.999), it is unlikely that any of the entries will be zero (and even less likely that the whole first column will consist of zeros). This means that we will usually be able to apply Steps 2 and 3 of the Gauss-Jordan algorithm to turn the first column into $\begin{bmatrix} 1 \\ 0 \\ 0 \end{bmatrix}$; this is indeed possible in our example: $\begin{bmatrix} 0.141 & 0.592 & 0.653 \\ 0.589 & 0.793 & 0.238 \\ 0.462 & 0.643 & 0.383 \end{bmatrix} \to \begin{bmatrix} 1 & 4.199 & 4.631 \\ 0 & -1.680 & -2.490 \\ 0 & -1.297 & -1.757 \end{bmatrix}$

Again, it is unlikely that any entries in the second column of the new matrix will be zero. Therefore, we can turn the second column into $\begin{bmatrix} 0 \\ 1 \\ 0 \end{bmatrix}$.

Likewise, we will be able to clear up the third column, so that $\text{rref}(A) = \begin{bmatrix} 1 & 0 & 0 \\ 0 & 1 & 0 \\ 0 & 0 & 1 \end{bmatrix}$.

We summarize:
As we apply Gauss-Jordan elimination to a random matrix A (of any size), it is unlikely that we will ever encounter a zero on the diagonal. Therefore, rref(A) is likely to have all ones along the diagonal.

39. We will usually get $\text{rref}(A) = \begin{bmatrix} 1 & 0 & 0 & a \\ 0 & 1 & 0 & b \\ 0 & 0 & 1 & c \end{bmatrix}$, where a, b, and c are arbitrary.

(Compare with the summary at the end of Exercise 38.)

40. We will usually have $\text{rref}(A) = \begin{bmatrix} 1 & 0 & 0 \\ 0 & 1 & 0 \\ 0 & 0 & 1 \\ 0 & 0 & 0 \end{bmatrix}$.

(Compare with the summary to Exercise 38.)

41. If $A\vec{x} = \vec{b}$ is a "random" system, then rref(A) will usually be $\begin{bmatrix} 1 & 0 & 0 \\ 0 & 1 & 0 \\ 0 & 0 & 1 \end{bmatrix}$ (by Exercise 38), so that we will have a unique solution.

42. If $A\vec{x} = \vec{b}$ is a "random" system of three equations with four unknowns, then rref(A) will usually be $\begin{bmatrix} 1 & 0 & 0 & a \\ 0 & 1 & 0 & b \\ 0 & 0 & 1 & c \end{bmatrix}$ (by Exercise 39), so that the system will have infinitely many solutions (x_4 is a nonleading variable).

43. If $A\vec{x} = \vec{b}$ is a "random" system of four equations with three unknowns, then $\text{rref}\left[A \mid b\right]$ will usually be $\left[\begin{array}{ccc|c} 1 & 0 & 0 & 0 \\ 0 & 1 & 0 & 0 \\ 0 & 0 & 1 & 0 \\ 0 & 0 & 0 & 1 \end{array}\right]$ (see the summary to Exercise 38), so that the system is inconsistent.

44. This holds by Example 13, since rank(A) $\le n < m$.

45. Write $A = \left[\vec{v}_1 \quad \vec{v}_2 \quad \dots \quad \vec{v}_n\right]$ and $\vec{x} = \begin{bmatrix} x_1 \\ \dots \\ x_n \end{bmatrix}$. Then $A(k\vec{x}) = \left[\vec{v}_1 \quad \dots \quad \vec{v}_n\right]\begin{bmatrix} kx_1 \\ \dots \\ kx_n \end{bmatrix} = kx_1\vec{v}_1 + \cdots + kx_n\vec{v}_n$ and

$k(A\vec{x}) = k(x_1\vec{v}_1 + \cdots + x_n\vec{v}_n) = kx_1\vec{v}_1 + \cdots + kx_n\vec{v}_n$.

The two results agree, as claimed.

46. Since a, d, and f are all nonzero, we can divide the first row by a, the second row by d, and the third row by f to obtain $\begin{bmatrix} 1 & \frac{b}{a} & \frac{c}{a} \\ 0 & 1 & \frac{e}{d} \\ 0 & 0 & 1 \end{bmatrix}$.

It follows that the rank of the matrix is 3.

47. **a.** $\vec{x} = \vec{0}$ is a solution.

b. This holds by part (a) and Fact 1.3.3.

c. If $\vec{x}_1$ and $\vec{x}_2$ are solutions, then $A\vec{x}_1 = \vec{0}$ and $A\vec{x}_2 = \vec{0}$.

Therefore, $A(\vec{x}_1 + \vec{x}_2) = A\vec{x}_1 + A\vec{x}_2 = \vec{0} + \vec{0} = \vec{0}$, so that $\vec{x}_1 + \vec{x}_2$ is a solution as well. Note that we have used Fact 1.3.7a.

d. $A(k\vec{x}) = k(A\vec{x}) = k\vec{0} = \vec{0}$

We have used Fact 1.3.7b

48. The fact that $\vec{x}_1$ is a solution of $A\vec{x} = \vec{b}$ means that $A\vec{x}_1 = \vec{b}$.

a. $A(\vec{x}_1 + \vec{x}_h) = A\vec{x}_1 + A\vec{x}_h = \vec{b} + \vec{0} = \vec{b}$

b. $A(\vec{x}_2 - \vec{x}_1) = A\vec{x}_2 - A\vec{x}_1 = \vec{b} - \vec{b} = \vec{0}$

c. Parts (a) and (b) show that the solutions of $A\vec{x} = \vec{b}$ are exactly the vectors of the form $\vec{x}_1 + \vec{x}_h$, where $\vec{x}_h$ is a solution of $A\vec{x} = \vec{0}$; indeed, if $\vec{x}_2$ is a solution of $A\vec{x} = \vec{b}$, we can write $\vec{x}_2 = \vec{x}_1 + (\vec{x}_2 - \vec{x}_1)$, and $\vec{x}_2 - \vec{x}_1$ will be a solution of $A\vec{x} = \vec{0}$, by part (b).
Geometrically, the vectors of the form $\vec{x}_1 + \vec{x}_h$ are those whose tips are on the line L sketched below; the line L runs through the tip of $\vec{x}_1$ and is parallel to the given line consisting of the solutions of $A\vec{x} = \vec{0}$.

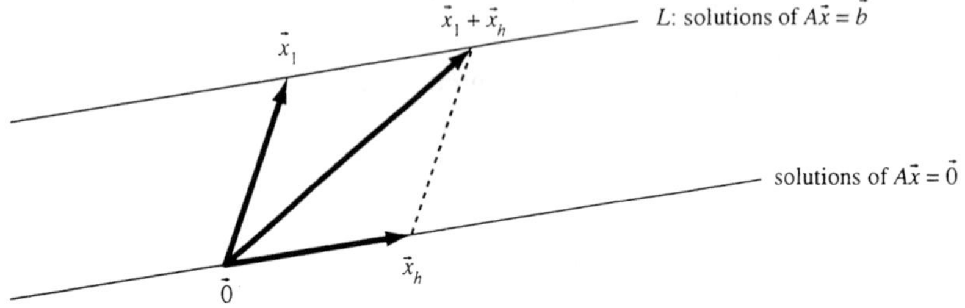

49. a. This system has either infinitely many solutions (if the right-most column of rref$[A \mid b]$ does *not* contain a leading one), or no solutions (if the right-most column *does* contain a leading one).

b. This system has either a unique solution (if rank$[A \mid b] = 3$), or no solution (if rank$[A \mid \vec{b}] = 4$).

c. The right-most column of rref$[A \mid b]$ must contain a leading one, so that the system has no solutions.

d. This system has infinitely many solutions, since there is one non-leading variable.

50. For right-most column of rref$[A \mid \vec{b}]$ must contain a leading one, so that the system has no solutions.

51. For $B\vec{x}$ to be defined, the number of columns of B, which is s, must equal the number of components of $\vec{x}$, which is p, so that we must have $s = p$. Then $B\vec{x}$ will be a vector in $\mathbb{R}^r$; for $A(B\vec{x})$ to be defined we must have $n = r$.
Summary: We must have $s = p$ and $n = r$.

52. $A(B\vec{x}) = A\left(\begin{bmatrix} 0 & -1 \\ 1 & 0 \end{bmatrix}\begin{bmatrix} x_1 \\ x_2 \end{bmatrix}\right) = \begin{bmatrix} 1 & 0 \\ 1 & 2 \end{bmatrix}\begin{bmatrix} -x_2 \\ x_1 \end{bmatrix} = \begin{bmatrix} -x_2 \\ 2x_1 - x_2 \end{bmatrix} = \begin{bmatrix} 0 & -1 \\ 2 & -1 \end{bmatrix}\begin{bmatrix} x_1 \\ x_2 \end{bmatrix}$, so that $C = \begin{bmatrix} 0 & -1 \\ 2 & -1 \end{bmatrix}$.

53. Yes; write $A = \begin{bmatrix} \vec{v}_1 & \dots & \vec{v}_n \end{bmatrix}$, $B = \begin{bmatrix} \vec{w}_1 & \dots & \vec{w}_n \end{bmatrix}$, and $\vec{x} = \begin{bmatrix} x_1 \\ \dots \\ x_n \end{bmatrix}$.

Then $(A+B)\vec{x} = \begin{bmatrix} \vec{v}_1 + \vec{w}_1 & \dots & \vec{v}_n + \vec{w}_n \end{bmatrix} \begin{bmatrix} x_1 \\ \dots \\ x_n \end{bmatrix} = x_1(\vec{v}_1 + \vec{w}_1) + \cdots + x_n(\vec{v}_n + \vec{w}_n)$ and

$A\vec{x} + B\vec{x} = \begin{bmatrix} \vec{v}_1 & \dots & \vec{v}_n \end{bmatrix} \begin{bmatrix} x_1 \\ \dots \\ x_n \end{bmatrix} + \begin{bmatrix} \vec{w}_1 & \dots & \vec{w}_n \end{bmatrix} \begin{bmatrix} x_1 \\ \dots \\ x_2 \end{bmatrix} = x_1\vec{v}_1 + \cdots + x_n\vec{v}_n + x_1\vec{w}_1 + \cdots + x_n\vec{w}_2$.

The two results agree, as claimed.

54. The vectors of the form $c_1\vec{v}_2 + c_2\vec{v}_2$ form a plane through the origin containing $\vec{v}_1$ and $\vec{v}_2$; in the figure below we draw a typical vector in this plane.

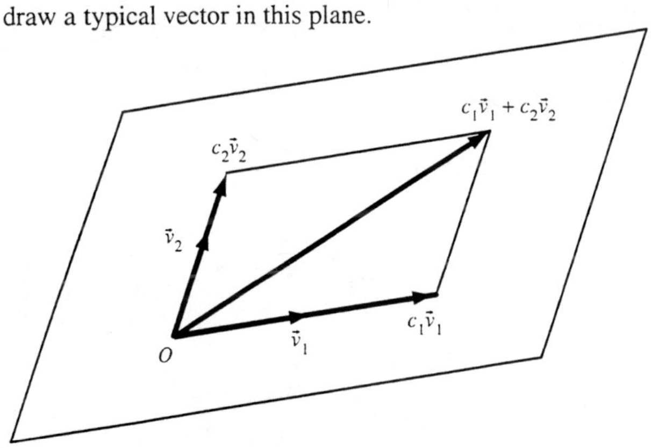

55. We are looking for constants a and b such that $a\begin{bmatrix} 1 \\ 2 \\ 3 \end{bmatrix} + b\begin{bmatrix} 4 \\ 5 \\ 6 \end{bmatrix} = \begin{bmatrix} 7 \\ 8 \\ 9 \end{bmatrix}$.

The resulting system $\left| \begin{matrix} a + 4b = 7 \\ 2a + 5b = 8 \\ 3a + 6b = 9 \end{matrix} \right|$ has the unique solution $a = -1$, $b = 2$, so that $\begin{bmatrix} 7 \\ 8 \\ 9 \end{bmatrix}$ is indeed a linear combination of the vectors $\begin{bmatrix} 1 \\ 2 \\ 3 \end{bmatrix}$ and $\begin{bmatrix} 4 \\ 5 \\ 6 \end{bmatrix}$.

56. We can use technology to determine that the system $\begin{bmatrix} 30 \\ -1 \\ 38 \\ 56 \\ 62 \end{bmatrix} = x_1 \begin{bmatrix} 1 \\ 7 \\ 1 \\ 9 \\ 4 \end{bmatrix} + x_2 \begin{bmatrix} 5 \\ 6 \\ 3 \\ 2 \\ 8 \end{bmatrix} + x_3 \begin{bmatrix} 9 \\ 2 \\ 3 \\ 5 \\ 2 \end{bmatrix} + x_4 \begin{bmatrix} -2 \\ -5 \\ 4 \\ 7 \\ 9 \end{bmatrix}$ is inconsistent;

therefore, the vector $\begin{bmatrix} 30 \\ -1 \\ 38 \\ 56 \\ 62 \end{bmatrix}$ is not a linear combination of the other four vectors.

57. Pick a vector on each line, say $\begin{bmatrix} 2 \\ 1 \end{bmatrix}$ on $y = \frac{x}{2}$ and $\begin{bmatrix} 1 \\ 3 \end{bmatrix}$ on $y = 3x$.

Then write $\begin{bmatrix} 7 \\ 11 \end{bmatrix}$ as a linear combination of $\begin{bmatrix} 2 \\ 1 \end{bmatrix}$ and $\begin{bmatrix} 1 \\ 3 \end{bmatrix}$: $a\begin{bmatrix} 2 \\ 1 \end{bmatrix} + b\begin{bmatrix} 1 \\ 3 \end{bmatrix} = \begin{bmatrix} 7 \\ 11 \end{bmatrix}$.

The unique solution is $a = 2$, $b = 3$, so that the desired representation is $\begin{bmatrix} 7 \\ 11 \end{bmatrix} = \begin{bmatrix} 4 \\ 2 \end{bmatrix} + \begin{bmatrix} 3 \\ 9 \end{bmatrix}$.

$\begin{bmatrix} 4 \\ 2 \end{bmatrix}$ is on line $y = \frac{x}{2}$; $\begin{bmatrix} 3 \\ 9 \end{bmatrix}$ is on line $y = 3x$.

Chapter 2

2.1

1. Not a linear transformation, since $y_2 = x_2 + 2$ is not linear in our sense.

2. Linear, with matrix $\begin{bmatrix} 0 & 2 & 0 \\ 0 & 0 & 3 \\ 1 & 0 & 0 \end{bmatrix}$.

3. Not linear, since $y_2 = x_1 x_3$ is nonlinear.

4. $A = \begin{bmatrix} 9 & 3 & -3 \\ 2 & -9 & 1 \\ 4 & -9 & -2 \\ 5 & 1 & 5 \end{bmatrix}$

5. By Fact 2.1.2, the three columns of the 2×3 matrix A are $T(\vec{e}_1)$, $T(\vec{e}_2)$, and $T(\vec{e}_3)$, so that $A = \begin{bmatrix} 7 & 6 & -13 \\ 11 & 9 & 17 \end{bmatrix}$.

6. Note that $x_1 \begin{bmatrix} 1 \\ 2 \\ 3 \end{bmatrix} + x_2 \begin{bmatrix} 4 \\ 5 \\ 6 \end{bmatrix} = \begin{bmatrix} 1 & 4 \\ 2 & 5 \\ 3 & 6 \end{bmatrix} \begin{bmatrix} x_1 \\ x_2 \end{bmatrix}$, so that T is indeed linear, with matrix $\begin{bmatrix} 1 & 4 \\ 2 & 5 \\ 3 & 6 \end{bmatrix}$.

7. Note that $x_1\vec{v}_1 + \cdots + x_n\vec{v}_n = \begin{bmatrix} \vec{v}_1 & \cdots & \vec{v}_n \end{bmatrix} \begin{bmatrix} x_1 \\ \ldots \\ x_n \end{bmatrix}$, so that T is indeed linear, with matrix $\begin{bmatrix} \vec{v}_1 & \vec{v}_2 & \cdots & \vec{v}_n \end{bmatrix}$.

8. Reducing the system $\left| \begin{array}{l} x_1 + 7x_2 = y_1 \\ 3x_1 + 20x_2 = y_2 \end{array} \right|$, we obtain $\left| \begin{array}{rcrcr} x_1 & = & -20y_1 & + & 7y_2 \\ x_2 & = & 3y_1 & - & y_2 \end{array} \right|$.

9. We have to attempt to solve the equation $\begin{bmatrix} y_1 \\ y_2 \end{bmatrix} = \begin{bmatrix} 2 & 3 \\ 6 & 9 \end{bmatrix} \begin{bmatrix} x_1 \\ x_2 \end{bmatrix}$ for x_1 and x_2. Reducing the system

$\left| \begin{array}{rcrcrcr} 2x_1 & + & 3x_2 & = & y_1 & & \\ 6x_1 & + & 9x_2 & = & & & y_2 \end{array} \right|$ we obtain $\left| \begin{array}{rcrcrcr} x_1 & + & 1.5x_2 & = & 0.5y_1 & & \\ & & 0 & = & -3y_1 & + & y_2 \end{array} \right|$.

No unique solution (x_1, x_2) can be found for a given (y_1, y_2); the matrix is noninvertible.

10. We have to attempt to solve the equation $\begin{bmatrix} y_1 \\ y_2 \end{bmatrix} = \begin{bmatrix} 1 & 2 \\ 4 & 9 \end{bmatrix} \begin{bmatrix} x_1 \\ x_2 \end{bmatrix}$ for x_1 and x_2. Reducing the system

$\left| \begin{array}{rcrcrcr} x_1 & + & 2x_2 & = & y_1 & & \\ 4x_1 & + & 9x_2 & = & & & y_2 \end{array} \right|$ we find that $\left| \begin{array}{rcrcr} x_1 & = & 9y_1 & + & 2y_2 \\ x_2 & = & -4y_1 & + & y_2 \end{array} \right|$ or $\begin{bmatrix} x_1 \\ x_2 \end{bmatrix} = \begin{bmatrix} 9 & -2 \\ -4 & 1 \end{bmatrix} \begin{bmatrix} y_1 \\ y_2 \end{bmatrix}$. The inverse matrix is $\begin{bmatrix} 9 & -2 \\ -4 & 1 \end{bmatrix}$.

11. We have to attempt to solve the equation $\begin{bmatrix} y_1 \\ y_2 \end{bmatrix} = \begin{bmatrix} 1 & 2 \\ 3 & 9 \end{bmatrix}\begin{bmatrix} x_1 \\ x_2 \end{bmatrix}$ for x_1 and x_2. Reducing the system

$\left| \begin{array}{lllll} x_1 & + & 2x_2 & = & y_1 \\ 3x_1 & + & 9x_2 & = & y_2 \end{array} \right|$ we find that $\left| \begin{array}{lllll} x_1 & = & 3y_1 & - & \frac{2}{3}y_2 \\ x_2 & = & -y_1 & + & \frac{1}{3}y_2 \end{array} \right|$.

The inverse matrix is $\begin{bmatrix} 3 & -\frac{2}{3} \\ -1 & \frac{1}{3} \end{bmatrix}$.

12. Reducing the system $\left| \begin{array}{lllll} x_1 & + & kx_2 & = & y_1 \\ & & x_2 & = & y_2 \end{array} \right|$ we find that $\left| \begin{array}{lllll} x_1 & = & y_1 & - & ky_2 \\ x_2 & = & & & y_2 \end{array} \right|$.

The inverse matrix is $\begin{bmatrix} 1 & -k \\ 0 & 1 \end{bmatrix}$.

13. **a.** First suppose that $a \neq 0$. We have to attempt to solve the equation $\begin{bmatrix} y_1 \\ y_2 \end{bmatrix} = \begin{bmatrix} a & b \\ c & d \end{bmatrix}\begin{bmatrix} x_1 \\ x_2 \end{bmatrix}$ for x_1 and x_2.

$$\left| \begin{array}{lllll} ax_1 & + & bx_2 & = & y_1 \\ cx_1 & + & dx_2 & = & y_2 \end{array} \right| \begin{array}{l} \div a \\ \\ \end{array} \rightarrow \left| \begin{array}{lllll} x_1 & + & \frac{b}{a}x_2 & = & \frac{1}{a}y_1 \\ cx_1 & + & dx_2 & = & y_2 \end{array} \right| \begin{array}{l} \\ -c(\mathrm{I}) \end{array}$$

$$\left| \begin{array}{lllll} x_1 & + & \frac{b}{a}x_2 & = & \frac{1}{a}y_1 \\ & & \left(d - \frac{bc}{a}\right)x_2 & = & -\frac{c}{a}y_1 + y_2 \end{array} \right|$$

$$\left| \begin{array}{lllll} x_1 & + & \frac{b}{a}x_2 & = & \frac{1}{a}y_1 \\ & & \left(\frac{ad-bc}{a}\right)x_2 & = & -\frac{c}{a}y_1 + y_2 \end{array} \right|$$

We can solve this system for x_1 and x_2 if (and only if) $ad - bc \neq 0$, as claimed.

If $a = 0$, then we have to consider the system $\left| \begin{array}{lllll} & & bx_2 & = & y_1 \\ cx_1 & + & dx_2 & = & y_2 \end{array} \right| \xrightarrow{\mathrm{I} \leftrightarrow \mathrm{II}} \left| \begin{array}{lllll} cx_1 & + & dx_2 & = & y_2 \\ & & bx_2 & = & y_1 \end{array} \right|$

We can solve for x_1 and x_2 provided that both b and c are nonzero, that is if $bc \neq 0$. Since $a = 0$, this means that $ad - bc \neq 0$, as claimed.

b. First suppose that $ad - bc \neq 0$ and $a \neq 0$. Let $D = ad - bc$ for simplicity. We continue our work in part (a):

$$\left| \begin{array}{lllll} x_1 & + & \frac{b}{a}x_2 & = & \frac{1}{a}y_1 \\ & & \frac{D}{a}x_2 & = & -\frac{c}{a}y_1 + y_2 \end{array} \right| \begin{array}{l} \\ \cdot\frac{a}{D} \end{array}$$

$$\left| \begin{array}{lllll} x_1 & + & \frac{b}{a}x_2 & = & \frac{1}{a}y_1 \\ & & x_2 & = & -\frac{c}{D}y_1 + \frac{a}{D}y_2 \end{array} \right| \begin{array}{l} -\frac{b}{a}(\mathrm{II}) \\ \\ \end{array}$$

$$\left| \begin{array}{lll} x_1 & = & \left(\frac{1}{a} + \frac{bc}{aD}\right)y_1 - \frac{b}{D}y_2 \\ x_2 & = & -\frac{c}{D}y_1 + \frac{a}{D}y_2 \end{array} \right|$$

$$\left| \begin{array}{lll} x_1 & = & \frac{d}{D}y_1 - \frac{b}{D}y_2 \\ x_2 & = & -\frac{c}{D}y_1 + \frac{a}{D}y_2 \end{array} \right|$$

$\left(\text{Note that } \frac{1}{a} + \frac{bc}{aD} = \frac{D+bc}{aD} = \frac{ad}{aD} = \frac{d}{D}.\right)$

It follows that $\begin{bmatrix} a & b \\ c & d \end{bmatrix}^{-1} = \frac{1}{ad-bc}\begin{bmatrix} d & -b \\ -c & a \end{bmatrix}$, as claimed. If $ad - bc \neq 0$ and $a = 0$, then we have to solve the system

$$\left|\begin{array}{rcrcr} cx_1 & + & dx_2 & = & y_2 \\ & & bx_2 & = y_1 & \end{array}\right| \begin{array}{l} \div c \\ \div b \end{array}$$

$$\left|\begin{array}{rcrcr} x_1 & + & \frac{d}{c}x_2 & = & \frac{1}{c}y_2 \\ & & x_2 & = \frac{1}{b}y_1 & \end{array}\right| \begin{array}{l} -\frac{d}{c}(\text{II}) \\ \\ \end{array}$$

$$\left|\begin{array}{rclcl} x_1 & & = -\frac{d}{bc}y_1 & + & \frac{1}{c}y_2 \\ & x_2 & = \frac{1}{b}y_1 & & \end{array}\right|$$

It follows that $\begin{bmatrix} a & b \\ c & d \end{bmatrix}^{-1} = \begin{bmatrix} -\frac{d}{bc} & \frac{1}{c} \\ \frac{1}{b} & 0 \end{bmatrix} = \frac{1}{ad-bc}\begin{bmatrix} d & -b \\ -c & a \end{bmatrix}$ (recall that $a = 0$), as claimed.

14. a. By Exercise 13.1, $\begin{bmatrix} 2 & 3 \\ 5 & k \end{bmatrix}$ is invertible if (and only if) $2k - 15 \neq 0$, or $k \neq 7.5$.

b. By Exercise 13 (b), $\begin{bmatrix} 2 & 3 \\ 5 & k \end{bmatrix}^{-1} = \frac{1}{2k-15}\begin{bmatrix} k & -3 \\ -5 & 2 \end{bmatrix}$.

If all entries of this inverse are integers, then $\frac{3}{2k-15} - \frac{2}{2k-15} = \frac{1}{2k-15}$ is a (nonzero) integer n, so that $2k - 15 = \frac{1}{n}$ or $k = 7.5 + \frac{1}{2n}$. Since $\frac{k}{2k-15} = kn = 7.5n + \frac{1}{2}$ is an integer as well, n must be odd.

We have shown: If all entries of the inverse are integers, then $k = 7.5 + \frac{1}{2n}$, where n is an odd integer. The converse is true as well: If k is chosen in this way, then the entries of $\begin{bmatrix} 2 & 3 \\ 5 & k \end{bmatrix}^{-1}$ will be integers.

15. By Exercise 13(a), the matrix $\begin{bmatrix} a & -b \\ b & a \end{bmatrix}$ is invertible if (and only if) $a^2 + b^2 \neq 0$, which is the case unless $a = b = 0$. If $\begin{bmatrix} a & -b \\ b & a \end{bmatrix}$ is invertible, then its inverse is $\frac{1}{a^2+b^2}\begin{bmatrix} a & b \\ -b & a \end{bmatrix}$, by Exercise 13(b).

16. If $A = \begin{bmatrix} 2 & 0 \\ 0 & 2 \end{bmatrix}$, then $A\vec{x} = 2\vec{x}$ for all $\vec{x}$ in $\mathbb{R}^2$, so that A represents a dilation by the factor of 2. Its inverse is a dilation by the factor of $\frac{1}{2}$ (that is, a contraction): $A^{-1} = \begin{bmatrix} \frac{1}{2} & 0 \\ 0 & \frac{1}{2} \end{bmatrix}$.

17. If $A = \begin{bmatrix} -1 & 0 \\ 0 & -1 \end{bmatrix}$, then $A\vec{x} = -\vec{x}$ for all $\vec{x}$ in $\mathbb{R}^2$, so that A represents a reflection about the origin. This transformation is its own inverse: $A^{-1} = A$.

18. Compare with Exercise 16: This matrix represents a dilation by the factor of $\frac{1}{2}$; the inverse is a dilation by 2.

19. If $A = \begin{bmatrix} 1 & 0 \\ 0 & 0 \end{bmatrix}$, then $A\begin{bmatrix} x_1 \\ x_2 \end{bmatrix} = \begin{bmatrix} x_1 \\ 0 \end{bmatrix}$, so that A represents the orthogonal projection onto the $\vec{e}_1$ axis. This transformation is not invertible, since the equation $A\vec{x} = \begin{bmatrix} 1 \\ 0 \end{bmatrix}$ has infinitely many solutions $\vec{x}$.

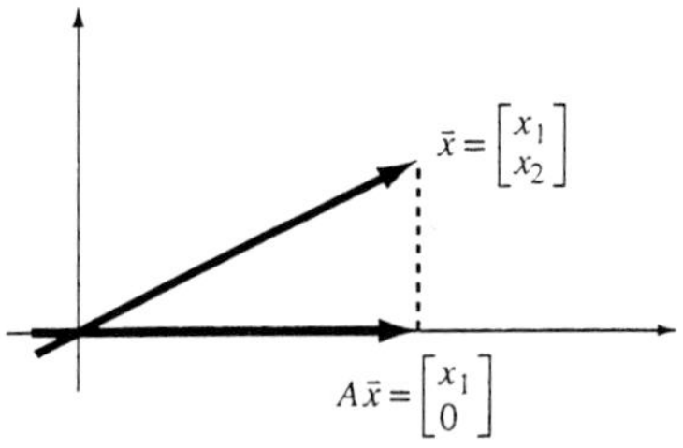

20. If $A = \begin{bmatrix} 0 & 1 \\ 1 & 0 \end{bmatrix}$, then $A\begin{bmatrix} x_1 \\ x_2 \end{bmatrix} = \begin{bmatrix} x_2 \\ x_1 \end{bmatrix}$, so that A represents the reflection in the line $x_2 = x_1$. This transformation is its own inverse: $A^{-1} = A$.

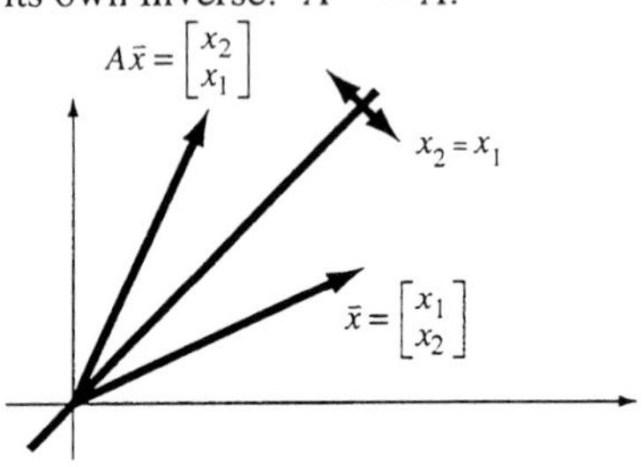

21. Compare with Example 5.

If $A = \begin{bmatrix} 0 & 1 \\ -1 & 0 \end{bmatrix}$, then $A\begin{bmatrix} x_1 \\ x_2 \end{bmatrix} = \begin{bmatrix} x_2 \\ -x_1 \end{bmatrix}$. Note that the vectors $\vec{x}$ and $A\vec{x}$ are perpendicular and have the same length. If $\vec{x}$ is in the first quadrant, then $A\vec{x}$ is in the fourth. Therefore, A represents the rotation through an angle of 90° in the clockwise direction. The inverse $A^{-1} = \begin{bmatrix} 0 & -1 \\ 1 & 0 \end{bmatrix}$ represents the rotation through 90° in the counterclockwise direction.

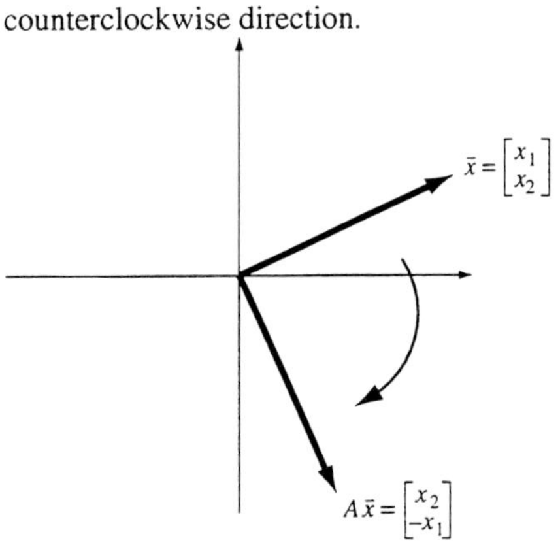

22. If $A=\begin{bmatrix}1 & 0\\ 0 & -1\end{bmatrix}$, then $A\begin{bmatrix}x_1\\ x_2\end{bmatrix}=\begin{bmatrix}x_1\\ -x_2\end{bmatrix}$, so that A represents the reflection in the $\vec{e}_1$ axis. This transformation is its own inverse: $A^{-1}=A$.

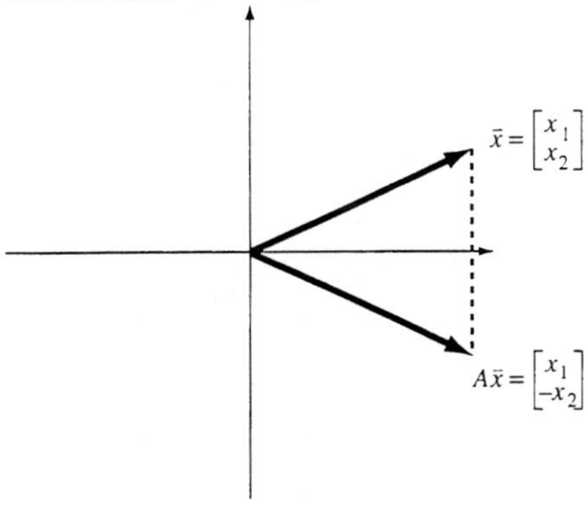

23. Compare with Exercise 21.

Note that $A=2\begin{bmatrix}0 & 1\\ -1 & 0\end{bmatrix}$, so that A represents a rotation through an angle of 90° in the clockwise direction, followed by a dilation by the factor of 2.

The inverse $A^{-1}=\begin{bmatrix}0 & -\frac{1}{2}\\ \frac{1}{2} & 0\end{bmatrix}$ represents a rotation through an angle of 90° in the counterclockwise direction, followed by a dilation by the factor of $\frac{1}{2}$.

24. Compare with Example 5.

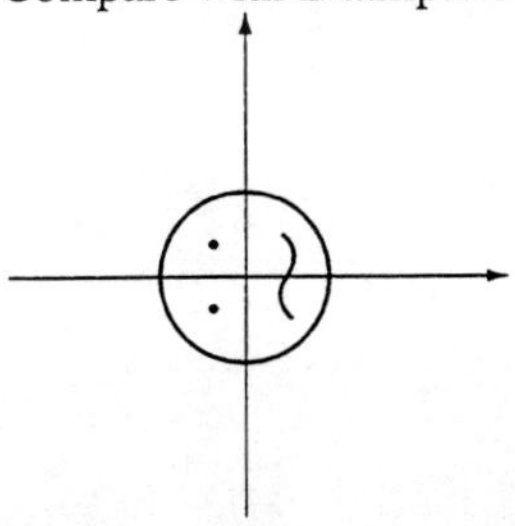

25. The matrix represents a dilation by the factor of 2.

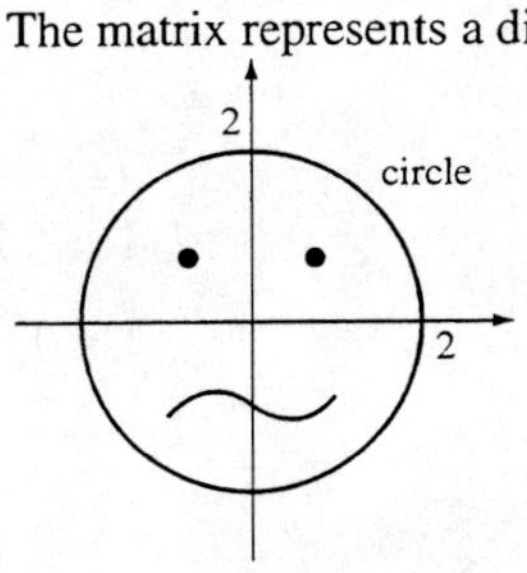

26. Matrix represents a reflection in the line $x_2 = x_1$.

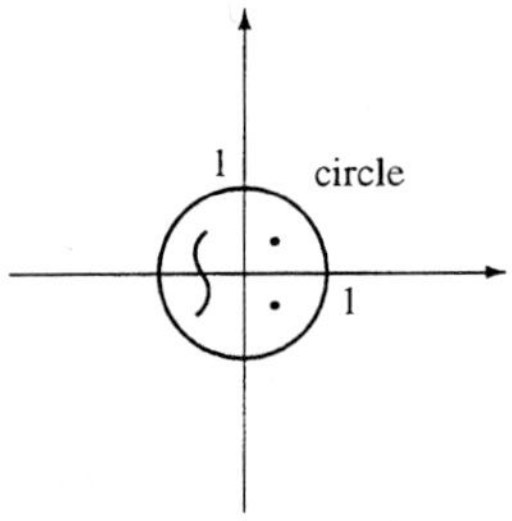

27. Matrix represents a reflection in the $\vec{e}_1$ axis.

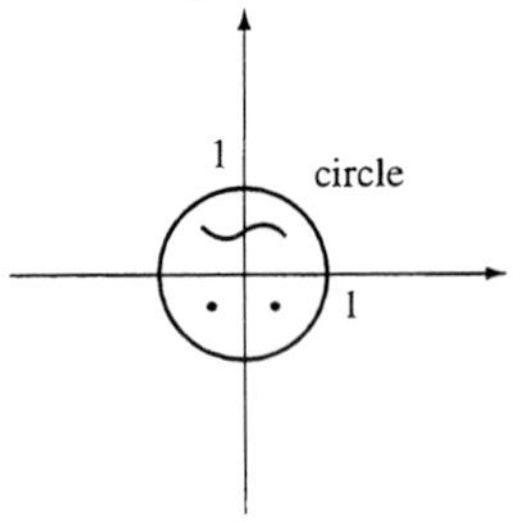

28. If $A = \begin{bmatrix} 1 & 0 \\ 0 & 2 \end{bmatrix}$, then $A\begin{bmatrix} x_1 \\ x_2 \end{bmatrix} = \begin{bmatrix} x_1 \\ 2x_2 \end{bmatrix}$, so that the x_2 component is multiplied by 2, while the x_1 component remains unchanged.

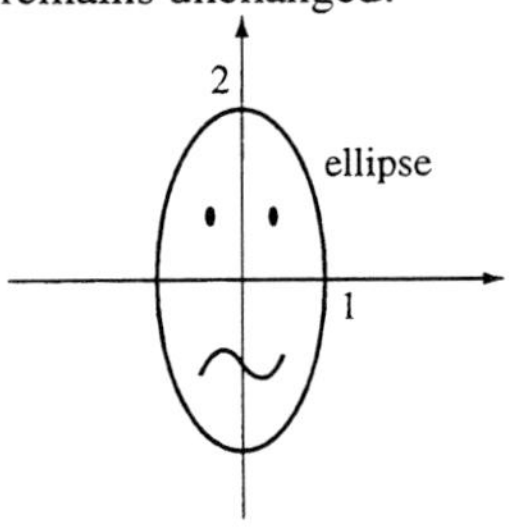

29. Matrix represents a reflection in the origin. Compare with Exercise 17.

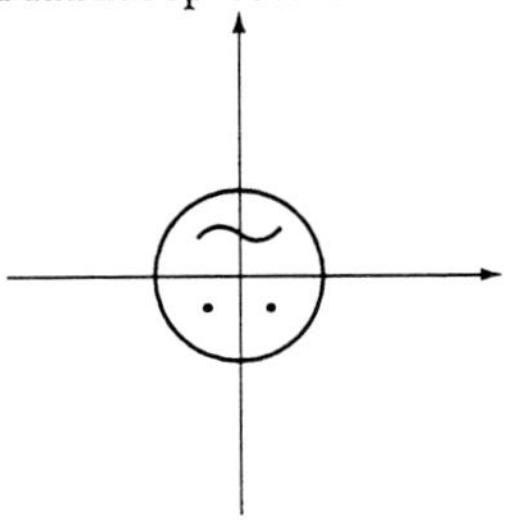

30. If $A=\begin{bmatrix}0&0\\0&1\end{bmatrix}$, then $A\begin{bmatrix}x_1\\x_2\end{bmatrix}=\begin{bmatrix}0\\x_2\end{bmatrix}$, so that A represents the projection onto the $\vec{e}_2$ axis.

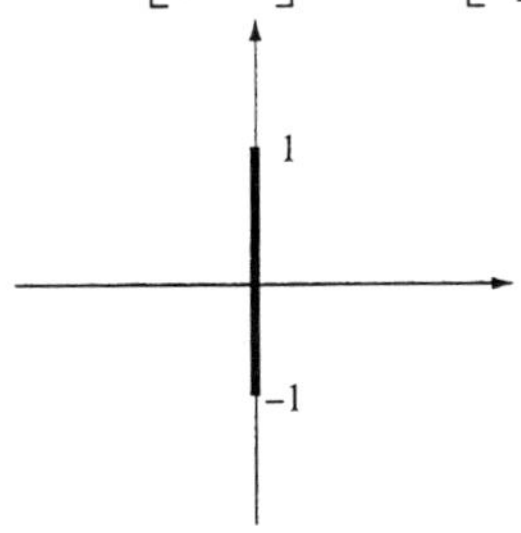

31. The image must be reflected in the $\vec{e}_2$ axis, that is, $\begin{bmatrix}x_1\\x_2\end{bmatrix}$ must be transformed into $\begin{bmatrix}-x_1\\x_2\end{bmatrix}$: This can be accomplished by means of the linear transformation $T(\vec{x})=\begin{bmatrix}-1&0\\0&1\end{bmatrix}\vec{x}$.

32. Using Fact 2.1.2, we find $A=\begin{bmatrix}3&0&\cdots&0\\0&3&\cdots&0\\\vdots&\vdots&\ddots&\vdots\\0&0&\cdots&3\end{bmatrix}$. This matrix has 3's on the diagonal and 0's everywhere else.

33. By Fact 2.1.2, $A=\begin{bmatrix}T\begin{bmatrix}1\\0\end{bmatrix} & T\begin{bmatrix}0\\1\end{bmatrix}\end{bmatrix}$. Now consider the figure below.

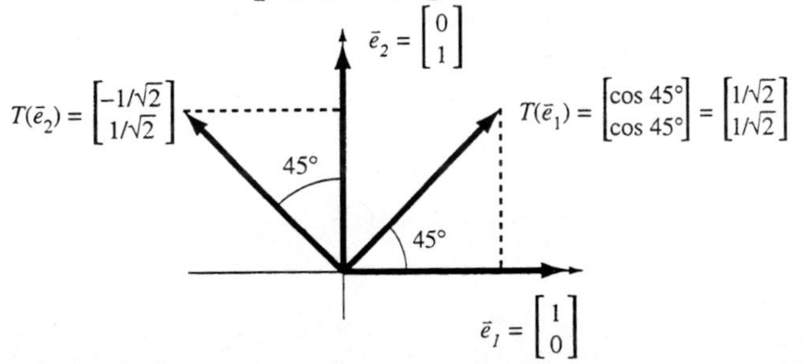

Therefore, $A=\begin{bmatrix}\frac{1}{\sqrt{2}}&-\frac{1}{\sqrt{2}}\\\frac{1}{\sqrt{2}}&\frac{1}{\sqrt{2}}\end{bmatrix}$.

34. As in Exercise 33, we find $T(\vec{e}_1)$ and $T(\vec{e}_2)$; then by Fact 2.1.2, $A=\begin{bmatrix}T(\vec{e}_1) & T(\vec{e}_2)\end{bmatrix}$.

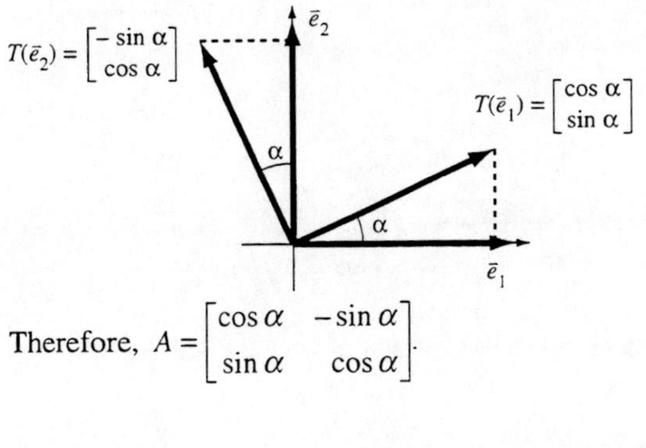

Therefore, $A=\begin{bmatrix}\cos\alpha&-\sin\alpha\\\sin\alpha&\cos\alpha\end{bmatrix}$.

35. We want to find a matrix $A = \begin{bmatrix} a & b \\ c & d \end{bmatrix}$ such that $A\begin{bmatrix} 5 \\ 42 \end{bmatrix} = \begin{bmatrix} 89 \\ 52 \end{bmatrix}$ and $A\begin{bmatrix} 6 \\ 41 \end{bmatrix} = \begin{bmatrix} 88 \\ 53 \end{bmatrix}$. This amounts to solving the system $\begin{vmatrix} 5a+42b & & =89 \\ 6a+41b & & =88 \\ & 5c+42d & =52 \\ & 6c+41d & =53 \end{vmatrix}$.

(Here we really have two systems with two unknowns each.)

The unique solution is $a = 1$, $b = 2$, $c = 2$, and $d = 1$, so that $A = \begin{bmatrix} 1 & 2 \\ 2 & 1 \end{bmatrix}$.

36. By Definition 2.1.1, there is an $m \times n$ matrix A such that $T(\vec{x}) = A\vec{x}$ for all $\vec{x}$ in $\mathbb{R}^n$.
Then $T(\vec{v}+\vec{w}) = A(\vec{v}+\vec{w}) = A\vec{v} + A\vec{w} = T(\vec{v}) + T(\vec{w})$, by Fact 1.3.7a, and $T(k\vec{v}) = A(k\vec{v}) = k(A\vec{v}) = kT(\vec{v})$, by Fact 1.3.7b.

37. Since $\vec{x} = \vec{v} + k(\vec{w}-\vec{v})$, we have $T(\vec{x}) = T(\vec{v}+k(\vec{w}-\vec{v})) = T(\vec{v}) + k(T(\vec{w}) - T(\vec{v}))$, by Exercise 36.
Since k is between 0 and 1, the tip of this vector $T(\vec{x})$ is on the line segment connecting the tips of $T(\vec{v})$ and $T(\vec{w})$.

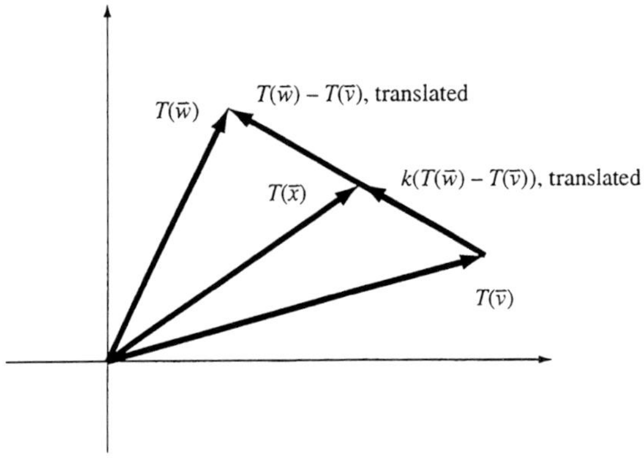

38. $T\begin{bmatrix} 2 \\ -1 \end{bmatrix} = \begin{bmatrix} \vec{v}_1 & \vec{v}_2 \end{bmatrix}\begin{bmatrix} 2 \\ -1 \end{bmatrix} = 2\vec{v}_1 - \vec{v}_2 = 2\vec{v}_1 + (-\vec{v}_2)$

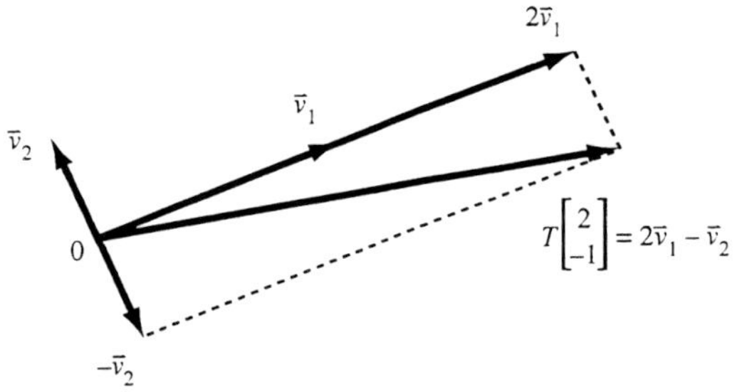

39. True, by Fact 2.1.2: $T\begin{bmatrix} x_1 \\ \dots \\ x_n \end{bmatrix} = \begin{bmatrix} T(\vec{e}_1) & \cdots & T(\vec{e}_n) \end{bmatrix}\begin{bmatrix} x_1 \\ \dots \\ x_n \end{bmatrix} = x_1T(\vec{e}_1) + \cdots + x_nT(\vec{e}_n)$

40. These linear transformations are of the form $[y] = [a][x]$, or $y = ax$. The graph of such a function is a line through the origin.

41. These linear transformations are of the form $[y]=[a \quad b]\begin{bmatrix} x_1 \\ x_2 \end{bmatrix}$, or $y = ax_1 + bx_2$. The graph of such a function is a plane through the origin.

42. a.

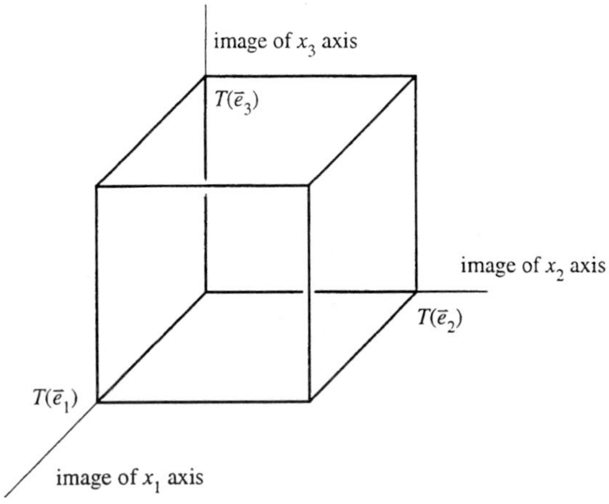

b. Solve the equation $\begin{bmatrix} -\frac{1}{2} & 1 & 0 \\ -\frac{1}{2} & 0 & 1 \end{bmatrix}\begin{bmatrix} x_1 \\ x_2 \\ x_3 \end{bmatrix} = \begin{bmatrix} 0 \\ 0 \end{bmatrix}$, or $\left| \begin{matrix} -\frac{1}{2}x_1 & + & x_2 & & & = 0 \\ -\frac{1}{2}x_1 & & & + & x_3 & = 0 \end{matrix} \right|$.

The solutions are of the form $\begin{bmatrix} x_1 \\ x_2 \\ x_3 \end{bmatrix} = \begin{bmatrix} 2t \\ t \\ t \end{bmatrix}$, where t is an arbitrary real number. These points are on the line through the origin and the observer's eye.

43. a. $T(\vec{x}) = \begin{bmatrix} 2 \\ 3 \\ 4 \end{bmatrix} \cdot \begin{bmatrix} x_1 \\ x_2 \\ x_3 \end{bmatrix} = 2x_1 + 3x_2 + 4x_3 = [2 \quad 3 \quad 4]\begin{bmatrix} x_1 \\ x_2 \\ x_3 \end{bmatrix}$

The transformation is indeed linear, with matrix [2 3 4].

b. If $\vec{v} = \begin{bmatrix} v_1 \\ v_2 \\ v_3 \end{bmatrix}$, then T is linear with matrix $[v_1 \quad v_2 \quad v_3]$, as in part (a).

c. Let $[a \quad b \quad c]$ be the matrix of T. Then $T\begin{bmatrix} x_1 \\ x_2 \\ x_3 \end{bmatrix} = [a \quad b \quad c]\begin{bmatrix} x_1 \\ x_2 \\ x_3 \end{bmatrix} = ax_1 + bx_2 + cx_3 = \begin{bmatrix} a \\ b \\ c \end{bmatrix} \cdot \begin{bmatrix} x_1 \\ x_2 \\ x_3 \end{bmatrix}$, so that $\vec{v} = \begin{bmatrix} a \\ b \\ c \end{bmatrix}$ does the job.

44. $T\begin{bmatrix} x_1 \\ x_2 \\ x_3 \end{bmatrix} = \begin{bmatrix} v_1 \\ v_2 \\ v_3 \end{bmatrix} \times \begin{bmatrix} x_1 \\ x_2 \\ x_3 \end{bmatrix} = \begin{bmatrix} v_2x_3 - v_3x_2 \\ v_3x_1 - v_1x_3 \\ v_1x_2 - v_2x_1 \end{bmatrix} = \begin{bmatrix} 0 & -v_3 & v_2 \\ v_3 & 0 & -v_1 \\ -v_2 & v_1 & 0 \end{bmatrix}\begin{bmatrix} x_1 \\ x_2 \\ x_3 \end{bmatrix}$, so that T is linear, with matrix

$$\begin{bmatrix} 0 & -v_3 & v_2 \\ v_3 & 0 & -v_1 \\ -v_2 & v_1 & 0 \end{bmatrix}.$$

45. False; as a counterexample consider the linear transformation $T\begin{bmatrix} x_1 \\ x_2 \\ x_3 \end{bmatrix} = \begin{bmatrix} x_1 \\ x_2 \\ 0 \end{bmatrix}$, the orthogonal projection onto the $\vec{e}_1 - \vec{e}_2$ plane.
Let $\vec{v} = \vec{e}_1$ and $\vec{w} = \vec{e}_2$.
Then $T(\vec{v} \times \vec{w}) = T(\vec{e}_1 \times \vec{e}_2) = T(\vec{e}_3) = \vec{0}$, but $T(\vec{v}) \times T(\vec{w}) = T(\vec{e}_1) \times T(\vec{e}_2) = \vec{e}_1 \times \vec{e}_2 = \vec{e}_3$.

2.2

1. See Example 2.

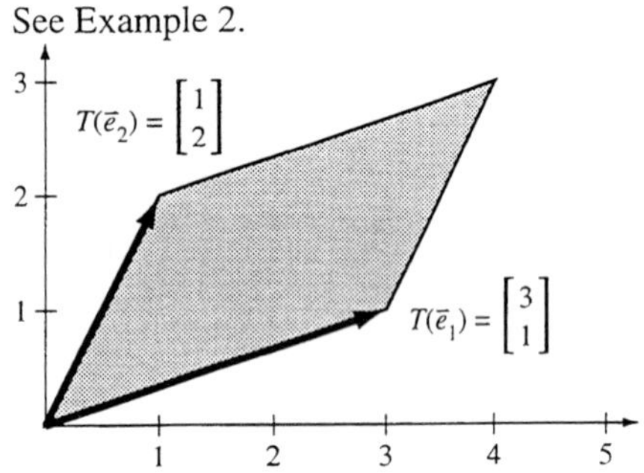

2. By Fact 2.2.2, this matrix is $\begin{bmatrix} \cos(60°) & -\sin(60°) \\ \sin(60°) & \cos(60°) \end{bmatrix} = \begin{bmatrix} \frac{1}{2} & -\frac{\sqrt{3}}{2} \\ \frac{\sqrt{3}}{2} & \frac{1}{2} \end{bmatrix}$.

3. Compare with Example 2.
If $\vec{x}$ is in the unit square in $\mathbb{R}^2$, then $\vec{x} = x_1\vec{e}_1 + x_2\vec{e}_2$ with $0 \le x_1, x_2 \le 1$, so that
$T(\vec{x}) = T(x_1\vec{e}_1 + x_2\vec{e}_2) = x_1T(\vec{e}_1) + x_2T(\vec{e}_2)$.
The image of the unit square is a parallelogram in $\mathbb{R}^3$; two of its sides are $T(\vec{e}_1)$ and $T(\vec{e}_2)$, and the origin is one of its vertices.

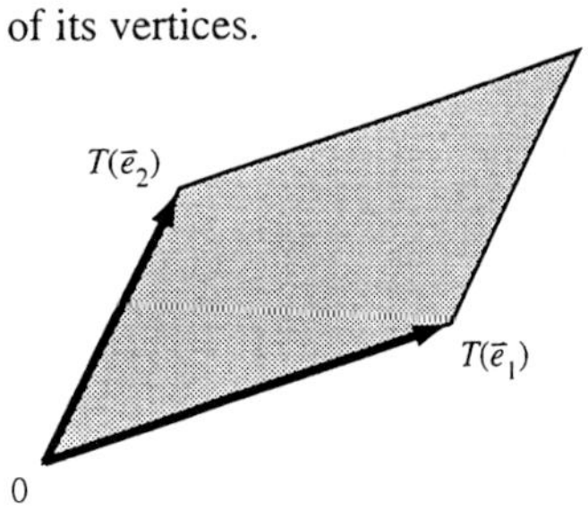

4. $\begin{bmatrix} 1 & -1 \\ 1 & 1 \end{bmatrix} = \sqrt{2}\begin{bmatrix} \frac{1}{\sqrt{2}} & -\frac{1}{\sqrt{2}} \\ \frac{1}{\sqrt{2}} & \frac{1}{\sqrt{2}} \end{bmatrix}$

By Fact 2.2.3, this matrix represents a rotation-dilation: a rotation by 45° followed by a dilation by $\sqrt{2}$.

5.

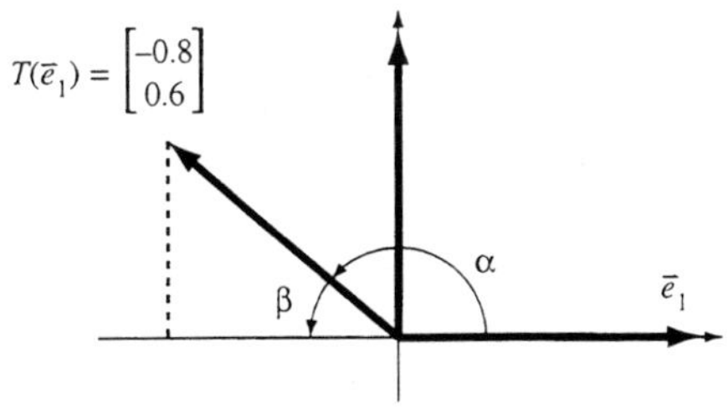

Note that $\sin \beta = 0.6$ so that $\beta = \arcsin(0.6) \approx 0.6435$, and $a = \pi - \beta \approx 2.498$.

6. By Fact 2.2.5, $\text{proj}_L \begin{bmatrix} 1 \\ 1 \\ 1 \end{bmatrix} = \left(\vec{u} \cdot \begin{bmatrix} 1 \\ 1 \\ 1 \end{bmatrix} \right) \vec{u}$, where $\vec{u}$ is a unit vector on L. To get $\vec{u}$, we normalize $\begin{bmatrix} 2 \\ 1 \\ 2 \end{bmatrix}$:

$\vec{u} = \frac{1}{3}\begin{bmatrix} 2 \\ 1 \\ 2 \end{bmatrix}$, so that $\text{proj}_L \begin{bmatrix} 1 \\ 1 \\ 1 \end{bmatrix} = \frac{5}{3} \cdot \frac{1}{3} \begin{bmatrix} 2 \\ 1 \\ 2 \end{bmatrix} = \begin{bmatrix} \frac{10}{9} \\ \frac{5}{9} \\ \frac{10}{9} \end{bmatrix}$.

7. By Fact 2.2.7, $\text{ref}_L \begin{bmatrix} 1 \\ 1 \\ 1 \end{bmatrix} = 2\left(\vec{u} \cdot \begin{bmatrix} 1 \\ 1 \\ 1 \end{bmatrix} \right) \vec{u} - \begin{bmatrix} 1 \\ 1 \\ 1 \end{bmatrix}$, where $\vec{u}$ is a unit vector in L. To get $\vec{u}$, we normalize $\begin{bmatrix} 2 \\ 1 \\ 2 \end{bmatrix}$:

$\vec{u} = \frac{1}{3}\begin{bmatrix} 2 \\ 1 \\ 2 \end{bmatrix}$, so that $\text{ref}_L \begin{bmatrix} 1 \\ 1 \\ 1 \end{bmatrix} = 2\frac{5}{3} \cdot \frac{1}{3} \begin{bmatrix} 2 \\ 1 \\ 2 \end{bmatrix} - \begin{bmatrix} 1 \\ 1 \\ 1 \end{bmatrix} = \begin{bmatrix} \frac{11}{9} \\ \frac{1}{9} \\ \frac{11}{9} \end{bmatrix}$.

8. Examine the effect this transformation has on the unit square (compare with Example 6):

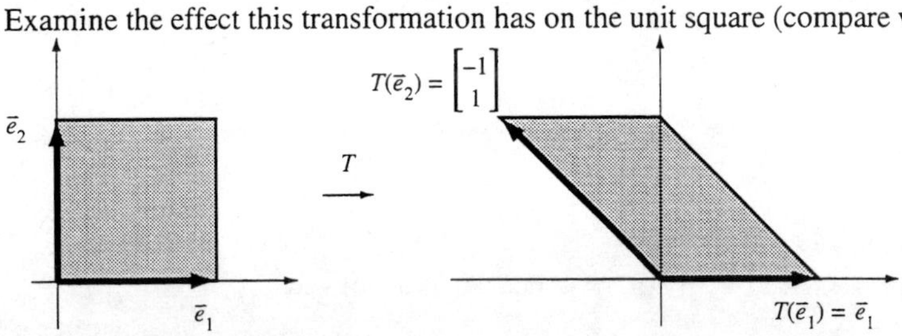

This is a shear parallel to the $\vec{e}_1$ axis.

9. Examine the effect this transformation has on the unit square (compare with Example 6):

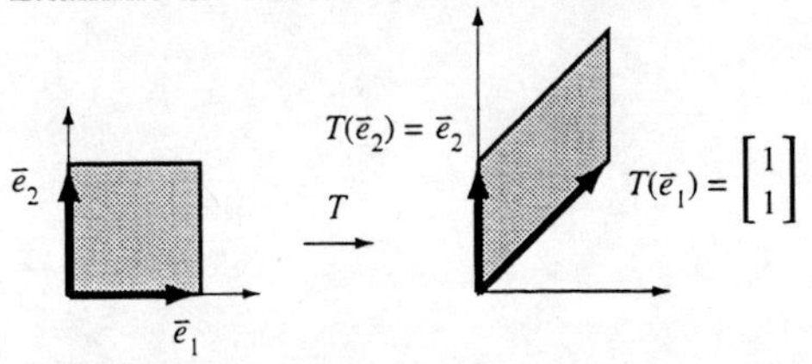

This is a shear parallel to the $\vec{e}_2$ axis.

10. By Fact 2.2.5, $\text{proj}_L\vec{x} = (\vec{u}\cdot\vec{x})\vec{u}$, where $\vec{u}$ is a unit vector on L. We can choose $\vec{u} = \frac{1}{5}\begin{bmatrix}4\\3\end{bmatrix} = \begin{bmatrix}0.8\\0.6\end{bmatrix}$. Then

$$\text{proj}_L\begin{bmatrix}x_1\\x_2\end{bmatrix} = \left(\begin{bmatrix}0.8\\0.6\end{bmatrix}\cdot\begin{bmatrix}x_1\\x_2\end{bmatrix}\right)\begin{bmatrix}0.8\\0.6\end{bmatrix} = (0.8x_1+0.6x_2)\begin{bmatrix}0.8\\0.6\end{bmatrix} = \begin{bmatrix}0.64x_1+0.48x_2\\0.48x_1+0.36x_2\end{bmatrix} = \begin{bmatrix}0.64 & 0.48\\0.48 & 0.36\end{bmatrix}\begin{bmatrix}x_1\\x_2\end{bmatrix}.$$

The matrix is $A = \begin{bmatrix}0.64 & 0.48\\0.48 & 0.36\end{bmatrix}$.

11. In Exercise 10 we found the matrix $A = \begin{bmatrix}0.64 & 0.48\\0.48 & 0.36\end{bmatrix}$ of the projection onto the line L. By Fact 2.2.6,

$\text{ref}_L\vec{x} = 2(\text{proj}_L\vec{x}) - \vec{x} = 2A\vec{x} - \vec{x} = (2A - I_2)\vec{x}$, so that the matrix of the reflection is $2A - I_2 = \begin{bmatrix}0.28 & 0.96\\0.96 & -0.28\end{bmatrix}$.

12. By Fact 2.2.5, $\text{proj}_L\begin{bmatrix}x_1\\x_2\end{bmatrix} = \left(\begin{bmatrix}u_1\\u_2\end{bmatrix}\cdot\begin{bmatrix}x_1\\x_2\end{bmatrix}\right)\begin{bmatrix}u_1\\u_2\end{bmatrix} = (u_1x_1+u_2x_2)\begin{bmatrix}u_1\\u_2\end{bmatrix} = \begin{bmatrix}u_1^2x_1+u_1u_2x_2\\u_1u_2x_1+u_2^2x_2\end{bmatrix} = \begin{bmatrix}u_1^2 & u_1u_2\\u_1u_2 & u_2^2\end{bmatrix}\begin{bmatrix}x_1\\x_2\end{bmatrix}$.

The matrix is $\begin{bmatrix}u_1^2 & u_1u_2\\u_1u_2 & u_2^2\end{bmatrix}$.

13. By Fact 2.2.6, $\text{ref}_L\begin{bmatrix}x_1\\x_2\end{bmatrix} = 2\left(\begin{bmatrix}u_1\\u_2\end{bmatrix}\cdot\begin{bmatrix}x_1\\x_2\end{bmatrix}\right)\begin{bmatrix}u_1\\u_2\end{bmatrix} - \begin{bmatrix}x_1\\x_2\end{bmatrix} = 2(u_1x_1+u_2x_2)\begin{bmatrix}u_1\\u_2\end{bmatrix} - \begin{bmatrix}x_1\\x_2\end{bmatrix} = \begin{bmatrix}(2u_1^2-1)x_1+2u_1u_2x_2\\2u_1u_2x_1+(2u_2^2-1)x_2\end{bmatrix}$.

The matrix is $\begin{bmatrix}2u_1^2-1 & 2u_1u_2\\2u_1u_2 & 2u_2^2-1\end{bmatrix}$.

14. a. Proceeding as in Exercise 12, we find that A is the matrix whose ijth entry is u_iu_j:

$$A = \begin{bmatrix}u_1^2 & u_1u_2 & \cdots & u_1u_n\\u_2u_1 & u_2^2 & \cdots & u_2u_n\\\vdots & \vdots & \ddots & \vdots\\u_nu_1 & u_nu_2 & \cdots & u_n^2\end{bmatrix}$$

b. The sum of the diagonal entries is $u_1^2 + u_2^2 + \cdots + u_n^2 = 1$, since $\vec{u}$ is a unit vector.

15. Proceeding as in Exercise 13, we find that A is the matrix whose ijth entry is $2u_iu_j$ if $i \neq j$ and $u_i^2 - 1$ if $i = j$:

$$A = \begin{bmatrix}u_1^2-1 & 2u_1u_2 & \cdots & 2u_1u_n\\2u_2u_1 & u_2^2-1 & \cdots & 2u_2u_n\\\vdots & \vdots & \ddots & \vdots\\2u_nu_1 & 2u_nu_2 & \cdots & u_n^2-1\end{bmatrix}$$

16. a.

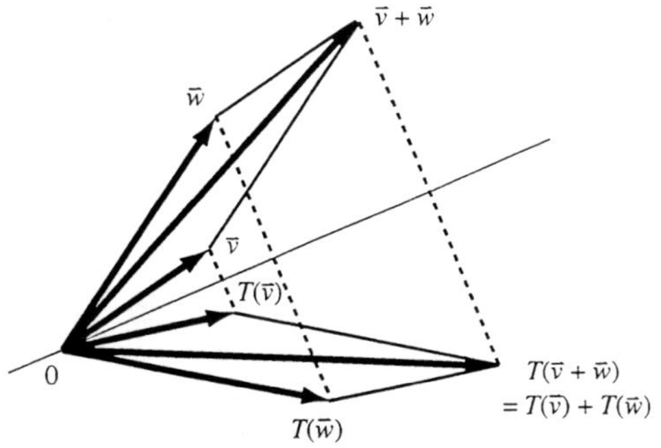

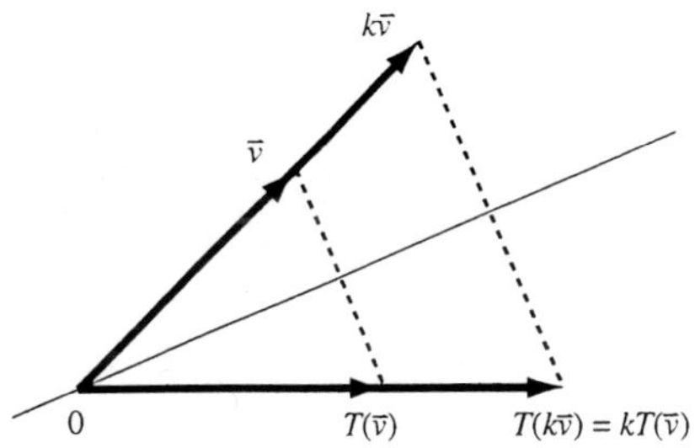

b. By Fact 2.1.2, the matrix of T is $\begin{bmatrix} T(\vec{e}_1) & T(\vec{e}_2) \end{bmatrix}$.

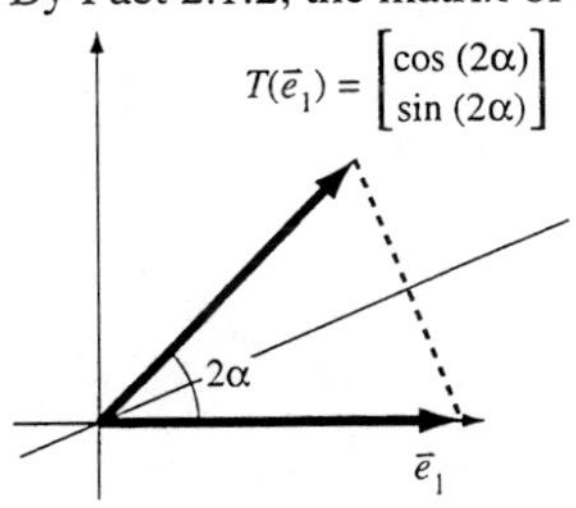

$T(\vec{e}_2)$ is the unit vector in the fourth quadrant perpendicular to $T(\vec{e}_1) = \begin{bmatrix} \cos(2\alpha) \\ \sin(2\alpha) \end{bmatrix}$, so that $T(\vec{e}_2) = \begin{bmatrix} \sin(2\alpha) \\ -\cos(2\alpha) \end{bmatrix}$. The matrix of T is therefore $\begin{bmatrix} \cos(2\alpha) & \sin(2\alpha) \\ \sin(2\alpha) & -\cos(2\alpha) \end{bmatrix}$.

Alternatively, we can use the result of Exercise 13, with $\begin{bmatrix} u_1 \\ u_2 \end{bmatrix} = \begin{bmatrix} \cos\alpha \\ \sin\alpha \end{bmatrix}$ to find the matrix

$$\begin{bmatrix} 2\cos^2\alpha - 1 & 2\cos\alpha\sin\alpha \\ 2\cos\alpha\sin\alpha & 2\sin^2\alpha - 1 \end{bmatrix}.$$

You can use trigonometric identities to show that the two results agree.

17. a.

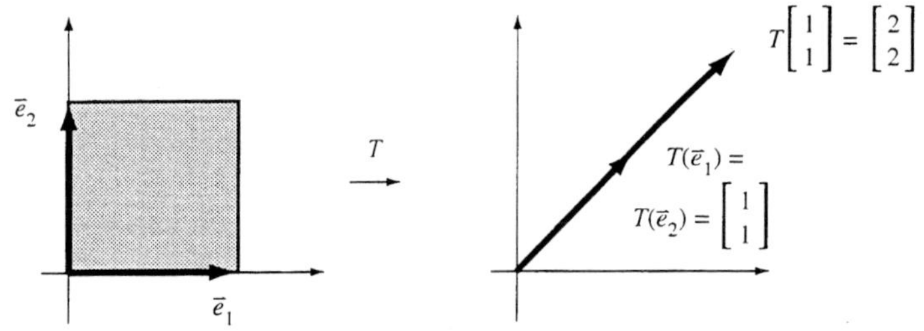

The image is a "collapsed parallelogram"; it is just the line segment from $\begin{bmatrix}0\\0\end{bmatrix}$ to $\begin{bmatrix}2\\2\end{bmatrix}$.

b. The figure below shows that the matrix of the projection onto the line through $\begin{bmatrix}1\\1\end{bmatrix}$ is $\begin{bmatrix}\frac{1}{2} & \frac{1}{2}\\ \frac{1}{2} & \frac{1}{2}\end{bmatrix}$.

Therefore, $\begin{bmatrix}1 & 1\\ 1 & 1\end{bmatrix}=2\begin{bmatrix}\frac{1}{2} & \frac{1}{2}\\ \frac{1}{2} & \frac{1}{2}\end{bmatrix}$ represents this projection followed by a dilation by 2.

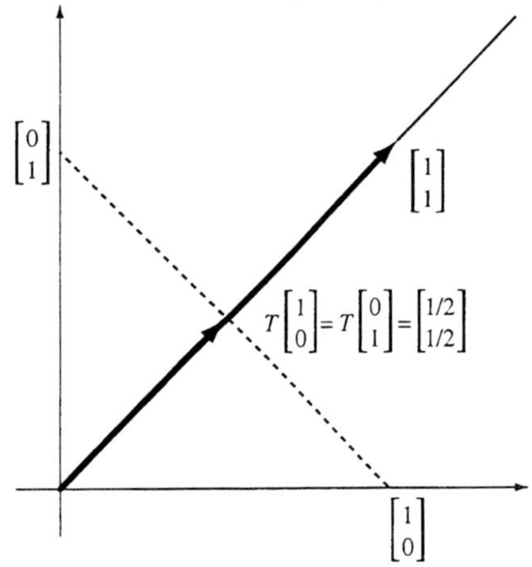

18. If T is indeed a reflection-dilation, then the factor of dilation is $\|T(\vec{e}_1)\|=\left\|\begin{bmatrix}3\\4\end{bmatrix}\right\|=5$, since a reflection leaves the length of a vector unchanged.

Now write $\begin{bmatrix}3 & 4\\ 4 & -3\end{bmatrix}=5\begin{bmatrix}0.6 & 0.8\\ 0.8 & -0.6\end{bmatrix}$; we need to verify that the matrix $\begin{bmatrix}0.6 & 0.8\\ 0.8 & -0.6\end{bmatrix}$ represents a reflection. The result of Exercise 13 shows that this is indeed the case, with $\begin{bmatrix}u_1\\u_2\end{bmatrix}=\begin{bmatrix}\frac{2}{\sqrt{5}}\\ \frac{1}{\sqrt{5}}\end{bmatrix}$.

19. $T(\vec{e}_1)=\vec{e}_1$, $T(\vec{e}_2)=\vec{e}_2$, and $T(\vec{e}_3)=\vec{0}$, so that the matrix is $\begin{bmatrix}1 & 0 & 0\\ 0 & 1 & 0\\ 0 & 0 & 0\end{bmatrix}$.

20. $T(\vec{e}_1) = \vec{e}_1$, $T(\vec{e}_2) = -\vec{e}_2$, and $T(\vec{e}_3) = \vec{e}_3$, so that the matrix is $\begin{bmatrix} 1 & 0 & 0 \\ 0 & -1 & 0 \\ 0 & 0 & 1 \end{bmatrix}$.

21. $T(\vec{e}_1) = \vec{e}_2$, $T(\vec{e}_2) = -\vec{e}_1$, and $T(\vec{e}_3) = \vec{e}_3$, so that the matrix is $\begin{bmatrix} 0 & -1 & 0 \\ 1 & 0 & 0 \\ 0 & 0 & 1 \end{bmatrix}$.

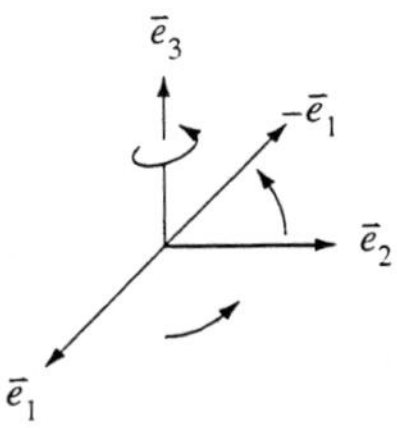

22. Sketch the $\vec{e}_1 - \vec{e}_3$ plane, as viewed from the positive $\vec{e}_2$ axis.

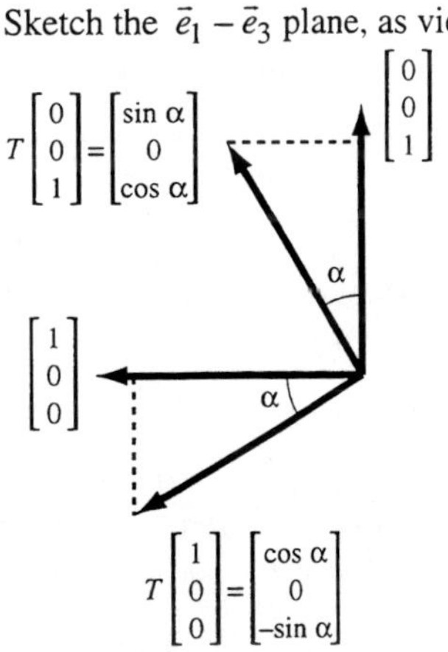

Since $T(\vec{e}_2) = \vec{e}_2$, the matrix is $\begin{bmatrix} \cos\alpha & 0 & \sin\alpha \\ 0 & 1 & 0 \\ -\sin\alpha & 0 & \cos\alpha \end{bmatrix}$.

23. $T(\vec{e}_1) = \vec{e}_3$, $T(\vec{e}_2) = \vec{e}_2$, and $T(\vec{e}_3) = \vec{e}_1$, so that the matrix is $\begin{bmatrix} 0 & 0 & 1 \\ 0 & 1 & 0 \\ 1 & 0 & 0 \end{bmatrix}$.

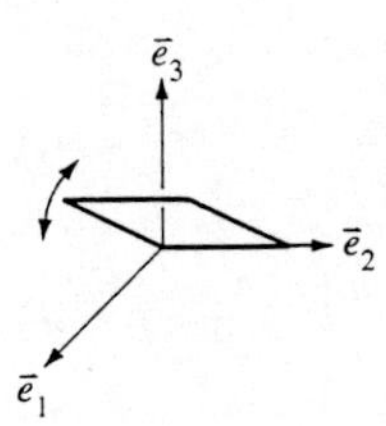

24. a. Compare with Example 6.

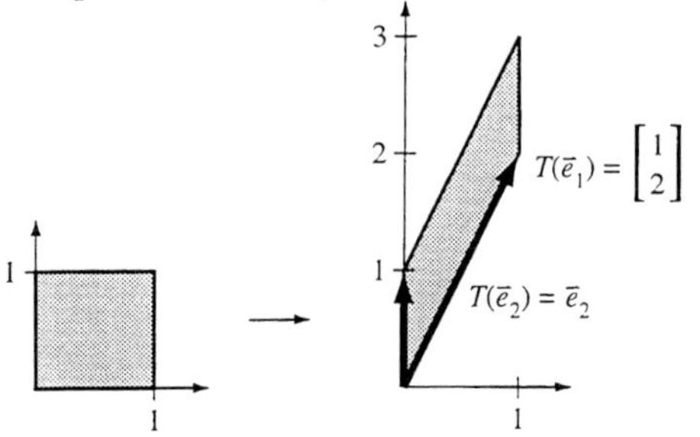

b. We claim that T is a shear parallel to the $\vec{e}_2$ axis L. We need to check the two parts of Definition 2.2.4:
$A\vec{v}=\begin{bmatrix}1&0\\2&1\end{bmatrix}\begin{bmatrix}0\\v_2\end{bmatrix}=\begin{bmatrix}0\\v_2\end{bmatrix}=\vec{v}$, for all $\vec{v}$ in L, and $A\vec{x}-\vec{x}=\begin{bmatrix}1&0\\2&1\end{bmatrix}\begin{bmatrix}x_1\\x_2\end{bmatrix}-\begin{bmatrix}x_1\\x_2\end{bmatrix}=\begin{bmatrix}x_1\\2x_1+x_2\end{bmatrix}-\begin{bmatrix}x_1\\x_2\end{bmatrix}=\begin{bmatrix}0\\2x_2\end{bmatrix}$ is in L for all $\vec{x}$ in $\mathbb{R}^2$.

c. The matrix of the inverse transformation is $\begin{bmatrix}1&0\\-2&1\end{bmatrix}$, representing a shear parallel to the $\vec{e}_2$ axis as well.

25. The matrix $A=\begin{bmatrix}1&k\\0&1\end{bmatrix}$ represents a shear parallel to the $\vec{e}_1$ axis, and its inverse $A^{-1}=\begin{bmatrix}1&-k\\0&1\end{bmatrix}$ represents such a shear as well, but "the other way."

26. In view of part a of Definition 2.2.4, we will first find the vectors $\vec{v}=\begin{bmatrix}v_1\\v_2\end{bmatrix}$ such that $T(\vec{v})=\vec{v}$. We need to solve the system $\begin{vmatrix}-v_1+4v_2=v_1\\-v_1+3v_2=v_2\end{vmatrix}$ or $\begin{vmatrix}-2v_1+4v_2=0\\-v_1+2v_2=0\end{vmatrix}$, with solutions $\begin{bmatrix}v_1\\v_2\end{bmatrix}=\begin{bmatrix}2t\\t\end{bmatrix}$, where t is an arbitrary real number.

To show that T is a shear parallel to the line L consisting of all vectors of the form $\begin{bmatrix}2t\\t\end{bmatrix}=t\begin{bmatrix}2\\1\end{bmatrix}$, we will verify part b of Definition 2.2.4:
$T(\vec{x})-\vec{x}=\begin{bmatrix}-1&4\\-1&3\end{bmatrix}\begin{bmatrix}x_1\\x_2\end{bmatrix}-\begin{bmatrix}x_1\\x_2\end{bmatrix}=\begin{bmatrix}-x_1+4x_2\\-x_1+3x_2\end{bmatrix}-\begin{bmatrix}x_1\\x_2\end{bmatrix}=\begin{bmatrix}-2x_1+4x_2\\-x_1+2x_2\end{bmatrix}=(-x_1+2x_2)\begin{bmatrix}2\\1\end{bmatrix}$ is indeed in L, for all $\vec{x}$ in $\mathbb{R}^2$.

27. Yes; let us verify the two requirements expressed in Fact 2.2.1. Write $F(\vec{x})=L(T(\vec{x}))$.

a. $F(\vec{v}+\vec{w})=L(T(\vec{v}+\vec{w}))=L(T(\vec{v})+T(\vec{w}))=L(T(\vec{v}))+L(T(\vec{w}))=F(\vec{v})+F(\vec{w})$

b. $F(k\vec{v})=L(T(k\vec{v}))=L(kT(\vec{v}))=kL(T(\vec{v}))=kF(\vec{v})$

28. $T(\vec{e}_1)=B(A\vec{e}_1)=\begin{bmatrix}p&q\\r&s\end{bmatrix}\begin{bmatrix}a\\c\end{bmatrix}=\begin{bmatrix}pa+qc\\ra+sc\end{bmatrix}$, and $T(\vec{e}_2)=B(Ae_2)=\begin{bmatrix}p&q\\r&s\end{bmatrix}\begin{bmatrix}b\\d\end{bmatrix}=\begin{bmatrix}pb+qd\\rb+sd\end{bmatrix}$, so that the desired matrix is $\begin{bmatrix}pa+qc&pb+qd\\ra+sc&rb+sd\end{bmatrix}$.

29. To check that L is linear, we verify the two parts of Fact 2.2.1.

a. Use the hint and apply L on both sides of the equation $\vec{x}+\vec{y}=T(L(\vec{x})+L(\vec{y}))$:
$L(\vec{x}+\vec{y})=L(T(L(\vec{x})+L(\vec{y})))=L(\vec{x})+L(\vec{y})$, as claimed.

b. $L(k\vec{x})=L(kT(L(\vec{x}))=L(T(kL(\vec{x})))=kL(\vec{x})$, as claimed.

$\uparrow$ $\vec{x}=T(L(\vec{x}))$ $\quad$ $\uparrow$ T is linear.

30. Write $A=\begin{bmatrix}\vec{v}_1 & \vec{v}_2\end{bmatrix}$; then $A\vec{x}=\begin{bmatrix}\vec{v}_1 & \vec{v}_2\end{bmatrix}\begin{bmatrix}x_1\\x_2\end{bmatrix}=x_1\vec{v}_1+x_2\vec{v}_2$. We must choose $\vec{v}_1$ and $\vec{v}_2$ in such a way that $x_1\vec{v}_1+x_2\vec{v}_2$ is a scalar multiple of the vector $\begin{bmatrix}1\\2\end{bmatrix}$, for all x_1 and x_2. This is the case if (and only if) both $\vec{v}_1$ and $\vec{v}_2$ are scalar multiples of $\begin{bmatrix}1\\2\end{bmatrix}$.

For example, choose $\vec{v}_1=\begin{bmatrix}1\\2\end{bmatrix}$ and $\vec{v}_2=\begin{bmatrix}0\\0\end{bmatrix}$, so that $A=\begin{bmatrix}1 & 0\\2 & 0\end{bmatrix}$.

31. Write $A=\begin{bmatrix}\vec{v}_1 & \vec{v}_2 & \vec{v}_3\end{bmatrix}$; then $A\vec{x}=\begin{bmatrix}\vec{v}_1 & \vec{v}_2 & \vec{v}_3\end{bmatrix}\begin{bmatrix}x_1\\x_2\\x_3\end{bmatrix}=x_1\vec{v}_1+x_2\vec{v}_2+x_3\vec{v}_3$.

We must choose $\vec{v}_1$, $\vec{v}_2$, and $\vec{v}_3$ in such a way that $x_1\vec{v}_1+x_2\vec{v}_2+x_3\vec{v}_3$ is perpendicular to $\vec{w}=\begin{bmatrix}1\\2\\3\end{bmatrix}$ for all x_1, x_2, and x_3. This is the case if (and only if) all the vectors $\vec{v}_1$, $\vec{v}_2$, and $\vec{v}_3$ are perpendicular to $\vec{w}$, that is, if $\vec{v}_1\cdot\vec{w}=\vec{v}_2\cdot\vec{w}=\vec{v}_3\cdot\vec{w}=0$.

For example, we can choose $\vec{v}_1=\begin{bmatrix}-2\\1\\0\end{bmatrix}$ and $\vec{v}_2=\vec{v}_3=\vec{0}$, so that $A=\begin{bmatrix}-2 & 0 & 0\\1 & 0 & 0\\0 & 0 & 0\end{bmatrix}$.

32. a.

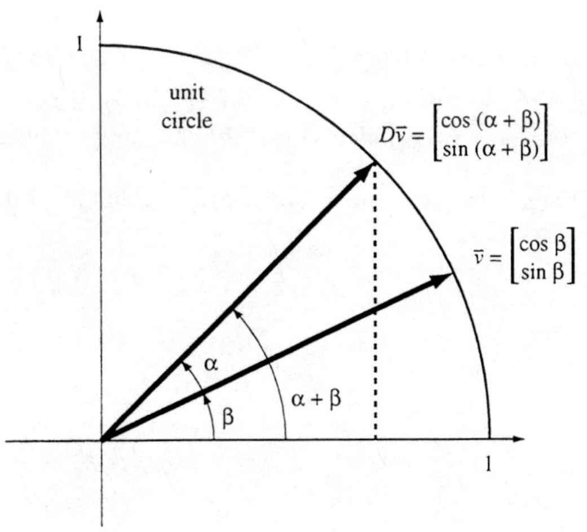

b. Compute $D\vec{v} = \begin{bmatrix} \cos\alpha & -\sin\alpha \\ \sin\alpha & \cos\alpha \end{bmatrix}\begin{bmatrix} \cos\beta \\ \sin\beta \end{bmatrix} = \begin{bmatrix} \cos\alpha\cos\beta - \sin\alpha\sin\beta \\ \sin\alpha\cos\beta + \cos\alpha\sin\beta \end{bmatrix}$.

Comparing this result with our finding in part (a), we get the addition theorems
$\cos(\alpha+\beta) = \cos\alpha\cos\beta - \sin\alpha\sin\beta$
$\sin(\alpha+\beta) = \sin\alpha\cos\beta + \cos\alpha\sin\beta$

33. Geometrically, we can find the representation $\vec{v} = \vec{v}_1 + \vec{v}_2$ by means of a parallelogram:

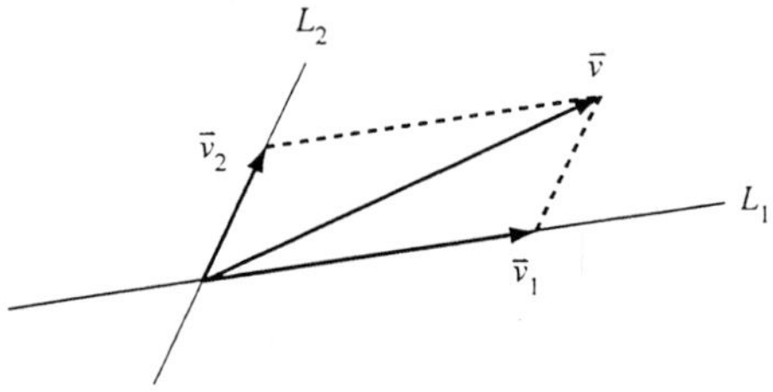

To show the existence and uniqueness of this representation algebraically, choose a nonzero vector $\vec{w}_1$ in L_1 and a nonzero $\vec{w}_2$ in L_2. Then the system $x_1\vec{w}_1 + x_2\vec{w}_2 = \vec{0}$ or $\begin{bmatrix} \vec{w}_1 & \vec{w}_2 \end{bmatrix}\begin{bmatrix} x_1 \\ x_2 \end{bmatrix} = \vec{0}$ has only the solution $x_1 = x_2 = 0$ (if $x_1\vec{w}_1 + x_2\vec{w}_2 = \vec{0}$ then $x_1\vec{w}_1 = -x_2\vec{w}_2$ is both in L_1 and in L_2, so that it must be the zero vector).

Therefore the system $x_1\vec{w}_1 + x_2\vec{w}_2 = \vec{v}$ or $\begin{bmatrix} \vec{w}_1 & \vec{w}_2 \end{bmatrix}\begin{bmatrix} x_1 \\ x_2 \end{bmatrix} = \vec{v}$ has a unique solution x_1, x_2 for all $\vec{v}$ in $\mathbb{R}^2$ (by Fact 1.3.4). Now set $\vec{v}_1 = x_1\vec{w}_1$ and $\vec{v}_2 = x_2\vec{w}_2$ to obtain the desired representation $\vec{v} = \vec{v}_1 + \vec{v}_2$.
(Compare with Exercise 1.3.57.)
To show that the transformation $T(\vec{v}) = \vec{v}_1$ is linear, we will verify the two parts of Fact 2.2.1.
Let $\underset{\text{in } L_1}{\underset{\uparrow}{\vec{v} = \vec{v}_1}} + \underset{\text{in } L_2}{\underset{\uparrow}{\vec{v}_2}}$, $\underset{\text{in } L_1}{\underset{\uparrow}{\vec{w} = \vec{w}_1}} + \underset{\text{in } L_2}{\underset{\uparrow}{\vec{w}_2}}$, so that $\vec{v} + \vec{w} = \underset{\text{in } L_1}{\underset{\uparrow}{(\vec{v}_1 + \vec{w}_1)}} + \underset{\text{in } L_2}{\underset{\uparrow}{(\vec{v}_2 + \vec{w}_2)}}$ and $k\vec{v} = \underset{\text{in } L_1}{\underset{\uparrow}{k\vec{v}_1}} + \underset{\text{in } L_2}{\underset{\uparrow}{k\vec{v}_2}}$.

a. $T(\vec{v} + \vec{w}) = \vec{v}_1 + \vec{w}_1 = T(\vec{v}) + T(\vec{w})$, and

b. $T(k\vec{v}) = k\vec{v}_1 = kT(\vec{v})$, as claimed.

34. Keep in mind that the columns of the matrix of a linear transformation T are $T(\vec{e}_1)$, $T(\vec{e}_2)$, and $T(\vec{e}_3)$.

If T is the orthogonal projection onto a line L, then $T(\vec{x})$ will be on L for all $\vec{x}$ in $\mathbb{R}^3$; in particular, the three columns of the matrix of T will be on L, and therefore pairwise parallel. This is the case only for matrix B:
B represents an orthogonal projection onto a line.
A reflection transforms orthogonal vectors into orthogonal vectors; therefore, the three columns of its matrix must be pairwise orthogonal. This is the case only for matrix E:
E represents the reflection in a line.

35. If the vectors $\vec{v}_1$ and $\vec{v}_2$ are defined as shown in the figure below, then the parallelogram *P* consists of all vectors of the form $\vec{v} = c_1\vec{v}_1 + c_2\vec{v}_2$, where $0 \le c_1, c_2 \le 1$.
The image of *P* consists of all vectors of the form $T(\vec{v}) = T(c_1\vec{v}_1 + c_2\vec{v}_2) = c_1T(\vec{v}_1) + c_2T(\vec{v}_2)$.
These vectors form the parallelogram shown below.

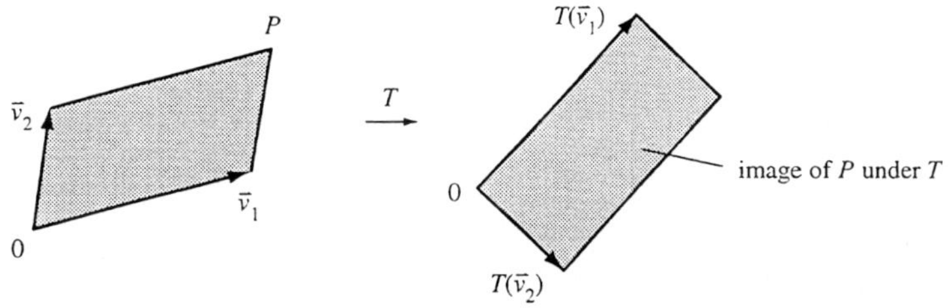

36. If the vectors $\vec{v}_0$, $\vec{v}_1$, and $\vec{v}_2$ are defined as shown in the figure below, then the parallelogram *P* consists of all vectors $\vec{v}$ of the form $\vec{v} = \vec{v}_0 + c_1\vec{v}_1 + c_2\vec{v}_2$, where $0 \le c_1, c_2 \le 1$.
The image of *P* consists of all vectors of the form $T(\vec{v}) = T(\vec{v}_0 + c_1\vec{v}_1 + c_2\vec{v}_2) = T(\vec{v}_0) + c_1T(\vec{v}_1) + c_2T(\vec{v}_2)$.
These vectors form the parallelogram shown below.

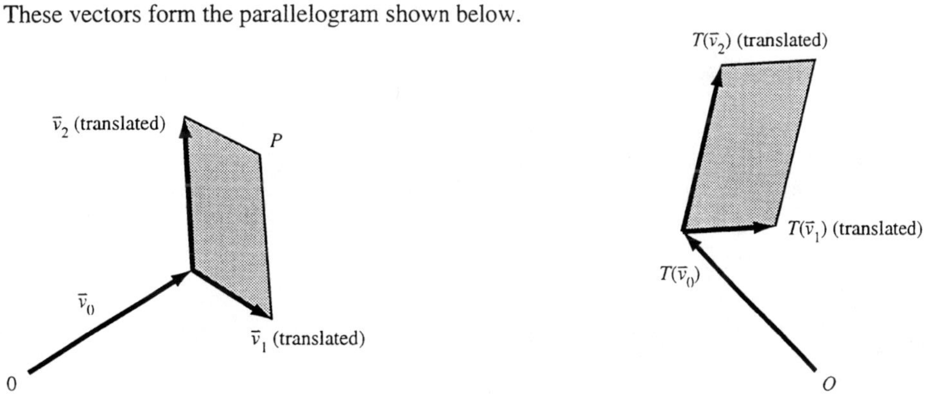

37. Write $\vec{w}$ as a linear combination of $\vec{v}_1$ and $\vec{v}_2$: $\vec{w} = c_1\vec{v}_1 + c_2\vec{v}_2$.

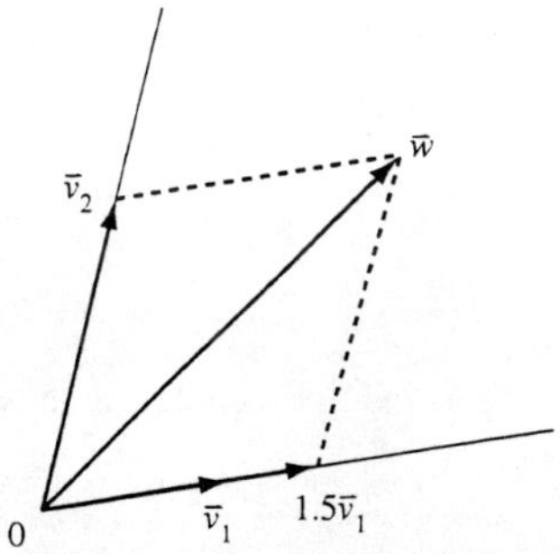

Measurements show that we have *roughly* $\vec{w} = 1.5\vec{v}_1 + \vec{v}_2$.
Therefore, by linearity, $T(\vec{w}) = T(1.5\vec{v}_1 + \vec{v}_2) = 1.5T(\vec{v}_1) + T(\vec{v}_2)$.

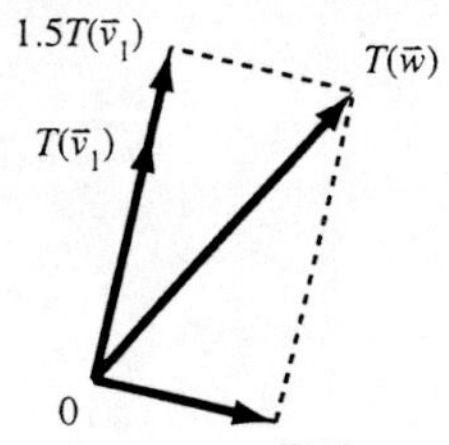

38. Write $\vec{w}$ as a linear combination of $\vec{v}_1$ and $\vec{v}_2$: $\vec{w} = c_1\vec{v}_1 + c_2\vec{v}_2$.
Then, by linearity, $T(\vec{w}) = T(c_1\vec{v}_1 + c_2\vec{v}_2) = c_1T(\vec{v}_1) + c_2T(\vec{v}_2) = c_1\vec{v}_1 + 3c_2\vec{v}_2$.

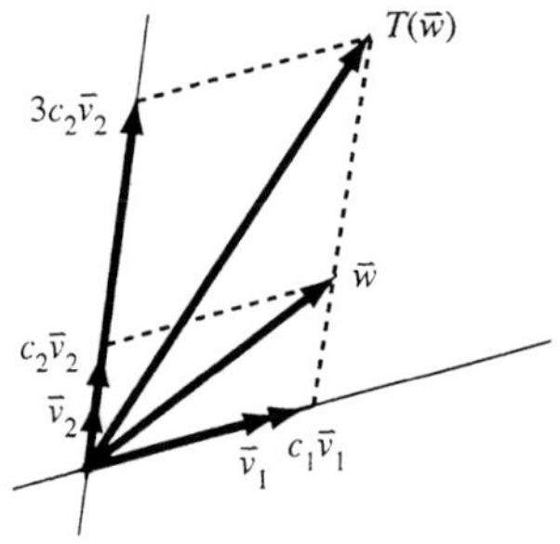

39. $T\begin{bmatrix}x_1\\x_2\end{bmatrix} = \begin{bmatrix}x_1\\x_2\end{bmatrix} + (ax_1 + bx_2)\begin{bmatrix}-b\\a\end{bmatrix} = \begin{bmatrix}(1-ab)x_1 - b^2x_2\\a^2x_1 + (ab+1)x_2\end{bmatrix} = \begin{bmatrix}1-ab & -b^2\\a^2 & 1+ab\end{bmatrix}\begin{bmatrix}x_1\\x_2\end{bmatrix}$

The matrix of T is $\begin{bmatrix}1-ab & -b^2\\a^2 & 1+ab\end{bmatrix}$.

40.

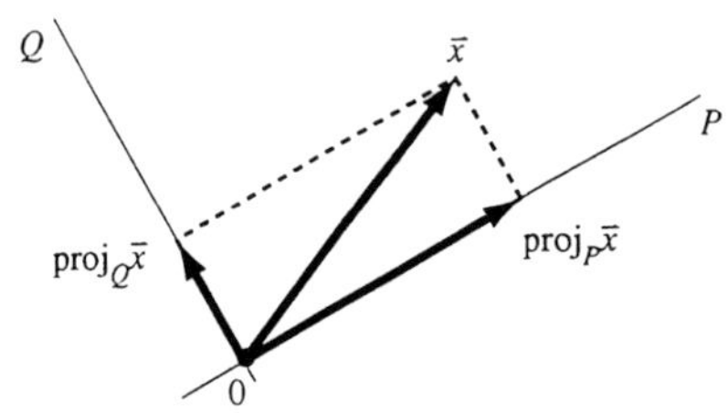

$\vec{x} = \text{proj}_P\vec{x} + \text{proj}_Q\vec{x}$

41.

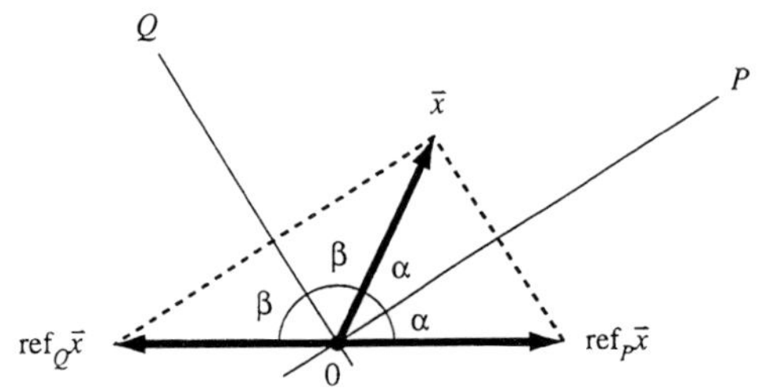

$\text{ref}_Q\vec{x} = -\text{ref}_P\vec{x}$, since $\text{ref}_Q\vec{x}$, $\text{ref}_P\vec{x}$, and $\vec{x}$ all have the same length, and $\text{ref}_Q\vec{x}$ and $\text{ref}_P\vec{x}$ enclose an angle of $2\alpha + 2\beta = 2(\alpha + \beta) = \pi$.

42. $T(\vec{x}) = T(T(\vec{x}))$ since $T(\vec{x})$ is on L hence the projection of $T(\vec{x})$ onto L is $T(\vec{x})$ itself.

43. Since $\vec{y} = A\vec{x}$ is obtained from $\vec{x}$ by a rotation through α in the counterclockwise direction, $\vec{x}$ is obtained from $\vec{y}$ by a rotation through α in the *clockwise* direction, that is, a rotation through $-\alpha$.

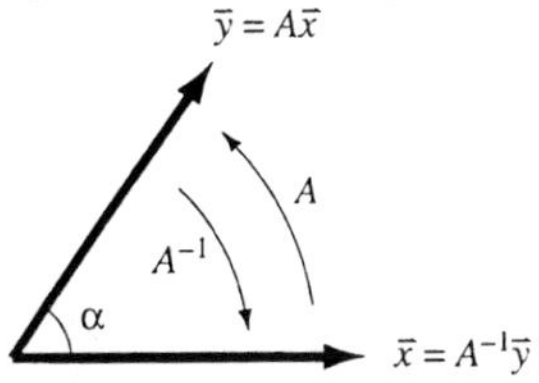

Therefore, the matrix of the inverse transformation is $A^{-1} = \begin{bmatrix} \cos(-\alpha) & -\sin(-\alpha) \\ \sin(-\alpha) & \cos(-\alpha) \end{bmatrix} = \begin{bmatrix} \cos\alpha & \sin\alpha \\ -\sin\alpha & \cos\alpha \end{bmatrix}$.

You can use the formula in Exercise 2.1.13b to check this result.

44. By Exercise 1.1.13b, $A^{-1} = \begin{bmatrix} a & -b \\ b & a \end{bmatrix}^{-1} = \dfrac{1}{a^2+b^2}\begin{bmatrix} a & b \\ -b & a \end{bmatrix}$.

If A represents a rotation through α followed by a dilation by r, then A^{-1} represents a rotation through $-\alpha$ followed by a dilation by $\dfrac{1}{r}$.

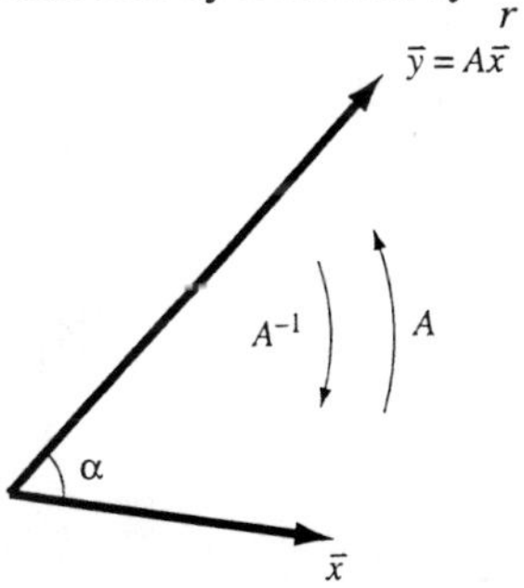

45. Since $\vec{v}_1$ and $\vec{v}_2$ are not parallel, any vector $\vec{x}$ in $\mathbb{R}^2$ can be written as a linear combination of $\vec{v}_1$ and $\vec{v}_2$: $\vec{x} = c_1\vec{v}_1 + c_2\vec{v}_2$.
Then $T(\vec{x}) = T(c_1\vec{v}_1 + c_2\vec{v}_2) = c_1T(\vec{v}_1) + c_2T(\vec{v}_2) = c_1L(\vec{v}_1) + c_2L(\vec{v}_2) = L(c_1\vec{v}_1 + c_2\vec{v}_2) = L(\vec{x})$, as claimed.

46. Suppose T is a shear parallel to the line L. Let $\vec{v}$ be a unit vector perpendicular to L, and set $\vec{w} = T(\vec{v}) - \vec{v}$; by definition of a shear, $\vec{w}$ will be in L and thus perpendicular to $\vec{v}$.

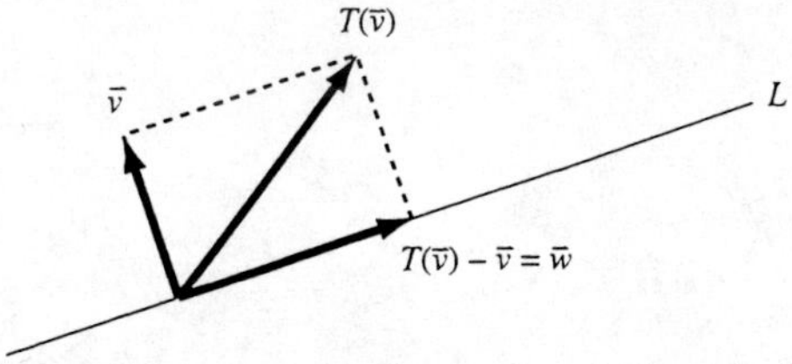

Define the linear transformation $L(\vec{x}) = \vec{x} + (\vec{v}\cdot\vec{x})\vec{w}$. We will show that $L(\vec{v}) = T(\vec{v})$ and $L(\vec{w}) = T(\vec{w})$; then $L(\vec{x}) = T(\vec{x})$ for all $\vec{x}$ in $\mathbb{R}^2$, as claimed (by Exercise 45):
$L(\vec{v}) = \vec{v} + (\vec{v}\cdot\vec{v})\vec{w} = \vec{v} + \vec{w} = T(\vec{v})$, and $L(\vec{w}) = \vec{w} + (\vec{v}\cdot\vec{w})\vec{w} = \vec{w} = T(\vec{w})$.

47. Write $T\begin{bmatrix} x_1 \\ x_2 \end{bmatrix} = \begin{bmatrix} a & b \\ c & d \end{bmatrix}\begin{bmatrix} x_1 \\ x_2 \end{bmatrix} = \begin{bmatrix} ax_1 + bx_2 \\ cx_1 + dx_2 \end{bmatrix}$.

a. $f(t) = \left(T\begin{bmatrix}\cos t\\ \sin t\end{bmatrix}\right)\cdot\left(T\begin{bmatrix}-\sin t\\ \cos t\end{bmatrix}\right) = \begin{bmatrix}a\cos t + b\sin t\\ c\cos t + d\sin t\end{bmatrix}\cdot\begin{bmatrix}-a\sin t + b\cos t\\ -c\sin t + d\cos t\end{bmatrix}$

$= (a\cos t + b\sin t)(-a\sin t + b\cos t) + (c\cos t + e\sin t)(-c\sin t + d\cos t)$

This function $f(t)$ is continuous, since $\cos(t)$, $\sin(t)$, and constant functions are continuous, and sums and products of continuous functions are continuous.

b. $f\left(\frac{\pi}{2}\right) = T\begin{bmatrix}0\\1\end{bmatrix}\cdot T\begin{bmatrix}-1\\0\end{bmatrix} = -\left(T\begin{bmatrix}0\\1\end{bmatrix}\cdot T\begin{bmatrix}1\\0\end{bmatrix}\right)$

↑

T is linear.

$f(0) = T\begin{bmatrix}1\\0\end{bmatrix}\cdot T\begin{bmatrix}0\\1\end{bmatrix} = T\begin{bmatrix}0\\1\end{bmatrix}\cdot T\begin{bmatrix}1\\0\end{bmatrix}$

The claim follows.

c. By part (b), the numbers $f(0)$ and $f\left(\frac{\pi}{2}\right)$ have different signs (one is positive and the other negative), or they are both zero. Since $f(t)$ is continuous, by part (a), we can apply the intermediate value theorem.

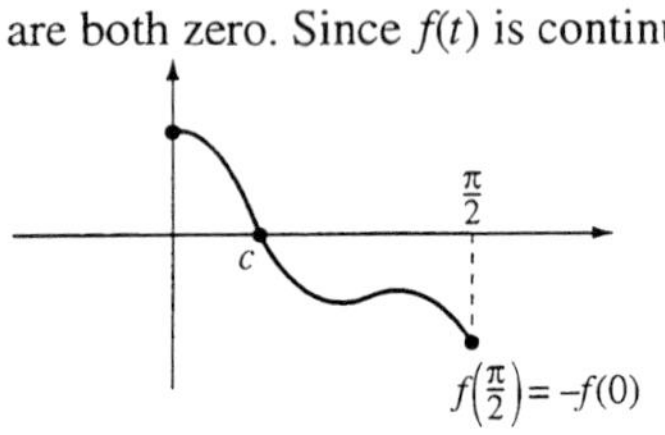

d. Note that $\begin{bmatrix}\cos(t)\\ \sin(t)\end{bmatrix}$ and $\begin{bmatrix}-\sin(t)\\ \cos(t)\end{bmatrix}$ are perpendicular unit vectors, for any t. If we set $\vec{v}_1 = \begin{bmatrix}\cos(c)\\ \sin(c)\end{bmatrix}$, $\vec{v}_2 = \begin{bmatrix}-\sin(c)\\ \cos(c)\end{bmatrix}$, with the number c we found in part (c), then $f(c) = T(\vec{v}_1)\cdot T(\vec{v}_2) = 0$, so that $T(\vec{v}_1)$ and $T(\vec{v}_2)$ are perpendicular, as claimed. Note that $T(\vec{v}_1)$ or $T(\vec{v}_2)$ may be zero.

48. We find $f(t) = \left(\begin{bmatrix}0 & 4\\ 5 & -3\end{bmatrix}\begin{bmatrix}\cos(t)\\ \sin(t)\end{bmatrix}\right)\cdot\left(\begin{bmatrix}0 & 4\\ 5 & -3\end{bmatrix}\begin{bmatrix}-\sin(t)\\ \cos(t)\end{bmatrix}\right) = \begin{bmatrix}4\sin(t)\\ 5\cos(t) - 3\sin(t)\end{bmatrix}\cdot\begin{bmatrix}4\cos(t)\\ -5\sin(t) - 3\cos(t)\end{bmatrix}$

$= 15(\sin^2 t - \cos^2 t) = 15(2\sin^2 t - 1)$

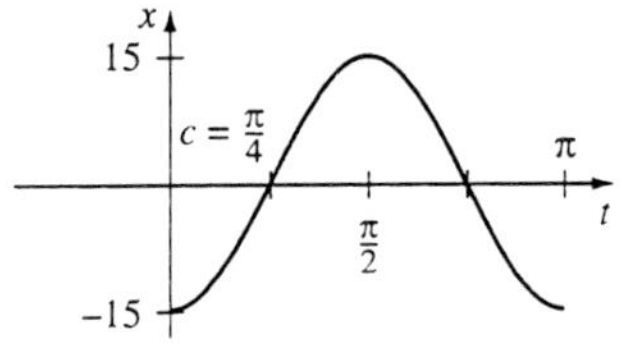

The only zero of $f(t)$ between 0 and $\frac{\pi}{2}$ is at $c = \frac{\pi}{4}$.

Therefore, $\vec{v}_1 = \begin{bmatrix}\cos\left(\frac{\pi}{4}\right)\\ \sin\left(\frac{\pi}{4}\right)\end{bmatrix} = \begin{bmatrix}\frac{\sqrt{2}}{2}\\ \frac{\sqrt{2}}{2}\end{bmatrix}$ and $\vec{v}_2 = \begin{bmatrix}-\sin\left(\frac{\pi}{4}\right)\\ \cos\left(\frac{\pi}{4}\right)\end{bmatrix} = \begin{bmatrix}-\frac{\sqrt{2}}{2}\\ \frac{\sqrt{2}}{2}\end{bmatrix}$ work. Note that $T(\vec{v}_1) = \frac{1}{\sqrt{2}}\begin{bmatrix}4\\2\end{bmatrix}$ and $T(\vec{v}_2) = \frac{1}{\sqrt{2}}\begin{bmatrix}4\\-8\end{bmatrix}$ are indeed perpendicular.

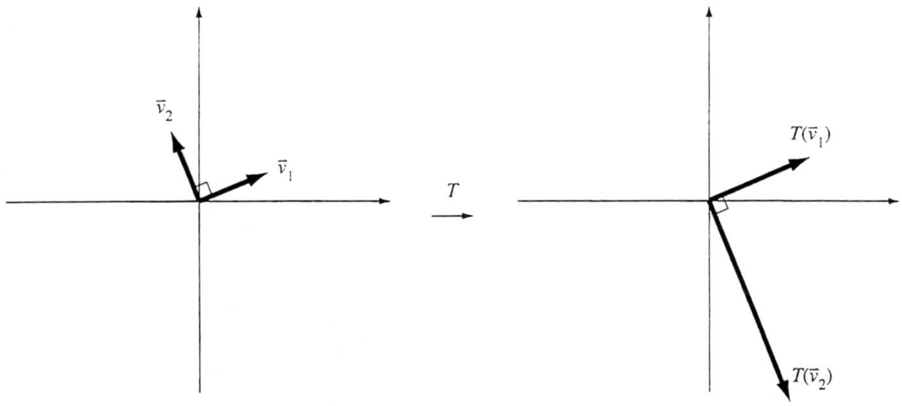

49. If $\vec{x}=\begin{bmatrix}\cos(t)\\ \sin(t)\end{bmatrix}$ then $T(\vec{x})=\begin{bmatrix}5 & 0\\ 0 & 2\end{bmatrix}\begin{bmatrix}\cos(t)\\ \sin(t)\end{bmatrix}=\begin{bmatrix}5\cos(t)\\ 2\sin(t)\end{bmatrix}=\cos(t)\begin{bmatrix}5\\ 0\end{bmatrix}+\sin(t)\begin{bmatrix}0\\ 2\end{bmatrix}$.

These vectors form an ellipse; consider the characterization of an ellipse given in the footnote on page 85, with $\vec{w}_1=\begin{bmatrix}5\\ 0\end{bmatrix}$ and $\vec{w}_2=\begin{bmatrix}0\\ 2\end{bmatrix}$.

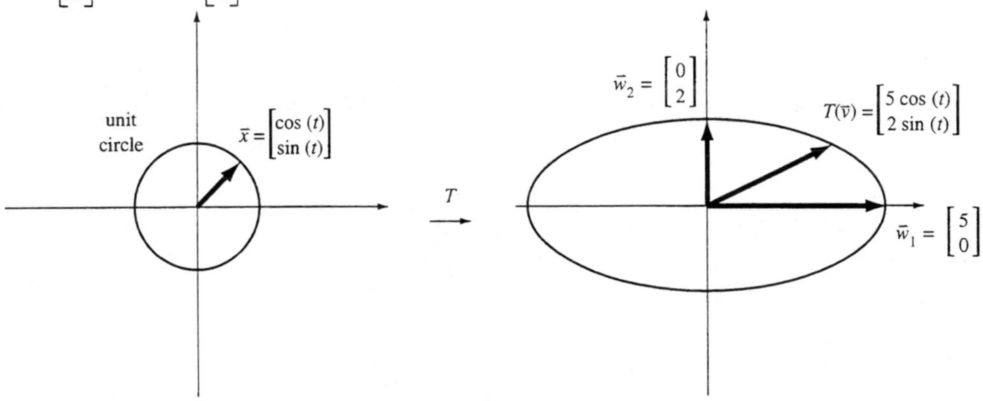

50. Use the hint: Since the vectors on the unit circle are of the form $\vec{v}=\cos(t)\vec{v}_1+\sin(t)\vec{v}_2$, the image of the unit circle consists of the vectors of the form $T(\vec{v})=T(\cos(t)\vec{v}_1+\sin(t)\vec{v}_2)=\cos(t)T(\vec{v}_1)+\sin(t)T(\vec{v}_2)$.

These vectors form an ellipse: Consider the characterization of an ellipse given in the footnote, with $\vec{w}_1=T(\vec{v}_1)$ and $\vec{w}_2=T(\vec{v}_2)$. The key point is that $T(\vec{v}_1)$ and $T(\vec{v}_2)$ are perpendicular.

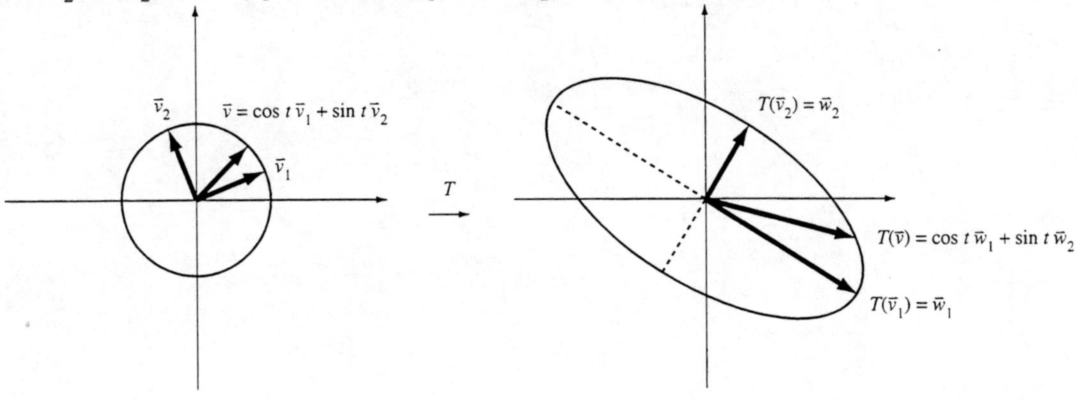

51. Consider the linear transformation T with matrix $A = [\vec{w}_1 \quad \vec{w}_2]$, that is,

$$T\begin{bmatrix} x_1 \\ x_2 \end{bmatrix} = A\begin{bmatrix} x_1 \\ x_2 \end{bmatrix} = [\vec{w}_1 \quad \vec{w}_2]\begin{bmatrix} x_1 \\ x_2 \end{bmatrix} = x_1\vec{w}_1 + x_2\vec{w}_2.$$

The curve C is the image of the unit circle under the transformation T: if $\vec{v} = \begin{bmatrix} \cos(t) \\ \sin(t) \end{bmatrix}$ is on the unit circle, then $T(\vec{v}) = \cos(t)\vec{w}_1 + \sin(t)\vec{w}_2$ is on the curve C. Therefore C is an ellipse, by Exercise 50.

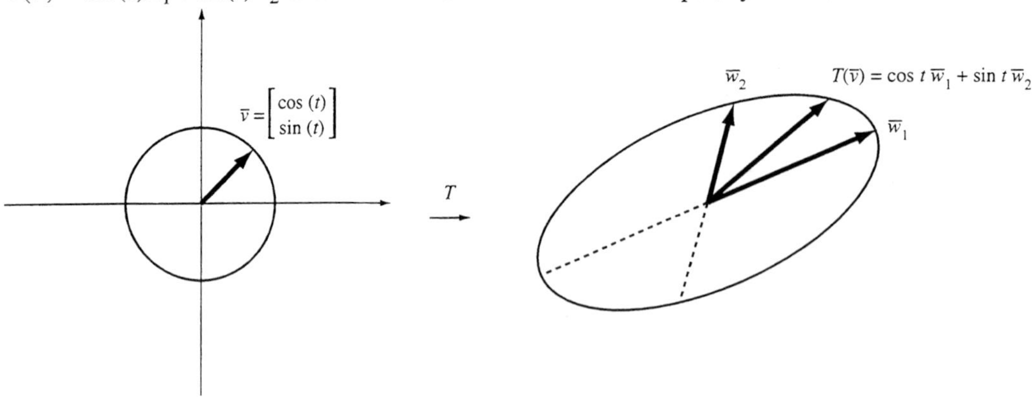

52. By definition, the vectors $\vec{v}$ on an ellipse E are of the form $\vec{v} = \cos(t)\vec{v}_1 + \sin(t)\vec{v}_2$, for some perpendicular vectors $\vec{v}_1$ and $\vec{v}_2$. Then the vectors on the image C of E are of the form $T(\vec{v}) = \cos(t)T(\vec{v}_1) + \sin(t)T(\vec{v}_2)$. These vectors form an ellipse, by Exercise 51 (with $\vec{w}_1 = T(\vec{v}_1)$ and $\vec{w}_2 = T(\vec{v}_2)$).

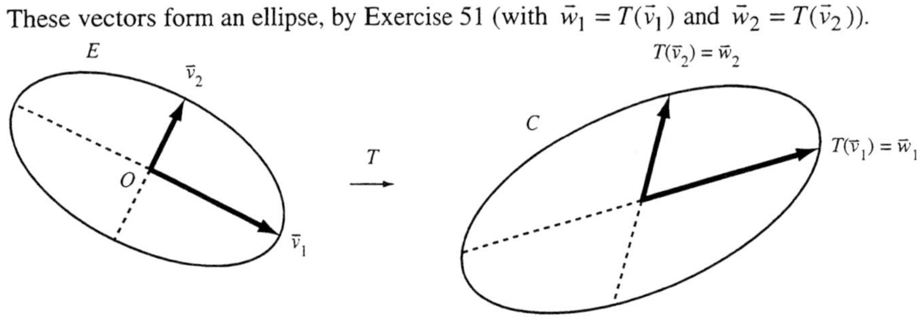

2.3

1. $\text{rref}\left[\begin{array}{cc|cc} 2 & 3 & 1 & 0 \\ 5 & 8 & 0 & 1 \end{array}\right] = \left[\begin{array}{cc|cc} 1 & 0 & 8 & -3 \\ 0 & 1 & -5 & 2 \end{array}\right]$, so that $\begin{bmatrix} 2 & 3 \\ 5 & 8 \end{bmatrix}^{-1} = \begin{bmatrix} 8 & -3 \\ -5 & 2 \end{bmatrix}$.

2. $\text{rref}\left[\begin{array}{cc|cc} 1 & 1 & 1 & 0 \\ 1 & 1 & 0 & 1 \end{array}\right] = \left[\begin{array}{cc|cc} 1 & 1 & 0 & 1 \\ 0 & 0 & 1 & -1 \end{array}\right]$, so that $\begin{bmatrix} 1 & 1 \\ 1 & 1 \end{bmatrix}$ is not invertible.

3. $\text{rref}\left[\begin{array}{cc|cc} 0 & 2 & 1 & 0 \\ 1 & 1 & 0 & 1 \end{array}\right] = \left[\begin{array}{cc|cc} 1 & 0 & -\frac{1}{2} & 1 \\ 0 & 1 & \frac{1}{2} & 0 \end{array}\right]$, so that $\begin{bmatrix} 0 & 2 \\ 1 & 1 \end{bmatrix}^{-1} = \begin{bmatrix} -\frac{1}{2} & 1 \\ \frac{1}{2} & 0 \end{bmatrix}$.

4. Use Fact 2.3.5; the inverse is $\begin{bmatrix} \frac{3}{2} & -1 & \frac{1}{2} \\ \frac{1}{2} & 0 & -\frac{1}{2} \\ -\frac{3}{2} & 1 & \frac{1}{2} \end{bmatrix}$.

5. $\text{rref}\begin{bmatrix} 1 & 2 & 2 \\ 1 & 3 & 1 \\ 1 & 1 & 3 \end{bmatrix} = \begin{bmatrix} 1 & 0 & 4 \\ 0 & 1 & -1 \\ 0 & 0 & 0 \end{bmatrix}$, so that the matrix is not invertible, by Fact 2.3.3.

6. Use Fact 2.3.5; the inverse is $\begin{bmatrix} 1 & -2 & 1 \\ 0 & 1 & -2 \\ 0 & 0 & 1 \end{bmatrix}$.

7. $\text{rref}\begin{bmatrix} 1 & 2 & 3 \\ 0 & 0 & 2 \\ 0 & 0 & 3 \end{bmatrix} = \begin{bmatrix} 1 & 2 & 0 \\ 0 & 0 & 1 \\ 0 & 0 & 0 \end{bmatrix}$, so that the matrix is not invertible, by Fact 2.3.3.

8. Use Fact 2.3.5; the inverse is $\begin{bmatrix} 0 & 0 & 1 \\ 0 & 1 & 0 \\ 1 & 0 & 0 \end{bmatrix}$.

9. $\text{rref}\begin{bmatrix} 1 & 1 & 1 \\ 1 & 1 & 1 \\ 1 & 1 & 1 \end{bmatrix} = \begin{bmatrix} 1 & 1 & 1 \\ 0 & 0 & 0 \\ 0 & 0 & 0 \end{bmatrix}$, so that the matrix is not invertible, by Fact 2.3.3.

10. Use Fact 2.3.5; the inverse is $\begin{bmatrix} 3 & -3 & 1 \\ -3 & 5 & -2 \\ 1 & -2 & 1 \end{bmatrix}$.

11. Use Fact 2.3.5; the inverse is $\begin{bmatrix} 1 & 0 & -1 \\ 0 & 1 & 0 \\ 0 & 0 & 1 \end{bmatrix}$.

12. Use Fact 2.3.5; the inverse is $\begin{bmatrix} 5 & -20 & -2 & -7 \\ 0 & -1 & 0 & 0 \\ -2 & 6 & 1 & 2 \\ 0 & 3 & 0 & 1 \end{bmatrix}$.

13. Use Fact 2.3.5; the inverse is $\begin{bmatrix} 1 & 0 & 0 & 0 \\ -2 & 1 & 0 & 0 \\ 1 & -2 & 1 & 0 \\ 0 & 1 & -2 & 1 \end{bmatrix}$.

14. Use Fact 2.3.5; the inverse is $\begin{bmatrix} 3 & -5 & 0 & 0 \\ -1 & 2 & 0 & 0 \\ 0 & 0 & 5 & -2 \\ 0 & 0 & -2 & 1 \end{bmatrix}$.

15. Use Fact 2.3.5; the inverse is $\begin{bmatrix} -6 & 9 & -5 & 1 \\ 9 & -1 & -5 & 2 \\ -5 & -5 & 9 & -3 \\ 1 & 2 & -3 & 1 \end{bmatrix}$.

16. Solving for x_1 and x_2 in terms of y_1 and y_2 we find that

$$\begin{aligned} x_1 &= -8y_1 + 5y_2 \\ x_2 &= \ \ 5y_1 - 3y_2 \end{aligned}.$$

17. We make an attempt to solve for x_1 and x_2 in terms of y_1 and y_2:

$$\left| \begin{array}{rcrcr} x_1 & + & 2x_2 & = & y_1 \\ 4x_1 & + & 8x_2 & = & y_2 \end{array} \right| \xrightarrow[-4(\mathrm{I})]{} \left| \begin{array}{rcrcrcr} x_1 & + & 2x_2 & = & y_1 & & \\ & & 0 & = & -4y_1 & + & y_2 \end{array} \right|.$$

This system has no solutions (x_1, x_2) for some (y_1, y_2), and infinitely many solutions for others; the transformation is not invertible.

18. Solving for x_1, x_2, and x_3 in terms of y_1, y_2, and y_3 we find that

$$\begin{aligned} x_1 &= y_3 \\ x_2 &= y_1 \\ x_3 &= y_2 \end{aligned}.$$

19. Solving for x_1, x_2, and x_3 in terms of y_1, y_2, and y_3 we find that

$$\begin{aligned} x_1 &= 3y_1 - \frac{5}{2}y_2 + \frac{1}{2}y_3 \\ x_2 &= -3y_1 + 4y_2 - y_3 \\ x_3 &= y_1 - \frac{3}{2}y_2 + \frac{1}{2}y_3 \end{aligned}.$$

20. Solving for x_1, x_2, and x_3 in terms of y_1, y_2, and y_3 we find that

$$\begin{aligned} x_1 &= -8y_1 - 15y_2 + 12y_3 \\ x_2 &= \ \ 4y_1 + 6y_2 - \ \ 5y_3 \\ x_3 &= - \ \ y_1 - \ \ y_2 + \ \ \ \ y_3 \end{aligned}.$$

21. $f(x) = x^2$ is not invertible, since the equation $f(x) = x^2 = 1$ has two solutions $x_{1,2} = \pm 1$.

22. $f(x) = 2^x$ is not invertible, since the equation $f(x) = 2^x = 0$ has no solution x.

23. Note that $f'(x) = 3x^2 + 1$ is always positive; this implies that the function $f(x) = x^3 + x$ is increasing throughout. Therefore, the equation $f(x) = b$ has *at most* one solution x for all b.
Now observe that $\lim_{x \to \infty} f(x) = \infty$ and $\lim_{x \to -\infty} f(x) = -\infty$; this implies that the equation $f(x) = b$ has at least one solution x for a given b (for a careful proof, use the intermediate value theorem; compare with Exercise 2.2.47c).

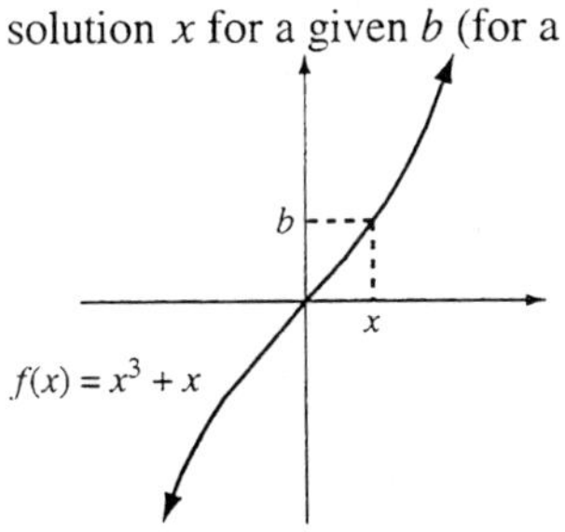

24. We can write $f(x) = x^3 - x = x(x^2 - 1) = x(x-1)(x+1)$.
The equation $f(x) = 0$ has three solutions, $x = 0, 1, -1$, so that $f(x)$ is not invertible.

25. Invertible, with inverse $\begin{bmatrix} x_1 \\ x_2 \end{bmatrix} = \begin{bmatrix} \sqrt[3]{y_1} \\ y_2 \end{bmatrix}$

26. Invertible, with inverse $\begin{bmatrix} x_1 \\ x_2 \end{bmatrix} = \begin{bmatrix} \sqrt[3]{y_2 - y_1} \\ y_1 \end{bmatrix}$

27. Not invertible, since the equation $\begin{bmatrix} x_1 + x_2 \\ x_1 x_2 \end{bmatrix} = \begin{bmatrix} 0 \\ 1 \end{bmatrix}$ has no solutions.

28. We are asked to find the inverse of the matrix $A = \begin{bmatrix} 22 & 13 & 8 & 3 \\ -16 & -3 & -2 & -2 \\ 8 & 9 & 7 & 2 \\ 5 & 4 & 3 & 1 \end{bmatrix}$.

We find that $A^{-1} = \begin{bmatrix} 1 & -2 & 9 & -25 \\ -2 & 5 & -22 & 60 \\ 4 & -9 & 41 & -112 \\ -9 & 17 & 80 & 222 \end{bmatrix}$.

T^{-1} is the transformation from $\mathbb{R}^4$ to $\mathbb{R}^4$ with matrix A^{-1}.

29. Use Fact 2.3.3:

$$\begin{bmatrix} 1 & 1 & 1 \\ 1 & 2 & k \\ 1 & 4 & k^2 \end{bmatrix} \begin{matrix} \to \\ -\text{I} \\ -\text{I} \end{matrix} \begin{bmatrix} 1 & 1 & 1 \\ 0 & 1 & k-1 \\ 0 & 3 & k^2-1 \end{bmatrix} \begin{matrix} -\text{II} \\ \to \\ -3(\text{II}) \end{matrix} \begin{bmatrix} 1 & 0 & 2-k \\ 0 & 1 & k-1 \\ 0 & 0 & k^2-3k+2 \end{bmatrix}$$

The matrix is invertible if (and only if) $k^2 - 3k + 2 = (k-2)(k-1) \neq 0$, in which case we can further reduce it to I_3. Therefore, the matrix is invertible if $k \neq 1$ and $k \neq 2$.

30. Use Fact 2.3.3:

$$\begin{bmatrix} 0 & 1 & b \\ -1 & 0 & c \\ -b & -c & 0 \end{bmatrix} \xrightarrow{\text{I} \leftrightarrow \text{II}} \begin{bmatrix} -1 & 0 & c \\ 0 & 1 & b \\ -b & -c & 0 \end{bmatrix} \begin{matrix} \div(-1) \\ \\ \\ \end{matrix} \longrightarrow \begin{bmatrix} 1 & 0 & -c \\ 0 & 1 & b \\ -b & -c & 0 \end{bmatrix} \begin{matrix} \\ \\ +b(\text{I})+c(\text{II}) \end{matrix} \rightarrow \begin{bmatrix} 1 & 0 & -c \\ 0 & 1 & b \\ 0 & 0 & 0 \end{bmatrix}$$

This matrix is noninvertible, regardless of the values of b and c.

31. Use Fact 2.3.3; first assume that $a \neq 0$.

$$\begin{bmatrix} 0 & a & b \\ -a & 0 & c \\ -b & -c & 0 \end{bmatrix} \xrightarrow{\text{I} \leftrightarrow \text{II}} \begin{bmatrix} -a & 0 & c \\ 0 & a & b \\ -b & -c & 0 \end{bmatrix} \begin{matrix} \div(-a) \\ \\ \\ \end{matrix} \rightarrow \begin{bmatrix} 1 & 0 & -\frac{c}{a} \\ 0 & a & b \\ -b & -c & 0 \end{bmatrix} \begin{matrix} \\ \\ +b(\text{I}) \end{matrix} \rightarrow \begin{bmatrix} 1 & 0 & -\frac{c}{a} \\ 0 & a & b \\ 0 & -c & -\frac{bc}{a} \end{bmatrix} \begin{matrix} \\ \div a \\ \\ \end{matrix} \rightarrow \begin{bmatrix} 1 & 0 & -\frac{c}{a} \\ 0 & 1 & \frac{b}{a} \\ 0 & -c & -\frac{bc}{a} \end{bmatrix} \begin{matrix} \\ \\ +c(\text{II}) \end{matrix} \rightarrow$$

$$\begin{bmatrix} 1 & 0 & -\frac{c}{a} \\ 0 & 1 & \frac{b}{a} \\ 0 & 0 & 0 \end{bmatrix}$$

Now consider the case when $a = 0$:

$$\begin{bmatrix} 0 & 0 & b \\ 0 & 0 & c \\ -b & -c & 0 \end{bmatrix} \xrightarrow{\text{I} \leftrightarrow \text{III}} \begin{bmatrix} -b & -c & 0 \\ 0 & 0 & c \\ 0 & 0 & b \end{bmatrix}$$: The second entry on the diagonal of rref will be 0.

It follows that the matrix $\begin{bmatrix} 0 & a & b \\ -a & 0 & c \\ -b & -c & 0 \end{bmatrix}$ is noninvertible, regardless of the values of a, b, and c.

32. Use Fact 2.3.6.

If $A = \begin{bmatrix} a & b \\ c & d \end{bmatrix}$ is a matrix such that $ad - bc = 1$ and $A^{-1} = A$, then

$$A^{-1} = \frac{1}{ad-bc}\begin{bmatrix} d & -b \\ -c & a \end{bmatrix} = \begin{bmatrix} d & -b \\ -c & a \end{bmatrix} = \begin{bmatrix} a & b \\ c & d \end{bmatrix}$$, so that $b = 0$, $c = 0$, and $a = d$.

The condition $ad - bc = a^2 = 1$ now implies that $a = d = 1$ or $a = d = -1$.
This leaves only two matrices A, namely, I_2 and $-I_2$. Check that these two matrices do indeed satisfy the given requirements.

33. Use Fact 2.3.6.

The requirement $A^{-1} = A$ means that $-\frac{1}{a^2+b^2}\begin{bmatrix} -a & -b \\ -b & a \end{bmatrix} = \begin{bmatrix} a & b \\ b & -a \end{bmatrix}$. This is the case if (and only if) $a^2 + b^2 = 1$.

34. a. By Fact 2.3.3, A is invertible if (and only if) a, b, and c are all nonzero. In this case, $A^{-1} = \begin{bmatrix} \frac{1}{a} & 0 & 0 \\ 0 & \frac{1}{b} & 0 \\ 0 & 0 & \frac{1}{c} \end{bmatrix}$.

b. In general, a diagonal matrix is invertible if (and only if) all of its diagonal entries are nonzero.

35. a. A is invertible if (and only if) all its diagonal entries, a, d, and f, are nonzero.

b. As in part (a): if all the diagonal entries are nonzero.

c. Yes, A^{-1} will be upper triangular as well; as you construct $\text{rref}[A \mid I_n]$, you will perform only the following row operations:
· divide rows by scalars
· subtract a multiple of the jth row from the ith row, where $j > i$.
Applying these operations to I_n, you end up with an upper triangular matrix.

d. As in part (b): if all diagonal entries are nonzero.

36. If a matrix A can be transformed into B by elementary row operations, then A is invertible if (and only if) B is invertible. The claim now follows from Exercise 35, where we show that a triangular matrix is invertible if (and only if) its diagonal entries are nonzero.

37. Make an attempt to solve the linear transformation $\vec{y} = (cA)\vec{x} = c(A\vec{x})$ for $\vec{x}$:

$$A\vec{x} = \frac{1}{c}\vec{y}, \text{ so that } \vec{x} = A^{-1}\left(\frac{1}{c}\vec{y}\right) = \left(\frac{1}{c}A^{-1}\right)\vec{y}.$$

This shows that cA is indeed invertible, with $(cA)^{-1} = \frac{1}{c}A^{-1}$.

38. Use Fact 2.3.6; $A^{-1} = \frac{1}{-1}\begin{bmatrix} -1 & -k \\ 0 & 1 \end{bmatrix} = \begin{bmatrix} 1 & k \\ 0 & -1 \end{bmatrix} (= A)$.

39. Suppose the ijth entry of M is k, and all other entries are as in the identity matrix. Then we can find $\text{rref}[M \mid I_n]$ by subtracting k times the jth row from the ith row. Therefore, M is indeed invertible, and M^{-1} differs from the identity matrix only at the ijth entry; that entry is $-k$.

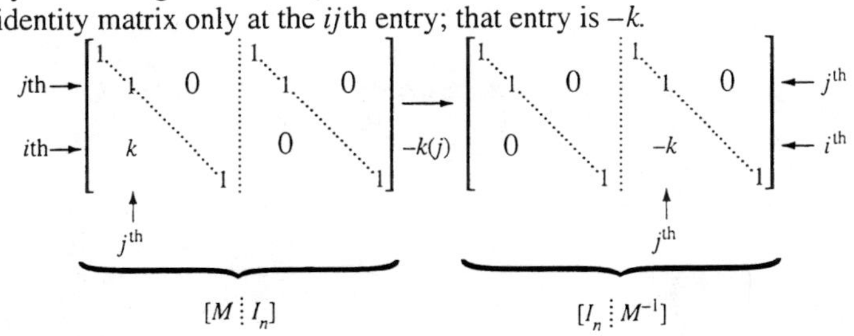

40. If you apply an elementary row operation to a matrix with two equal columns, then the resulting matrix will also have two equal columns. Therefore, rref(A) has two equal columns, so that $\text{rref}(A) \neq I_n$. Now use Fact 2.3.3.

41. a. Invertible: the transformation is its own inverse.

b. Not invertible: the equation $T(\vec{x}) = \vec{b}$ has infinitely many solutions if $\vec{b}$ is on the plane, and none otherwise.

c. Invertible: The inverse is a dilation by $\frac{1}{5}$ (that is, a contraction by 5). If $\vec{y} = 5\vec{x}$, then $\vec{x} = \frac{1}{5}\vec{y}$.

d. Invertible: The inverse is a rotation about the same axis through the same angle in the opposite direction.

42. Permutation matrices are invertible since they row reduce to I_n in an obvious way, just by row swaps. The inverse of a permutation matrix A is also a permutation matrix since $\text{rref}[A \mid I_n] = [I_n \mid A^{-1}]$ is obtained from $[A \mid I_n]$ by a sequence of row swaps.

43. We make an attempt to solve the equation $\vec{y} = A(B\vec{x})$ for $\vec{x}$:
$B\vec{x} = A^{-1}\vec{y}$, so that $\vec{x} = B^{-1}(A^{-1}\vec{y})$.

44. a. $\text{rref}(M_4) = \begin{bmatrix} 1 & 0 & -1 & -2 \\ 0 & 1 & 2 & 3 \\ 0 & 0 & 0 & 0 \\ 0 & 0 & 0 & 0 \end{bmatrix}$, so that $\text{rank}(M_4) = 2$.

b. To simplify the notation, we introduce the row vectors $\vec{v} = [1 \;\; 1 \;\; \cdots \;\; 1]$ and $\vec{w} = [0 \;\; n \;\; 2n \;\; \cdots \;\; (n-1)n]$ with n components. Then we can write M_n in terms of its rows as $M_n = \begin{bmatrix} \vec{v}+\vec{w} \\ 2\vec{v}+\vec{w} \\ \cdots \\ n\vec{v}+\vec{w} \end{bmatrix} \begin{matrix} \\ -2(\text{I}) \\ \vdots \\ -n(\text{I}) \end{matrix}$.

Applying the Gauss-Jordan algorithm to the first column we get $\begin{bmatrix} \vec{v}+\vec{w} \\ -\vec{w} \\ -2\vec{w} \\ \cdots \\ -(n-1)\vec{w} \end{bmatrix}$.

All the rows below the second are scalar multiples of the second; therefore, $\text{rank}(M_n) = 2$.

c. By part (b), the matrix M_n is invertible only if $n = 1$ or $n = 2$.

45. a. Each of the three row divisions requires 3 operations, and each of the six row subtractions requires three operations as well; altogether, we have $3 \cdot 3 + 6 \cdot 3 = 9 \cdot 3 = 3^3 = 27$ operations.

b. Suppose we have already taken care of the first m columns: $[A \mid I_n]$ has been reduced to

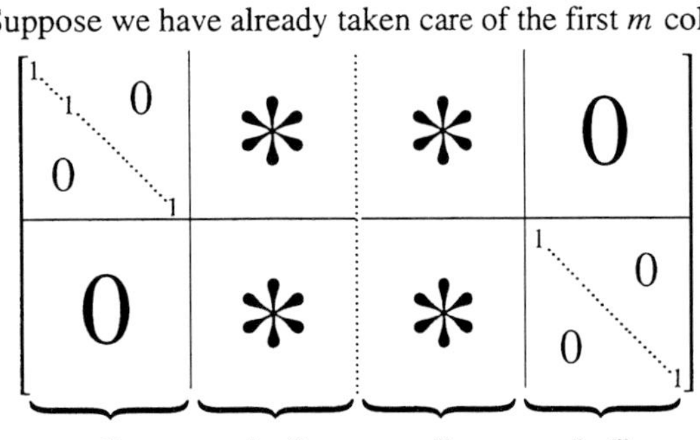

Here, the stars represent arbitrary entries.
Suppose the $(m + 1)$th entry on the diagonal is k. Dividing the $(m + 1)$th row by k requires n operations: $n - m - 1$ to the left of the dotted line $\left(\text{not counting the computation } \frac{k}{k} = 1\right)$, and $m + 1$ to the right of the dotted line $\left(\text{including } \frac{1}{k}\right)$. Now the matrix has the form

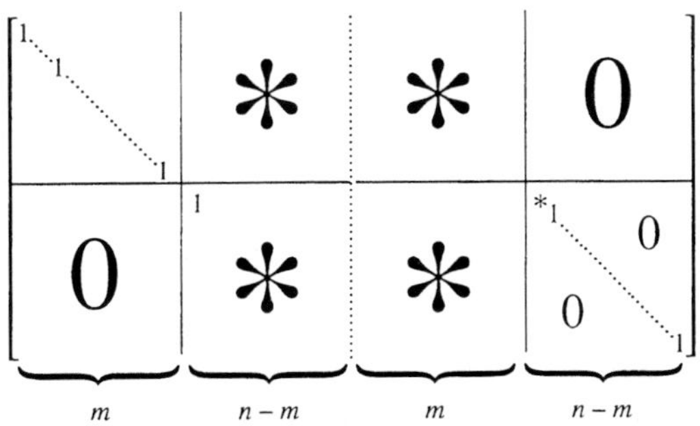

Eliminating each of the other $n-1$ components of the $(m+1)$th column now requires n operations $(n-m-1$ to the left of the dotted line, and $m+1$ to the right). Altogether, it requires $n+(n-1)n=n^2$ operations to process the mth column. To process all n columns requires $n \cdot n^2 = n^3$ operations.

c. The inversion of a 12×12 matrix requires $12^3 = 4^3 3^3 = 64 \cdot 3^3$ operations, that is, 64 times as much as the inversion of a 3×3 matrix. If the inversion of a 3×3 matrix takes one second, then the inversion of a 12×12 matrix takes 64 seconds.

46. Computing $A^{-1}\vec{b}$ requires n^3+n^2 operations: First we need n^3 operations to find A^{-1} (see Exercise 45b) and then n^2 operations to compute $A^{-1}\vec{b}$ (n multiplications for each component).
How many operations are required to perform Gauss-Jordan elimination on $\left[A \mid \vec{b}\right]$? Let us count these operations "column by column." If m columns of the coefficient matrix are left, then processing the next column requires nm operations (compare with Exercise 45(b)). To process all the columns requires

$$n \cdot n + n(n-1) + \cdots + n \cdot 2 + n \cdot 1 = n(n+n-1+\cdots+2+1) = n\frac{n(n+1)}{2} = \frac{n^3+n^2}{2} \text{ operations,}$$

only half of what was required to compute $A^{-1}\vec{b}$.
We mention in passing that one can reduce the number of operations further (by about 50% for large matrices) by performing the steps of the row reduction in a different order.

47. Let $f(x) = x^2$; the equation $f(x) = 0$ has the unique solution $x = 0$.

48. Let $A = \begin{bmatrix} 1 & 0 \\ 0 & 1 \\ 0 & 0 \end{bmatrix}$ and $\vec{b} = \begin{bmatrix} 0 \\ 0 \\ 0 \end{bmatrix}$. The equation $A\vec{x} = \vec{b}$ has the unique solution $\vec{x} = \begin{bmatrix} 0 \\ 0 \end{bmatrix}$. Note that Fact 2.3.4 applies to *square* matrices only.

49. a. $A = \begin{bmatrix} 0.293 & 0 & 0 \\ 0.014 & 0.207 & 0.017 \\ 0.044 & 0.01 & 0.216 \end{bmatrix}$, $I_3 - A = \begin{bmatrix} 0.707 & 0 & 0 \\ -0.014 & 0.793 & -0.017 \\ -0.044 & -0.01 & 0.784 \end{bmatrix}$

$(I_3 - A)^{-1} \approx \begin{bmatrix} 1.41 & 0 & 0 \\ 0.0267 & 1.26 & 0.0274 \\ 0.0797 & 0.0161 & 1.28 \end{bmatrix}$

b. We have $\vec{b} = \begin{bmatrix} 1 \\ 0 \\ 0 \end{bmatrix}$, so that $\vec{x} = (I_3 - A)^{-1}\vec{e}_1$ = first column of $(I_3 - A)^{-1} \approx \begin{bmatrix} 1.41 \\ 0.0267 \\ 0.0797 \end{bmatrix}$

c. As illustrated in part (b), the ith column of $(I_3 - A)^{-1}$ gives the output vector required to satisfy a consumer demand of 1 unit on industry i, in the absence of any other consumer demands. In particular, the ith diagonal entry of $(I_3 - A)^{-1}$ gives the output of industry i required to satisfy this demand. Since industry i has to satisfy the consumer demand of 1 as well as the interindustry demand, its total output will be at least 1.

d. Suppose the consumer demand increases from $\vec{b}$ to $\vec{b} + \vec{e}_2$ (that is, the demand on manufacturing increases by one unit). Then the output must change from $(I_3 - A)^{-1}\vec{b}$ to
$(I_3 - A)^{-1}(\vec{b} + \vec{e}_2) = (I_3 - A)^{-1}\vec{b} + (I_3 - A)^{-1}\vec{e}_2 = (I_3 - A)^{-1}\vec{b} + (\text{second column of } (I_3 - A)^{-1})$.
The components of the second column of $(I_3 - A)^{-1}$ tell us by how much each industry has to increase its output.

e. The ijth entry of $(I_n - A)^{-1}$ gives the required increase of the output x_i of industry i to satisfy an increase of the consumer demand b_j on industry j by one unit. In the language of multivariable calculus, this quantity is $\dfrac{\partial x_i}{\partial b_j}$.

50. Recall that $1 + k + k^2 + \cdots = \dfrac{1}{1-k}$.

The top left entry of $I_3 - A$ is $1 - k$, and the top left entry of $(I_3 - A)^{-1}$ will therefore be $\dfrac{1}{1-k}$, as claimed:

$$\left[\begin{array}{ccc|ccc} 1-k & 0 & 0 & 1 & 0 & 0 \\ * & * & * & 0 & 1 & 0 \\ * & * & * & 0 & 0 & 1 \end{array}\right] \xrightarrow{\div(1-k)} \left[\begin{array}{ccc|ccc} 1 & 0 & 0 & \frac{1}{1-k} & 0 & 0 \\ * & * & * & 0 & 1 & 0 \\ * & * & * & 0 & 0 & 1 \end{array}\right]$$

$\rightarrow \ldots$ (first row will remain unchanged)

In terms of economics, we can explain this fact as follows: The top left entry of $(I_3 - A)^{-1}$ is the output of industry 1 (Agriculture) required to satisfy a consumer demand of 1 unit on industry 1. Producing this one unit to satisfy the consumer demand will generate an extra demand of $k = 0.293$ units on industry 1. Producing these k units in turn will generate an extra demand of $k \cdot k = k^2$ units, and so forth. We are faced with an infinite series of (ever smaller) demands, $1 + k + k^2 + \cdots$.

2.4

1. $\begin{bmatrix} 4 & 6 \\ 3 & 4 \end{bmatrix}$

2. $\begin{bmatrix} 4 & 4 \\ -8 & -8 \end{bmatrix}$

3. Undefined

4. $\begin{bmatrix} 2 & 2 \\ 2 & 0 \\ 7 & 4 \end{bmatrix}$

5. $\begin{bmatrix} a & b \\ c & d \\ 0 & 0 \end{bmatrix}$

6. $\begin{bmatrix} ad-bc & 0 \\ 0 & ad-bc \end{bmatrix}$

7. $\begin{bmatrix} -1 & 1 & 0 \\ 5 & 3 & 4 \\ -6 & -2 & -4 \end{bmatrix}$

8. $\begin{bmatrix} 0 & 0 \\ 0 & 0 \end{bmatrix}$

9. $\begin{bmatrix} 0 & 0 \\ 0 & 0 \end{bmatrix}$

10. $[0 \quad 1]$

11. $[10]$

12. $\begin{bmatrix} 1 & 2 & 3 \\ 2 & 4 & 6 \\ 3 & 6 & 9 \end{bmatrix}$

13. $[h]$

14. $A^2 = \begin{bmatrix} 2 & 2 \\ 2 & 2 \end{bmatrix}$, $BC = [14 \quad 8 \quad 2]$, $BD = [6]$, $C^2 = \begin{bmatrix} -2 & -2 & -2 \\ 4 & 1 & -2 \\ 10 & 4 & -2 \end{bmatrix}$, $CD = \begin{bmatrix} 0 \\ 3 \\ 6 \end{bmatrix}$, $DB = \begin{bmatrix} 1 & 2 & 3 \\ 1 & 2 & 3 \\ 1 & 2 & 3 \end{bmatrix}$, $DE = \begin{bmatrix} 5 \\ 5 \\ 5 \end{bmatrix}$,

$EB = [5 \quad 10 \quad 15]$, $E^2 = [25]$

15. $\begin{bmatrix} 1 & 0 \\ 0 & 1 \end{bmatrix}$; Fact 2.4.9 applies to square matrices only.

16. True; $(I_n - A)(I_n + A) = I_n^2 + A - A - A^2 = I_n - A^2$.

17. Not necessarily true; $(A+B)^2 = (A+B)(A+B) = A^2 + AB + BA + B^2 \neq A^2 + 2AB + B^2$ if $AB \neq BA$.

18. True; apply Fact 2.4.8 to $B = A$.

19. Not necessarily true; consider the case $A = I_n$ and $B = -I_n$.

20. Not necessarily true; $(A-B)(A+B) = A^2 + AB - BA - B^2 \neq A^2 - B^2$ if $AB \neq BA$.

21. True; $ABB^{-1}A^{-1} = AI_nA^{-1} = AA^{-1} = I_n$.

22. Not necessarily true; the equation $ABA^{-1} = B$ is equivalent to $AB = BA$ (multiply by A from the right), which is not true in general.

23. True; $(ABA^{-1})^3 = ABA^{-1}ABA^{-1}ABA^{-1} = AB^3A^{-1}$

24. True; $(I_n + A)(I_n + A^{-1}) = I_n^2 + A + A^{-1} + AA^{-1} = 2I_n + A + A^{-1}$.

25. True; $(A^{-1}B)^{-1} = B^{-1}(A^{-1})^{-1} = B^{-1}A$ (use Fact 2.4.8).

26. $$\begin{bmatrix} \begin{bmatrix}1&0\\0&1\end{bmatrix}\begin{bmatrix}1&2\\3&4\end{bmatrix}+\begin{bmatrix}1&0\\0&1\end{bmatrix}\begin{bmatrix}0&0\\0&0\end{bmatrix} & \begin{bmatrix}1&0\\0&1\end{bmatrix}\begin{bmatrix}2&3\\4&5\end{bmatrix}+\begin{bmatrix}1&0\\0&1\end{bmatrix}\begin{bmatrix}1&2\\3&4\end{bmatrix} \\ \begin{bmatrix}0&0\\0&0\end{bmatrix}\begin{bmatrix}1&2\\3&4\end{bmatrix}+\begin{bmatrix}1&0\\0&1\end{bmatrix}\begin{bmatrix}0&0\\0&0\end{bmatrix} & \begin{bmatrix}0&0\\0&0\end{bmatrix}\begin{bmatrix}2&3\\4&5\end{bmatrix}+\begin{bmatrix}1&0\\0&1\end{bmatrix}\begin{bmatrix}1&2\\3&4\end{bmatrix} \end{bmatrix} = \begin{bmatrix} \begin{bmatrix}1&2\\3&4\end{bmatrix} & \begin{bmatrix}3&5\\7&9\end{bmatrix} \\ \begin{bmatrix}0&0\\0&0\end{bmatrix} & \begin{bmatrix}1&2\\3&4\end{bmatrix} \end{bmatrix} = \begin{bmatrix}1&2&3&5\\3&4&7&9\\0&0&1&2\\0&0&3&4\end{bmatrix}$$

27. $$\begin{bmatrix} \begin{bmatrix}1&0\\0&1\end{bmatrix}\begin{bmatrix}1\\2\end{bmatrix}+\begin{bmatrix}0\\0\end{bmatrix}[3] & \begin{bmatrix}1&0\\0&1\end{bmatrix}\begin{bmatrix}0\\0\end{bmatrix}+\begin{bmatrix}0\\0\end{bmatrix}[4] \\ [1\ \ 3]\begin{bmatrix}1\\2\end{bmatrix}+[4][3] & [1\ \ 3]\begin{bmatrix}0\\0\end{bmatrix}+[4][4] \end{bmatrix} = \begin{bmatrix} \begin{bmatrix}1\\2\end{bmatrix} & \begin{bmatrix}0\\0\end{bmatrix} \\ [19] & [16] \end{bmatrix} = \begin{bmatrix}1&0\\2&0\\19&16\end{bmatrix}$$

28. $A = \begin{bmatrix}0&1\\0&0\end{bmatrix}$ is one such matrix.

29. The columns of B must be solutions of the system $\begin{bmatrix}1&3\\2&6\end{bmatrix}\vec{x} = \begin{bmatrix}0\\0\end{bmatrix}$. One possible solution is $B = \begin{bmatrix}3&0\\-1&0\end{bmatrix}$.

30. Yes; by Fact 2.3.4b, the equation $A\vec{x} = \vec{0}$ has a nonzero solution $\vec{v}$. Let $B = [\vec{v}\ \ \vec{v}\ \ \cdots\ \ \vec{v}]$.
Then $AB = [A\vec{v}\ \ A\vec{v}\ \ \cdots\ \ A\vec{v}] = 0$.

31. The two columns of A must be solutions of the linear systems $B\vec{x} = \begin{bmatrix}1\\0\end{bmatrix}$ and $B\vec{x} = \begin{bmatrix}0\\1\end{bmatrix}$, respectively. Each of these systems has *infinitely many solutions*. One possible solution is $A = \begin{bmatrix}2&-1\\-1&1\\0&0\end{bmatrix}$.

32. By Fact 1.3.3, there is a nonzero $\vec{x}$ such that $B\vec{x} = \vec{0}$ and therefore $AB\vec{x} = \vec{0}$. This implies that $AB \neq I_3$, since $I_3\vec{x} = \vec{x} \neq \vec{0}$.

33. By Fact 1.3.3, there is a nonzero $\vec{x}$ such that $B\vec{x} = \vec{0}$ and therefore $AB\vec{x} = \vec{0}$. By Fact 2.3.4b, the 3×3 matrix AB is not invertible.

34. We can write $AB(AB)^{-1} = A(B(AB)^{-1}) = I_n$ and $(AB)^{-1}AB = ((AB)^{-1}A)B = I_n$.
By Fact 2.4.9, A and B are invertible.

35. **a.** Consider a solution $\vec{x}$ of the equation $A\vec{x} = \vec{0}$.
Multiply both sides by B from the left: $BA\vec{x} = B\vec{0} = \vec{0}$, so that $\vec{x} = \vec{0}$ (since $BA = I_m$).
It follows that $\vec{x} = \vec{0}$ is the only solution of $A\vec{x} = \vec{0}$.

b. $\vec{x} = A\vec{b}$ is a solution, since $B\vec{x} = BA\vec{b} = \vec{b}$ (because $BA = I_m$).

c. rank(A) = m, by part (a) (all variables are leading).
rank(B) = m, by part (b) (compare with Example 13, Section 1.3).

d. $m = \text{rank}(B) \leq (\text{number of columns of } B) = n$

36. The columns of X must be solutions of the system $A\vec{x} = \vec{0}$. These solutions are of the form $\begin{bmatrix} -2t \\ t \end{bmatrix}$, where t is arbitrary. Therefore, $X = \begin{bmatrix} -2s & -2t \\ s & t \end{bmatrix}$, where s and t are arbitrary.

37. We need to find all matrices $X = \begin{bmatrix} a & b \\ c & d \end{bmatrix}$ such that $\begin{bmatrix} a & b \\ c & d \end{bmatrix}\begin{bmatrix} 1 & 0 \\ 0 & 2 \end{bmatrix} = \begin{bmatrix} 1 & 0 \\ 0 & 2 \end{bmatrix}\begin{bmatrix} a & b \\ c & d \end{bmatrix}$, or, $\begin{bmatrix} a & 2b \\ c & 2d \end{bmatrix} = \begin{bmatrix} a & b \\ 2c & 2d \end{bmatrix}$.

This is the case if $b = 0$ and $c = 0$. This means that X must be a *diagonal matrix*, $X = \begin{bmatrix} a & 0 \\ 0 & d \end{bmatrix}$.

38. We need to find all matrices $X = \begin{bmatrix} a & b \\ c & d \end{bmatrix}$ such that $\begin{bmatrix} a & b \\ c & d \end{bmatrix}\begin{bmatrix} 1 & 2 \\ 0 & 3 \end{bmatrix} = \begin{bmatrix} 1 & 2 \\ 0 & 3 \end{bmatrix}\begin{bmatrix} a & b \\ c & d \end{bmatrix}$, or

$\begin{bmatrix} a & 2a+3b \\ c & 2c+3d \end{bmatrix} = \begin{bmatrix} a+2c & b+2d \\ 3c & 3d \end{bmatrix}$.

The condition $c = 3c$ implies that $c = 0$. There is one other constraint: $2a + 3b = b + 2d$, or, $d = a + b$.

This means that X must be of the form $X = \begin{bmatrix} a & b \\ 0 & a+b \end{bmatrix}$, where a and b are arbitrary.

39. By Exercise 37, X must be diagonal, of the form $X = \begin{bmatrix} a & 0 \\ 0 & d \end{bmatrix}$. X must also commute with $\begin{bmatrix} 0 & 1 \\ 0 & 0 \end{bmatrix}$, so that

$\begin{bmatrix} a & 0 \\ 0 & d \end{bmatrix}\begin{bmatrix} 0 & 1 \\ 0 & 0 \end{bmatrix} = \begin{bmatrix} 0 & a \\ 0 & 0 \end{bmatrix}$ must equal $\begin{bmatrix} 0 & 1 \\ 0 & 0 \end{bmatrix}\begin{bmatrix} a & 0 \\ 0 & d \end{bmatrix} = \begin{bmatrix} 0 & d \\ 0 & 0 \end{bmatrix}$. It follows that we must have $a = d$, so that X must be of the form $\begin{bmatrix} a & 0 \\ 0 & a \end{bmatrix}$, that is, a scalar multiple of the identity matrix.

40. $A = (AB)B^{-1} = ((AB)^{-1})^{-1}B^{-1} = \begin{bmatrix} 1 & 3 \\ 2 & 5 \end{bmatrix}^{-1}\begin{bmatrix} 1 & 2 \\ 3 & 5 \end{bmatrix} = \begin{bmatrix} -5 & 3 \\ 2 & -1 \end{bmatrix}\begin{bmatrix} 1 & 2 \\ 3 & 5 \end{bmatrix} = \begin{bmatrix} 4 & 5 \\ -1 & -1 \end{bmatrix}$

41. a. $D_\alpha D_\beta$ and $D_\beta D_\alpha$ are the same transformation, namely, a rotation by $\alpha + \beta$.

b. $D_\alpha D_\beta = \begin{bmatrix} \cos\alpha & -\sin\alpha \\ \sin\alpha & \cos\alpha \end{bmatrix}\begin{bmatrix} \cos\beta & -\sin\beta \\ \sin\beta & \cos\beta \end{bmatrix} = \begin{bmatrix} \cos\alpha\cos\beta - \sin\alpha\sin\beta & -\cos\alpha\sin\beta - \sin\alpha\cos\beta \\ \sin\alpha\cos\beta + \cos\alpha\sin\beta & -\sin\alpha\sin\beta + \cos\alpha\cos\beta \end{bmatrix}$

$= \begin{bmatrix} \cos(\alpha+\beta) & -\sin(\alpha+\beta) \\ \sin(\alpha+\beta) & \cos(\alpha+\beta) \end{bmatrix}$

$D_\beta D_\alpha$ yields the same answer.

42. **a.**

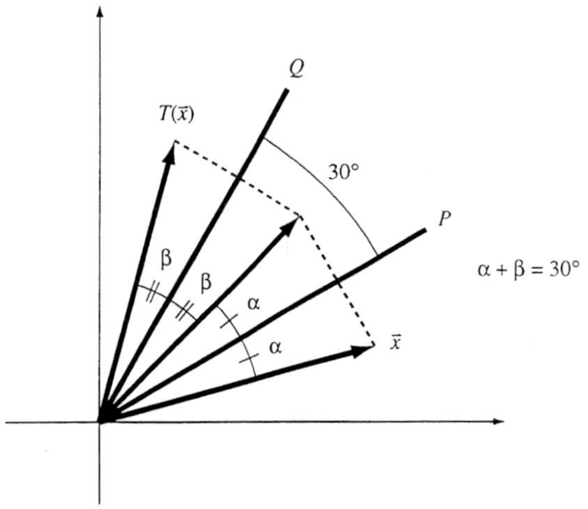

The vectors $\vec{x}$ and $T(\vec{x})$ have the same length (since reflections leave the length unchanged), and they enclose an angle of $2(\alpha+\beta) = 2 \cdot 30° = 60°$.

b. Based on the answer in part (a), we conclude that T is a rotation by 60°.

c. The matrix of T is $\begin{bmatrix} \cos(60°) & -\sin(60°) \\ \sin(60°) & \cos(60°) \end{bmatrix} = \begin{bmatrix} \frac{1}{2} & -\frac{\sqrt{3}}{2} \\ \frac{\sqrt{3}}{2} & \frac{1}{2} \end{bmatrix}$.

43. Let A represent the rotation by 120°; then A^3 represents the rotation by 360°, that is $A^3 = I_2$.

$$A = \begin{bmatrix} \cos(120°) & -\sin(120°) \\ \sin(120°) & \cos(120°) \end{bmatrix} = \begin{bmatrix} -\frac{1}{2} & -\frac{\sqrt{3}}{2} \\ \frac{\sqrt{3}}{2} & -\frac{1}{2} \end{bmatrix}$$

44. We want A such that $A\begin{bmatrix} 1 & 2 \\ 2 & 5 \end{bmatrix} = \begin{bmatrix} 2 & 1 \\ 1 & 3 \end{bmatrix}$, so that $A = \begin{bmatrix} 2 & 1 \\ 1 & 3 \end{bmatrix}\begin{bmatrix} 1 & 2 \\ 2 & 5 \end{bmatrix}^{-1} = \begin{bmatrix} 8 & -3 \\ -1 & 1 \end{bmatrix}$.

45. We want A such that $A\vec{v}_i = \vec{w}_i$, for $i = 1, 2, \ldots, n$, or $A[\vec{v}_1 \;\; \vec{v}_2 \;\; \cdots \;\; \vec{v}_n] = [\vec{w}_1 \;\; \vec{w}_2 \;\; \cdots \;\; \vec{w}_n]$, or, $AS = B$. Multiplying by S^{-1} from the right we find the unique solution $A = BS^{-1}$.

46. Use the result of Exercise 45, with $S = \begin{bmatrix} 1 & 2 \\ 2 & 5 \end{bmatrix}$ and $B = \begin{bmatrix} 7 & 1 \\ 5 & 2 \\ 3 & 3 \end{bmatrix}$:

$$A = BS^{-1} = \begin{bmatrix} 33 & -13 \\ 21 & -8 \\ 9 & -3 \end{bmatrix}$$

47. Use the result of Exercise 45, with $S = \begin{bmatrix} 3 & 1 \\ 1 & 2 \end{bmatrix}$ and $B = \begin{bmatrix} 6 & 3 \\ 2 & 6 \end{bmatrix}$:

$$A = BS^{-1} = \frac{1}{5}\begin{bmatrix} 9 & 3 \\ -2 & 16 \end{bmatrix}$$

48. $P_0 \xrightarrow{T} P_1$, $P_1 \xrightarrow{T} P_3$, $P_2 \xrightarrow{T} P_2$, $P_3 \xrightarrow{T} P_0$

$P_0 \xrightarrow{L} P_0$, $P_1 \xrightarrow{L} P_2$, $P_2 \xrightarrow{L} P_1$, $P_3 \xrightarrow{L} P_3$

a. T^{-1} is the rotation about the axis through O and P_2 that transforms P_3 into P_1.

b. $L^{-1} = L$

c. $T^2 = T^{-1}$ (See part (a).)

d. $P_0 \xrightarrow{T\circ L} P_1$, $P_1 \longrightarrow P_2$, $P_2 \longrightarrow P_3$, $P_3 \longrightarrow P_0$

$P_0 \xrightarrow{L\circ T} P_2$, $P_1 \longrightarrow P_3$, $P_2 \longrightarrow P_1$, $P_3 \longrightarrow P_0$

The transformations $T \circ L$ and $L \circ T$ are not the same.

e. $P_0 \xrightarrow{L\circ T\circ L} P_2$

$P_1 \longrightarrow P_1$

$P_2 \longrightarrow P_3$

$P_3 \longrightarrow P_0$

This is the rotation about the axis through O and P_1 that sends P_0 to P_2.

49. Let A be the matrix of T and C the matrix of L. We want that $AP_0 = P_1$, $AP_1 = P_3$, and $AP_2 = P_2$. We can use the result of Exercise 45, with $S = \begin{bmatrix} 1 & 1 & -1 \\ 1 & -1 & 1 \\ 1 & -1 & -1 \end{bmatrix}$ and $B = \begin{bmatrix} 1 & -1 & -1 \\ -1 & -1 & 1 \\ -1 & 1 & -1 \end{bmatrix}$.

Then $A = BS^{-1} = \begin{bmatrix} 0 & 0 & 1 \\ -1 & 0 & 0 \\ 0 & -1 & 0 \end{bmatrix}$.

Using an analogous approach, we find that $C = \begin{bmatrix} 0 & 1 & 0 \\ 1 & 0 & 0 \\ 0 & 0 & 1 \end{bmatrix}$.

50. **a.** $EA = \begin{bmatrix} a & b & c \\ d-3a & e-3b & f-3c \\ g & h & k \end{bmatrix}$

The matrix EA is obtained from A by an elementary row operation: subtract three times the first row from the second.

b. $EA = \begin{bmatrix} a & b & c \\ \frac{1}{4}d & \frac{1}{4}e & \frac{1}{4}f \\ g & h & k \end{bmatrix}$

T he matrix EA is obtained from A by dividing the second row of A by 4 (an elementary row operation).

c. If we set $E = \begin{bmatrix} 1 & 0 & 0 \\ 0 & 0 & 1 \\ 0 & 1 & 0 \end{bmatrix}$ then $\begin{bmatrix} 1 & 0 & 0 \\ 0 & 0 & 1 \\ 0 & 1 & 0 \end{bmatrix}\begin{bmatrix} a & b & c \\ d & e & f \\ g & h & k \end{bmatrix} = \begin{bmatrix} a & b & c \\ g & h & k \\ d & e & f \end{bmatrix}$, as desired.

d. An elementary $n \times n$ matrix E has the same form as I_n except that either

- $e_{ij} = k(\neq 0)$ for some $i \neq j$ (as in part (a)), or
- $e_{ii} = k(\neq 0, 1)$ for some i (as in part (b)), or
- $e_{ij} = e_{ji} = 1$, $e_{ii} = e_{jj} = 0$ for some $i \neq j$ (as in part (c)).

51. Let E be an elementary $n \times n$ matrix (obtained from I_n by a certain elementary row operation), and let F be the elementary matrix obtained from I_n by the reversed row operation. Our work in Exercise 50 (parts (a) to (c)) shows that $EF = I_n$, so that E is indeed invertible, and $E^{-1} = F$ is an elementary matrix as well.

52. a. The matrix rref(A) is obtained from A by performing a sequence of p elementary row operations. By Exercise 50 (parts (a) to (c)) each of these operations can be represented by the left multiplication with an elementary matrix, so that $\text{rref}(A) = E_1E_2\cdots E_pA$.

b. $A = \begin{bmatrix} 0 & 2 \\ 1 & 3 \end{bmatrix}$ swap rows 1 and 2, represented by $\begin{bmatrix} 0 & 1 \\ 1 & 0 \end{bmatrix}$

$\downarrow$

$\begin{bmatrix} 1 & 3 \\ 0 & 2 \end{bmatrix} \div 2$, represented by $\begin{bmatrix} 1 & 0 \\ 0 & \frac{1}{2} \end{bmatrix}$

$\downarrow$

$\begin{bmatrix} 1 & 3 \\ 0 & 1 \end{bmatrix} -3(\text{II})$, represented by $\begin{bmatrix} 1 & -3 \\ 0 & 1 \end{bmatrix}$

$\downarrow$

$\text{rref}(A) = \begin{bmatrix} 1 & 0 \\ 0 & 1 \end{bmatrix}$

Therefore, $\text{rref}(A) = \begin{bmatrix} 1 & 0 \\ 0 & 1 \end{bmatrix} = \begin{bmatrix} 1 & -3 \\ 0 & 1 \end{bmatrix}\begin{bmatrix} 1 & 0 \\ 0 & \frac{1}{2} \end{bmatrix}\begin{bmatrix} 0 & 1 \\ 1 & 0 \end{bmatrix}\begin{bmatrix} 0 & 2 \\ 1 & 3 \end{bmatrix} = E_1E_2E_3A$.

53. a. Let $S = E_1E_2\cdots E_p$ in Exercise 52(a).

By Exercise 51, the elementary matrices E_i are invertible; now use Fact 2.4.8 repeatedly to see that S is invertible.

b. $A = \begin{bmatrix} 2 & 4 \\ 4 & 8 \end{bmatrix} \div 2$, represented by $\begin{bmatrix} \frac{1}{2} & 0 \\ 0 & 1 \end{bmatrix}$

$\begin{bmatrix} 1 & 2 \\ 4 & 8 \end{bmatrix} -4(\text{I})$, represented by $\begin{bmatrix} 1 & 0 \\ -4 & 1 \end{bmatrix}$

$\text{rref}(A) = \begin{bmatrix} 1 & 2 \\ 0 & 0 \end{bmatrix}$

Therefore, $\text{rref}(A) = \begin{bmatrix} 1 & 2 \\ 0 & 0 \end{bmatrix} = \begin{bmatrix} 1 & 0 \\ -4 & 1 \end{bmatrix}\begin{bmatrix} \frac{1}{2} & 0 \\ 0 & 1 \end{bmatrix}\begin{bmatrix} 2 & 4 \\ 4 & 8 \end{bmatrix} = E_1E_2A = SA$, where $S = \begin{bmatrix} 1 & 0 \\ -4 & 1 \end{bmatrix}\begin{bmatrix} \frac{1}{2} & 0 \\ 0 & 1 \end{bmatrix} = \begin{bmatrix} \frac{1}{2} & 0 \\ -2 & 1 \end{bmatrix}$.

(There are other correct answers.)

54. a. By Exercise 52(a), $I_n = \text{rref}(A) = E_1E_2 \ldots E_pA$, for some elementary matrices $E_1, \ldots, E_p$. By Exercise 51, the E_i are invertible, and their inverses are elementary as well. Therefore, $A = (E_1E_2 \ldots E_p)^{-1} = E_p^{-1} \ldots E_2^{-1}E_1^{-1}$ expresses A as a product of elementary matrices.

b. We can use out work in Exercise 52(b):

$$\begin{bmatrix} 0 & 2 \\ 1 & 3 \end{bmatrix} = \left(\begin{bmatrix} 1 & -3 \\ 0 & 1 \end{bmatrix}\begin{bmatrix} 1 & 0 \\ 0 & \frac{1}{2} \end{bmatrix}\begin{bmatrix} 0 & 1 \\ 1 & 0 \end{bmatrix}\right)^{-1} = \begin{bmatrix} 0 & 1 \\ 1 & 0 \end{bmatrix}^{-1}\begin{bmatrix} 1 & 0 \\ 0 & \frac{1}{2} \end{bmatrix}^{-1}\begin{bmatrix} 1 & -3 \\ 0 & 1 \end{bmatrix}^{-1} = \begin{bmatrix} 0 & 1 \\ 1 & 0 \end{bmatrix}\begin{bmatrix} 1 & 0 \\ 0 & 2 \end{bmatrix}\begin{bmatrix} 1 & 3 \\ 0 & 1 \end{bmatrix}$$

55. $\begin{bmatrix} 1 & k \\ 0 & 1 \end{bmatrix}$ represents a shear parallel to the $\vec{e}_1$ axis, $\begin{bmatrix} 1 & 0 \\ k & 1 \end{bmatrix}$ represents a shear parallel to the $\vec{e}_2$ axis,

$\begin{bmatrix} k & 0 \\ 0 & 1 \end{bmatrix}$ represents a "dilation in $\vec{e}_1$ direction" (leaving the $\vec{e}_2$ component unchanged),

$\begin{bmatrix} 1 & 0 \\ 0 & k \end{bmatrix}$ represents a "dilation in $\vec{e}_2$ direction" (leaving the $\vec{e}_1$ component unchanged), and

$\begin{bmatrix} 0 & 1 \\ 1 & 0 \end{bmatrix}$ represents the reflection in the line spanned by $\begin{bmatrix} 1 \\ 1 \end{bmatrix}$.

56. Performing a sequence of p elementary row operations on a matrix A amounts to multiplying A with $E_1E_2 \cdots E_p$ from the left, where the E_i are elementary matrices. If $I_n = E_1E_2 \cdots E_pA$, then $E_1E_2 \cdots E_p = A^{-1}$, so that

a. $E_1E_2 \cdots E_pAB = B$, and

b. $E_1E_2 \cdots E_pI_n = A^{-1}$.

57. Let A and B be two lower triangular $n \times n$ matrices. We need to show that the ijth entry of AB is 0 whenever $i < j$.

This entry is the dot product of the ith row of A and the jth column of B, $[a_{i1} \quad a_{i2} \cdots a_{ii} \quad 0 \cdots 0] \cdot \begin{bmatrix} 0 \\ \vdots \\ 0 \\ b_{jj} \\ \vdots \\ b_{nj} \end{bmatrix}$, which is indeed 0 if $i < j$.

58. a. $\begin{bmatrix} 1 & 2 & 3 \\ 2 & 6 & 7 \\ 2 & 2 & 4 \end{bmatrix} \begin{matrix} \\ -2\text{I} \\ -2\text{I} \end{matrix}$, represented by $\begin{bmatrix} 1 & 0 & 0 \\ 0 & 1 & 0 \\ -2 & 0 & 1 \end{bmatrix}\begin{bmatrix} 1 & 0 & 0 \\ -2 & 1 & 0 \\ 0 & 0 & 1 \end{bmatrix}$

$\downarrow$

$\begin{bmatrix} 1 & 2 & 3 \\ 0 & 2 & 1 \\ 0 & -2 & -2 \end{bmatrix} \begin{matrix} \\ \\ +\text{II} \end{matrix}$, represented by $\begin{bmatrix} 1 & 0 & 0 \\ 0 & 1 & 0 \\ 0 & 1 & 1 \end{bmatrix}$

$\downarrow$

$\begin{bmatrix} 1 & 2 & 3 \\ 0 & 2 & 1 \\ 0 & 0 & -1 \end{bmatrix}$, so that

$$\begin{bmatrix}1&2&3\\0&2&1\\0&0&-1\end{bmatrix}=\begin{bmatrix}1&0&0\\0&1&0\\0&1&1\end{bmatrix}\begin{bmatrix}1&0&0\\0&1&0\\-2&0&1\end{bmatrix}\begin{bmatrix}1&0&0\\-2&1&0\\0&0&1\end{bmatrix}\begin{bmatrix}1&2&3\\2&6&7\\2&2&4\end{bmatrix}$$
$$\uparrow U \quad \uparrow E_3 \quad \uparrow E_2 \quad \uparrow E_1 \quad \uparrow A$$

b. $A=(E_3E_2E_1)^{-1}U=E_1^{-1}E_2^{-1}E_3^{-1}U=\begin{bmatrix}1&0&0\\2&1&0\\0&0&1\end{bmatrix}\begin{bmatrix}1&0&0\\0&1&0\\2&0&1\end{bmatrix}\begin{bmatrix}1&0&0\\0&1&0\\0&-1&1\end{bmatrix}\begin{bmatrix}1&2&3\\0&2&1\\0&0&-1\end{bmatrix}$
$$\uparrow M_1 \quad \uparrow M_2 \quad \uparrow M_3 \quad \uparrow U$$

c. Let $L=M_1M_2M_3$ in part (b); we compute $L=\begin{bmatrix}1&0&0\\2&1&0\\2&-1&1\end{bmatrix}$.

Then $\begin{bmatrix}1&2&3\\2&6&7\\2&2&4\end{bmatrix}=\begin{bmatrix}1&0&0\\2&1&0\\2&-1&1\end{bmatrix}\begin{bmatrix}1&2&3\\0&2&1\\0&0&-1\end{bmatrix}$
$$\uparrow A \quad \uparrow L \quad \uparrow U$$

d. We can use the matrix L we found in part (c), but U needs to be modified. Let $D=\begin{bmatrix}1&0&0\\0&2&0\\0&0&-1\end{bmatrix}$.

(Take the diagonal entries of the matrix U in part (c)).

Then $\begin{bmatrix}1&2&3\\2&6&7\\2&2&4\end{bmatrix}=\begin{bmatrix}1&0&0\\2&1&0\\2&-1&1\end{bmatrix}\begin{bmatrix}1&0&0\\0&2&0\\0&0&-1\end{bmatrix}\begin{bmatrix}1&2&3\\0&1&\frac{1}{2}\\0&0&1\end{bmatrix}$.
$$\uparrow A \quad \uparrow L \quad \uparrow D \quad \uparrow U$$

59. a. Write the system $L\vec{y}=\vec{b}$ in components:

$\begin{vmatrix}y_1 & & & & & & & =-3\\ -3y_1 &+& y_2 & & & & & =14\\ y_1 &+& 2y_2 &+& y_3 & & & =9\\ -y_1 &+& 8y_2 &-& 5y_3 &+& y_4 & =33\end{vmatrix}$, so that $y_1=-3,\ y_2=14+3y_1=5,\ y_3=9-y_1-2y_2=2$, and $y_4=33+y_1-8y_2+5y_3=0$:

$\vec{y}=\begin{bmatrix}-3\\5\\2\\0\end{bmatrix}$.

b. Proceeding as in part (a) we find that $\vec{x} = \begin{bmatrix} 1 \\ -1 \\ 2 \\ 0 \end{bmatrix}$.

60. We try to find matrices $L = \begin{bmatrix} a & 0 \\ b & c \end{bmatrix}$ and $U = \begin{bmatrix} d & e \\ 0 & f \end{bmatrix}$ such that

$\begin{bmatrix} 0 & 1 \\ 1 & 0 \end{bmatrix} = \begin{bmatrix} a & 0 \\ b & c \end{bmatrix}\begin{bmatrix} d & e \\ 0 & f \end{bmatrix} = \begin{bmatrix} ad & ae \\ bd & be+cf \end{bmatrix}$.

Note that the equations $ad = 0$, $ae = 1$, and $bd = 1$ cannot be solved simultaneously: If $ad = 0$ then a or d is 0 so that ae or bd is zero.

Therefore, the matrix $\begin{bmatrix} 0 & 1 \\ 1 & 0 \end{bmatrix}$ does not have an LU factorization.

61. a. Write $L = \begin{bmatrix} L^{(m)} & 0 \\ L_3 & L_4 \end{bmatrix}$ and $U = \begin{bmatrix} U^{(m)} & U_2 \\ 0 & U_4 \end{bmatrix}$. Then $A = LU = \begin{bmatrix} L^{(m)}U^{(m)} & L^{(m)}U_2 \\ L_3U^{(m)} & L_3U_2 + L_4U_4 \end{bmatrix}$, so that $A^{(m)} = L^{(m)}U^{(m)}$, as claimed.

b. By Exercise 34, the matrices L and U are both invertible. By Exercise 2.3.35, the diagonal entries of L and U are all nonzero. For any m, the matrices $L^{(m)}$ and $U^{(m)}$ are triangular, with nonzero diagonal entries so that they are invertible. By Fact 2.4.8, the matrix $A^{(m)} = L^{(m)}U^{(m)}$ is invertible as well.

c. Use the hint, we write $A = \begin{bmatrix} A^{(n-1)} & \vec{v} \\ \vec{w} & k \end{bmatrix} = \begin{bmatrix} L' & 0 \\ \vec{x} & t \end{bmatrix}\begin{bmatrix} U' & \vec{y} \\ 0 & s \end{bmatrix}$.

We are looking for a column vector $\vec{y}$, a row vector $\vec{x}$, and scalars t and s satisfying these equations. The following equations need to be satisfied: $\vec{v} = L'\vec{y}$, $\vec{w} = \vec{x}U'$, and $k = \vec{x}\vec{y} + ts$.

We find that $\vec{y} = (L')^{-1}\vec{v}$, $\vec{x} = \vec{w}(U')^{-1}$, and $ts = k - \vec{w}(U')^{-1}(L')^{-1}\vec{v}$.

We can choose, for example, $s = 1$ and $t = k - \vec{w}(U')^{-1}(L')^{-1}\vec{v}$, proving that A does indeed have an LU factorization.

Alternatively, one can show that if all principal submatrices are invertible then no row swaps are required in the Gauss-Jordan algorithm. In this case, we can find an LU-factorization as outlined in Exercise 58.

62. a. If $A = LU$ is an LU factorization, then the diagonal entries of L and U are nonzero (compare with Exercise 61). Let D_1 and D_2 be the diagonal matrices whose diagonal entries are the same as those of L and U, respectively. Then $A = \underbrace{(LD_1^{-1})}_{\text{new } L}\underbrace{(D_1D_2)}_{D}\underbrace{(D_2^{-1}U)}_{\text{new } U}$ is the desired factorization (verify that LD_1^{-1} and $D_2^{-1}U$ are of the required form).

b. If $A = L_1D_1U_1 = L_2D_2U_2$ and A is invertible, then L_1, D_1, U_1, L_2, D_2, U_2 are all invertible, so we can multiply the above equation by $D_2^{-1}L_2^{-1}$ from the left and by U_1^{-1} from the right:

$D_2^{-1}L_2^{-1}L_1D_1 = U_2U_1^{-1}$

Since products and inverses of upper triangular matrices are upper triangular (and likewise for lower triangular matrices), the matrix $D_2^{-1}L_2^{-1}L_1D_1 = U_2U_1^{-1}$ is both upper and lower triangular, that is, it is diagonal. Since the diagonal entries of U_2 and U_1 are all 1, so are the diagonal entries of $U_2U_1^{-1}$, that is $U_2U_1^{-1} = I_n$, and thus $U_2 = U_1$.

Now $L_1D_1 = L_2D_2$, so that $L_2^{-1}L_1 = D_2D_1^{-1}$ is diagonal. As above, we have in fact $L_2^{-1}L_1 = I_n$ and therefore $L_2 = L_1$.

63. We will prove that $A(C+D) = AC + AD$, repeatedly using Fact 1.3.7a: $A(\vec{x}+\vec{y}) = A\vec{x} + A\vec{y}$.
Write $B = [\vec{v}_1 \ \cdots \ \vec{v}_n]$ and $C = [\vec{w}_1 \ \cdots \ \vec{w}_n]$. Then
$A(C+D) = A[\vec{v}_1 + \vec{w}_1 \ \cdots \ \vec{v}_n + \vec{w}_n] = [A\vec{v}_1 + A\vec{w}_1 \ \cdots \ A\vec{v}_n + A\vec{w}_n]$, and
$AC + AD = A[\vec{v}_1 \ \cdots \ \vec{v}_n] + A[\vec{w}_1 \ \cdots \ \vec{w}_n] = [A\vec{v}_1 + A\vec{w}_1 \ \cdots \ A\vec{v}_n + A\vec{w}_n]$.
The results agree.

64. The ijth entries of the three matrices are $\sum_{h=1}^{n}(ka_{ih})b_{hj}$, $\sum_{h=1}^{n}a_{ih}(kb_{hj})$, and $k\left(\sum_{h=1}^{n}a_{ih}b_{hj}\right)$.

The three results agree.

65. Suppose A_{11} is a $p \times p$ matrix and A_{22} is a $q \times q$ matrix. For B to be the inverse of A we must have $AB = I_{p+q}$.
Let us partition B the same way as A:

$B = \begin{bmatrix} B_{11} & B_{12} \\ B_{21} & B_{22} \end{bmatrix}$, where B_{11} is $p \times p$ and B_{22} is $q \times q$.

Then $AB = \begin{bmatrix} A_{11} & 0 \\ 0 & A_{22} \end{bmatrix}\begin{bmatrix} B_{11} & B_{12} \\ B_{21} & B_{22} \end{bmatrix} = \begin{bmatrix} A_{11}B_{11} & A_{11}B_{12} \\ A_{22}B_{21} & A_{22}B_{22} \end{bmatrix} = \begin{bmatrix} I_p & 0 \\ 0 & I_q \end{bmatrix}$ means that

$A_{11}B_{11} = I_p$, $A_{22}B_{22} = I_q$, $A_{11}B_{12} = 0$, $A_{22}B_{21} = 0$.

This implies that A_{11} and A_{22} are invertible, and $B_{11} = A_{11}^{-1}$, $B_{22} = A_{22}^{-1}$.
This in turn implies that $B_{12} = 0$ and $B_{21} = 0$.

We summarize: A is invertible if (and only if) both A_{11} and A_{22} are invertible; in this case $A^{-1} = \begin{bmatrix} A_{11}^{-1} & 0 \\ 0 & A_{22}^{-1} \end{bmatrix}$.

66. This exercise is very similar to Example 4 in the text. We outline the solution:

$\begin{bmatrix} A_{11} & 0 \\ A_{21} & A_{22} \end{bmatrix}\begin{bmatrix} B_{11} & B_{12} \\ B_{21} & B_{22} \end{bmatrix} = \begin{bmatrix} I_p & 0 \\ 0 & I_q \end{bmatrix}$ means that

$A_{11}B_{11} = I_q$, $A_{11}B_{12} = 0$, $A_{21}B_{11} + A_{22}B_{21} = 0$, $A_{21}B_{12} + A_{22}B_{22} = I_q$.

This implies that A_{11} is invertible, and $B_{11} = A_{11}^{-1}$. Multiplying the second equation with A_{11}^{-1}, we conclude that $B_{12} = 0$. Then the last equation simplifies to $A_{22}B_{22} = I_q$, so that $B_{22} = A_{22}^{-1}$.

Finally, $B_{21} = -A_{22}^{-1}A_{21}B_{11} = -A_{22}^{-1}A_{21}A_{11}^{-1}$.
We summarize: A is invertible if (and only if) both A_{11} and A_{22} are invertible. In this case,

$$A^{-1} = \begin{bmatrix} A_{11}^{-1} & 0 \\ -A_{22}^{-1}A_{21}A_{11}^{-1} & A_{22}^{-1} \end{bmatrix}.$$

67. Write A in terms of its rows: $A = \begin{bmatrix} \vec{w}_1 \\ \vec{w}_2 \\ \cdots \\ \vec{w}_n \end{bmatrix}$ (suppose A is $n \times m$).

We can think of this as a partition into n $1 \times m$ matrices. Now $AB = \begin{bmatrix} \vec{w}_1 \\ \vec{w}_2 \\ \cdots \\ \vec{w}_n \end{bmatrix} B = \begin{bmatrix} \vec{w}_1 B \\ \vec{w}_2 B \\ \cdots \\ \vec{w}_n B \end{bmatrix}$ (a product of partitioned matrices).

We see that the ith row of AB is the product of the ith row of A and the matrix B.

68. By Exercise 65 or by Example 4 in the text, we find that $S^{-1} = \begin{bmatrix} 1 & 0 \\ 0 & R^{-1} \end{bmatrix}$. Then

$$S^{-1}AS = \begin{bmatrix} 1 & 0 \\ 0 & R^{-1} \end{bmatrix} \begin{bmatrix} k & \vec{v} \\ 0 & B \end{bmatrix} \begin{bmatrix} 1 & 0 \\ 0 & R \end{bmatrix} = \begin{bmatrix} 1 & 0 \\ 0 & R^{-1} \end{bmatrix} \begin{bmatrix} k & \vec{v}R \\ 0 & BR \end{bmatrix} = \begin{bmatrix} k & \vec{v}R \\ 0 & R^{-1}BR \end{bmatrix}.$$

69. Suppose A_{11} is a $p \times p$ matrix. Since A_{11} is invertible, $\text{rref}(A) = \begin{bmatrix} I_p & A_{12} & * \\ 0 & 0 & \text{rref}(A_{23}) \end{bmatrix}$, so that

$\text{rank}(A) = p + \text{rank}(A_{23}) = \text{rank}(A_{11}) + \text{rank}(A_{23})$.

70. Try to find a matrix $B = \begin{bmatrix} X & \vec{x} \\ \vec{y} & t \end{bmatrix}$ (where X is $n \times n$) such that

$$AB = \begin{bmatrix} I_n & \vec{v} \\ \vec{w} & 1 \end{bmatrix} \begin{bmatrix} X & \vec{x} \\ \vec{y} & t \end{bmatrix} = \begin{bmatrix} X + \vec{v}\vec{y} & \vec{x} + t\vec{v} \\ \vec{w}X + \vec{y} & \vec{w}\vec{x} + t \end{bmatrix} = \begin{bmatrix} I_n & 0 \\ 0 & 1 \end{bmatrix}.$$

We want $X + \vec{v}\vec{y} = I_n$, $\vec{x} + t\vec{v} = \vec{0}$, $\vec{w}X + \vec{y} = \vec{0}$, and $\vec{w}\vec{x} + t = 1$.

Substituting $\vec{x} = -t\vec{v}$ into the last equation we find $-t\vec{w}\vec{v} + t = 1$ or $t(1 - \vec{w}\vec{v}) = 1$.

This can be solved only if $\vec{w}\vec{v} \neq 1$, in which case $t = \dfrac{1}{1 - \vec{w}\vec{v}}$. Now substituting $X = I_n - \vec{v}\vec{y}$ into the third equation

we find $\vec{w} - \vec{w}\vec{v}\vec{y} + \vec{y} = \vec{0}$ or $\vec{y} = -\dfrac{1}{1 - \vec{w}\vec{v}}\vec{w} = -t\vec{w}$.

We summarize: A is invertible if (and only if) $\vec{w}\vec{v} \neq 1$. In this case, $A^{-1} = \begin{bmatrix} I_n + t\vec{v}\vec{w} & -t\vec{v} \\ -t\vec{w} & t \end{bmatrix}$, where $t = \dfrac{1}{1 - \vec{w}\vec{v}}$.

The same result can be found (perhaps more easily) by working with $\text{rref}\left[A \mid I_{n+1}\right]$, rather than partitioned matrices.

71. Multiplying both sides with A^{-1} we find that $A = I_n$: The identity matrix is the only invertible matrix with this property.

72. Suppose the entries of A are all a. Then the entries of A^2 are all na^2. The equation $na^2 = a$ is satisfied if $a = 0$ and if $a = \dfrac{1}{n}$. The two possible solutions are therefore $A = \begin{bmatrix} 0 & 0 & 0 & \cdots & 0 \\ 0 & 0 & 0 & \cdots & 0 \\ & & & \ddots & \\ 0 & 0 & 0 & \cdots & 0 \end{bmatrix}$ and $A = \begin{bmatrix} \frac{1}{n} & \frac{1}{n} & \cdots & \frac{1}{n} \\ \frac{1}{n} & \frac{1}{n} & \cdots & \frac{1}{n} \\ & & \ddots & \\ \frac{1}{n} & \frac{1}{n} & \cdots & \frac{1}{n} \end{bmatrix}$.

73. The ijth entry of AB is $\sum_{k=1}^{n} a_{ik}b_{kj}$. Then $\sum_{k=1}^{n} a_{ik}b_{kj} \le \sum_{k=1}^{n} sb_{kj} = s\left(\sum_{k=1}^{n} b_{kj}\right) \le sr$.

↑ since $a_{ik} \le s$ ↑ this is $\le r$, as it is the jth column sum of B.

74. The notation is confusing. We should say: Show that the entries of A^m are less than or equal to r^m, etc.

a. We proceed by induction on m_1. Since the column sums of A are $\le r$, the entries of $A^1 = A$ are also $\le r^1 = r$, so that the claim holds for $m = 1$. Suppose the claim holds for some fixed m. Now write $A^{m+1} = A^m A$; since the entries of A^m are $\le r^m$ and the column sums of A are $\le r$, we can conclude that the entries of A^{m+1} are $\le r^m r = r^{m+1}$, by Exercise 73.

b. For a fixed i and j, let b_m be the ijth entry of A^m. In part (a) we have seen that $0 \le b_m \le r^m$. Note that $\lim_{m\to\infty} r^m = 0$ (since $r < 1$), so that $\lim_{m\to\infty} b_m = 0$ as well (this follows from what some calculus texts call the "squeeze theorem").

c. For a fixed i and j, let c_m be the ijth entry of the matrix $I_n + A + A^2 + \cdots + A^m$. By part (a),
$c_m \le 1 + r + r^2 + \cdots + r^m < \frac{1}{1-r}$.
Since the c_m form an increasing bounded sequence, $\lim_{m\to\infty} c_m$ exists (this is a fundamental fact of calculus).

d. $(I_n - A)(I_n + A + A^2 + \cdots + A^m) = I_n + A + A^2 + \cdots A^m - A - A^2 - \cdots - A^m - A^{m+1} = I_n - A^{m+1}$
Now let m go to infinity; use parts (b) and (c). $(I_n - A)(I_n + A + A^2 + \cdots + A^m + \cdots) = I_n$, so that $(I_n - A)^{-1} = I_n + A + A^2 + \cdots + A^m + \cdots$.

75. a. The components of the jth column of the technology matrix A give the demands industry J_j makes on the other industries, per unit output of J_j. The fact that the jth column sum is less than 1 means that industry J_j *adds value* to the products it produces.

b. A productive economy can satisfy any consumer demand $\vec{b}$, since the equation $(I_n - A)\vec{x} = \vec{b}$ can be solved for the output vector $\vec{x}$: $\vec{x} = (I_n - A)^{-1}\vec{b}$ (compare with Exercise 2.3.49).

c. The output $\vec{x}$ required to satisfy a consumer demand $\vec{b}$ is
$\vec{x} = (I_n - A)^{-1}\vec{b} = (I_n + A + A^2 + \cdots + A^m + \cdots)\vec{b} = \vec{b} + A\vec{b} + A^2\vec{b} + \cdots + A^m\vec{b} + \cdots$.
To interpret the terms in this series, keep in mind that whatever output $\vec{v}$ the industries produce generates an interindustry demand of $A\vec{v}$.
The industries first need to satisfy the consumer demand, $\vec{b}$. Producing the output $\vec{b}$ will generate an interindustry demand, $A\vec{b}$. Producing $A\vec{b}$ in turn generates an extra interindustry demand, $A(A\vec{b}) = A^2\vec{b}$, and so forth.
For a simple example, see Exercise 2.3.50; also read the discussion of "chains of interindustry demands" in the footnote to Exercise 2.3.49.

76. a. $\begin{matrix} I = \frac{1}{3}R + \frac{1}{3}G + \frac{1}{3}B \\ L = R - G \\ S = -\frac{1}{2}R - \frac{1}{2}G + B \end{matrix}$, so the matrix is $P = \begin{bmatrix} \frac{1}{3} & \frac{1}{3} & \frac{1}{3} \\ 1 & -1 & 0 \\ -\frac{1}{2} & -\frac{1}{2} & 1 \end{bmatrix}$.

b. $\begin{bmatrix} R \\ G \\ B \end{bmatrix}$ is transformed into $\begin{bmatrix} R \\ G \\ 0 \end{bmatrix}$, with matrix $A = \begin{bmatrix} 1 & 0 & 0 \\ 0 & 1 & 0 \\ 0 & 0 & 0 \end{bmatrix}$.

c. This matrix is $PA = \begin{bmatrix} \frac{1}{3} & \frac{1}{3} & 0 \\ 1 & -1 & 0 \\ -\frac{1}{2} & -\frac{1}{2} & 0 \end{bmatrix}$ (we apply first A, then P.)

d.

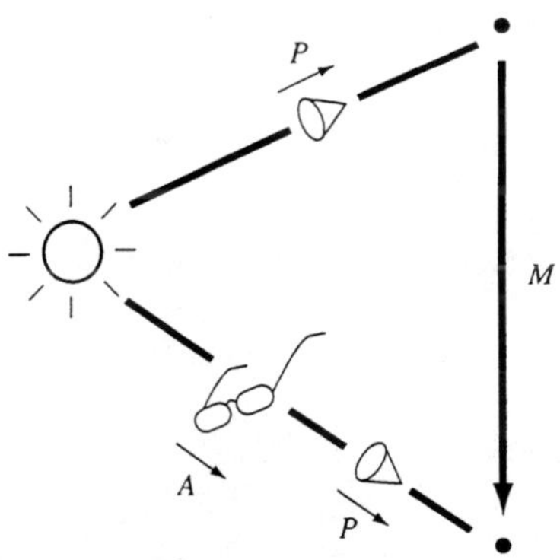

A "diagram chase" shows that $M = PAP^{-1} = \begin{bmatrix} \frac{2}{3} & 0 & -\frac{2}{9} \\ 0 & 1 & 0 \\ -1 & 0 & \frac{1}{3} \end{bmatrix}$.

77. a. $A^{-1} = \begin{bmatrix} 0 & 0 & 1 \\ 1 & 0 & 0 \\ 0 & 1 & 0 \end{bmatrix}$ and $B^{-1} = \begin{bmatrix} 1 & 0 & 0 \\ 0 & 0 & 1 \\ 0 & 1 & 0 \end{bmatrix}$.

Matrix A^{-1} transforms a wife's clan into her husband's clan, and B^{-1} transforms a child's clan into the mother's clan.

b. B^2 transforms a women's clan into the clan of a child of her daughter.

c. AB transforms a woman's clan into the clan of her daughter-in-law (her son's wife), while BA transforms a man's clan into the clan of his children. The two transformations are different.

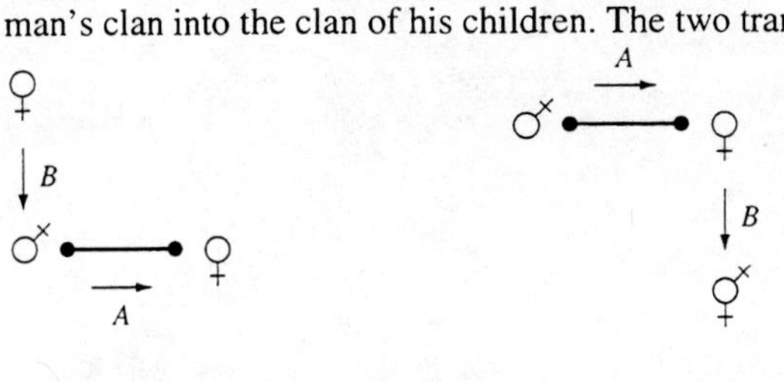

d. The matrices for the four given diagrams (in the same order) are $BB^{-1} = I_3$,

$$BAB^{-1} = \begin{bmatrix} 0 & 0 & 1 \\ 1 & 0 & 0 \\ 0 & 1 & 0 \end{bmatrix},\ B(BA)^{-1} = \begin{bmatrix} 0 & 1 & 0 \\ 0 & 0 & 1 \\ 1 & 0 & 0 \end{bmatrix},\ BA(BA)^{-1} = I_3.$$

e. Yes; since $BAB^{-1} = A^{-1} = \begin{bmatrix} 0 & 0 & 1 \\ 1 & 0 & 0 \\ 0 & 1 & 0 \end{bmatrix}$, in the second case in part (d) the cousin belongs to Bueya's husband's clan.

78. a. We need 8 multiplications: 2 to compute each of the four entries of the product.

b. We need n multiplications to compute each of the mp entries of the product, mnp multiplications altogether.

79. $g(f(x)) = x$, for all x, so that $g \circ f$ is the identity, but $f(g(x)) = \begin{cases} x & \text{if } x \text{ is even} \\ x+1 & \text{if } x \text{ is odd} \end{cases}$.

80. a. The formula $\begin{bmatrix} y \\ n \end{bmatrix} = \begin{bmatrix} 1-Rk & L+R-kLR \\ -k & 1-kL \end{bmatrix}\begin{bmatrix} x \\ m \end{bmatrix}$ is given, which implies that $y = (1-Rk)x + (L+R-kLR)m$.

In order for y to be independent of x it is required that $1 - Rk = 0$, or $k = \frac{1}{R} = 40$ (diopters).

$\frac{1}{k}$ then equals R, which is the distance between the plane of the lens and the plane on which parallel incoming rays focus at a point; thus the term "focal length" for $\frac{1}{k}$.

b. Now we want y to be independent of the slope m (it must depend on x alone). In view of the formula above, this is the case if $L + R - kLR = 0$, or, $k = \frac{L+R}{LR} = \frac{1}{R} + \frac{1}{L} = 40 + \frac{10}{3} \approx 43.3$ (diopters).

c. Here the transformation is $\begin{bmatrix} y \\ n \end{bmatrix} = \begin{bmatrix} 1 & 0 \\ -k_1 & 1 \end{bmatrix}\begin{bmatrix} 1 & D \\ 0 & 1 \end{bmatrix}\begin{bmatrix} 1 & 0 \\ -k_1 & 1 \end{bmatrix}\begin{bmatrix} x \\ m \end{bmatrix} = \begin{bmatrix} 1-k_1D & D \\ k_1k_2D - k_1 - k_2 & 1-k_2D \end{bmatrix}\begin{bmatrix} x \\ m \end{bmatrix}$.

We want the slope n of the outgoing rays to depend on the slope m of the incoming rays alone, and not on x; this forces $k_1k_2D - k_1 - k_2 = 0$, or, $D = \frac{k_1+k_2}{k_1k_2} = \frac{1}{k_1} + \frac{1}{k_2}$, the sum of the focal lengths of the two lenses.

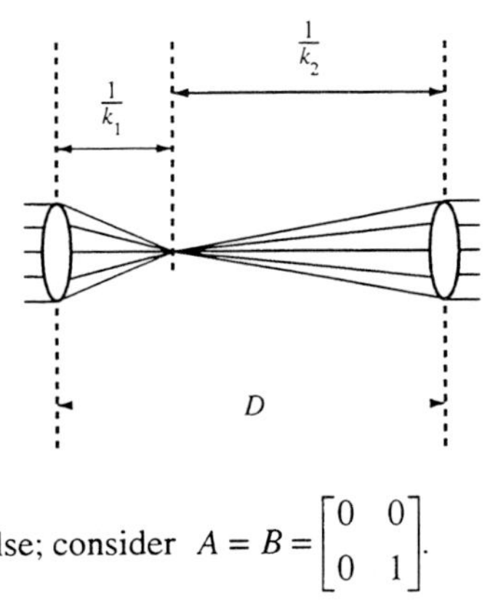

81. False; consider $A = B = \begin{bmatrix} 0 & 0 \\ 0 & 1 \end{bmatrix}$.

Chapter 3

3.1

1. Find all $\vec{x}$ such that $A\vec{x} = \vec{0}$:

$$\left[\begin{array}{cc|c} 1 & 2 & 0 \\ 3 & 4 & 0 \end{array}\right] \to \left[\begin{array}{cc|c} 1 & 0 & 0 \\ 0 & 1 & 0 \end{array}\right], \text{ so that } x_1 = x_2 = 0.$$

$\ker(A) = \{\vec{0}\}$

2. Find all $\vec{x}$ such that $A\vec{x} = \vec{0}$:

$$\left[\begin{array}{cc|c} 2 & 3 & 0 \\ 6 & 9 & 0 \end{array}\right] \to \left[\begin{array}{cc|c} 1 & \frac{3}{2} & 0 \\ 0 & 0 & 0 \end{array}\right], \text{ so that } \begin{bmatrix} x_1 \\ x_2 \end{bmatrix} = \begin{bmatrix} -\frac{3t}{2} \\ t \end{bmatrix}$$

Setting $t = 2$ we find $\ker(A) = \operatorname{span}\begin{bmatrix} -3 \\ 2 \end{bmatrix}$.

3. Find all $\vec{x}$ such that $A\vec{x} = \vec{0}$; note that all $\vec{x}$ in $\mathbb{R}^2$ satisfy the equation, so that $\ker(A) = \mathbb{R}^2 = \operatorname{span}(\vec{e}_1, \vec{e}_2)$.

4. Find all $\vec{x}$ such that $A\vec{x} = \vec{0}$, or $x_1 + 2x_2 + 3x_3 = 0$.

The solutions are of the form $\begin{bmatrix} x_1 \\ x_2 \\ x_2 \end{bmatrix} = \begin{bmatrix} -2t-3r \\ t \\ r \end{bmatrix} = t\begin{bmatrix} -2 \\ 1 \\ 0 \end{bmatrix} + r\begin{bmatrix} -3 \\ 0 \\ 1 \end{bmatrix}$, so that $\ker(A) = \operatorname{span}\left(\begin{bmatrix} -2 \\ 1 \\ 0 \end{bmatrix}, \begin{bmatrix} -3 \\ 0 \\ 1 \end{bmatrix}\right)$

5. Find all $\vec{x}$ such that $A\vec{x} = \vec{0}$.

$$\left[\begin{array}{ccc|c} 1 & 1 & 1 & 0 \\ 1 & 2 & 3 & 0 \\ 1 & 3 & 5 & 0 \end{array}\right] \to \left[\begin{array}{ccc|c} 1 & 0 & -1 & 0 \\ 0 & 1 & 2 & 0 \\ 0 & 0 & 0 & 0 \end{array}\right]; \begin{array}{l} x_1 = x_3 \\ x_2 = -2x_3 \end{array}; \begin{bmatrix} x_1 \\ x_2 \\ x_3 \end{bmatrix} = \begin{bmatrix} t \\ -2t \\ t \end{bmatrix}$$

$$\ker(A) = \operatorname{span}\begin{bmatrix} 1 \\ -2 \\ 1 \end{bmatrix}$$

6. Find all $\vec{x}$ such that $A\vec{x} = \vec{0}$.

$$\left[\begin{array}{ccc|c} 1 & 1 & 1 & 0 \\ 1 & 1 & 1 & 0 \\ 1 & 1 & 1 & 0 \end{array}\right] \to \left[\begin{array}{ccc|c} 1 & 1 & 1 & 0 \\ 0 & 0 & 0 & 0 \\ 0 & 0 & 0 & 0 \end{array}\right]; x_1 + x_2 + x_3 = 0$$

$$\begin{bmatrix} x_1 \\ x_2 \\ x_3 \end{bmatrix} = \begin{bmatrix} -r-t \\ r \\ t \end{bmatrix} = r\begin{bmatrix} -1 \\ 1 \\ 0 \end{bmatrix} + t\begin{bmatrix} -1 \\ 0 \\ 1 \end{bmatrix}$$

$$\ker(A) = \operatorname{span}\left(\begin{bmatrix} -1 \\ 1 \\ 0 \end{bmatrix}, \begin{bmatrix} -1 \\ 0 \\ 1 \end{bmatrix}\right)$$

7. Find all $\vec{x}$ such that $A\vec{x} = \vec{0}$. Since $\operatorname{rref}(A) = I_3$ we have $\ker(A) = \{\vec{0}\}$.

8. Find all $\vec{x}$ such that $A\vec{x} = \vec{0}$. Solving this system yields $\ker(A) = \text{span}\begin{bmatrix} 1 \\ -2 \\ 1 \end{bmatrix}$.

9. Find all $\vec{x}$ such that $A\vec{x} = \vec{0}$. Solving this system yields $\ker(A) = \{\vec{0}\}$.

10. Solving the system $A\vec{x} = \vec{0}$ we find that $\ker(A) = \text{span}\begin{bmatrix} 1 \\ -2 \\ 1 \\ 0 \end{bmatrix}$.

11. Solving the system $A\vec{x} = \vec{0}$ we find that $\ker(A) = \text{span}\begin{bmatrix} -2 \\ 3 \\ 1 \\ 0 \end{bmatrix}$.

12. Solving the system $A\vec{x} = \vec{0}$ we find that $\ker(A) = \text{span}\left(\begin{bmatrix} 1 \\ 1 \\ 0 \\ 0 \\ 0 \end{bmatrix}, \begin{bmatrix} -2 \\ 0 \\ -1 \\ 1 \\ 0 \end{bmatrix}\right)$.

13. Solving the system $A\vec{x} = \vec{0}$ we find that $\ker(A) = \text{span}\left(\begin{bmatrix} -2 \\ 1 \\ 0 \\ 0 \\ 0 \\ 0 \end{bmatrix}, \begin{bmatrix} -3 \\ 0 \\ -2 \\ -1 \\ 1 \\ 0 \end{bmatrix}, \begin{bmatrix} 0 \\ 0 \\ 0 \\ 0 \\ 0 \\ 1 \end{bmatrix}\right)$.

14. By Fact 3.1.3, the image of A is the span of the columns of A:

$$\text{im}(A) = \text{span}\left(\begin{bmatrix} 1 \\ 1 \\ 1 \\ 1 \end{bmatrix}, \begin{bmatrix} 1 \\ 2 \\ 3 \\ 4 \end{bmatrix}\right).$$

15. By Fact 3.1.3, the image of A is the span of the columns of A:

$$\text{im(A)} = \text{span}\left(\begin{bmatrix} 1 \\ 1 \end{bmatrix}, \begin{bmatrix} 1 \\ 2 \end{bmatrix}, \begin{bmatrix} 1 \\ 3 \end{bmatrix}, \begin{bmatrix} 1 \\ 4 \end{bmatrix}\right).$$

Since any two of these vectors span all of $\mathbb{R}^2$ already, we can write

$$\text{im}(A) = \text{span}\left(\begin{bmatrix} 1 \\ 1 \end{bmatrix}, \begin{bmatrix} 1 \\ 2 \end{bmatrix}\right).$$

16. By Fact 3.1.3, the image of A is the span of the columns of A:

$$\text{im}(A) = \text{span}\left(\begin{bmatrix}1\\1\\1\end{bmatrix}, \begin{bmatrix}2\\2\\2\end{bmatrix}, \begin{bmatrix}3\\3\\3\end{bmatrix}\right).$$

Since these three vectors are parallel, we need only one of them to span the image:

$$\text{im}(A) = \text{span}\begin{bmatrix}1\\1\\1\end{bmatrix}.$$

17. By Fact 3.1.3, $\text{im}(A) = \text{span}\left(\begin{bmatrix}1\\3\end{bmatrix}, \begin{bmatrix}2\\4\end{bmatrix}\right) = \mathbb{R}^2$ (the whole plane).

18. By Fact 3.1.3, $\text{im}(A) = \text{span}\left(\begin{bmatrix}1\\3\end{bmatrix}, \begin{bmatrix}4\\12\end{bmatrix}\right) = \text{span}\begin{bmatrix}1\\3\end{bmatrix}$ (a line in $\mathbb{R}^2$).

19. Since the four column vectors of A are parallel, we have $\text{im}(A) = \text{span}\begin{bmatrix}1\\-2\end{bmatrix}$, a line in $\mathbb{R}^2$.

20. Since the three column vectors of A are parallel, we have $\text{im}(A) = \text{span}\begin{bmatrix}1\\1\\1\end{bmatrix}$, a line in $\mathbb{R}^3$.

21. By Fact 3.1.3, $\text{im}(A) = \text{span}\left(\begin{bmatrix}4\\1\\5\end{bmatrix}, \begin{bmatrix}7\\9\\6\end{bmatrix}, \begin{bmatrix}3\\2\\8\end{bmatrix}\right)$.

It is hard to tell by inspection whether this span is a plane in $\mathbb{R}^3$ or all of $\mathbb{R}^3$; we need to find out whether the third column vector is a linear combination of the first two.

$$\left[\begin{array}{cc|c}4 & 7 & 3\\1 & 9 & 2\\5 & 6 & 8\end{array}\right] \to \left[\begin{array}{cc|c}1 & 0 & 0\\0 & 1 & 0\\0 & 0 & 1\end{array}\right]$$

This shows that the third column vector is not contained in the span of the first two, so that $\text{im}(A) = \mathbb{R}^3$.

22. Compare with the solution to Exercise 21.

$$\left[\begin{array}{cc|c}2 & 1 & 3\\3 & 4 & 2\\6 & 5 & 7\end{array}\right] \to \left[\begin{array}{cc|c}1 & 0 & 2\\0 & 1 & -1\\0 & 0 & 0\end{array}\right]$$

This computation shows that the third column vector of A, $\vec{v}_3$, is a linear combination of the first two,

$\vec{v}_3 = 2\vec{v}_1 - \vec{v}_2$, so that $\text{im}(A) = \text{span}\left(\begin{bmatrix}2\\3\\6\end{bmatrix}, \begin{bmatrix}1\\4\\5\end{bmatrix}, \begin{bmatrix}3\\2\\7\end{bmatrix}\right) = \text{span}\left(\begin{bmatrix}2\\3\\6\end{bmatrix}, \begin{bmatrix}1\\4\\5\end{bmatrix}\right)$, a plane in $\mathbb{R}^3$.

23. $\text{im}(T) = \mathbb{R}^2$ and $\ker(T) = \{\vec{0}\}$, since T is invertible (see Summary 3.1.8).

24. im(T) is the plane $x+2y+3z=0$, and ker(T) is the line perpendicular to this plane, spanned by the vector $\begin{bmatrix}1\\2\\3\end{bmatrix}$ (compare with Examples 6 and 10).

25. $\text{im}(T)=\mathbb{R}^2$ and $\ker(T)=\{\vec{0}\}$, since T is invertible (see Summary 3.1.8).

26. Since $\lim\limits_{t\to\infty} f(t)=\infty$ and $\lim\limits_{t\to-\infty} f(t)=-\infty$, we have $\text{im}(f)=\mathbb{R}$.
A careful proof involves the intermediate value theorem (see Exercise 2.2.47c).

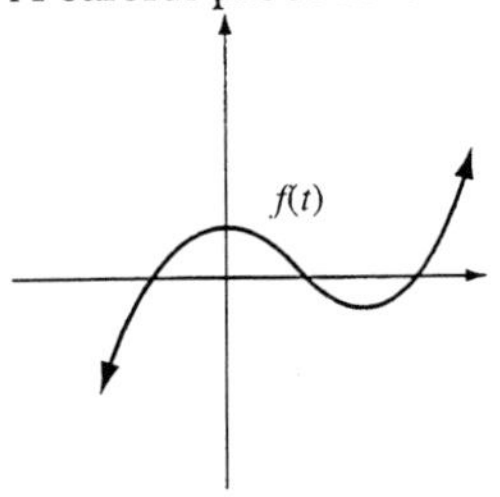

Any horizontal line intersects this graph at least once (compare with Example 3).

27. Let $f(x)=x^3-x=x(x^2-1)=x(x-1)(x+1)$.
Then im(f) = $\mathbb{R}$, by Exercise 26, but the function fails to be invertible since the equation $f(x)=0$ has three solutions, $x=0$, 1, and -1.

28. This ellipse can be obtained from the unit circle by means of the linear transformation with matrix $\begin{bmatrix}1&0\\0&2\end{bmatrix}$, as shown in the figure below (compare with Exercise 2.2.49).

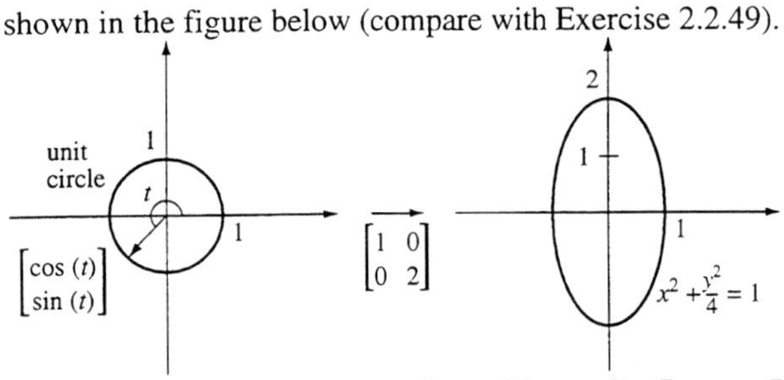

We obtain the parametrization $\begin{bmatrix}1&0\\0&2\end{bmatrix}\begin{bmatrix}\cos(t)\\\sin(t)\end{bmatrix}=\begin{bmatrix}\cos(t)\\2\sin(t)\end{bmatrix}$ for the ellipse.

We can check that $x^2+\dfrac{y^2}{4}=\cos^2(t)+\dfrac{4\sin^2(t)}{4}=1$.

29. Use spherical coordinates (see any good text in multivariable calculus): $f\begin{bmatrix}\phi\\\theta\end{bmatrix}=\begin{bmatrix}\sin(\phi)\cos(\theta)\\\sin(\phi)\sin(\theta)\\\cos(\phi)\end{bmatrix}$

30. By Fact 3.1.3, $A=\begin{bmatrix}1\\5\end{bmatrix}$ does the job. There are many other possible answers: any nonzero $2\times n$ matrix A whose column vectors are scalar multiples of vector $\begin{bmatrix}1\\5\end{bmatrix}$.

31. The plane $x+3y+2z=0$ is spanned by the two vectors $\begin{bmatrix}-2\\0\\1\end{bmatrix}$ and $\begin{bmatrix}-3\\1\\0\end{bmatrix}$, for example. Therefore, $A=\begin{bmatrix}-2&-3\\0&1\\1&0\end{bmatrix}$ does the job. There are many other correct answers.

32. By Fact 3.1.3, $A=\begin{bmatrix}7\\6\\5\end{bmatrix}$ does the job. There are many other correct answers: any nonzero $3\times n$ matrix A whose column vectors are scalar multiples of $\begin{bmatrix}7\\6\\5\end{bmatrix}$.

33. The plane is the kernel of the linear transformation $T\begin{bmatrix}x\\y\\z\end{bmatrix}=x+2y+3z$ from $\mathbb{R}^3$ to $\mathbb{R}$.

34. To describe a subset of $\mathbb{R}^3$ as a kernel means to describe it as an intersection of planes (think about it).
By inspection, the given line is the intersection of the planes

$x+y \quad =0$ and

$2x \quad +z=0.$

This means that the line is the kernel of the linear transformation $T\begin{bmatrix}x\\y\\z\end{bmatrix}=\begin{bmatrix}x+y\\2x\quad+z\end{bmatrix}$ from $\mathbb{R}^3$ to $\mathbb{R}^2$.

35. kernel$(T)=\{\vec{x}\text{: } T(\vec{x})=\vec{v}\cdot\vec{x}=0\}$ = the plane whose normal vector is $\vec{v}$.
im$(T)=\mathbb{R}$, since for every real number k there is a vector $\vec{x}$ such that $T(\vec{x})=k$, namely, $\vec{x}=\frac{k}{\vec{v}\cdot\vec{v}}\vec{v}$.

36. kernel$(T)=\{\vec{x}\text{: } T(\vec{x})=\vec{v}\times\vec{x}=\vec{0}\}$ = the line spanned by $\vec{v}$ (see Fact A.10b)
im(T) = the plane whose normal vector is $\vec{v}$
By Fact A.10a, $T(\vec{x})=\vec{v}\times\vec{x}$ is in this plane, for all $\vec{x}$ in $\mathbb{R}^3$. Conversely, for every vector $\vec{w}$ in this plane there is an $\vec{x}$ in $\mathbb{R}^3$ such that $T(\vec{x})=\vec{w}$, namely $\vec{x}=-\frac{1}{\vec{v}\cdot\vec{v}}T(\vec{w})$ (verify this!).

37. $A=\begin{bmatrix}0&1&0\\0&0&1\\0&0&0\end{bmatrix}$, $A^2=\begin{bmatrix}0&0&1\\0&0&0\\0&0&0\end{bmatrix}$, $A^3=\begin{bmatrix}0&0&0\\0&0&0\\0&0&0\end{bmatrix}$, so that
$\ker(A)=\text{span}(\vec{e}_1)$, $\ker(A^2)=\text{span}(\vec{e}_1,\vec{e}_2)$, $\ker(A^3)=\mathbb{R}^3$, and
$\text{im}(A)=\text{span}(\vec{e}_1,\vec{e}_2)$, $\text{im}(A^2)=\text{span}(\vec{e}_1)$, $\text{im}(A^3)=\{\vec{0}\}$.

38. a. If a vector $\vec{x}$ is in $\ker(A^k)$, that is, $A^k\vec{x}=\vec{0}$, then $\vec{x}$ is also in $\ker(A^{k+1})$, since $A^{k+1}\vec{x}=AA^k\vec{x}=A\vec{0}=\vec{0}$.
Therefore, $\ker(A)\subseteq\ker(A^2)\subseteq\ker(A^3)\ldots$
Exercise 37 shows that these kernels need not be equal.

b. If a vector $\vec{y}$ is in $\text{im}(A^{k+1})$, that is, $\vec{y} = A^{k+1}\vec{x}$ for some $\vec{x}$, then $\vec{y}$ is also in $\text{im}(A^k)$, since we can write $\vec{y} = A^k(A\vec{x})$. Therefore, $\ldots\text{im}(A^3) \subseteq \text{im}(A^2) \subseteq \text{im}(A)$.
Exercise 37 shows that these images need not be equal.

39. a. If a vector $\vec{x}$ is in ker(B), that is, $B\vec{x} = \vec{0}$, then $\vec{x}$ is also in ker(AB), since $AB\vec{x} = A(B\vec{x}) = A\vec{0} = \vec{0}$:
ker(B) $\subseteq$ ker(AB).
Exercise 37 (with $A = B$) illustrates that these kernels need not be equal.

b. If a vector $\vec{y}$ is in im(AB), that is, $\vec{y} = AB\vec{x}$ for some $\vec{x}$, then $\vec{y}$ is also in im(A), since we can write $\vec{y} = A(B\vec{x})$:
im(AB) $\subseteq$ im(A).
Exercise 37 (with $A = B$) illustrates that these images need not be equal.

40. For any $\vec{x}$ in $\mathbb{R}^p$, the vector $B\vec{x}$ is in im(B) = ker(A), so that $AB\vec{x} = \vec{0}$. If we apply this fact to $\vec{x} = \vec{e}_1, \vec{e}_2, \ldots, \vec{e}_p$, we find that all the columns of the matrix AB are zero, so that $AB = 0$.

41. a. $\text{rref}(A) = \begin{bmatrix} 1 & \frac{4}{3} \\ 0 & 0 \end{bmatrix}$, so that $\ker(A) = \text{span}\begin{bmatrix} -4 \\ 3 \end{bmatrix}$.

$\text{im}(A) = \text{span}\begin{bmatrix} 0.36 \\ 0.48 \end{bmatrix} = \text{span}\begin{bmatrix} 3 \\ 4 \end{bmatrix}$

Note that im(A) and ker(A) are perpendicular lines.

b. $A^2 = A$
If $\vec{v}$ is in im(A) with $\vec{v} = A\vec{x}$, then $A\vec{v} = A^2\vec{x} = A\vec{x} = \vec{v}$.

c.

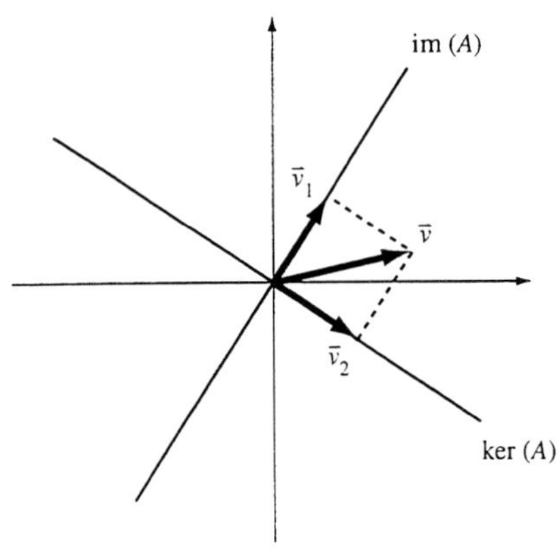

Any vector $\vec{v}$ in $\mathbb{R}^2$ can be written uniquely as $\vec{v} = \vec{v}_1 + \vec{v}_2$, where $\vec{v}_1$ is in im(A) and $\vec{v}_2$ is in ker(A). Then $A\vec{v} = A\vec{v}_1 + A\vec{v}_2 = \vec{v}_1$ ($A\vec{v}_1 = \vec{v}_1$ by part b, $A\vec{v}_2 = \vec{0}$ since $\vec{v}_2$ is in ker(A)), so that A represents the *orthogonal projection* onto $\text{im}(A) = \text{span}\begin{bmatrix} 3 \\ 4 \end{bmatrix}$.

42. Using the hint, we see that the vector $y = \begin{bmatrix} y_1 \\ y_2 \\ y_3 \\ y_4 \end{bmatrix}$ is in the image of A if

$$\begin{aligned} y_1 \quad - 3y_3 + 2y_4 &= 0 \text{ and} \\ y_2 - 2y_3 + \ y_4 &= 0. \end{aligned}$$

This means that im(A) is the kernel of the matrix $\begin{bmatrix} 1 & 0 & -3 & 2 \\ 0 & 1 & -2 & 1 \end{bmatrix}$.

43. Using our work in Exercise 42 as a guide, we come up with the following procedure to express the image of an $m \times n$ matrix A as the kernel of a matrix B:
If rank(A) = m, let B be the $m \times m$ zero matrix.
If r = rank(A) < m, let B be the $(m - r) \times m$ matrix obtained by omitting the first r rows and the first n columns of $\text{rref}\left[A \mid I_m\right]$.

44. a. Yes; by construction of the echelon form, the systems $A\vec{x} = \vec{0}$ and $B\vec{x} = \vec{0}$ have the same solutions (it is the whole point of Gaussian elimination not to change the solutions of a system).

b. No; as a counterexample, consider $A = \begin{bmatrix} 0 & 0 \\ 1 & 0 \end{bmatrix}$, with $\text{im(A)} = \text{span}(\vec{e}_2)$, but $B = \text{rref}(A) = \begin{bmatrix} 1 & 0 \\ 0 & 0 \end{bmatrix}$, with $\text{im}(B) = \text{span}(\vec{e}_1)$.

45. As we solve the system $A\vec{x} = \vec{0}$, we obtain r leading variables and $n - r$ nonleading variables. The "general vector" in ker(A) can be written as a linear combination of $n - r$ vectors, with the nonleading variables as coefficients. (See Example 12, where $n - r = 5 - 3 = 2$.)

46. If rank(A) = r, then $\text{im}(A) = \text{span}(\vec{e}_1, \ldots, \vec{e}_r)$.

rank(A) = 0 rank(A) = 1 rank(A) = 2 rank(A) = 3

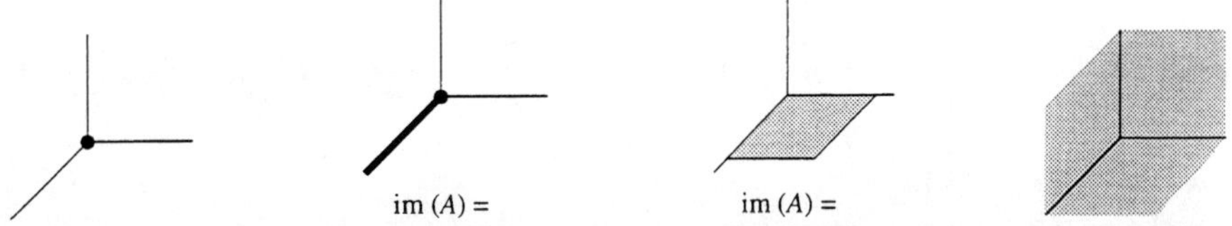

im (A) = $\{\vec{0}\}$ im (A) = span ($\vec{e}_1$) im (A) = span ($\vec{e}_1, \vec{e}_2$) im (A) = $\mathbb{R}^3$

47. $\text{im}(T) = L_1$ and $\text{ker}(T) = L_2$.

48. a. $\vec{w} = A\vec{x}$, for some $\vec{x}$, so that $A\vec{w} = A^2\vec{x} = A\vec{x} = \vec{w}$.

b. If rank(A) = 2, then A is invertible, and the equation $A^2 = A$ implies that $A = I_2$ (multiply by A^{-1}).

If rank(A) = 0 then $A = \begin{bmatrix} 0 & 0 \\ 0 & 0 \end{bmatrix}$.

c. First note that im(A) and ker(A) are lines (there is one nonleading variable).

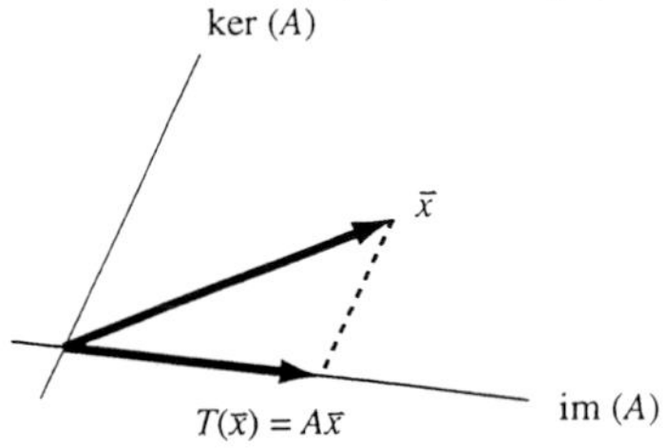

By definition of a projection, we need to verify that $\vec{x} - A\vec{x}$ is in ker(A). This is indeed the case, since $A(\vec{x} - A\vec{x}) = A\vec{x} - A^2\vec{x} = A\vec{x} - A\vec{x} = \vec{0}$ (we are told that $A^2 = A$).

49. If $\vec{v}$ and $\vec{w}$ are in ker(T), then $T(\vec{v} + \vec{w}) = T(\vec{v}) + T(\vec{w}) = \vec{0} + \vec{0} = \vec{0}$, so that $\vec{v} + \vec{w}$ is in ker(T) as well.
If $\vec{v}$ is in ker(T) and k is an arbitrary scalar, then $T(k\vec{v}) = kT(\vec{v}) = k\vec{0} = \vec{0}$, so that $k\vec{v}$ is in ker(T) as well.

50. From Exercise 38 we know that $\ker(A^3) \subseteq \ker(A^4)$. Conversely, if $\vec{x}$ is in $\ker(A^4)$, then $A^4\vec{x} = A^3(A\vec{x}) = \vec{0}$, so that $A\vec{x}$ is in $\ker(A^3) = \ker(A^2)$, which implies that $A^2(A\vec{x}) = A^3\vec{x} = \vec{0}$, that is, $\vec{x}$ is in $\ker(A^3)$. We have shown that $\ker(A^3) = \ker(A^4)$.

51. We need to find all $\vec{x}$ such that $AB\vec{x} = \vec{0}$. If $AB\vec{x} = \vec{0}$, then $B\vec{x}$ is in $\ker(A) = \{\vec{0}\}$, so that $B\vec{x} = \vec{0}$.
Since $\ker(B) = \{\vec{0}\}$, we can conclude that $\vec{x} = \vec{0}$. It follows that $\ker(AB) = \{\vec{0}\}$.

52. Since $C\vec{x} = \begin{bmatrix} A \\ B \end{bmatrix}\vec{x} = \begin{bmatrix} A\vec{x} \\ B\vec{x} \end{bmatrix}$, we can conclude that $C\vec{x} = \vec{0}$ if (and only if) both $A\vec{x} = \vec{0}$ and $B\vec{x} = \vec{0}$. It follows that ker(C) is the intersection of ker(A) and ker(B): ker(C) = ker(A) $\cap$ ker(B)

53. a. Using the equation 1 + 1 = 0 (or −1 = 1), we can write the general vector $\vec{x}$ in ker(H) as

$$\vec{x} = \begin{bmatrix} x_1 \\ x_2 \\ x_3 \\ x_4 \\ x_5 \\ x_6 \\ x_7 \end{bmatrix} = \begin{bmatrix} p+r+s \\ p+q+s \\ p+q+r \\ p \\ q \\ r \\ s \end{bmatrix} = p\underset{\vec{v}_1}{\underset{\uparrow}{\begin{bmatrix} 1 \\ 1 \\ 1 \\ 1 \\ 0 \\ 0 \\ 0 \end{bmatrix}}} + q\underset{\vec{v}_2}{\underset{\uparrow}{\begin{bmatrix} 0 \\ 1 \\ 1 \\ 0 \\ 1 \\ 0 \\ 0 \end{bmatrix}}} + r\underset{\vec{v}_3}{\underset{\uparrow}{\begin{bmatrix} 1 \\ 0 \\ 1 \\ 0 \\ 0 \\ 1 \\ 0 \end{bmatrix}}} + s\underset{\vec{v}_4}{\underset{\uparrow}{\begin{bmatrix} 1 \\ 1 \\ 0 \\ 0 \\ 0 \\ 0 \\ 1 \end{bmatrix}}}$$

b. $\ker(H) = \text{span}(\vec{v}_1, \vec{v}_2, \vec{v}_3, \vec{v}_4)$ by part (a), and $\text{im}(M) = \text{span}(\vec{v}_1, \vec{v}_2, \vec{v}_3, \vec{v}_4)$ by Fact 3.1.3, so that im(M) = ker(H). $M\vec{x}$ is in im(M) = ker(H), so that $H(M\vec{x}) = \vec{0}$.

54. a. If no error occurred, then $\vec{w} = \vec{v} = M\vec{u}$, and $H\vec{w} = H(M\vec{u}) = \vec{0}$, by Exercise 53b.
If an error occurred in the ith component, then $\vec{w} = \vec{v} + \vec{e}_i = M\vec{u} + \vec{e}_i$, so that
$H\vec{w} = H(M\vec{u}) + H\vec{e}_i$ = ith column of H.
Since the columns of H are all different, this method allows us to find out where an error occurred.

b. $H\vec{w} = \begin{bmatrix} 1 \\ 1 \\ 0 \end{bmatrix}$ = seventh column of H: an error occurred in the seventh component of $\vec{v}$.

Therefore $\vec{v} = \vec{w} + \vec{e}_7 = \begin{bmatrix} 1 \\ 0 \\ 1 \\ 0 \\ 1 \\ 0 \\ 1 \end{bmatrix}$ and $\vec{u} = \begin{bmatrix} 0 \\ 1 \\ 0 \\ 1 \end{bmatrix}$.

3.2

1. Not a subspace, since W does not contain the zero vector.

2. Not a subspace, since W contains the vector $\vec{v} = \begin{bmatrix} 1 \\ 2 \\ 3 \end{bmatrix}$ but not the vector $(-1)\vec{v} = \begin{bmatrix} -1 \\ -2 \\ -3 \end{bmatrix}$.

3. $W = \text{im}\begin{bmatrix} 1 & 2 & 3 \\ 4 & 5 & 6 \\ 7 & 8 & 9 \end{bmatrix}$ is a subspace of $\mathbb{R}^3$, by Fact 3.2.2.

4. $\text{span}(\vec{v}_1, \ldots, \vec{v}_m) = \text{im}[\vec{v}_1 \ \cdots \ \vec{v}_m]$ is a subspace of $\mathbb{R}^n$, by Fact 3.2.2.

5. We have subspaces $\{\vec{0}\}$, $\mathbb{R}^3$, and all lines and planes (through the origin). To prove this, mimic the reasoning in Example 3.

6. a. Yes!

- The zero vector is in $V \cap W$, since $\vec{0}$ is in both V and W.
- If $\vec{x}$ and $\vec{y}$ are in $V \cap W$, then both $\vec{x}$ and $\vec{y}$ are in V, so that $\vec{x} + \vec{y}$ is in V as well, since V is a subspace of $\mathbb{R}^n$. Likewise, $\vec{x} + \vec{y}$ is in W, so that $\vec{x} + \vec{y}$ is in $V \cap W$.
- If $\vec{x}$ is in $V \cap W$ and k is an arbitrary scalar, then $k\vec{x}$ is in both V and W, since they are subspaces of $\mathbb{R}^n$. Therefore, $k\vec{x}$ is in $V \cap W$.

b. No; as a counterexample consider $V = \text{span}(\vec{e}_1)$ and $W = \text{span}(\vec{e}_2)$ in $\mathbb{R}^2$.

7. Yes; we need to show that W contains the zero vector. We are told that W is nonempty, so that it contains some vector $\vec{v}$. Since W is closed under scalar multiplication, it will contain the vector $0\vec{v} = \vec{0}$, as claimed.

8. We need to solve the system $c_1\begin{bmatrix}1\\2\end{bmatrix}+c_2\begin{bmatrix}2\\3\end{bmatrix}+c_3\begin{bmatrix}3\\4\end{bmatrix}=\begin{bmatrix}1&2&3\\2&3&4\end{bmatrix}\begin{bmatrix}c_1\\c_2\\c_3\end{bmatrix}=\begin{bmatrix}0\\0\end{bmatrix}$.

The general solution is $\begin{bmatrix}c_1\\c_2\\c_3\end{bmatrix}=\begin{bmatrix}t\\-2t\\t\end{bmatrix}$.

Picking $t=1$ we find the nontrivial relation $1\begin{bmatrix}1\\2\end{bmatrix}-2\begin{bmatrix}2\\3\end{bmatrix}+1\begin{bmatrix}3\\4\end{bmatrix}=\begin{bmatrix}0\\0\end{bmatrix}$.

9. These vectors are linearly dependent, since $\vec{v}_m=0\vec{v}_1+0\vec{v}_2+\cdots+0\vec{v}_{m-1}$.

10. Linearly dependent, since $\begin{bmatrix}6\\3\end{bmatrix}=3\begin{bmatrix}2\\1\end{bmatrix}$.

11. Linearly independent, since the two vectors are not parallel.

12. Linearly dependent, since $\begin{bmatrix}0\\0\end{bmatrix}=0\begin{bmatrix}7\\11\end{bmatrix}$.

13. Linearly dependent, since $\begin{bmatrix}1\\2\end{bmatrix}=1\begin{bmatrix}1\\2\end{bmatrix}$.

14. Linearly independent, since $\text{rref}\begin{bmatrix}1&1&1\\0&2&2\\0&0&3\end{bmatrix}=I_3$ (use Fact 3.2.6).

15. Linearly dependent, since $\ker\begin{bmatrix}1&2&3\\2&3&4\end{bmatrix}\neq\{\vec{0}\}$ (by Fact 1.3.3).

16. Linearly dependent, since $\text{rref}\begin{bmatrix}1&4&7\\2&5&8\\3&6&9\end{bmatrix}=\begin{bmatrix}1&0&-1\\0&1&2\\0&0&0\end{bmatrix}$ (use Fact 3.2.6).

17. Linearly independent, since $\text{rref}\begin{bmatrix}1&1&1\\1&2&3\\1&3&6\end{bmatrix}=I_3$ (use Fact 3.2.6).

18. Linearly dependent, since $\text{rref}\begin{bmatrix}1&1&1\\1&2&4\\1&3&7\\1&4&10\end{bmatrix}=\begin{bmatrix}1&0&-2\\0&1&3\\0&0&0\\0&0&0\end{bmatrix}$.

19. Linearly dependent, since $\ker\begin{bmatrix}1 & 1 & 5 & 1 & 1\\ 2 & 5 & 4 & 8 & 1\\ 9 & 1 & 9 & 1 & 1\\ 1 & 5 & 1 & 5 & 1\end{bmatrix} \neq \{\vec{0}\}$ (by Fact 1.3.3).

20. Linearly dependent, since $\operatorname{rref}\begin{bmatrix}1 & 5 & 9\\ 2 & 6 & 10\\ 3 & 7 & 11\\ 4 & 8 & 12\end{bmatrix} = \begin{bmatrix}1 & 0 & -1\\ 0 & 1 & 2\\ 0 & 0 & 0\\ 0 & 0 & 0\end{bmatrix}$.

21. Linearly dependent, since rref does not have a leading one in the fourth column.

22. If a, c, and f are nonzero, then $\operatorname{rref}\begin{bmatrix}a & b & d\\ 0 & c & e\\ 0 & 0 & f\\ 0 & 0 & 0\end{bmatrix} = \begin{bmatrix}1 & 0 & 0\\ 0 & 1 & 0\\ 0 & 0 & 1\\ 0 & 0 & 0\end{bmatrix}$, and the three vectors are linearly independent, by Fact 3.2.6. If at least one of the constants a, c, or f is zero, then at least one column of rref will not contain a leading one, so that the three vectors are linearly dependent.

23. The zero vector is in $V^\perp$, since $\vec{0}\cdot\vec{v} = 0$ for all $\vec{v}$ in V.
If $\vec{w}_1$ and $\vec{w}_2$ are both in $V^\perp$, then $(\vec{w}_1 + \vec{w}_2)\cdot\vec{v} = \vec{w}_1\cdot\vec{v} + \vec{w}_2\cdot\vec{v} = 0 + 0 = 0$ for all $\vec{v}$ in V, so that $\vec{w}_1 + \vec{w}_2$ is in $V^\perp$ as well.
If $\vec{w}$ is in $V^\perp$ and k is an arbitrary constant, then $(k\vec{w})\cdot\vec{v} = k(\vec{w}\cdot\vec{v}) = k0 = 0$ for all $\vec{v}$ in V, so that $k\vec{w}$ is in $V^\perp$ as well.

24. We need to find all vectors $\begin{bmatrix}x\\ y\\ z\end{bmatrix}$ in $\mathbb{R}^3$ such that $\begin{bmatrix}x\\ y\\ z\end{bmatrix}\cdot\begin{bmatrix}1\\ 2\\ 3\end{bmatrix} = x + 2y + 3z = 0$.

These vectors have the form $\begin{bmatrix}x\\ y\\ z\end{bmatrix} = \begin{bmatrix}-2s-3t\\ s\\ t\end{bmatrix} = s\begin{bmatrix}-2\\ 1\\ 0\end{bmatrix} + t\begin{bmatrix}-3\\ 0\\ 1\end{bmatrix}$.

Therefore, $\begin{bmatrix}-2\\ 1\\ 0\end{bmatrix}, \begin{bmatrix}-3\\ 0\\ 1\end{bmatrix}$ is a basis of $L^\perp$.

25. We need to find all vectors $\vec{x}$ in $\mathbb{R}^5$ such that $\begin{bmatrix} x_1 \\ x_2 \\ x_3 \\ x_4 \\ x_5 \end{bmatrix} \cdot \begin{bmatrix} 1 \\ 2 \\ 3 \\ 4 \\ 5 \end{bmatrix} = x_1 + 2x_2 + 3x_3 + 4x_4 + 5x_5 = 0.$

These vectors are of the form $\begin{bmatrix} x_1 \\ x_2 \\ x_3 \\ x_4 \\ x_5 \end{bmatrix} = \begin{bmatrix} -2a-3b-4c-5d \\ a \\ b \\ c \\ d \end{bmatrix} = a\begin{bmatrix} -2 \\ 1 \\ 0 \\ 0 \\ 0 \end{bmatrix} + b\begin{bmatrix} -3 \\ 0 \\ 1 \\ 0 \\ 0 \end{bmatrix} + c\begin{bmatrix} -4 \\ 0 \\ 0 \\ 1 \\ 0 \end{bmatrix} + d\begin{bmatrix} -5 \\ 0 \\ 0 \\ 0 \\ 1 \end{bmatrix}.$

The four vectors to the right form a basis of $L^{\perp}$; they span $L^{\perp}$, by construction, and none of them is a linear combination of the three others (since each of them has a one as a component where the other three have a zero).

26. Consider a linear relation $c_1\vec{v}_1 + c_2\vec{v}_2 + c_3\vec{v}_3 + c_4\vec{v}_4 = \vec{0}$ among the four given vectors. The last component of the vector on the left hand side is c_3, so that $c_3 = 0$. Now the fifth component on the left is c_1, so that $c_1 = 0$. The third component is now c_4, so $c_4 = 0$. It follows that $c_2 = 0$ as well.
We have shown that there is only the trivial relation among the given vectors, so that they are linearly independent, regardless of the values of the constants $a, b, \ldots, m$.

27. A basis of im(A) is $\begin{bmatrix} 1 \\ 1 \\ 1 \end{bmatrix}, \begin{bmatrix} 1 \\ 2 \\ 3 \end{bmatrix}$, by Fact 3.1.3.

28. The three column vectors are linearly independent, since $\text{rref}\begin{bmatrix} 1 & 1 & 1 \\ 1 & 2 & 5 \\ 1 & 3 & 7 \end{bmatrix} = I_3$.

Therefore, the three columns form a basis of $\text{im}(A)$ $(= \mathbb{R}^3)$:

$\begin{bmatrix} 1 \\ 1 \\ 1 \end{bmatrix}, \begin{bmatrix} 1 \\ 2 \\ 3 \end{bmatrix}, \begin{bmatrix} 1 \\ 5 \\ 7 \end{bmatrix}.$

Another sensible choice for a basis of im(A) is $\vec{e}_1, \vec{e}_2, \vec{e}_3$.

29. The three column vectors of A span all of $\mathbb{R}^2$, so that $\text{im}(A) = \mathbb{R}^2$. We can choose any two of the columns of A to form a basis of im(A); another sensible choice is $\vec{e}_1, \vec{e}_2$.

30. $\text{im}(A) = \text{span}(\vec{e}_1, \vec{e}_2)$
We can choose $\vec{e}_1, \vec{e}_2$ as a basis of im(A).

31. The two column vectors of the given matrix A are linearly independent (they are not parallel), so that they form a basis of im(A).

32. The two column vectors $\vec{v}_1$ and $\vec{v}_2$ of the given matrix A are parallel, with $\vec{v}_2 = 2\vec{v}_1$, so that $\text{im}(A) = \text{span}(\vec{v}_1, \vec{v}_2) = \text{span}(\vec{v}_1)$.

Therefore, $\vec{v}_1 = \begin{bmatrix} 1 \\ 2 \\ 3 \\ 4 \end{bmatrix}$ is a basis of $\text{im}(A)$.

33. $\text{im}(A) = \text{span}(\vec{e}_1, \vec{e}_2, \vec{e}_3)$, so that $\vec{e}_1, \vec{e}_2, \vec{e}_3$ is a basis of $\text{im}(A)$.

34. The fact that $\begin{bmatrix} 1 \\ 2 \\ 3 \\ 4 \end{bmatrix}$ is in $\ker(A)$ means that

$$A\begin{bmatrix} 1 \\ 2 \\ 3 \\ 4 \end{bmatrix} = \begin{bmatrix} \vec{v}_1 & \vec{v}_2 & \vec{v}_3 & \vec{v}_4 \end{bmatrix}\begin{bmatrix} 1 \\ 2 \\ 3 \\ 4 \end{bmatrix} = \vec{v}_1 + 2\vec{v}_2 + 3\vec{v}_3 + 4\vec{v}_4 = \vec{0}, \text{ so that}$$

$$\vec{v}_4 = -\frac{1}{4}\vec{v}_1 - \frac{1}{2}\vec{v}_2 - \frac{3}{4}\vec{v}_3.$$

35. Consider a nontrivial relation $c_1\vec{v}_1 + c_2\vec{v}_2 + \cdots + c_m\vec{v}_m = \vec{0}$.
Let i be the highest index such that $c_i \neq 0$; since $\vec{v}_1 \neq \vec{0}$ we know that $i > 1$. Now we have $c_1\vec{v}_1 + c_2\vec{v}_2 + \cdots + c_i\vec{v}_i = \vec{0}$.
We can solve this relation for $\vec{v}_i$ and thus express $\vec{v}_i$ as a linear combination of $\vec{v}_1, \vec{v}_2, \ldots, \vec{v}_{i-1}$:

$$\vec{v}_i = -\frac{c_1}{c_i}\vec{v}_1 - \frac{c_2}{c_i}\vec{v}_2 - \cdots - \frac{c_{i-1}}{c_i}\vec{v}_{i-1}.$$

36. Yes; we know that there is a nontrivial relation $c_1\vec{v}_1 + c_2\vec{v}_2 + \cdots + c_m\vec{v}_m = \vec{0}$.
Now apply the transformation T to the vectors on both sides, and use linearity:
$T(c_1\vec{v}_1 + c_2\vec{v}_2 + \cdots + c_m\vec{v}_m) = T(\vec{0})$, so that $c_1T(\vec{v}_1) + c_2T(\vec{v}_2) + \cdots + c_mT(\vec{v}_m) = \vec{0}$.
This is a nontrivial relation among the vectors $T(\vec{v}_1), \ldots, T(\vec{v}_m)$, so that these vectors are linearly dependent, as claimed.

37. No; as a counter example, consider the extreme case when T is the zero transformation, that is, $T(\vec{x}) = \vec{0}$ for all $\vec{x}$. Then the vectors $T(\vec{v}_1), \ldots, T(\vec{v}_m)$ will all be zero, so that they are linearly dependent.

38. **a.** Consider vectors $\vec{v}_1, \ldots, \vec{v}_m$ in $\mathbb{R}^n$, where $m > n$. The rank of the $n \times m$ matrix $\begin{bmatrix} \vec{v}_1 & \vec{v}_2 & \cdots & \vec{v}_m \end{bmatrix}$ is at most n, so that there is at least one column without a leading one. By Fact 3.2.6, the vectors $\vec{v}_1, \vec{v}_2, \ldots, \vec{v}_m$ will be linearly dependent.

b. Using the terminology introduced in the exercise, we need to show that any vector $\vec{v}$ in V is a linear combination of $\vec{v}_1, \ldots, \vec{v}_m$. Choose a specific vector $\vec{v}$ in V. Since we can find no more than m linearly independent vectors in V, the $m + 1$ vectors $\vec{v}_1, \ldots, \vec{v}_m, \vec{v}$ will be linearly dependent, that is, there will be a nontrivial relation $c_1\vec{v}_1 + \cdots + c_m\vec{v}_m + c\vec{v} = \vec{0}$.
We must have $c \neq 0$ (if c were zero, we could drop the term $c\vec{v}$ in the relation above to get a nontrivial

relation among $\vec{v}_1, \ldots, \vec{v}_m$, contradicting the linear independence of these vectors.) Now $\vec{v} = -\frac{c_1}{c}\vec{v}_1 - \cdots - \frac{c_m}{c}\vec{v}_m$. We were able to express $\vec{v}$ as a linear combination of $\vec{v}_1, \ldots, \vec{v}_m$, as claimed.

39. Yes; we can use Exercise 35 here. None of the vectors in the sequence $\vec{v}_1, \ldots, \vec{v}_m, \vec{v}$ is a linear combination of the previous vectors in that sequence.

40. Yes; by Fact 3.2.6, $\ker(A) = \{\vec{0}\}$ and $\ker(B) = \{\vec{0}\}$. Then $\ker(AB) = \{\vec{0}\}$ by Exercise 3.1.51, so that the columns of AB are linearly independent, by Fact 3.2.6.

41. To show that the columns of B are linearly independent, we show that $\ker(B) = \{\vec{0}\}$. Indeed, if $B\vec{x} = \vec{0}$, then $AB\vec{x} = A\vec{0} = \vec{0}$, so that $\vec{x} = \vec{0}$ (since $AB = I_m$).
By Fact 3.2.6, rank(B) = # columns = m, so that $m \le n$ and in fact $m < n$ (we are told that $m \ne n$). This implies that the rank of the $m \times n$ matrix A is less than n, so that the columns of A are linearly dependent (by Fact 3.2.6).

42. We can use the hint and form the dot product of $\vec{v}_i$ and both sides of the relation
$c_1\vec{v}_1 + \cdots + c_i\vec{v}_i + \cdots + c_m\vec{v}_m = \vec{0}$:
$(c_1\vec{v}_1 + \cdots + c_i\vec{v}_i + \cdots + c_m\vec{v}_m)\cdot\vec{v}_i = \vec{0}\cdot\vec{v}_i$, so that $c_1(\vec{v}_1\cdot\vec{v}_i) + \cdots + c_i(\vec{v}_i\cdot\vec{v}_i) + \cdots + c_m(\vec{v}_m\cdot\vec{v}_i) = 0$.
Since $\vec{v}_i$ is perpendicular to all the other $\vec{v}_j$, we will have $\vec{v}_i\cdot\vec{v}_j = 0$ whenever $j \ne i$; since $\vec{v}_i$ is a unit vector, we will have $\vec{v}_i\cdot\vec{v}_i = 1$. Therefore, the equation above simplifies to $c_i = 0$.
Since this reasoning applies to all $i = 1, \ldots, m$, we have only the trivial relation among the vectors $\vec{v}_1, \vec{v}_2, \ldots, \vec{v}_m$, so that these vectors are linearly independent, as claimed.

43. Consider a linear relation $c_1\vec{v}_1 + c_2(\vec{v}_1 + \vec{v}_2) + c_3(\vec{v}_1 + \vec{v}_2 + \vec{v}_3) = \vec{0}$, or, $(c_1 + c_2 + c_3)\vec{v}_1 + (c_2 + c_3)\vec{v}_2 + c_3\vec{v}_3 = \vec{0}$.
Since there is only the trivial relation among the vectors $\vec{v}_1, \vec{v}_2, \vec{v}_3$, we must have $c_1 + c_2 + c_3 = c_2 + c_3 = c_3 = 0$, so that $c_3 = 0$ and then $c_2 = 0$ and then $c_1 = 0$, as claimed.

44. Yes; this is a special case of Exercise 40 (recall that $\ker(A) = \{\vec{0}\}$, by Fact 3.1.7).

45. Yes; if A is invertible, then $\text{rref}(A) = I_n$, so that the columns of A are linearly independent, by Fact 3.2.6.

46. Solve the system $\left| \begin{array}{ccccccc} x_1 & + & 2x_2 & & + & 3x_4 & + & 5x_5 & = & 0 \\ & & & x_3 & + & 4x_4 & + & 6x_5 & = & 0 \end{array} \right|$.

The solutions are of the form $\begin{bmatrix} x_1 \\ x_2 \\ x_3 \\ x_4 \\ x_5 \end{bmatrix} = \begin{bmatrix} -2s - 3t - 5r \\ s \\ -4t - 6r \\ t \\ r \end{bmatrix} = s\begin{bmatrix} -2 \\ 1 \\ 0 \\ 0 \\ 0 \end{bmatrix} + t\begin{bmatrix} -3 \\ 0 \\ -4 \\ 1 \\ 0 \end{bmatrix} + r\begin{bmatrix} -5 \\ 0 \\ -6 \\ 0 \\ 1 \end{bmatrix}$.

The three vectors $\begin{bmatrix} -2 \\ 1 \\ 0 \\ 0 \\ 0 \end{bmatrix}, \begin{bmatrix} -3 \\ 0 \\ -4 \\ 1 \\ 0 \end{bmatrix}, \begin{bmatrix} -5 \\ 0 \\ -6 \\ 0 \\ 1 \end{bmatrix}$ span the kernel, by construction, and none of them is a linear combination of the two others (since each has a one as a component where the other two have a zero; consider the second, fourth, and fifth components). Therefore, the three vectors form a basis of the kernel.

47. By Fact 3.2.6, $\text{rref}(A) = \begin{bmatrix} 1 & 0 & 0 \\ 0 & 1 & 0 \\ 0 & 0 & 1 \\ 0 & 0 & 0 \end{bmatrix}$.

48. We can write $3x_1 + 4x_2 + 5x_3 = [3 \quad 4 \quad 5] \begin{bmatrix} x_1 \\ x_2 \\ x_3 \end{bmatrix} = 0$, so that $E = \ker[3 \quad 4 \quad 5]$.

To express E as an image, choose a basis in E, for example, $\begin{bmatrix} 4 \\ -3 \\ 0 \end{bmatrix}, \begin{bmatrix} 0 \\ 5 \\ -4 \end{bmatrix}$.

Then, $E = \text{im}\begin{bmatrix} 4 & 0 \\ -3 & 5 \\ 0 & -4 \end{bmatrix}$.

There are other solutions.

49. $L = \text{im}\begin{bmatrix} 1 \\ 1 \\ 1 \end{bmatrix}$

To write L as a kernel, think of L as the intersection of the planes $x = y$ and $y = z$, that is, as the solution set of the system $\begin{vmatrix} x & - & y & & & = & 0 \\ & & y & - & z & = & 0 \end{vmatrix}$.

Therefore, $L = \ker\begin{bmatrix} 1 & -1 & 0 \\ 0 & 1 & -1 \end{bmatrix}$.

There are other solutions.

50. The verification of the three properties listed in Definition 3.2.1 is straightforward. Alternatively, we can choose a basis $\vec{v}_1, \ldots, \vec{v}_p$ of V and a basis $\vec{w}_1, \ldots, \vec{w}_q$ of W (see Exercise 38b) and show that $V + W = \text{span}(\vec{v}_1, \ldots, \vec{v}_p, \vec{w}_1, \ldots, \vec{w}_q)$ (compare with Exercise 4).

Indeed, if $\vec{v} + \vec{w}$ is in $V + W$, then $\vec{v}$ is a linear combination of the $\vec{v}_i$ and $\vec{w}$ is a linear combination of the $\vec{w}_i$, so that $\vec{v} + \vec{w}$ is a linear combination of $\vec{v}_1, \ldots, \vec{v}_p, \vec{w}_1, \ldots, \vec{w}_q$. Conversely, if $\vec{x}$ is in $\text{span}(\vec{v}_1, \ldots, \vec{v}_p, \vec{w}_1, \ldots, \vec{w}_q)$, then $\vec{x} = (c_1\vec{v}_1 + \cdots + c_p\vec{v}_p) + (d_1\vec{w}_1 + \cdots + d_q\vec{w}_q)$, so that $\vec{x}$ is in $V + W$.

If V and W are distinct lines in $\mathbb{R}^3$ (spanned by $\vec{v}$ and $\vec{w}$, respectively), then $V + W$ is the plane spanned by $\vec{v}$ and $\vec{w}$.

51. **a.** Consider a relation $c_1\vec{v}_1 + \cdots + c_p\vec{v}_p + d_1\vec{w}_1 + \cdots + d_q\vec{w}_q = \vec{0}$.

Then the vector $c_1\vec{v}_1 + \cdots + c_p\vec{v}_p = -d_1\vec{w}_1 - \cdots - d_q\vec{w}_q$ is both in V and in W, so that this vector is $\vec{0}$: $c_1\vec{v}_1 + \cdots + c_p\vec{v}_p = \vec{0}$ and $d_1\vec{w}_1 + \cdots + d_q\vec{w}_q = \vec{0}$.

Now the c_i are all zero (since the $\vec{v}_i$ are linearly independent) and the d_j are zero (since the $\vec{w}_j$ are linearly independent).

Since there is only the trivial relation among the vectors $\vec{v}_1, \ldots, \vec{v}_p, \vec{w}_1, \ldots, \vec{w}_q$, they are linearly independent.

b. In Exercise 50 we show that $V + W = \text{span}(\vec{v}_1, \ldots, \vec{v}_p, \vec{w}_1, \ldots, \vec{w}_q)$, and in part (a) we show that these vectors are linearly independent.

52. True; by Exercise 38(b), there is a basis $\vec{v}_1, \ldots, \vec{v}_m$ of V, where $m \le n$. Now let A be the $n \times n$ matrix whose first m columns are $\vec{v}_1, \ldots, \vec{v}_m$, with the remaining $n - m$ columns being zero: $A = \begin{bmatrix} \vec{v}_1 & \vec{v}_2 \ldots \vec{v}_m & \vec{0} \ldots \vec{0} \end{bmatrix}$.
Then, $\text{im}(A) = \text{span}(\vec{v}_1, \ldots, \vec{v}_m) = V$.

3.3

1. $\text{rref}(A) = \begin{bmatrix} 1 & 2 \\ 0 & 0 \end{bmatrix}$

A basis of ker(A) is $\begin{bmatrix} -2 \\ 1 \end{bmatrix}$; dim(ker($A$)) = 1.

2. $\text{rref}(A) = \begin{bmatrix} 1 & 1 & 1 \\ 0 & 0 & 0 \\ 0 & 0 & 0 \end{bmatrix}$

A basis of ker(A) is $\begin{bmatrix} -1 \\ 1 \\ 0 \end{bmatrix}, \begin{bmatrix} -1 \\ 0 \\ 1 \end{bmatrix}$. Dimension is 2.

3. $\text{rref}(A) = \begin{bmatrix} 1 & 0 & 5 \\ 0 & 1 & -1 \\ 0 & 0 & 0 \end{bmatrix}$

A basis of ker(A) is $\begin{bmatrix} -5 \\ 1 \\ 1 \end{bmatrix}$, so that dim(ker($A$)) = 1.

4. $\text{rref}(A) = \begin{bmatrix} 1 & 0 \\ 0 & 1 \\ 0 & 0 \end{bmatrix}$

The basis of ker(A) is $\varnothing$ (the empty set), so that dim(ker(A)) = 0.

5. $\text{rref}(A) = \begin{bmatrix} 1 & -\frac{4}{5} & \frac{3}{5} \end{bmatrix}$

A basis of ker A is $\begin{bmatrix} 4 \\ 5 \\ 0 \end{bmatrix}, \begin{bmatrix} -3 \\ 0 \\ 5 \end{bmatrix}$, so that dim(ker($A$)) = 2.

6. $\text{rref}(A) = I_3$

The basis of ker(A) is $\varnothing$, so that dim(ker(A)) = 0.

7. $\text{rref}(A) = A$

A basis of ker(A) is $\begin{bmatrix}1\\0\\0\\0\\0\end{bmatrix}, \begin{bmatrix}0\\-2\\1\\0\\0\end{bmatrix}, \begin{bmatrix}0\\-3\\0\\-4\\1\end{bmatrix}$, so that dim(ker($A$)) = 3.

8. $\text{rref}(A) = \begin{bmatrix}1 & 2 & 0 & 5 & -2\\0 & 0 & 1 & -1 & 1\\0 & 0 & 0 & 0 & 0\\0 & 0 & 0 & 0 & 0\end{bmatrix}$

A basis of ker(A) is $\begin{bmatrix}-2\\1\\0\\0\\0\end{bmatrix}, \begin{bmatrix}-5\\0\\1\\1\\0\end{bmatrix}, \begin{bmatrix}2\\0\\-1\\0\\1\end{bmatrix}$, so that dim(ker($A$)) = 3.

9. $\text{rref}(A) = A$

A basis of ker(A) is $\begin{bmatrix}1\\1\\0\\0\\0\end{bmatrix}, \begin{bmatrix}-1\\0\\1\\0\\0\end{bmatrix}, \begin{bmatrix}1\\0\\0\\1\\0\end{bmatrix}, \begin{bmatrix}-1\\0\\0\\0\\1\end{bmatrix}$, so that dim(ker($A$)) = 4.

10. $\text{rref}(A) = \begin{bmatrix}1 & 0 & 2 & 0\\0 & 1 & -3 & 0\\0 & 0 & 0 & 1\\0 & 0 & 0 & 0\end{bmatrix}$

A basis of ker(A) is $\begin{bmatrix}-2\\3\\1\\0\end{bmatrix}$, so that dim(ker($A$)) = 1.

11. A basis of im(A) is $\begin{bmatrix}1\\2\end{bmatrix}$, so that dim(im($A$)) = 1 (see Exercise 1).

12. A basis of im(A) is $\begin{bmatrix}1\\1\\1\end{bmatrix}$, so that dim(im($A$)) = 1 (see Exercise 2).

13. A basis of im(A) is $\begin{bmatrix}1\\1\\1\end{bmatrix}, \begin{bmatrix}3\\2\\1\end{bmatrix}$, so that dim(im($A$)) = 2 (see Exercise 4).

14. A basis of im(A) is $\begin{bmatrix}1\\2\\3\end{bmatrix}, \begin{bmatrix}2\\3\\4\end{bmatrix}$, so that dim(im($A$)) = 2 (see Exercise 4).

15. A basis of im(A) is [5] (or [k] for any nonzero k), so that dim(im(A)) = 1.

16. A basis of im(A) is $\begin{bmatrix}1\\1\\1\end{bmatrix}, \begin{bmatrix}3\\2\\1\end{bmatrix}, \begin{bmatrix}2\\3\\3\end{bmatrix}$, so that dim(im($A$)) = 3 (see Exercise 6).

17. $\text{rref}(A) = \begin{bmatrix}1 & 2 & 0\\0 & 0 & 1\\0 & 0 & 0\\0 & 0 & 0\end{bmatrix}$, so that a basis of im($A$) is $\begin{bmatrix}1\\1\\1\\1\end{bmatrix}, \begin{bmatrix}1\\2\\3\\4\end{bmatrix}$, with dim(im($A$)) = 2.

18. $\text{rref}(A) = \begin{bmatrix}1 & 2 & 0 & 0 & 2\\0 & 0 & 1 & 0 & -3\\0 & 0 & 0 & 1 & 1\\0 & 0 & 0 & 0 & 0\end{bmatrix}$, so that a basis of im($A$) is $\begin{bmatrix}4\\3\\2\\1\end{bmatrix}, \begin{bmatrix}1\\1\\1\\1\end{bmatrix}, \begin{bmatrix}1\\2\\9\\1\end{bmatrix}$, with dim(im($A$)) = 3.

19. $\text{rref}(A) = \begin{bmatrix}1 & 0 & 2 & 0\\0 & 1 & -3 & 0\\0 & 0 & 0 & 1\\0 & 0 & 0 & 0\end{bmatrix}$, so that a basis of im($A$) is $\begin{bmatrix}1\\0\\3\\0\end{bmatrix}, \begin{bmatrix}0\\1\\4\\-1\end{bmatrix}, \begin{bmatrix}4\\-1\\8\\4\end{bmatrix}$, with dim(im($A$)) = 3.

20. A basis of im(A) is $\begin{bmatrix}1\\3\\1\\2\end{bmatrix}, \begin{bmatrix}3\\9\\4\\9\end{bmatrix}$, so that dim(im($A$)) = 2 (see Exercise 8).

21. $\text{rref}(A) = \begin{bmatrix}1 & -1 & 0 & 2 & 0\\0 & 0 & 1 & 1 & 0\\0 & 0 & 0 & 0 & 1\\0 & 0 & 0 & 0 & 0\end{bmatrix}$

A basis of im(A) is $\begin{bmatrix}1\\-1\\1\\2\end{bmatrix}, \begin{bmatrix}-1\\0\\-2\\-1\end{bmatrix}, \begin{bmatrix}1\\2\\3\\4\end{bmatrix}$, so that dim(im($A$)) = 3.

A basis of ker(A) is $\begin{bmatrix}1\\1\\0\\0\\0\end{bmatrix}, \begin{bmatrix}-2\\0\\-1\\1\\0\end{bmatrix}$, so that dim(ker($A$)) = 2.

dim(im(A)) + dim(ker(A)) = 5 = # of columns, in accordance with Fact 3.3.9.

22. $\text{rref}(A) = \begin{bmatrix} 1 & 2 & 0 & 3 & 0 \\ 0 & 0 & 1 & -1 & 0 \\ 0 & 0 & 0 & 0 & 1 \\ 0 & 0 & 0 & 0 & 0 \end{bmatrix}$

A basis of im(A) is $\begin{bmatrix} 1 \\ 1 \\ 2 \\ 0 \end{bmatrix}, \begin{bmatrix} 1 \\ 2 \\ 3 \\ 1 \end{bmatrix}, \begin{bmatrix} 1 \\ 2 \\ 3 \\ -1 \end{bmatrix}$, so that dim(im($A$)) = 3.

A basis of ker(A) is $\begin{bmatrix} -2 \\ 1 \\ 0 \\ 0 \\ 0 \end{bmatrix}, \begin{bmatrix} -3 \\ 0 \\ 1 \\ 1 \\ 0 \end{bmatrix}$, so that dim(ker($A$)) = 2.

dim(im(A)) + dim(ker(A)) = 5 = # of columns, in accordance with Fact 3.3.9.

23. A basis of im(A) is $\begin{bmatrix} 1 \\ 1 \\ 1 \\ 1 \\ 1 \end{bmatrix}$ and for ker(A) we have $\begin{bmatrix} -1 \\ 1 \\ 0 \\ 0 \\ 0 \end{bmatrix}, \begin{bmatrix} -1 \\ 0 \\ 1 \\ 0 \\ 0 \end{bmatrix}, \begin{bmatrix} -1 \\ 0 \\ 0 \\ 1 \\ 0 \end{bmatrix}, \begin{bmatrix} -1 \\ 0 \\ 0 \\ 0 \\ 1 \end{bmatrix}$.

dim(im(A)) + dim(ker(A)) = 1 + 4 = 5 = # of columns, in accordance with Fact 3.3.9.

24. We form a 5×5 matrix A with the given vectors as its columns, and we find a basis of im(A).

$\text{rref}(A) = \begin{bmatrix} 1 & 3 & 0 & -1 & 3 \\ 0 & 0 & 1 & 3 & -1 \\ 0 & 0 & 0 & 0 & 0 \\ 0 & 0 & 0 & 0 & 0 \\ 0 & 0 & 0 & 0 & 0 \end{bmatrix}$

A basis of im(A) is $\begin{bmatrix} 1 \\ 2 \\ 3 \\ 2 \\ 1 \end{bmatrix}, \begin{bmatrix} 3 \\ 2 \\ 4 \\ 1 \\ 2 \end{bmatrix}$.

25. We form a 5×9 matrix A with the given vectors as its columns.

$$\text{rref}(A) = \begin{bmatrix} 1 & 3 & 0 & -1 & 3 & 0 & 0 & -1 & 0 \\ 0 & 0 & 1 & 3 & -1 & 0 & 0 & 3 & 0 \\ 0 & 0 & 0 & 0 & 0 & 1 & 0 & -1 & 0 \\ 0 & 0 & 0 & 0 & 0 & 0 & 1 & 0 & 0 \\ 0 & 0 & 0 & 0 & 0 & 0 & 0 & 0 & 1 \end{bmatrix}$$

Since rank$(A) = 5$ we have $\text{im}(A) = \mathbb{R}^5$, and we can use Fact 3.3.7 to pick a basis of $\text{im}(A) = \mathbb{R}^5$ from among the given vectors: $\begin{bmatrix} 1 \\ 2 \\ 3 \\ 2 \\ 1 \end{bmatrix}, \begin{bmatrix} 3 \\ 2 \\ 4 \\ 1 \\ 2 \end{bmatrix}, \begin{bmatrix} 4 \\ 3 \\ 2 \\ 1 \\ 4 \end{bmatrix}, \begin{bmatrix} 2 \\ 3 \\ 5 \\ 7 \\ 5 \end{bmatrix}, \begin{bmatrix} 1 \\ 2 \\ 9 \\ 1 \\ 8 \end{bmatrix}$.

26. Find a basis $\vec{v}_1, \vec{v}_2, \vec{v}_3$ of ker(A) and let $B = \begin{bmatrix} \vec{v}_1 & \vec{v}_2 & \vec{v}_3 \end{bmatrix}$.

$$B = \begin{bmatrix} 1 & 0 & 0 \\ 0 & -2 & -3 \\ 0 & 1 & 0 \\ 0 & 0 & -4 \\ 0 & 0 & 1 \end{bmatrix}$$

27. Form a 4×4 matrix A with the given vectors as its columns. We find that $\text{rref}(A) = I_4$, so that the vectors do indeed form a basis of $\mathbb{R}^4$, by Fact 3.3.10.

28. Form a 4×4 matrix A with the given vectors as its columns. The matrix A reduces to $\begin{bmatrix} 1 & 0 & 0 & 2 \\ 0 & 1 & 0 & 3 \\ 0 & 0 & 1 & 4 \\ 0 & 0 & 0 & k-29 \end{bmatrix}$.

This matrix can be reduced further to I_4 if (and only if) $k - 29 \neq 0$, that is, if $k \neq 29$.
By Fact 3.3.10, the four given vectors form a basis of $\mathbb{R}^4$ unless $k = 29$.

29. $x_1 = -\frac{3}{2}x_2 - \frac{1}{2}x_3$; let $x_2 = s$ and $x_3 = t$. Then the solutions are of the form

$$\begin{bmatrix} x_1 \\ x_2 \\ x_3 \end{bmatrix} = \begin{bmatrix} -\frac{3}{2}s - \frac{1}{2}t \\ s \\ t \end{bmatrix} = s\begin{bmatrix} -\frac{3}{2} \\ 1 \\ 0 \end{bmatrix} + t\begin{bmatrix} -\frac{1}{2} \\ 0 \\ 1 \end{bmatrix}.$$

Multiplying the two vectors by 2 to simplify, we obtain the basis $\begin{bmatrix} -3 \\ 2 \\ 0 \end{bmatrix}, \begin{bmatrix} -1 \\ 0 \\ 2 \end{bmatrix}$.

30. Proceeding as in Exercise 29, we find the basis $\begin{bmatrix} 1 \\ 2 \\ 0 \\ 0 \end{bmatrix}, \begin{bmatrix} -1 \\ 0 \\ 1 \\ 0 \end{bmatrix}, \begin{bmatrix} -2 \\ 0 \\ 0 \\ 1 \end{bmatrix}$.

31. Proceeding as in Exercise 29, we can find the following basis of V: $\begin{bmatrix}1\\1\\0\\0\end{bmatrix}, \begin{bmatrix}-2\\0\\1\\0\end{bmatrix}, \begin{bmatrix}-4\\0\\0\\1\end{bmatrix}$.

Now let A be the 4×3 matrix with these three vectors as its columns. Then im(A) = V by Fact 3.1.3, and $\ker(A) = \{\vec{0}\}$ by Fact 3.2.6, so that A does the job.

$$A = \begin{bmatrix}1 & -2 & -4\\1 & 0 & 0\\0 & 1 & 0\\0 & 0 & 1\end{bmatrix}$$

32. We need to find all vectors $\vec{x}$ in $\mathbb{R}^4$ such that $\begin{bmatrix}x_1\\x_2\\x_3\\x_4\end{bmatrix} \cdot \begin{bmatrix}1\\0\\-1\\1\end{bmatrix} = 0$ and $\begin{bmatrix}x_1\\x_2\\x_3\\x_4\end{bmatrix} \cdot \begin{bmatrix}0\\1\\2\\3\end{bmatrix} = 0$.

This amounts to solving the system $\left| \begin{array}{ccccccc} x_1 & & - & x_3 & + & x_4 & = 0 \\ & x_2 & + & 2x_3 & + & 3x_4 & = 0 \end{array} \right|$, which in turn amounts to finding the kernel of $\begin{bmatrix}1 & 0 & -1 & 1\\0 & 1 & 2 & 3\end{bmatrix}$.

Using the standard approach, we find the basis $\begin{bmatrix}1\\-2\\1\\0\end{bmatrix}, \begin{bmatrix}-1\\-3\\0\\1\end{bmatrix}$.

33. We can write $V = \ker(A)$, where A is the $1 \times n$ matrix $A = [c_1 \quad c_2 \quad \cdots \quad c_n]$.
Since at least one of the c_i is nonzero, the rank of A is 1, so that dim(V) = dim(ker(A)) = n – rank(A) = n – 1, by Fact 3.3.5.
A "hyperplane" in $\mathbb{R}^2$ is a line, and a "hyperplane" in $\mathbb{R}^3$ is just a plane.

34. We can write $V = \ker(A)$, where A is the $m \times n$ matrix with entries a_{ij}. Note that rank(A) ≤ m. Therefore, dim(V) = dim(ker(A)) = $n - \text{rank}(A) \geq n - m$, by Fact 3.3.5.

35. We need to find all vectors $\vec{x}$ in $\mathbb{R}^n$ such that $\vec{v} \cdot \vec{x} = 0$, or $v_1x_1 + v_2x_2 + \cdots + v_nx_n = 0$, where the v_i are the components of the vector $\vec{v}$. These vectors form a hyperplane in $\mathbb{R}^n$ (see Exercise 33), so that the dimension of the space is $n - 1$.

36. No; if im(A) = ker(A) for an $n \times n$ matrix A, then n = dim(ker(A)) + dim(im(A)) = 2dim(im(A)), so that n is an even number.

37. Since dim(ker(A)) = 5 – rank(A), any 4×5 matrix with rank 2 will do; for example, $A = \begin{bmatrix}1 & 0 & 0 & 0 & 0\\0 & 1 & 0 & 0 & 0\\0 & 0 & 0 & 0 & 0\\0 & 0 & 0 & 0 & 0\end{bmatrix}$.

38. **a.** The rank of a 3×5 matrix A is 0, 1, 2, or 3, so that dim(ker(A)) = 5 – rank(A) is 2, 3, 4, or 5.

b. The rank of a 7×4 matrix A is at most 4, so that dim(im(A)) = rank(A) is 0, 1, 2, 3, or 4.

39. Note that $\ker(C) \neq \{\vec{0}\}$, by Fact 1.3.3, and $\ker(C) \subseteq \ker(A)$. Therefore, $\ker(A) \neq \{\vec{0}\}$, so that A is not invertible.

40. We can choose a basis $\vec{v}_1, \ldots, \vec{v}_p$ in V, where $p = \dim(V)$. Then $\vec{v}_1, \ldots, \vec{v}_p$ are linearly independent vectors in W, so that $\dim(V) = p \leq \dim(W)$, by Fact 3.3.4a, as claimed.

41. We can choose a basis $\vec{v}_1, \ldots, \vec{v}_p$ in V, where $p = \dim(V) = \dim(W)$. Then $\vec{v}_1, \ldots, \vec{v}_p$ is a basis of W as well, by Fact 3.3.4c, so that $V = W = \text{span}(\vec{v}_1, \ldots, \vec{v}_p)$, as claimed.

42. Consider a basis $\vec{v}_1, \ldots, \vec{v}_n$ of V. Since the $\vec{v}_i$ are n linearly independent vectors in $\mathbb{R}^n$, they form a basis of $\mathbb{R}^n$ (by parts (vii) and (ix) of Summary 3.3.11), so that $V = \text{span}(\vec{v}_1, \ldots, \vec{v}_n) = \mathbb{R}^n$, as claimed. (Note that Exercise 42 is a special case of Exercise 41.)

43. $\dim(V + W) = \dim(V) + \dim(W)$, by Exercise 3.2.51b.

44. Suppose that $V \cap W = \{\vec{0}\}$ and $\dim(V) + \dim(W) = n$.
Choose a basis $\vec{v}_1, \ldots, \vec{v}_p$ of V and a basis $\vec{w}_1, \ldots, \vec{w}_q$ in W; note that $p + q = n$. By Exercise 3.2.51b, the n vectors $\vec{v}_1, \ldots, \vec{v}_p, \vec{w}_1, \ldots, \vec{w}_q$ in $\mathbb{R}^n$ are linearly independent, so that they form a basis of $\mathbb{R}^n$ (by parts (vii) and (ix) of Summary 3.3.11). By Fact 3.2.7, any vector $\vec{x}$ can be written uniquely as
$\vec{x} = (c_1\vec{v}_1 + \cdots + c_p\vec{v}_p) + (d_1\vec{w}_1 + \cdots + d_q\vec{w}_q)$, with $\vec{v} = c_1\vec{v}_1 + \cdots + c_p\vec{v}_p$ in V and $\vec{w} = d_1\vec{w}_1 + \cdots + d_q\vec{w}_q$ in W
which gives the desired representation.
Conversely, suppose V and W are complements. Let us first show that $V \cap W = \{\vec{0}\}$ in this case. Indeed, if $\vec{x}$ is in $V \cap W$, then we can write

$$\underset{\substack{\uparrow\\ \text{in}\\ V}}{\vec{x}} = \underset{\substack{\uparrow\\ \text{in}\\ W}}{\vec{x}} + \underset{}{\vec{0}} = \ldots$$

$$\vec{x} = \underbrace{\vec{x}}_{\text{in } V} + \underbrace{\vec{0}}_{\text{in } W} = \underbrace{\vec{0}}_{\text{in } V} + \underbrace{\vec{x}}_{\text{in } W}$$

Since this representation is unique (by definition of complements), we must have $\vec{x} = \vec{0}$, so that $V \cap W = \{\vec{0}\}$. By definition of complements, we have $\mathbb{R}^n = V + W$, so that $n = \dim(V + W) = \dim(V) + \dim(W)$, by Exercise 43.

45. Note that $\text{im}(A) = \text{span}(\vec{v}_1, \ldots, \vec{v}_p, \vec{w}_1, \ldots, \vec{w}_q) = V$, since the $\vec{w}_j$ alone span V.
Note further that the first p columns of rref(A) will contain leading ones (since the $\vec{v}_i$ are linearly independent); therefore the $\vec{v}_i$ are pivot columns of A.
If we use Fact 3.3.7 to find a basis of $V = \text{im}(A)$, this basis will contain all of the $\vec{v}_i$ (since they are pivot columns) and possibly some of the $\vec{w}_j$.

46. Use Exercise 45 with $\vec{v}_1 = \begin{bmatrix} 1 \\ 2 \\ 3 \\ 4 \end{bmatrix}$, $\vec{v}_2 = \begin{bmatrix} 1 \\ 4 \\ 6 \\ 8 \end{bmatrix}$, and $\vec{w}_i = \vec{e}_i$ for $i = 1, 2, 3, 4$.

Now $\text{rref}\begin{bmatrix} 1 & 1 & 1 & 0 & 0 & 0 \\ 2 & 4 & 0 & 1 & 0 & 0 \\ 3 & 6 & 0 & 0 & 1 & 0 \\ 4 & 8 & 0 & 0 & 0 & 1 \end{bmatrix} = \begin{bmatrix} 1 & 0 & 2 & 0 & 0 & -\frac{1}{4} \\ 0 & 1 & -1 & 0 & 0 & \frac{1}{4} \\ 0 & 0 & 0 & 1 & 0 & -\frac{1}{2} \\ 0 & 0 & 0 & 0 & 1 & -\frac{3}{4} \end{bmatrix}$.

Picking the pivot columns gives the basis $\begin{bmatrix} 1 \\ 2 \\ 3 \\ 4 \end{bmatrix}, \begin{bmatrix} 1 \\ 4 \\ 6 \\ 8 \end{bmatrix}, \begin{bmatrix} 0 \\ 1 \\ 0 \\ 0 \end{bmatrix}, \begin{bmatrix} 0 \\ 0 \\ 1 \\ 0 \end{bmatrix}$.

47. Using the terminology suggested in the hint, we need to show that $\vec{u}_1, \ldots, \vec{u}_m, \vec{v}_1, \ldots, \vec{v}_p, \vec{w}_1, \ldots, \vec{w}_q$ is a basis of $V + W$. Then $\dim(V + W) + \dim(V \cap W) = (m + p + q) + m = (m + p) + (m + q) = \dim(V) + \dim(W)$, as claimed. Any vector $\vec{x}$ in $V + W$ can be written as $\vec{x} = \vec{v} + \vec{w}$, where $\vec{v}$ is in V and $\vec{w}$ is in W. Since $\vec{v}$ is a linear combination of the $\vec{u}_i$ and the $\vec{v}_j$, and $\vec{w}$ is a linear combination of the $\vec{u}_i$ and $\vec{w}_j$, $\vec{x}$ will be a linear combination of the $\vec{u}_i$, $\vec{v}_j$, and $\vec{w}_k$; this shows that the vectors $\vec{u}_1, \ldots, \vec{u}_m, \vec{v}_1, \ldots, \vec{v}_p, \vec{w}_1, \ldots, \vec{w}_q$ span $V + W$.

To show linear independence, consider the relation $a_1\vec{u}_1 + \cdots + a_m\vec{u}_m + b_1\vec{v}_1 + \cdots + b_p\vec{v}_p + c_1\vec{w}_1 + \cdots + c_q\vec{w}_q = \vec{0}$. Then the vector $a_1\vec{u}_1 + \cdots + a_m\vec{u}_m + b_1\vec{v}_1 + \cdots + b_p\vec{v}_p = -c_1\vec{w}_1 - \cdots - c_q\vec{w}_q$ is in $V \cap W$, so that it can be expressed uniquely as a linear combination of $\vec{u}_1, \ldots, \vec{u}_m$ alone; this implies that the b_i are all zero. Now our relation simplifies to $a_1\vec{u}_1 + \cdots + a_m\vec{u}_m + c_1\vec{w}_1 + \cdots + c_q\vec{w}_q = \vec{0}$, which implies that the a_i and the c_j are zero as well (since the vectors $\vec{u}_1, \ldots, \vec{u}_m, \vec{w}_1, \ldots, \vec{w}_q$ are linearly independent).

48. By Exercise 47, $\dim(V \cap W) = \dim(V) + \dim(W) - \dim(V + W) = 13 - \dim(V + W)$.
The dimension of $V + W$ is at least 7 (since $W \subseteq V + W$) and at most 10 (since $V + W \subseteq \mathbb{R}^{10}$); therefore the dimension of $V \cap W$ is at least 3 and at most 6.

49. The nonzero rows of E span the row space, and they are linearly independent (consider the leading ones), so that they form a basis of the row space: $[0\ \ 1\ \ 0\ \ 2\ \ 0]$, $[0\ \ 0\ \ 1\ \ 3\ \ 0]$, $[0\ \ 0\ \ 0\ \ 0\ \ 1]$.

50. As in Exercise 49, we observe that the nonzero rows of E form a basis of the row space, so that $\dim(\text{row space of } E) = \text{rank}(E)$.

51. **a.** All elementary row operations leave the row space unchanged, so that A and rref(A) have the same row space.

b. By part (a) and Exercise 50,
$\dim(\text{row space of } A) = \dim(\text{row space of rref}(A)) = \text{rank}(\text{rref}(A)) = \text{rank}(A)$.

52. $\text{rref}(A) = \begin{bmatrix} 1 & 0 & -1 & -2 \\ 0 & 1 & 2 & 3 \\ 0 & 0 & 0 & 0 \\ 0 & 0 & 0 & 0 \end{bmatrix}$

By Exercises 50 and 51a, $[1\ \ 0\ \ -1\ \ -2]$, $[0\ \ 1\ \ 2\ \ 3]$ is a basis of the row space of A.

53. Using the terminology suggested in the hint, we observe that the vectors $\vec{v}, A\vec{v}, \ldots, A^n\vec{v}$ are linearly dependent (by Fact 3.3.4a), so that there is a nontrivial relation $c_0\vec{v} + c_1A\vec{v} + \cdots + c_nA^n\vec{v} = \vec{0}$.
We can rewrite this relation in the form $(c_0I_n + c_1A + \cdots + c_nA^n)\vec{v} = \vec{0}$.
The nonzero vector $\vec{v}$ is in the kernel of the matrix $c_0I_n + c_1A + \cdots + c_nA^n$, so that this matrix is not invertible.

54. We can use the approach outlined in Exercise 53, with $\vec{v} = \begin{bmatrix} 1 \\ 0 \end{bmatrix}$, say.
Then $\vec{v} = \begin{bmatrix} 1 \\ 0 \end{bmatrix}$, $A\vec{v} = \begin{bmatrix} 1 \\ 2 \end{bmatrix}$, and $A^2\vec{v} = \begin{bmatrix} -3 & -4 \\ 4 & -3 \end{bmatrix}\begin{bmatrix} 1 \\ 0 \end{bmatrix} = \begin{bmatrix} -3 \\ 4 \end{bmatrix}$.
We find the relation $5\vec{v} - 2A\vec{v} + A^2\vec{v} = \vec{0}$, so that the matrix $5I_2 - 2A + A^2$ does the job.

55. If rank$(A) = m$, then the m pivot columns of A form a basis of $\text{im}(A) = \mathbb{R}^m$, so that the matrix formed by the pivot columns is invertible (by Fact 3.3.10).
Conversely, if A has an invertible $m \times m$ submatrix B, then the columns of B form a basis of $\mathbb{R}^m$ (again by Fact 3.3.10), so that $\text{im}(A) = \mathbb{R}^m$ and therefore rank(A) = dim(im(A)) = m.

56. Using the terminology suggested in the Exercise, we multiply the relation $c_0\vec{v} + c_1A\vec{v} + \cdots + c_{m-1}A^{m-1}\vec{v} = \vec{0}$ with A^{m-1} and obtain $c_0A^{m-1}\vec{v} = \vec{0}$ (all other terms vanish since $A^m = 0$).
Since the vector $A^{m-1}\vec{v}$ is nonzero (by construction), the scalar c_0 must be zero, and our relation simplifies to $c_1A\vec{v} + c_2A^2\vec{v} + \cdots + c_{m-1}A^{m-1}\vec{v} = \vec{0}$.
Now we multiply both sides with A^{m-2} and obtain $c_1A^{m-1}\vec{v} = \vec{0}$, so that $c_1 - 0$ as above. Continuing like this we conclude that all the c_i must be zero, as claimed.

57. As in Exercise 56, let m be the smallest positive integer such that $A^m = 0$. In Exercise 56 we construct m linearly independent vectors $\vec{v}, A\vec{v}, \ldots, A^{m-1}\vec{v}$ in $\mathbb{R}^n$; now $m \le n$ by Fact 3.3.4a.
Therefore $A^n = A^mA^{n-m} = 0A^{n-m} = 0$, as claimed.

58. If the vectors $\vec{w}_1, \ldots, \vec{w}_q$ span a d-dimensional space V (with basis $\vec{v}_1, \ldots, \vec{v}_d$), then $d \le q$ by Fact 3.3.1 (since the vectors $\vec{v}_i$ are linearly independent).

59. Prove Fact 3.3.4d: If d vectors $\vec{v}_1, \ldots, \vec{v}_d$ span a d-dimensional space V, then they form a basis of V. We need to show that the $\vec{v}_i$ are linearly independent. We will argue indirectly, assuming that the vectors are linearly dependent; this means that one of the vectors is a linear combination of the others, say
$\vec{v}_d = c_1\vec{v}_1 + c_2\vec{v}_2 + \cdots + c_{d-1}\vec{v}_{d-1}$.
But then $V = \text{span}(\vec{v}_1, \ldots, \vec{v}_d) = \text{span}(\vec{v}_1, \ldots, v_{d-1})$, contradicting Fact 3.3.4b (see Exercise 58).

60. im(A) is the plane onto which we project, so that rank(A) = dim(im(A)) = 2.

61. a. Note that rank$(B) \le 2$, so that dim(ker(B)) = 5 – rank$(B) \ge 3$ and dim(ker(AB)) ≥ 3 since ker$(B) \subseteq$ ker(AB).
Since ker(AB) is a subspace of $\mathbb{R}^5$, dim(ker(AB)) could be 3, 4, or 5. It is easy to give an example for each case; for example, if $A = \begin{bmatrix} 1 & 0 \\ 0 & 1 \\ 0 & 0 \\ 0 & 0 \end{bmatrix}$ and $B = \begin{bmatrix} 1 & 0 & 0 & 0 & 0 \\ 0 & 1 & 0 & 0 & 0 \end{bmatrix}$, then $AB = \begin{bmatrix} 1 & 0 & 0 & 0 & 0 \\ 0 & 1 & 0 & 0 & 0 \\ 0 & 0 & 0 & 0 & 0 \\ 0 & 0 & 0 & 0 & 0 \end{bmatrix}$ and dim(ker(AB)) = 3.

b. Since dim(im(AB)) = 5 – dim(ker(AB)), the possible values of dim(im(AB)) are 0, 1, and 2, by part a.

62. False: Argue indirectly and assume that $BA = 0$. Then $B(A\vec{x}) = \vec{0}$ for all $\vec{x}$ in $\mathbb{R}^6$, so that im(A) is contained in ker(B). We can conclude that rank(A) = dim(im(A)) ≤ dim(ker(B)) = 7 – rank(B), or rank(A) + rank(B) ≤ 7, contradicting the fact that rank(A) = rank(B) = 4.

63. a. By Exercise 3.1.39b, im(AB) ⊆ im(A), and therefore rank(AB) ≤ rank(A).

b. Write $B = [\vec{v}_1 \ \cdots \ \vec{v}_n]$ and $AB = [A\vec{v}_1 \ \cdots \ A\vec{v}_n]$. If r = rank(B), then the r pivot columns of B will span im(B), and the corresponding r columns of AB will span im(AB), by linearity of A. By Fact 3.3.4b, rank(AB) = dim(im(AB)) ≤ r = rank(B).
Summary: rank(AB) ≤ rank(A), and rank(AB) ≤ rank(B).

64. Write $A = [\vec{v}_1 \ \cdots \ \vec{v}_n]$ and $B = [\vec{w}_1 \ \cdots \ \vec{w}_n]$, so that $A + B = [\vec{v}_1 + \vec{w}_1 \ \cdots \ \vec{v}_n + \vec{w}_n]$. Any linear combination of the columns of $A + B$, $\vec{y} = c_1(\vec{v}_1 + \vec{w}_1) + \cdots + c_n(\vec{v}_n + \vec{w}_n)$, can be written as

$$\vec{y} = \underbrace{(c_1\vec{v}_1 + \cdots + c_n\vec{v}_n)}_{\text{in im}(A)} + \underbrace{(c_1\vec{w}_1 + \cdots + c_n\vec{w}_n)}_{\text{in im}(B)}$$

so that im($A + B$) ⊆ im(A) + im(B) (see Exercise 3.2.50). Since dim($V + W$) ≤ dim(V) + dim(W), by Exercise 3.3.47, we can conclude that rank($A + B$) = dim(im($A + B$)) ≤ dim(im(A)) + dim(im(B)) = rank(A) + rank(B).
Summary: rank($A + B$) ≤ rank(A) + rank(B).

65. Write $A = [\vec{v}_1 \ \vec{v}_2 \ \vec{v}_3 \ \vec{v}_4 \ \vec{v}_5]$ and $B = [\vec{w}_1 \ \vec{w}_2 \ \vec{w}_3 \ \vec{w}_4 \ \vec{w}_5]$. We will show "column by column" that $\vec{v}_i = \vec{w}_i$. Our work will be based on the following two properties:

(P) The relations among the $\vec{v}_i$ correspond to the relations among the $\vec{w}_i$ (see the proof of Fact 3.3.7).

(Q) If a matrix A is in rref, then a column of A contains a leading one if and only if that column is not a linear combination of the previous columns of the matrix.

$i = 1$: Since $1\vec{w}_1 = \vec{0}$, we have $1\vec{v}_1 = \vec{0}$ (by (P)), so that $\vec{v}_1 = \vec{w}_1 = \vec{0}$.

$i = 2$: Since $\vec{w}_2$ is not a multiple of $\vec{w}_1$, $\vec{v}_2$ is not a multiple of $\vec{v}_1$ (by P), so that $\vec{v}_2$ will contain a leading one (by Q).

$$\vec{v}_2 = \begin{bmatrix} 1 \\ 0 \\ 0 \\ 0 \end{bmatrix} = \vec{w}_2$$

$i = 3$: Since $\vec{w}_3$ is not a linear combination of $\vec{w}_1$ and $\vec{w}_2$, $\vec{v}_3$ is not a linear combination of $\vec{v}_1$ and $\vec{v}_2$ (by P), so that $\vec{v}_3$ will contain a leading 1 (by Q):

$$\vec{v}_3 = \begin{bmatrix} 0 \\ 1 \\ 0 \\ 0 \end{bmatrix} = \vec{w}_3.$$

$i = 4$: We have $\vec{w}_4 = 2\vec{w}_2 + 3\vec{w}_3$, so that $\vec{v}_4 = 2\vec{v}_2 + 3\vec{v}_3 = 2\vec{w}_2 + 3\vec{w}_3 = \vec{w}_4$, by (P).

$i = 5$: Arguing exactly as in the case of $i = 3$ we can show that $\vec{v}_5 = \vec{w}_5$.

66. Let $A=[\vec{v}_1 \;\cdots\; \vec{v}_n]$ and $B=[\vec{w}_1 \;\cdots\; \vec{w}_n]$. We will show "column by column" that $\vec{v}_i=\vec{w}_i$, using the principles (P) and (Q) established in the solution to Exercise 65.
In the case $i=1$, note that the first column of a matrix in rref is either $\vec{0}$ or $\vec{e}_1$. If $\vec{v}_1=\vec{0}$ then $1\vec{v}_1=\vec{0}$ and $1\vec{w}_1=\vec{0}$ (by P), so that $\vec{w}_1=\vec{0}$. Likewise, if $\vec{w}_1=\vec{0}$ then $\vec{v}_1=\vec{0}$, so that the vectors $\vec{v}_1$ and $\vec{w}_1$ are either both $\vec{0}$ or both $\vec{e}_1$.
Now suppose we have shown that $\vec{v}_1=\vec{w}_1$, $\vec{v}_2=\vec{w}_2$, ..., $\vec{v}_{i-1}=\vec{w}_{i-1}$, and we wish to show that $\vec{v}_i=\vec{w}_i$. Suppose that r of the vectors $\vec{v}_1,\ldots,\vec{v}_{i-1}$ contain a leading one. If $\vec{v}_i$ is not a linear combination of $\vec{v}_1,\ldots,\vec{v}_{i-1}$, then neither is $\vec{w}_i$ a linear combination of $\vec{w}_1,\ldots,\vec{w}_{i-1}$ (by P), so that $\vec{v}_i=\vec{w}_i=\vec{e}_{r+1}$ (by Q). If $\vec{v}_i=c_1\vec{v}_1+\cdots+c_{i-1}\vec{v}_{i-1}$, then $\vec{w}_i \underset{\text{by (P)}}{\overset{\uparrow}{=}} c_1\vec{w}_1+\cdots+c_{i-1}\vec{w}_{i-1}=c_1\vec{v}_1+\cdots+c_{i-1}\vec{v}_{i-1}=\vec{v}_i$, as claimed.
If you have trouble understanding this abstract reasoning, take another look at the example in Exercise 65.

67. This follows from Exercise 66, since both A and rref(M) are in rref, and they have the same kernel:
$\ker(A)=\ker(M)=\ker(\text{rref}(M))$
(since elementary row operations leave the kernel unchanged).

Chapter 4

4.1

1. $\|\vec{v}\| = \sqrt{7^2 + 11^2} = \sqrt{49 + 121} = \sqrt{170} \approx 13.04$

2. $\|\vec{v}\| = \sqrt{2^2 + 3^2 + 4^2} = \sqrt{4 + 9 + 16} = \sqrt{29} \approx 5.39$

3. $\|\vec{v}\| = \sqrt{2^2 + 3^2 + 4^2 + 5^2} = \sqrt{4 + 9 + 16 + 25} = \sqrt{54} \approx 7.35$

4. $\alpha = \arccos\frac{\vec{u}\cdot\vec{v}}{\|\vec{u}\|\|\vec{v}\|} = \arccos\frac{7+11}{\sqrt{2}\sqrt{170}} = \arccos\frac{18}{\sqrt{340}} \approx 0.219$ (radians)

5. $\alpha = \arccos\frac{\vec{u}\cdot\vec{v}}{\|\vec{u}\|\|\vec{v}\|} = \arccos\frac{2+6+12}{\sqrt{14}\sqrt{29}} \approx 0.122$ (radians)

6. $\alpha = \arccos\frac{\vec{u}\cdot\vec{v}}{\|\vec{u}\|\|\vec{v}\|} = \arccos\frac{2-3+8-10}{\sqrt{10}\sqrt{54}} \approx 1.700$ (radians)

7. Use the fact that $\vec{u}\cdot\vec{v} = \|\vec{u}\|\|\vec{v}\|\cos\alpha$, so that the angle is acute if $\vec{u}\cdot\vec{v} > 0$, and obtuse if $\vec{u}\cdot\vec{v} < 0$. Since $\vec{u}\cdot\vec{v} = 10 - 12 = -2$, the angle is obtuse.

8. Since $\vec{u}\cdot\vec{v} = 4 - 24 + 20 = 0$, the two vectors enclose a right angle.

9. Since $\vec{u}\cdot\vec{v} = 3 - 4 + 5 - 3 = 1$, the angle is acute (see Exercise 7).

10. $\vec{u}\cdot\vec{v} = 2 + 3k + 4 = 6 + 3k$
    The two vectors enclose a right angle if $\vec{u}\cdot\vec{v} = 6 + 3k = 0$, that is, if $k = -2$.

11. **a.** $\alpha_n = \arccos\frac{\vec{u}\cdot\vec{v}}{\|\vec{u}\|\|\vec{v}\|} = \arccos\frac{1}{\sqrt{n}}$

    $\alpha_2 = \arccos\frac{1}{\sqrt{2}} = \frac{\pi}{4}$ $(= 45°)$

    $\alpha_3 = \arccos\frac{1}{\sqrt{3}} \approx 0.955$ (radians)

    $a_4 = \arccos\frac{1}{2} = \frac{\pi}{3}$ $(= 60°)$

    b. Since $y = \arccos(x)$ is a continuous function, $\lim_{n\to\infty}\alpha_n = \arccos\left(\lim_{n\to\infty}\frac{1}{\sqrt{n}}\right) = \arccos(0) = \frac{\pi}{2}$ $(= 90°)$

12. $\|\vec{v} + \vec{w}\|^2 = (\vec{v} + \vec{w})\cdot(\vec{v} + \vec{w})$ (by hint)
    $= \|\vec{v}\|^2 + \|\vec{w}\|^2 + 2(\vec{v}\cdot\vec{w})$ (by definition of length)
    $\le \|\vec{v}\|^2 + \|\vec{w}\|^2 + 2\|\vec{v}\|\|\vec{w}\|$ (by Cauchy-Schwarz)
    $= (\|\vec{v}\| + \|\vec{w}\|)^2$, so that
    $\|\vec{v} + \vec{w}\|^2 \le (\|\vec{v}\| + \|\vec{w}\|)^2$
    Taking square roots of both sides, we find that $\|\vec{v} + \vec{w}\| \le \|\vec{v}\| + \|\vec{w}\|$, as claimed.

13. The figure below shows that $\left\|\vec{F}_2+\vec{F}_3\right\| = 2\cos\left(\frac{\alpha}{2}\right)\left\|\vec{F}_2\right\| = 20\cos\left(\frac{\alpha}{2}\right)$.

It is required that $\left\|\vec{F}_2+\vec{F}_3\right\| = 16$, so that $20\cos\left(\frac{\alpha}{2}\right) = 16$, or $\alpha = 2\arccos(0.8) \approx 74°$.

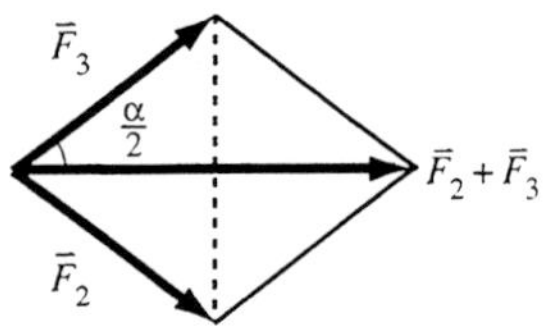

14. The horizontal components of $\vec{F}_1$ and $\vec{F}_2$ are $-\left\|\vec{F}_1\right\|\sin\beta$ and $\left\|\vec{F}_2\right\|\sin\alpha$, respectively (the horizontal component of $\vec{F}_3$ is zero).

Since the system is at rest, the horizontal components must add up to 0, so that $-\left\|\vec{F}_1\right\|\sin\beta + \left\|\vec{F}_2\right\|\sin\alpha = 0$ or

$\left\|\vec{F}_1\right\|\sin\beta = \left\|\vec{F}_2\right\|\sin\alpha$ or $\frac{\left\|\vec{F}_1\right\|}{\left\|\vec{F}_2\right\|} = \frac{\sin\alpha}{\sin\beta}$.

To find $\frac{\overline{EA}}{\overline{EB}}$, note that $\overline{EA} = \overline{ED}\tan\alpha$ and $\overline{EB} = \overline{ED}\tan\beta$ so that $\frac{\overline{EA}}{\overline{EB}} = \frac{\tan\alpha}{\tan\beta} = \frac{\sin\alpha}{\sin\beta}\cdot\frac{\cos\beta}{\cos\alpha} = \frac{\left\|\vec{F}_1\right\|}{\left\|\vec{F}_2\right\|}\frac{\cos\beta}{\cos\alpha}$.

Since α and β are two distinct acute angles, it follows that $\frac{\overline{EA}}{\overline{EB}} \neq \frac{\left\|\vec{F}_1\right\|}{\left\|\vec{F}_2\right\|}$, so that Leonardo was mistaken.

15. The subspace consists of all vectors $\vec{x}$ in $\mathbb{R}^4$ such that $\vec{x}\cdot\vec{v} = \begin{bmatrix} x_1 \\ x_2 \\ x_3 \\ x_4 \end{bmatrix}\cdot\begin{bmatrix} 1 \\ 2 \\ 3 \\ 4 \end{bmatrix} = x_1 + 2x_2 + 3x_3 + 4x_4 = 0$.

These are vectors of the form $\begin{bmatrix} -2r-3s-4t \\ r \\ s \\ t \end{bmatrix} = r\begin{bmatrix} -2 \\ 1 \\ 0 \\ 0 \end{bmatrix} + s\begin{bmatrix} -3 \\ 0 \\ 1 \\ 0 \end{bmatrix} + t\begin{bmatrix} -4 \\ 0 \\ 0 \\ 1 \end{bmatrix}$.

The three vectors to the right give the desired basis.

16. You may be able to find the solutions by educated guessing. Here is the systematic approach: we first find all vectors $\vec{x}$ that are orthogonal to $\vec{v}_1$, $\vec{v}_2$, and $\vec{v}_3$, then we identify the unit vectors among them.

Finding the vectors $\vec{x}$ with $\vec{x}\cdot\vec{v}_1 = \vec{x}\cdot\vec{v}_2 = \vec{x}\cdot\vec{v}_3 = 0$ amounts to solving the system

$$\begin{vmatrix} x_1 + x_2 + x_3 + x_4 = 0 \\ x_1 + x_2 - x_3 - x_4 = 0 \\ x_1 - x_2 + x_3 - x_4 = 0 \end{vmatrix}$$

$\left(\text{we can omit all the coefficients } \frac{1}{2}\right)$.

The solutions are of the form $\vec{x} = \begin{bmatrix} x_1 \\ x_2 \\ x_3 \\ x_4 \end{bmatrix} = \begin{bmatrix} t \\ -t \\ -t \\ t \end{bmatrix}$.

Since $\|\vec{x}\| = 2|t|$, we have a unit vector if $t = \frac{1}{2}$ or $t = -\frac{1}{2}$. Thus there are two possible choices for $\vec{v}_4$:

$$\begin{bmatrix} \frac{1}{2} \\ -\frac{1}{2} \\ -\frac{1}{2} \\ \frac{1}{2} \end{bmatrix} \text{ and } \begin{bmatrix} -\frac{1}{2} \\ \frac{1}{2} \\ \frac{1}{2} \\ -\frac{1}{2} \end{bmatrix}.$$

17. The orthogonal complement $W^{\perp}$ of W consists of the vectors $\vec{x}$ in $\mathbb{R}^4$ such that

$$\begin{bmatrix} x_1 \\ x_2 \\ x_3 \\ x_3 \end{bmatrix} \cdot \begin{bmatrix} 1 \\ 2 \\ 3 \\ 4 \end{bmatrix} = 0 \text{ and } \begin{bmatrix} x_1 \\ x_2 \\ x_3 \\ x_4 \end{bmatrix} \cdot \begin{bmatrix} 5 \\ 6 \\ 7 \\ 8 \end{bmatrix} = 0.$$

Finding these vectors amounts to solving the system $\left| \begin{matrix} x_1 + 2x_2 + 3x_3 + 4x_4 = 0 \\ 5x_1 + 6x_2 + 7x_3 + 8x_4 = 0 \end{matrix} \right|$.

The solutions are of the form

$$\begin{bmatrix} x_1 \\ x_2 \\ x_3 \\ x_4 \end{bmatrix} = \begin{bmatrix} s + 2t \\ -2s - 3t \\ s \\ t \end{bmatrix} = s\begin{bmatrix} 1 \\ -2 \\ 1 \\ 0 \end{bmatrix} + t\begin{bmatrix} 2 \\ -3 \\ 0 \\ 1 \end{bmatrix}.$$

The two vectors to the right give the desired basis.

18. a. $\|\vec{x}\|^2 = 1 + \frac{1}{4} + \frac{1}{16} + \frac{1}{64} + \cdots = \frac{1}{1 - \frac{1}{4}} = \frac{4}{3}$ $\left(\text{use the formula for a geometric series, with } a = \frac{1}{4}\right)$, so that $\|\vec{x}\| = \frac{2}{\sqrt{3}} \approx 1.155$.

b. If we let $\vec{u} = (1, 0, 0, \ldots)$ and $\vec{v} = \left(1, \frac{1}{2}, \frac{1}{4}, \ldots\right)$, then

$$\alpha = \arccos \frac{\vec{u} \cdot \vec{v}}{\|\vec{u}\|\|\vec{v}\|} = \arccos \frac{1}{\frac{2}{\sqrt{3}}} = \arccos \frac{\sqrt{3}}{2} = \frac{\pi}{6} \ (= 30°).$$

c. $\vec{x} = \left(1, \frac{1}{\sqrt{2}}, \frac{1}{\sqrt{3}}, \cdots, \frac{1}{\sqrt{n}}, \cdots\right)$ does the job, since the harmonic series $1 + \frac{1}{2} + \frac{1}{3} + \cdots$ diverges (a fact discussed in introductory calculus classes).

d. If we let $\vec{v} = (1, 0, 0, \ldots)$, $\vec{x} = \left(1, \frac{1}{2}, \frac{1}{4}, \ldots\right)$ and $\vec{u} = \frac{\vec{x}}{\|\vec{x}\|} = \frac{\sqrt{3}}{2}\left(1, \frac{1}{2}, \frac{1}{4}, \ldots\right)$ then

$$\text{proj}_L \vec{v} = (\vec{u} \cdot \vec{v})\vec{u} = \frac{3}{4}\left(1, \frac{1}{2}, \frac{1}{4}, \ldots\right).$$

19.

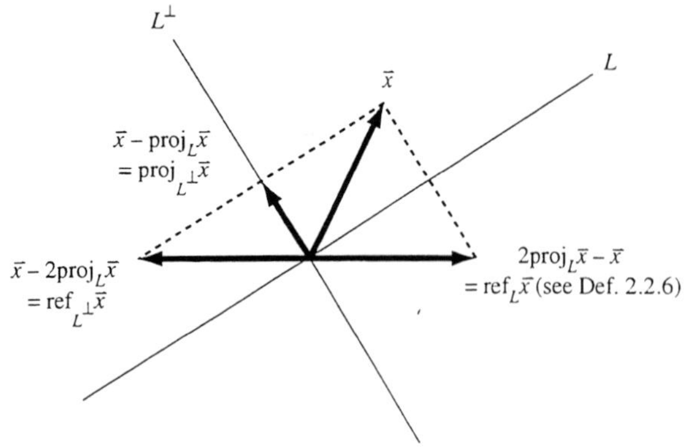

20. On the line L spanned by $\vec{x}$ we want to find the vector $m\vec{x}$ closest to $\vec{y}$ (that is, we want $\|m\vec{x} - \vec{y}\|$ to be minimal). We want $m\vec{x} - \vec{y}$ to be perpendicular to L (that is, to $\vec{x}$), which means that $\vec{x} \cdot (m\vec{x} - \vec{y}) = 0$ or $m(\vec{x} \cdot \vec{x}) - \vec{x} \cdot \vec{y} = 0$ or $m = \dfrac{\vec{x} \cdot \vec{y}}{\vec{x} \cdot \vec{x}} \approx \dfrac{4182.9}{198.53^2} \approx 0.106$.

Recall that the correlation coefficient r is $r = \dfrac{\vec{x} \cdot \vec{y}}{\|\vec{x}\|\|\vec{y}\|}$, so that $m = \dfrac{\|\vec{y}\|}{\|\vec{x}\|} r$.

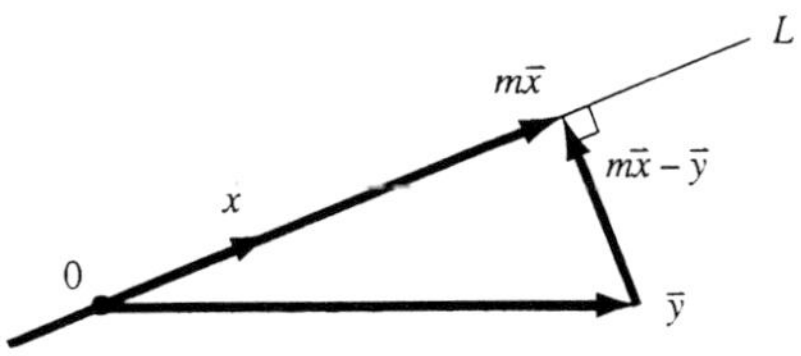

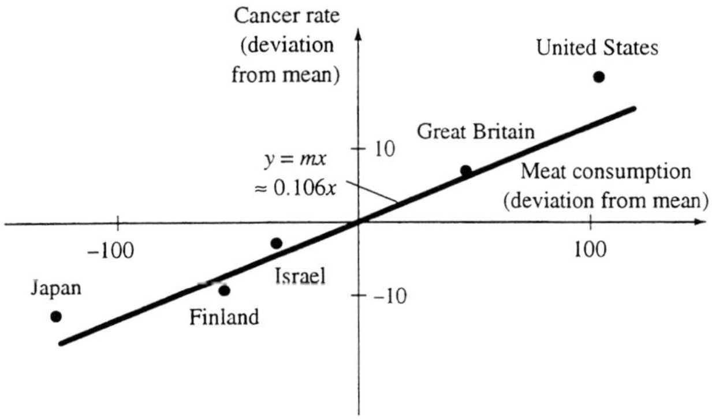

21. Call the three given vectors $\vec{v}_1$, $\vec{v}_2$, and $\vec{v}_3$. Since $\vec{v}_2$ is required to be a unit vector, we must have $b = g = 0$.
Now $\vec{v}_1 \cdot \vec{v}_2 = d$ must be zero, so that $d = 0$.
Likewise, $\vec{v}_2 \cdot \vec{v}_3 = e$ must be zero, so that $e = 0$.

Since $\vec{v}_3$ must be a unit vector, we have $\|\vec{v}_3\|^2 = c^2 + \dfrac{1}{4} = 1$, so that $c = \pm\dfrac{\sqrt{3}}{2}$.

Since we are asked to find just one solution, let us pick $c = \dfrac{\sqrt{3}}{2}$.

The condition $\vec{v}_1 \cdot \vec{v}_3 = 0$ now implies that $\dfrac{\sqrt{3}}{2}a + \dfrac{1}{2}f = 0$, or $f = -\sqrt{3}a$.

Finally, it is required that $\|\vec{v}_1\|^2 = a^2 + f^2 = a^2 + 3a^2 = 4a^2 = 1$, so that $a = \pm\frac{1}{2}$.

Let us pick $a = \frac{1}{2}$, so that $f = -\frac{\sqrt{3}}{2}$.

Summary:

$$\vec{v}_1 = \begin{bmatrix} \frac{1}{2} \\ 0 \\ -\frac{\sqrt{3}}{2} \end{bmatrix}, \vec{v}_2 = \begin{bmatrix} 0 \\ 1 \\ 0 \end{bmatrix}, \vec{v}_3 = \begin{bmatrix} \frac{\sqrt{3}}{2} \\ 0 \\ \frac{1}{2} \end{bmatrix}$$

There are other solutions; some components will have different signs.

22. Let $W = \{\vec{x} \text{ in } \mathbb{R}^n : \vec{x}\cdot\vec{v}_i = 0 \text{ for all } i = 1, \ldots, m\}$. We are asked to show that $V^\perp = W$, that is, any $\vec{x}$ in $V^\perp$ is in W, and vice versa.
If $\vec{x}$ is in $V^\perp$, then $\vec{x}\cdot\vec{v} = 0$ for all $\vec{v}$ in V; in particular, $x\cdot\vec{v}_i = 0$ for all i (since the $\vec{v}_i$ are in V), so that $\vec{x}$ is in W.
Conversely, consider a vector $\vec{x}$ in W. To show that $\vec{x}$ is in $V^\perp$, we have to verify that $\vec{x}\cdot\vec{v} = 0$ for all $\vec{v}$ in V. Pick a particular $\vec{v}$ in V. Since the $\vec{v}_i$ span V, we can write $\vec{v} = c_1\vec{v}_1 + \cdots + c_m\vec{v}_m$, for some scalars c_i. Then $\vec{x}\cdot\vec{v} = c_1(\vec{x}\cdot\vec{v}_1) + \cdots + c_m(\vec{x}\cdot\vec{v}_m) = 0$, as claimed.

23. We need to show that $V^\perp$ contains the zero vector and is closed under addition: $\vec{0}\cdot\vec{v} = 0$ for all $\vec{v}$ in V, so that $\vec{0}$ is in $V^\perp$.
Now suppose that $\vec{u}$ and $\vec{w}$ are in $V^\perp$. To show that $\vec{u}+\vec{w}$ is in $V^\perp$, we need to verify that $(\vec{u}+\vec{w})\cdot\vec{v} = 0$ for all $\vec{v}$ in V. Indeed, $(\vec{u}+\vec{w})\cdot\vec{v} = \vec{u}\cdot\vec{v} + \vec{w}\cdot\vec{v} = 0 + 0 = 0$, since $\vec{u}$ and $\vec{w}$ are in $V^\perp$.

24. Write $T(\vec{x}) = \text{proj}_V(\vec{x})$ for simplicity.

To prove the linearity of T we will use the definition of a projection: $T(\vec{x})$ is in V, and $\vec{x} - T(\vec{x})$ is in $V^\perp$.
To show that $T(\vec{x}+\vec{y}) = T(\vec{x}) + T(\vec{y})$, note that $T(\vec{x}) + T(\vec{y})$ is in V (since V is a subspace), and $\vec{x}+\vec{y} - (T(\vec{x}) + T(\vec{y})) = (\vec{x} - T(\vec{x})) + (\vec{y} - T(\vec{y}))$ is in $V^\perp$ (since $V^\perp$ is a subspace, by Fact 4.1.5).
To show that $T(k\vec{x}) = kT(\vec{x})$, note that $kT(\vec{x})$ is in V (since V is a subspace), and $k\vec{x} - kT(\vec{x}) = k(\vec{x} - T(\vec{x}))$ is in $V^\perp$ (since $V^\perp$ is a subspace).

25. **a.** $\|k\vec{v}\|^2 = (k\vec{v})\cdot(k\vec{v}) = k^2(\vec{v}\cdot\vec{v}) = k^2\|\vec{v}\|^2$

Now take square roots of both sides; note that $\sqrt{k^2} = |k|$, the absolute value of k (think about the case when k is negative).
$\|k\vec{v}\| = |k|\|\vec{v}\|$, as claimed.

b. $\|\vec{u}\| = \left\|\frac{1}{\|\vec{v}\|}\vec{v}\right\| \underset{\uparrow}{=} \frac{1}{\|\vec{v}\|}\|\vec{v}\| = 1$, as claimed.

by part a

26. The two given vectors spanning the subspace are orthogonal, but they are not unit vectors: both have length 7. To obtain an orthonormal basis $\vec{v}_1$, $\vec{v}_2$ of the subspace, we divide by 7:

$$\vec{v}_1 = \frac{1}{7}\begin{bmatrix} 2 \\ 3 \\ 6 \end{bmatrix}, \vec{v}_2 = \frac{1}{7}\begin{bmatrix} 3 \\ -6 \\ 2 \end{bmatrix}.$$

Now we can use Fact 4.1.6, with $\vec{x} = \begin{bmatrix} 49 \\ 49 \\ 49 \end{bmatrix}$:

$$\text{proj}_V\vec{x} = (\vec{v}_1 \cdot \vec{x})\vec{v}_1 + (\vec{v}_2 \cdot \vec{x})\vec{v}_2 = 11\begin{bmatrix} 2 \\ 3 \\ 6 \end{bmatrix} - \begin{bmatrix} 3 \\ -6 \\ 2 \end{bmatrix} = \begin{bmatrix} 19 \\ 39 \\ 64 \end{bmatrix}.$$

27. Since the two given vectors in the subspace are orthogonal, we have the orthonormal basis

$$\vec{v}_1 = \frac{1}{3}\begin{bmatrix} 2 \\ 2 \\ 1 \\ 0 \end{bmatrix}, \vec{v}_2 = \frac{1}{3}\begin{bmatrix} -2 \\ 2 \\ 0 \\ 1 \end{bmatrix}.$$

Now we can use Fact 4.1.6, with $\vec{x} = 9\vec{e}_1$: $\text{proj}_V\vec{x} = (\vec{v}_1 \cdot \vec{x})\vec{v}_1 + (\vec{v}_2 \cdot \vec{x})\vec{v}_2 = 2\begin{bmatrix} 2 \\ 2 \\ 1 \\ 0 \end{bmatrix} - 2\begin{bmatrix} -2 \\ 2 \\ 0 \\ 1 \end{bmatrix} = \begin{bmatrix} 8 \\ 0 \\ 2 \\ 2 \end{bmatrix}.$

28. Since the three given vectors in the subspace are orthogonal, we have the orthonormal basis

$$\vec{v}_1 = \frac{1}{2}\begin{bmatrix} 1 \\ 1 \\ 1 \\ 1 \end{bmatrix}, \vec{v}_2 = \frac{1}{2}\begin{bmatrix} 1 \\ 1 \\ -1 \\ -1 \end{bmatrix}, \vec{v}_3 = \frac{1}{2}\begin{bmatrix} 1 \\ -1 \\ -1 \\ 1 \end{bmatrix}.$$

Now we can use Fact 4.1.6, with $\vec{x} = \vec{e}_1$: $\text{proj}_V\vec{x} = (\vec{v}_1 \cdot \vec{x})\vec{v}_1 + (\vec{v}_2 \cdot \vec{x})\vec{v}_2 + (\vec{v}_3 \cdot \vec{x})\vec{v}_3 = \frac{1}{4}\begin{bmatrix} 3 \\ 1 \\ -1 \\ 1 \end{bmatrix}.$

29. By the Pythagorean theorem (Fact 4.1.8),

$$\begin{aligned}
\|\vec{x}\|^2 &= \|7\vec{v}_1 - 3\vec{v}_2 + 2\vec{v}_3 + \vec{v}_4 - \vec{v}_5\|^2 \\
&= \|7\vec{v}_1\|^2 + \|9\vec{v}_2\|^2 + \|2\vec{v}_3\|^2 + \|\vec{v}_4\|^2 + \|\vec{v}_5\|^2 \\
&= 49 + 9 + 4 + 1 + 1 \\
&= 64, \text{ so that } \|\vec{x}\| = 8.
\end{aligned}$$

30. Since $\vec{y} = \text{proj}_V \vec{x}$, the vector $\vec{x} - \vec{y}$ is orthogonal to $\vec{y}$, by definition of a projection (see Fact 4.1.6): $(\vec{x} - \vec{y}) \cdot \vec{y} = 0$ or $\vec{x} \cdot \vec{y} - \|\vec{y}\|^2 = 0$ or $\vec{x} \cdot \vec{y} = \|\vec{y}\|^2$.

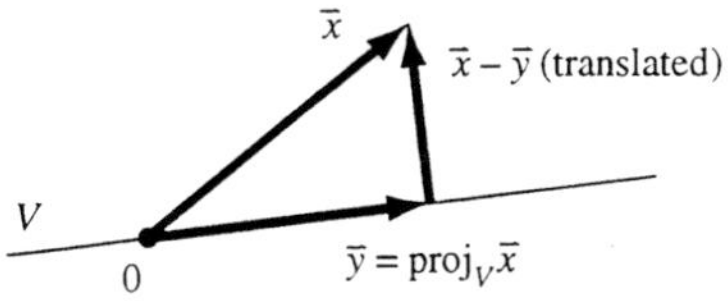

31. If $V = \text{span}(\vec{v}_1, \ldots, \vec{v}_m)$, then $\text{proj}_V \vec{x} = (\vec{v}_1 \cdot \vec{x})\vec{v}_1 + \cdots + (\vec{v}_m \cdot \vec{x})\vec{v}_m$, by Fact 4.1.6, and $\|\text{proj}_V \vec{x}\|^2 = (\vec{v}_1 \cdot \vec{x})^2 + \cdots + (\vec{v}_m \cdot \vec{x})^2 = p$, by Pythagorean theorem (Fact 4.1.8). Therefore $p \le \|\vec{x}\|^2$, by Fact 4.1.9.
The two quantities are equal if (and only if) $\vec{x}$ is in V.

32. By Fact 2.3.6a, the matrix G is invertible if (and only if) $(\vec{v}_1 \cdot \vec{v}_1)(\vec{v}_2 \cdot \vec{v}_2) - (\vec{v}_1 \cdot \vec{v}_2)^2 = \|\vec{v}_1\|^2 \|\vec{v}_2\|^2 - (\vec{v}_1 \cdot \vec{v}_2)^2 \neq 0$.
The Cauchy-Schwarz inequality (Fact 4.1.10) tells us that $\|\vec{v}_1\|^2 \|\vec{v}_2\|^2 - (\vec{v}_1 \cdot \vec{v}_2)^2 \ge 0$; equality holds if (and only if) $\vec{v}_1$ and $\vec{v}_2$ are parallel (that is, linearly dependent).

33. Using Definition 2.2.6 as a guide, we find that $\text{ref}_E \vec{x} = 2(\text{proj}_E \vec{x}) - \vec{x} = 2(\vec{v}_1 \cdot \vec{x})\vec{v}_1 + 2(\vec{v}_2 \cdot \vec{x})\vec{v}_2 - \vec{x}$.

34. Since $\vec{v}_1$ and $\vec{v}_2$ are unit vectors, the condition $\vec{v}_1 \cdot \vec{v}_2 = \|\vec{v}_1\| \|\vec{v}_2\| \cos(\alpha) = \cos(\alpha) = \frac{1}{2}$ implies that $\vec{v}_1$ and $\vec{v}_2$ enclose an angle of $60^\circ \left(= \frac{\pi}{3}\right)$. The vectors $\vec{v}_1$ and $\vec{v}_3$ enclose an angle of 60° as well.
In the case $n = 2$ there are two possible scenarios: either $\vec{v}_2 = \vec{v}_3$, or $\vec{v}_2$ and $\vec{v}_3$ enclose an angle of 120°.
Therefore, either $\vec{v}_2 \cdot \vec{v}_3 = 1$ or $\vec{v}_2 \cdot \vec{v}_3 = \cos(120^\circ) = -\frac{1}{2}$.
In the case $n = 3$, the vectors $\vec{v}_2$ and $\vec{v}_3$ could enclose any angle between 0° (if $\vec{v}_2 = \vec{v}_3$) and 120°, as illustrated in the figure below. We have $-\frac{1}{2} \le \vec{v}_2 \cdot \vec{v}_3 \le 1$.

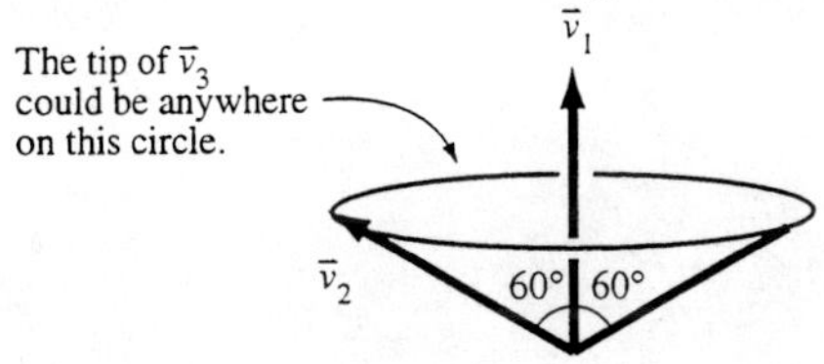

For example, consider $\vec{v}_1 = \begin{bmatrix} 0 \\ 0 \\ 1 \end{bmatrix}$, $\vec{v}_2 = \begin{bmatrix} 0 \\ \frac{\sqrt{3}}{2} \\ \frac{1}{2} \end{bmatrix}$, $\vec{v}_3 = \begin{bmatrix} \left(\frac{\sqrt{3}}{2}\right)\cos\theta \\ \left(\frac{\sqrt{3}}{2}\right)\sin\theta \\ \frac{1}{2} \end{bmatrix}$

Note that $\vec{v}_2 \cdot \vec{v}_3 = \left(\frac{3}{4}\right)\sin\theta + \frac{1}{4}$ could be anything between $-\frac{1}{2}$ (when $\sin\theta = -1$) and 1 (when $\sin\theta = 1$), as claimed.
If n exceeds three, we can consider the orthogonal projection $\vec{w}$ of $\vec{v}_3$ onto the plane E spanned by $\vec{v}_1$ and $\vec{v}_2$.
Since $\text{proj}_{\vec{v}_1} \vec{w} = (\vec{v}_1 \cdot \vec{w})\vec{v}_1 = \frac{1}{2}\vec{v}_1$, and since $\|\vec{w}\| \le \|\vec{v}_3\| = 1$, (by Fact 4.1.9), the tip of $\vec{w}$ will be on the line segment sketched below. Note that α is between 0° and 120°, so that $\cos\alpha$ is between $-\frac{1}{2}$ and 1.

Therefore, $\vec{v}_2 \cdot \vec{v}_3 = \vec{v}_2 \cdot \vec{w} = \|\vec{w}\|\cos\alpha$ is between $-\frac{1}{2}$ and 1.

This implies that $\angle(\vec{v}_2, \vec{v}_3)$ is between 0° and 120° as well. To see that all these values are attained, add $(n-3)$ zeros to the three vectors $\vec{v}_1, \vec{v}_2, \vec{v}_3$ in $\mathbb{R}^3$ given above.

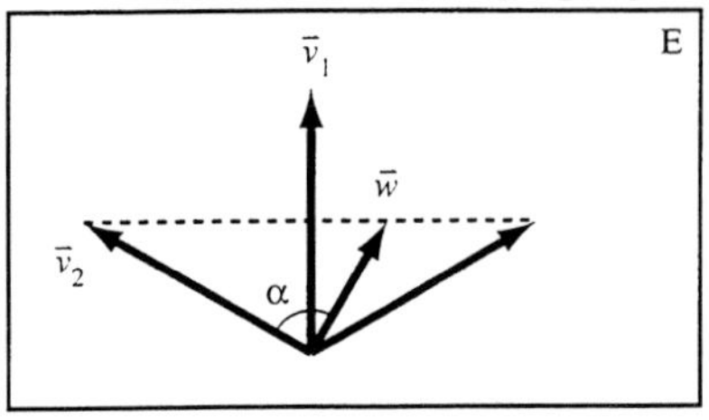

35. No! By definition of a projection, the vector $\vec{x} - \text{proj}_L\vec{x}$ is perpendicular to $\text{proj}_L\vec{x}$, so that $(\vec{x} - \text{proj}_L\vec{x})\cdot(\text{proj}_L\vec{x}) = \vec{x}\cdot\text{proj}_L\vec{x} - \|\text{proj}_L\vec{x}\|^2 = 0$ and $\vec{x}\cdot\text{proj}_L\vec{x} = \|\text{proj}_L\vec{x}\|^2 \geq 0$.

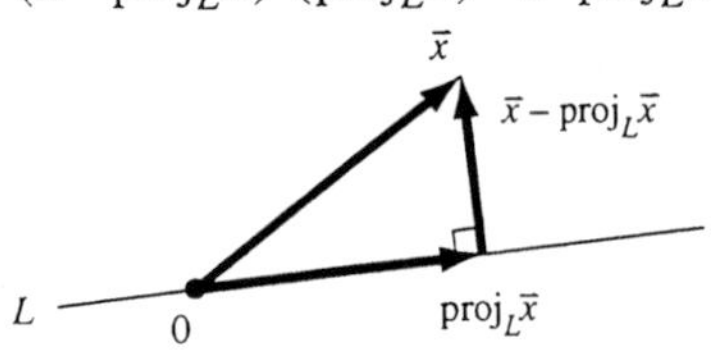

4.2

In Exercises 1–28, we will refer to the given vectors as $\vec{v}_1, \ldots, \vec{v}_m$, where $m = 1, 2,$ or 3.

1. $\vec{w}_1 = \frac{1}{\|\vec{v}_1\|}\vec{v}_1 = \frac{1}{3}\begin{bmatrix} 2 \\ 1 \\ -2 \end{bmatrix}$

2. $\vec{w}_1 = \frac{1}{\|\vec{v}_1\|}\vec{v}_1 = \frac{1}{7}\begin{bmatrix} 6 \\ 3 \\ 2 \end{bmatrix}$

$$\vec{w}_2 = \frac{\vec{v}_2 - (\vec{w}_1\cdot\vec{v}_2)\vec{w}_1}{\text{length}} = \frac{1}{7}\begin{bmatrix} 2 \\ -6 \\ 3 \end{bmatrix}$$

Note that $\vec{w}_1 \cdot \vec{v}_2 = 0$.

3. $\vec{w}_1 = \frac{1}{\|\vec{v}_1\|}\vec{v}_1 = \frac{1}{5}\begin{bmatrix} 4 \\ 0 \\ 3 \end{bmatrix}$

$$\vec{w}_2 = \frac{\vec{v}_2 - (\vec{w}_1\cdot\vec{v}_2)\vec{w}_1}{\text{length}} = \frac{1}{5}\begin{bmatrix} 3 \\ 0 \\ -4 \end{bmatrix}$$

4. $\vec{w}_1 = \frac{1}{5}\begin{bmatrix} 4 \\ 0 \\ 3 \end{bmatrix}$ and $\vec{w}_2 = \frac{1}{5}\begin{bmatrix} 3 \\ 0 \\ -4 \end{bmatrix}$ as in Exercise 3.

Since $\vec{v}_3$ is orthogonal to $\vec{w}_1$ and $\vec{w}_2$, $\vec{w}_3 = \frac{1}{\|\vec{v}_3\|}\vec{v}_3 = \begin{bmatrix} 0 \\ -1 \\ 0 \end{bmatrix}$.

5. $\vec{w}_1 = \frac{1}{\|\vec{v}_1\|}\vec{v}_1 = \frac{1}{3}\begin{bmatrix} 2 \\ 2 \\ 1 \end{bmatrix}$

$$\vec{w}_2 = \frac{\vec{v}_2 - (\vec{w}_1 \cdot \vec{v}_2)\vec{w}_1}{\text{length}} = \frac{1}{\sqrt{18}}\begin{bmatrix} -1 \\ -1 \\ 4 \end{bmatrix} = \frac{1}{3\sqrt{2}}\begin{bmatrix} -1 \\ -1 \\ 4 \end{bmatrix}$$

6. $\vec{w}_1 = \frac{1}{\|\vec{v}_1\|}\vec{v}_1 = \begin{bmatrix} 1 \\ 0 \\ 0 \end{bmatrix} = \vec{e}_1$

$$\vec{w}_2 = \frac{\vec{v}_2 - (\vec{w}_1 \cdot \vec{v}_2)\vec{w}_1}{\text{length}} = \begin{bmatrix} 0 \\ 1 \\ 0 \end{bmatrix} = \vec{e}_2$$

$$\vec{w}_3 = \frac{\vec{v}_3 - (\vec{w}_1 \cdot \vec{v}_3)\vec{w}_1 - (\vec{w}_2 \cdot \vec{v}_3)\vec{w}_2}{\text{length}} = \begin{bmatrix} 0 \\ 0 \\ 1 \end{bmatrix} = \vec{e}_3$$

7. Note that $\vec{v}_1$ and $\vec{v}_2$ are orthogonal, so that $\vec{w}_1 = \frac{1}{\|\vec{v}_1\|}\vec{v}_1 = \frac{1}{3}\begin{bmatrix} 2 \\ 2 \\ 1 \end{bmatrix}$ and $\vec{w}_2 = \frac{1}{\|\vec{v}_2\|}\vec{v}_2 = \frac{1}{3}\begin{bmatrix} -2 \\ 1 \\ 2 \end{bmatrix}$.

Then $\vec{w}_3 = \frac{\vec{v}_3 - (\vec{w}_1 \cdot \vec{v}_3)\vec{w}_1 - (\vec{w}_2 \cdot \vec{v}_2)\vec{w}_2}{\text{length}} = \frac{1}{\sqrt{36}}\begin{bmatrix} 2 \\ -4 \\ 4 \end{bmatrix} = \frac{1}{3}\begin{bmatrix} 1 \\ -2 \\ 2 \end{bmatrix}$.

8. $\vec{w}_1 = \frac{1}{\|\vec{v}_1\|}\vec{v}_1 = \frac{1}{7}\begin{bmatrix} 5 \\ 4 \\ 2 \\ 2 \end{bmatrix}$

$$\vec{w}_2 = \frac{\vec{v}_2 - (\vec{w}_1 \cdot \vec{v}_2)\vec{w}_1}{\text{length}} = \frac{1}{7}\begin{bmatrix} -2 \\ 2 \\ 5 \\ -4 \end{bmatrix}$$

9. $\vec{w}_1 = \dfrac{1}{\|\vec{v}_1\|}\vec{v}_1 = \dfrac{1}{2}\begin{bmatrix}1\\1\\1\\1\end{bmatrix}$

$\vec{w}_2 = \dfrac{\vec{v}_2 - (\vec{w}_1 \cdot \vec{v}_2)\vec{w}_1}{\text{length}} = \dfrac{1}{10}\begin{bmatrix}-1\\7\\-7\\1\end{bmatrix}$

10. $\vec{w}_1 = \dfrac{1}{\|\vec{v}_1\|}\vec{v}_1 = \dfrac{1}{2}\begin{bmatrix}1\\1\\1\\1\end{bmatrix}$

$\vec{w}_2 = \dfrac{\vec{v}_2 - (\vec{w}_1 \cdot \vec{v}_2)\vec{w}_1}{\text{length}} = \dfrac{1}{2}\begin{bmatrix}1\\-1\\1\\-1\end{bmatrix}$

11. $\vec{w}_1 = \dfrac{1}{\|\vec{v}_1\|}\vec{v}_1 = \dfrac{1}{5}\begin{bmatrix}4\\0\\0\\3\end{bmatrix}$

$\vec{w}_2 = \dfrac{\vec{v}_2 - (\vec{w}_1 \cdot \vec{v}_2)\vec{w}_1}{\text{length}} = \dfrac{1}{\sqrt{225}}\begin{bmatrix}-3\\2\\14\\4\end{bmatrix} = \dfrac{1}{15}\begin{bmatrix}-3\\2\\14\\4\end{bmatrix}$

12. $\vec{w}_1 = \dfrac{1}{7}\begin{bmatrix}2\\3\\0\\6\end{bmatrix}$

$\vec{w}_2 = \dfrac{\vec{v}_2 - (\vec{w}_1 \cdot \vec{v}_2)\vec{w}_1}{\text{length}} = \dfrac{1}{3}\begin{bmatrix}0\\-2\\2\\1\end{bmatrix}$

13. $\vec{w}_1 = \frac{1}{\|\vec{v}_1\|}\vec{v}_1 = \frac{1}{2}\begin{bmatrix}1\\1\\1\\1\end{bmatrix}$

$$\vec{w}_2 = \frac{\vec{v}_2 - (\vec{w}_1 \cdot \vec{v}_2)\vec{w}_1}{\text{length}} = \begin{bmatrix}\frac{1}{2}\\-\frac{1}{2}\\-\frac{1}{2}\\\frac{1}{2}\end{bmatrix}$$

$$\vec{w}_3 = \frac{\vec{v}_3 - (\vec{w}_1 \cdot \vec{v}_3)\vec{w}_1 - (\vec{w}_2 \cdot \vec{v}_3)\vec{w}_2}{\text{length}} = \begin{bmatrix}\frac{1}{2}\\\frac{1}{2}\\-\frac{1}{2}\\-\frac{1}{2}\end{bmatrix}$$

14. $\vec{w}_1 = \frac{1}{\|\vec{v}_1\|}\vec{v}_1 = \frac{1}{10}\begin{bmatrix}1\\7\\1\\7\end{bmatrix}$

$$\vec{w}_2 = \frac{\vec{v}_2 - (\vec{w}_1 \cdot \vec{v}_2)\vec{w}_1}{\text{length}} = \frac{1}{\sqrt{2}}\begin{bmatrix}-1\\0\\1\\0\end{bmatrix}$$

$$\vec{w}_3 = \frac{\vec{v}_3 - (\vec{w}_1 \cdot \vec{v}_3)\vec{w}_1 - (\vec{w}_2 \cdot \vec{v}_3)\vec{w}_2}{\text{length}} = \frac{1}{\sqrt{2}}\begin{bmatrix}0\\1\\0\\-1\end{bmatrix}$$

In Exercises 15–28, we will use the results of Exercises 1–14 (note that Exercise k, where $k = 1, \ldots, 14$, gives the QR factorization of Exercise $(k + 14)$). We can set $Q = [\vec{w}_1 \quad \ldots \quad \vec{w}_m]$; the entries of R are

$r_{11} = \|\vec{v}_1\|$

$r_{22} = \|\vec{v}_2 - (\vec{w}_1 \cdot \vec{v}_2)\vec{w}_1\|$

$r_{33} = \|\vec{v}_3 - (\vec{w}_1 \cdot \vec{v}_3)\vec{w}_1 - (\vec{w}_2 \cdot \vec{v}_3)\vec{w}_2\|$

$r_{ji} = \vec{w}_j \cdot \vec{v}_i$, where $j < i$.

15. $Q = \frac{1}{3}\begin{bmatrix}2\\1\\-2\end{bmatrix}$, $R = [3]$

16. $Q = \frac{1}{7}\begin{bmatrix}6 & 2\\3 & -6\\2 & 3\end{bmatrix}$, $R = \begin{bmatrix}7 & 0\\0 & 7\end{bmatrix}$

17. $Q = \frac{1}{5}\begin{bmatrix} 4 & 3 \\ 0 & 0 \\ 3 & -4 \end{bmatrix}, R = \begin{bmatrix} 5 & 5 \\ 0 & 35 \end{bmatrix}$

18. $Q = \frac{1}{5}\begin{bmatrix} 4 & 3 & 0 \\ 0 & 0 & -5 \\ 3 & -4 & 0 \end{bmatrix}, R = \begin{bmatrix} 5 & 5 & 0 \\ 0 & 35 & 0 \\ 0 & 0 & 2 \end{bmatrix}$

19. $Q = \frac{1}{3}\begin{bmatrix} 2 & -\frac{1}{\sqrt{2}} \\ 2 & -\frac{1}{\sqrt{2}} \\ 1 & \frac{4}{\sqrt{2}} \end{bmatrix}, R = 3\begin{bmatrix} 1 & 1 \\ 0 & \sqrt{2} \end{bmatrix}$

20. $Q = I_3,\ R = [\vec{v}_1 \quad \vec{v}_2 \quad \vec{v}_3] = \begin{bmatrix} 2 & 3 & 5 \\ 0 & 4 & 6 \\ 0 & 0 & 7 \end{bmatrix}$

21. $Q = \frac{1}{3}\begin{bmatrix} 2 & -2 & 1 \\ 2 & 1 & -2 \\ 1 & 2 & 2 \end{bmatrix}, R = \begin{bmatrix} 3 & 0 & 12 \\ 0 & 3 & -12 \\ 0 & 0 & 6 \end{bmatrix}$

22. $Q = \frac{1}{7}\begin{bmatrix} 5 & -2 \\ 4 & 2 \\ 2 & 5 \\ 2 & -4 \end{bmatrix}, R = \begin{bmatrix} 7 & 7 \\ 0 & 7 \end{bmatrix}$

23. $Q = \begin{bmatrix} 0.5 & -0.1 \\ 0.5 & 0.7 \\ 0.5 & -0.7 \\ 0.5 & 0.1 \end{bmatrix}, R = \begin{bmatrix} 2 & 4 \\ 0 & 10 \end{bmatrix}$

24. $Q = \frac{1}{2}\begin{bmatrix} 1 & 1 \\ 1 & -1 \\ 1 & 1 \\ 1 & -1 \end{bmatrix}, R = \begin{bmatrix} 2 & 10 \\ 0 & 2 \end{bmatrix}$

25. $Q = \frac{1}{15}\begin{bmatrix} 12 & -3 \\ 0 & 2 \\ 0 & 14 \\ 9 & 4 \end{bmatrix}, R = \begin{bmatrix} 5 & 10 \\ 0 & 15 \end{bmatrix}$

26. $Q=\begin{bmatrix}\frac{2}{7} & 0\\ \frac{3}{7} & -\frac{2}{3}\\ 0 & \frac{2}{3}\\ \frac{6}{7} & \frac{1}{3}\end{bmatrix},\ R=\begin{bmatrix}7 & 14\\ 0 & 3\end{bmatrix}$

27. $Q=\frac{1}{2}\begin{bmatrix}1 & 1 & 1\\ 1 & -1 & 1\\ 1 & -1 & -1\\ 1 & 1 & -1\end{bmatrix},\ R=\begin{bmatrix}2 & 1 & 1\\ 0 & 1 & -2\\ 0 & 0 & 1\end{bmatrix}$

28. $Q=\begin{bmatrix}\frac{1}{10} & -\frac{1}{\sqrt{2}} & 0\\ \frac{7}{10} & 0 & \frac{1}{\sqrt{2}}\\ \frac{1}{10} & \frac{1}{\sqrt{2}} & 0\\ \frac{7}{10} & 0 & -\frac{1}{\sqrt{2}}\end{bmatrix},\ R=\begin{bmatrix}10 & 10 & 10\\ 0 & \sqrt{2} & 0\\ 0 & 0 & \sqrt{2}\end{bmatrix}$

29. $\vec{w}_1=\frac{1}{\|\vec{v}_1\|}\vec{v}_1=\frac{1}{5}\begin{bmatrix}-3\\ 4\end{bmatrix}$

$\vec{w}_2=\frac{\vec{v}_2-(\vec{w}_1\cdot\vec{v}_2)\vec{w}_1}{\text{length}}=\frac{1}{5}\begin{bmatrix}4\\ 3\end{bmatrix}$

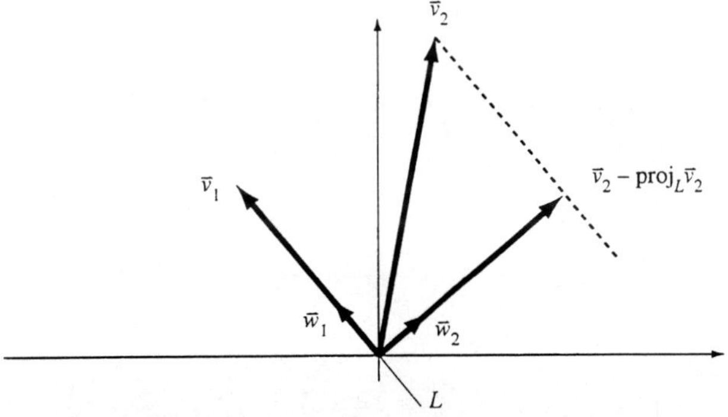

30.

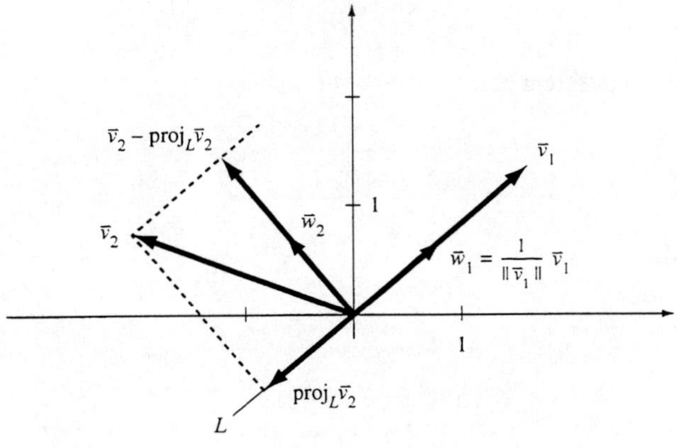

31. $\vec{w}_1 = \dfrac{1}{\|\vec{v}_1\|}\vec{v}_1 = \begin{bmatrix}1\\0\\0\end{bmatrix} = \vec{e}_1$

$$\underset{\uparrow}{\vec{v}_2 - \text{proj}_{V_1}\vec{v}_2} = \begin{bmatrix}b\\c\\0\end{bmatrix} - \begin{bmatrix}b\\0\\0\end{bmatrix} = \begin{bmatrix}0\\c\\0\end{bmatrix}, \text{ so that } \vec{w}_2 = \begin{bmatrix}0\\1\\0\end{bmatrix} = \vec{e}_2$$

Note that $V_1 = \text{span}(\vec{e}_1) = x$ axis

$$\underset{\uparrow}{\vec{v}_3 - \text{proj}_{V_2}\vec{v}_3} = \begin{bmatrix}d\\e\\f\end{bmatrix} - \begin{bmatrix}d\\e\\0\end{bmatrix} = \begin{bmatrix}0\\0\\f\end{bmatrix}, \text{ so that } \vec{w}_3 = \begin{bmatrix}0\\0\\1\end{bmatrix} = \vec{e}_3.$$

Note that $V_2 = \text{span}(\vec{e}_1, \vec{e}_2) = x\text{-}y$ plane

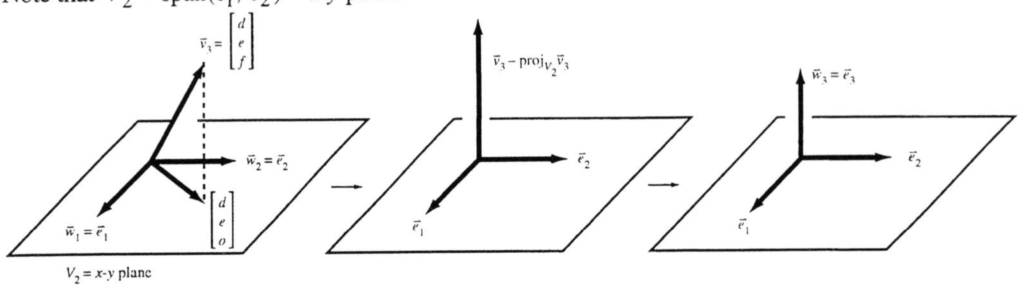

32. A basis of the plane is $\vec{v}_1 = \begin{bmatrix}-1\\1\\0\end{bmatrix}, \vec{v}_2 = \begin{bmatrix}-1\\0\\1\end{bmatrix}$.

Now apply the Gram-Schmidt process.

$$\vec{w}_1 = \frac{1}{\|\vec{v}_1\|}\vec{v}_1 = \frac{1}{\sqrt{2}}\begin{bmatrix}-1\\1\\0\end{bmatrix}$$

$$\vec{w}_2 = \frac{\vec{v}_2 - (\vec{w}_1 \cdot \vec{v}_2)\vec{w}_1}{\text{length}} = \frac{1}{\sqrt{6}}\begin{bmatrix}-1\\-1\\2\end{bmatrix}$$

Your solution may be different if you start with a different basis $\vec{v}_1, \vec{v}_2$ of the plane.

33. $\text{rref}(A) = \begin{bmatrix}1 & 0 & 0 & 1\\0 & 1 & 1 & 0\end{bmatrix}$

A basis of ker(A) is $\vec{v}_1 = \begin{bmatrix}-1\\0\\0\\1\end{bmatrix}, \vec{v}_2 = \begin{bmatrix}0\\-1\\1\\0\end{bmatrix}$.

Since $\vec{v}_1$ and $\vec{v}_2$ are orthogonal already, we obtain $\vec{w}_1 = \dfrac{1}{\sqrt{2}}\begin{bmatrix}-1\\0\\0\\1\end{bmatrix}, \vec{w}_2 = \dfrac{1}{\sqrt{2}}\begin{bmatrix}0\\-1\\1\\0\end{bmatrix}$.

34. $\text{rref}(A) = \begin{bmatrix} 1 & 0 & -1 & -2 \\ 0 & 1 & 2 & 3 \end{bmatrix}$

A basis of ker(A) is $\vec{v}_1 = \begin{bmatrix} 1 \\ -2 \\ 1 \\ 0 \end{bmatrix}, \vec{v}_2 = \begin{bmatrix} 2 \\ -3 \\ 0 \\ 1 \end{bmatrix}$.

We apply the Gram-Schmidt process and obtain

$$\vec{w}_1 = \frac{1}{\|\vec{v}_1\|}\vec{v}_1 = \frac{1}{\sqrt{6}}\begin{bmatrix} 1 \\ -2 \\ 1 \\ 0 \end{bmatrix}, \quad \vec{w}_2 = \frac{\vec{v}_2 - (\vec{w}_1 \cdot \vec{v}_2)\vec{w}_1}{\text{length}} = \frac{1}{\sqrt{30}}\begin{bmatrix} 2 \\ -1 \\ -4 \\ 3 \end{bmatrix}$$

35. $\text{rref}(A) = \begin{bmatrix} 1 & 0 & \frac{1}{3} \\ 0 & 1 & \frac{1}{3} \\ 0 & 0 & 0 \end{bmatrix}$

The pivot columns of A give us a basis of im(A):

$$\vec{v}_1 = \begin{bmatrix} 1 \\ 2 \\ 2 \end{bmatrix}, \vec{v}_2 = \begin{bmatrix} 2 \\ 1 \\ -2 \end{bmatrix}$$

Since $\vec{v}_1$ and $\vec{v}_2$ are orthogonal already, we obtain $\vec{w}_1 = \frac{1}{3}\begin{bmatrix} 1 \\ 2 \\ 2 \end{bmatrix}, \vec{w}_2 = \frac{1}{3}\begin{bmatrix} 2 \\ 1 \\ -2 \end{bmatrix}$.

36. Write $M = \underbrace{\frac{1}{2}\begin{bmatrix} 1 & 1 & 1 \\ 1 & -1 & -1 \\ 1 & -1 & 1 \\ 1 & 1 & -1 \end{bmatrix}}_{Q_0} \underbrace{\begin{bmatrix} 2 & 3 & 5 \\ 0 & -4 & 6 \\ 0 & 0 & 7 \end{bmatrix}}_{R_0}$

This is almost the QR factorization of M: the matrix Q_0 has orthonormal columns and R_0 is upper triangular; the only problem is the entry –4 on the diagonal of R_0. Keeping in mind how matrices are multiplied, we can change all the signs in the second column of Q_0 and in the second row of R_0 to fix this problem:

$$M = \underbrace{\frac{1}{2}\begin{bmatrix} 1 & -1 & 1 \\ 1 & 1 & -1 \\ 1 & 1 & -1 \\ 1 & -1 & 1 \end{bmatrix}}_{Q} \underbrace{\begin{bmatrix} 2 & 3 & 5 \\ 0 & 4 & -6 \\ 0 & 0 & 7 \end{bmatrix}}_{R}$$

37. Write $M = \frac{1}{2}\begin{bmatrix} 1 & 1 & 1 & 1 \\ 1 & -1 & -1 & 1 \\ 1 & -1 & 1 & -1 \\ 1 & 1 & -1 & -1 \end{bmatrix}\begin{bmatrix} 3 & 4 \\ 0 & 5 \\ 0 & 0 \\ 0 & 0 \end{bmatrix}$

$$\uparrow \qquad\qquad \uparrow$$
$$Q_0 \qquad\qquad R_0$$

Note that the last two columns of Q_0 and the last two rows of R_0 have no effect on the product Q_0R_0; if we drop them, we have the QR factorization of M:

$$M = \frac{1}{2}\begin{bmatrix} 1 & 1 \\ 1 & -1 \\ 1 & -1 \\ 1 & 1 \end{bmatrix}\begin{bmatrix} 3 & 4 \\ 0 & 5 \end{bmatrix}$$
$$\uparrow \qquad \uparrow$$
$$Q \qquad R$$

38. Since $\vec{v}_1 = 2\vec{e}_3$, $\vec{v}_2 = -3\vec{e}_1$ and $\vec{v}_3 = 4\vec{e}_4$ are orthogonal, we have

$$Q = \begin{bmatrix} \frac{\vec{v}_1}{\|\vec{v}_1\|} & \frac{\vec{v}_2}{\|\vec{v}_2\|} & \frac{\vec{v}_3}{\|\vec{v}_3\|} \end{bmatrix} = \begin{bmatrix} 0 & -1 & 0 \\ 0 & 0 & 0 \\ 1 & 0 & 0 \\ 0 & 0 & 1 \end{bmatrix} \text{ and } R = \begin{bmatrix} \|\vec{v}_1\| & 0 & 0 \\ 0 & \|\vec{v}_2\| & 0 \\ 0 & 0 & \|\vec{v}_3\| \end{bmatrix} = \begin{bmatrix} 2 & 0 & 0 \\ 0 & 3 & 0 \\ 0 & 0 & 4 \end{bmatrix}.$$

39. $\vec{w}_1 = \frac{1}{\sqrt{14}}\begin{bmatrix} 1 \\ 2 \\ 3 \end{bmatrix}$, $\vec{w}_2 = \frac{1}{\sqrt{3}}\begin{bmatrix} 1 \\ 1 \\ -1 \end{bmatrix}$, $\vec{w}_3 = \vec{w}_1 \times \vec{w}_2 = \frac{1}{\sqrt{42}}\begin{bmatrix} -5 \\ 4 \\ -1 \end{bmatrix}$

40. If $\vec{v}_1, \ldots, \vec{v}_n$ are the columns of A, then $Q = \begin{bmatrix} \frac{\vec{v}_1}{\|\vec{v}_1\|} & \cdots & \frac{\vec{v}_n}{\|\vec{v}_n\|} \end{bmatrix}$ and $R = \begin{bmatrix} \|\vec{v}_1\| & & 0 \\ & \ddots & \\ 0 & & \|\vec{v}_n\| \end{bmatrix}$.

(See Exercise 38 as an example.)

41. If all diagonal entries of A are positive, then we have $Q = I_n$ and $R = A$. A small modification is necessary if A has negative entries on the diagonal: if $a_{ii} < 0$ we let $r_{ij} = -a_{ij}$ for all j, and we let $q_{ii} = -1$; if $a_{ii} > 0$ we let $r_{ij} = a_{ij}$ and $q_{ii} = 1$. Furthermore, $q_{ij} = 0$ if $i \neq j$ (that is, Q is diagonal).

For example, $\begin{bmatrix} -1 & 2 & 3 \\ 0 & 4 & 5 \\ 0 & 0 & -6 \end{bmatrix} = \begin{bmatrix} -1 & 0 & 0 \\ 0 & 1 & 0 \\ 0 & 0 & -1 \end{bmatrix}\begin{bmatrix} 1 & -2 & -3 \\ 0 & 4 & 5 \\ 0 & 0 & 6 \end{bmatrix}$

$$\uparrow \qquad\qquad \uparrow \qquad\qquad \uparrow$$
$$A \qquad\qquad Q \qquad\qquad R$$

42. We have $r_{11} = \|\vec{v}_1\|$ and $r_{22} = \|\vec{v}_2 - \text{proj}_L\vec{v}_2\|$, so that $r_{11}r_{22}$ is the area of the parallelogram defined by $\vec{v}_1$ and $\vec{v}_2$.

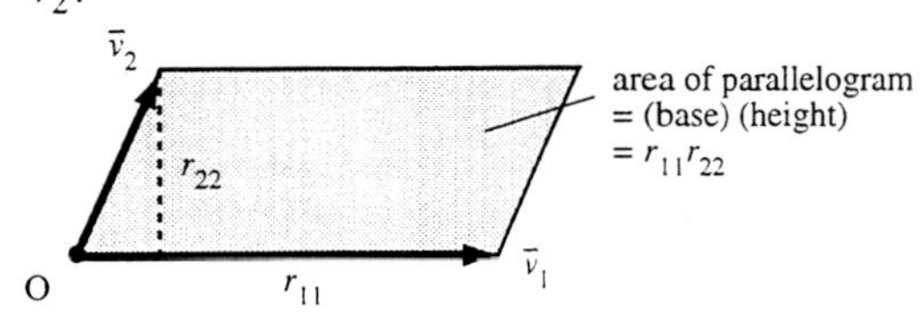

43. Partition the matrices Q and R in the QR factorization of A as follows:

$$[A_1 \quad A_2] = A = QR = [Q_1 \quad Q_2]\begin{bmatrix} R_1 & R_2 \\ 0 & R_3 \end{bmatrix} = [Q_1R_1 \quad Q_1R_2 + Q_2R_3],$$

where Q_1 is $n \times m_1$, Q_2 is $n \times m_2$, R_1 is $m_1 \times m_1$, and R_3 is $m_2 \times m_2$.
Then, $A_1 = Q_1R_1$ is the QR factorization of A_1: note that the columns of A_1 are orthonormal, and R_1 is upper triangular with positive diagonal entries.

44. No! If m exceeds n, then there is no $n \times m$ matrix Q with orthonormal columns (if the columns of a matrix are orthonormal, then they are linearly independent).

45. Yes. Let $A = [\vec{v}_1 \quad \cdots \quad \vec{v}_m]$. The idea is to perform the Gram-Schmidt process in reversed order, starting with $\vec{w}_m = \frac{1}{\|\vec{v}_m\|}\vec{v}_m$.

Then we can express $\vec{v}_j$ as a linear combination of $\vec{w}_j, \ldots, \vec{w}_m$, so that
$[\vec{v}_1 \quad \cdots \quad \vec{v}_j \quad \cdots \quad \vec{v}_m] = [\vec{w}_1 \quad \cdots \quad \vec{w}_j \quad \cdots \quad \vec{w}_m]L$ for some *lower* triangular matrix L, with

$$\vec{v}_j = [\vec{w}_1 \cdots \vec{w}_j \cdots \vec{w}_m]\begin{bmatrix} l_{1j} \\ \cdots \\ l_{jj} \\ \cdots \\ l_{mj} \end{bmatrix} = l_{jj}\vec{w}_j + \cdots + l_{mj}\vec{w}_m.$$

4.3

1. Using Facts 4.3.6 and 4.3.9a, we find that $(A\vec{v}) \cdot \vec{w} = (A\vec{v})^T\vec{w} = \vec{v}^TA^T\vec{w} = \vec{v} \cdot (A^T\vec{w})$, as claimed.

2. Write $L(\vec{x}) = A\vec{x}$; by Definition 4.3.1, A is an orthogonal $n \times n$ matrix, so that $A^TA = I_n$, by Fact 4.3.7.
Now $L(\vec{v}) \cdot L(\vec{w}) = (A\vec{v}) \cdot (A\vec{w}) = (A\vec{v})^TA\vec{w} = \vec{v}^TA^TA\vec{w} = \vec{v}^TI_n\vec{w} = \vec{v}^T\vec{w} = \vec{v} \cdot \vec{w}$, as claimed. Note that we have used Facts 4.3.6 and 4.3.9a.

3. We will use the fact that L preserves length (by Definition 4.3.1) and the dot product (by Exercise 2):

$$\angle(L(\vec{v}), L(\vec{w})) = \arccos\frac{L(\vec{v}) \cdot L(\vec{w})}{\|L(\vec{v})\|\|L(\vec{w})\|} = \arccos\frac{\vec{v} \cdot \vec{w}}{\|\vec{v}\|\|\vec{w}\|} = \angle(\vec{v}, \vec{w}).$$

4. If $L(\vec{x}) = \vec{0}$, then $\|L(\vec{x})\| = \|\vec{x}\| = 0$, so that $\vec{x} = \vec{0}$. Therefore, $\ker(L) = \{\vec{0}\}$.
By Fact 3.3.9, dim(im(L)) = n − dim(ker(L)) = n.
Since $\mathbb{R}^m$ has an n-dimensional subspace (namely, im(L)), the inequality $n \leq m$ holds.
The transformation L preserves right angles (the proof of Fact 4.3.2 applies), so that the columns of A are orthonormal (since they are $L(\vec{e}_1), \ldots, L(\vec{e}_n)$).
Therefore, we have $A^TA = I_n$ (the proof of Fact 4.3.7 applies).
Since the vectors $\vec{v}_1, \ldots, \vec{v}_n$ form an orthonormal basis of im(A), the matrix AA^T represents the orthogonal projection onto im(A), by Fact 4.3.10.

A simple example of such a transformation is $L(\vec{x}) = \begin{bmatrix} 1 & 0 \\ 0 & 1 \\ 0 & 0 \end{bmatrix}\vec{x}$, that is, $L\begin{bmatrix} x_1 \\ x_2 \end{bmatrix} = \begin{bmatrix} x_1 \\ x_2 \\ 0 \end{bmatrix}$.

5. Yes, since the product of orthogonal matrices is orthogonal, by Fact 4.3.4a.

6. Yes! By Facts 4.3.9b and 4.3.7, $(A^T)^{-1} = (A^{-1})^T = (A^T)^T$.
The equation $(A^T)^{-1} = (A^T)^T$ shows that A^T is orthogonal, again by Fact 4.3.7.

7. Yes! If A is orthogonal, then so is A^T, by Exercise 6. Since the columns of A are orthonormal, so are the rows of A^T.

8. a. No! As a counterexample, consider $A = \begin{bmatrix} 1 & 0 \\ 0 & 1 \\ 0 & 0 \end{bmatrix}$ (see Exercise 4).

b. Yes! More generally, if A and B are $n \times n$ matrices such that $BA = I_n$, then $AB = I_n$, by Fact 2.4.9c.

9. Write $A = [\vec{v}_1 \quad \vec{v}_2]$. The unit vector $\vec{v}_1$ can be expressed as $\vec{v} = \begin{bmatrix} \cos(\phi) \\ \sin(\phi) \end{bmatrix}$, for some ϕ. Then $\vec{v}_2$ will be one of the two unit vectors orthogonal to $\vec{v}_1$: $\vec{v}_2 = \begin{bmatrix} -\sin(\phi) \\ \cos(\phi) \end{bmatrix}$ or $\vec{v}_2 = \begin{bmatrix} \sin(\phi) \\ -\cos(\phi) \end{bmatrix}$.

Therefore, an orthogonal 2×2 matrix is either of the form $A = \begin{bmatrix} \cos(\phi) & -\sin(\phi) \\ \sin(\phi) & \cos(\phi) \end{bmatrix}$ or $A = \begin{bmatrix} \cos(\phi) & \sin(\phi) \\ \sin(\phi) & -\cos(\phi) \end{bmatrix}$, representing a rotation or a reflection.

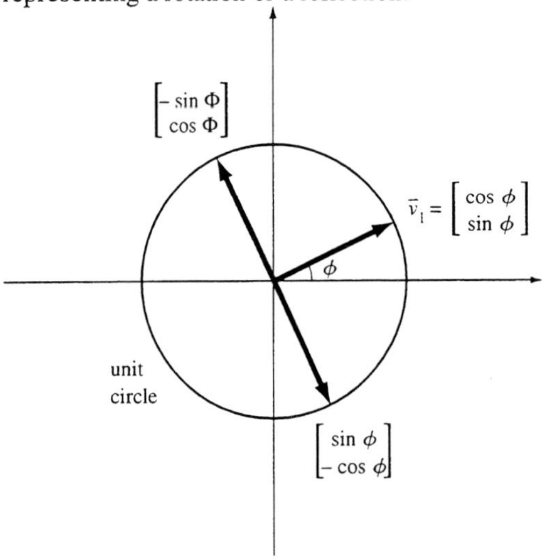

10. Since the first two columns are orthogonal to the third, we have $c = d = 0$. Then $\begin{bmatrix} a & b \\ e & f \end{bmatrix}$ is an orthogonal 2×2 matrix; By Exercise 9, the 3×3 matrix A is either of the form $A = \begin{bmatrix} \cos(\phi) & -\sin(\phi) & 0 \\ 0 & 0 & 1 \\ \sin(\phi) & \cos(\phi) & 0 \end{bmatrix}$ or

$$A = \begin{bmatrix} \cos(\phi) & \sin(\phi) & 0 \\ 0 & 0 & 1 \\ \sin(\phi) & -\cos(\phi) & 0 \end{bmatrix}.$$

11. Let us first think about the inverse $L = T^{-1}$ of T.

Write $L(\vec{x}) = A\vec{x} = [\vec{v}_1 \quad \vec{v}_2 \quad \vec{v}_3]\vec{x}$. It is required that $L(\vec{e}_3) = \vec{v}_3 = \begin{bmatrix} \frac{2}{3} \\ \frac{2}{3} \\ \frac{1}{3} \end{bmatrix}$.

Furthermore, the vectors $\vec{v}_1$, $\vec{v}_2$, $\vec{v}_3$ must form an orthonormal basis of $\mathbb{R}^3$. By inspection, we find $\vec{v}_1 = \begin{bmatrix} -\frac{2}{3} \\ \frac{1}{3} \\ \frac{2}{3} \end{bmatrix}$.

Then $\vec{v}_2 = \vec{v}_1 \times \vec{v}_3 = \begin{bmatrix} -\frac{1}{3} \\ \frac{2}{3} \\ -\frac{2}{3} \end{bmatrix}$ does the job. In summary, we have $L(\vec{x}) = \frac{1}{3}\begin{bmatrix} -2 & -1 & 2 \\ 1 & 2 & 2 \\ 2 & -2 & 1 \end{bmatrix}\vec{x}$.

Since the matrix of L is orthogonal, the matrix of $T = L^{-1}$ is the transpose of the matrix of L:

$$T(\vec{x}) = \frac{1}{3}\begin{bmatrix} -2 & 1 & 2 \\ -1 & 2 & -2 \\ 2 & 2 & 1 \end{bmatrix}\vec{x}.$$

There are many other answers (since there are many choices for the vector $\vec{v}_1$ above).

12. Let the third column be the cross product of the first two: $A = \begin{bmatrix} \frac{2}{3} & \frac{1}{\sqrt{2}} & \frac{1}{\sqrt{18}} \\ \frac{2}{3} & -\frac{1}{\sqrt{2}} & \frac{1}{\sqrt{18}} \\ \frac{1}{3} & 0 & -\frac{4}{\sqrt{18}} \end{bmatrix}$.

There is another solution, with the signs in the last column reversed.

13. No, since the vectors $\begin{bmatrix} 2 \\ 3 \\ 0 \end{bmatrix}$ and $\begin{bmatrix} -3 \\ 2 \\ 0 \end{bmatrix}$ are orthogonal, whereas $\begin{bmatrix} 3 \\ 0 \\ 2 \end{bmatrix}$ and $\begin{bmatrix} 2 \\ -3 \\ 0 \end{bmatrix}$ are not (see Fact 4.3.2).

14. By Fact 4.3.9a, $(A^TA)^T = A^T(A^T)^T = A^TA$.
The matrix AA^T is symmetric as well: $(AA^T)^T = (A^T)^TA^T = AA^T$.

15. No! If two symmetric matrices A and B do not commute, then $(AB)^T = B^TA^T = BA \neq AB$, so that AB is not symmetric.

Example: $A = \begin{bmatrix} 1 & 0 \\ 0 & 0 \end{bmatrix}$ and $B = \begin{bmatrix} 0 & 1 \\ 1 & 0 \end{bmatrix}$

16. Yes, since $(A^2)^T = (AA)^T = A^TA^T = AA = A^2$.

17. Yes! By Fact 4.3.9b, $(A^{-1})^T = (A^T)^{-1} = A^{-1}$.

18. a. Note that the diagonal entries must be zero, since $a_{ii} = -a_{ii}$.

Example: $A = \begin{bmatrix} 0 & 1 & 2 \\ -1 & 0 & -3 \\ -2 & 3 & 0 \end{bmatrix}$

b. By Fact 4.3.9a, $(A^2)^T = (AA)^T = A^TA^T = (-A)(-A) = A^2$, so that A^2 is symmetric.

19. By Fact 4.3.10, the matrix of the projection is $\vec{v}\vec{v}^T$; the ijth entry of this matrix is $v_i v_j$.

20. An orthonormal basis of W is $\vec{w}_1 = \begin{bmatrix} 0.5 \\ 0.5 \\ 0.5 \\ 0.5 \end{bmatrix}$, $\vec{w}_2 = \begin{bmatrix} -0.1 \\ 0.7 \\ -0.7 \\ 0.1 \end{bmatrix}$ (see Exercise 4.2.9).

By Fact 4.3.10, the matrix of the projection onto W is AA^T, where $A = [\vec{w}_1 \ \ \vec{w}_2]$.

$$AA^T = \frac{1}{100}\begin{bmatrix} 26 & 18 & 32 & 24 \\ 18 & 74 & -24 & 32 \\ 32 & -24 & 74 & 18 \\ 24 & 32 & 18 & 26 \end{bmatrix}$$

21. A unit vector on the line is $\vec{u} = \frac{1}{\sqrt{n}}\begin{bmatrix} 1 \\ \vdots \\ 1 \end{bmatrix}$.

The matrix of the orthogonal projection is $\vec{u}\vec{u}^T$, the $n \times n$ matrix whose entries are all $\frac{1}{n}$ (compare with Exercise 19).

22. a. Suppose we are projecting onto a subspace W of $\mathbb{R}^n$. Since $A\vec{x}$ is in W already, the orthogonal projection of $A\vec{x}$ onto W is just $A\vec{x}$ itself: $A(A\vec{x}) = A\vec{x}$, or $A^2\vec{x} = A\vec{x}$.
Since this equation holds for all $\vec{x}$, we have $A^2 = A$.

b. $A = MM^T$, for some matrix M with orthonormal columns $\vec{v}_1, \ldots, \vec{v}_m$. Note that $M^TM = I_m$, since the ijth entry of M^TM is $\vec{v}_i \cdot \vec{v}_j$. Then $A^2 = MM^TMM^T = M(M^TM)M^T = MI_mM^T = MM^T = A$.

23. Examine how A acts on $\vec{u}$, and on a vector $\vec{v}$ orthogonal to $\vec{u}$:
$A\vec{u} = (2\vec{u}\vec{u}^T - I_3)\vec{u} = 2\vec{u}\vec{u}^T\vec{u} - \vec{u} = \vec{u}$, since $\vec{u}^T\vec{u} = \vec{u}\cdot\vec{u} = \|\vec{u}\|^2 = 1$.
$A\vec{v} = (2\vec{u}\vec{u}^T - I_3)\vec{v} = 2\vec{u}\vec{u}^T\vec{v} - \vec{v} = -\vec{v}$, since $\vec{u}^T\vec{v} = \vec{u}\cdot\vec{v} = 0$.
Since A leaves the vectors in $L = \text{span}(\vec{u})$ unchanged and reverses the vectors in $V = L^\perp$, it represents the *reflection in L*.
Note that $B = -A$, so that B reverses the vectors in L and leaves the vectors in V unchanged; that is, B represents the reflection in V.

24. Note that A^T is an $n \times m$ matrix. By Facts 3.3.5 and 4.3.9c we have
$\dim(\ker(A^T)) = m - \text{rank}(A^T) = m - \text{rank}(A)$.
By Fact 3.3.8, $\dim(\text{im}(A)) = \text{rank}(A)$, so that $\dim(\text{im}(A)) + \dim(\ker(A^T)) = m$.

25. Note that A^T is an $n \times m$ matrix. By Facts 3.3.5 and 4.3.9c, we have
$\dim(\ker(A)) = n - \text{rank}(A)$ and $\dim(\ker(A^T)) = m - \text{rank}(A^T) = m - \text{rank}(A)$,
so that $\dim(\ker(A)) = \dim(\ker(A^T))$ if (and only if) A is a square matrix.

26. By Fact 4.2.2, the columns $\vec{w}_1, \ldots, \vec{w}_m$ of Q are orthonormal. Therefore, $Q^TQ = I_m$, since the ijth entry of Q^TQ is $\vec{w}_i \cdot \vec{w}_j$.
If we multiply the equation $M = QR$ by Q^T from the left then $Q^TM = Q^TQR = R$, as claimed.

27. By Fact 4.2.2, the columns $\vec{w}_1, \ldots, \vec{w}_m$ of Q are orthonormal. Therefore, $Q^TQ = I_m$, since the ijth entry of Q^TQ is $\vec{w}_i \cdot \vec{w}_j$.
By Fact 4.3.9a, we now have $A^TA = (QR)^TQR = R^TQ^TQR = R^TR$.

28. As suggested, we consider the QR factorization
$A^T = PR$
of A^T, where P is orthogonal and R is upper triangular with positive diagonal entries.
By Fact 4.3.9a, $A = (PR)^T = R^TP^T$.
Note that $L = R^T$ is lower triangular and $Q = P^T$ is orthogonal (see Exercise 6).

29. Yes! By Exercise 4.2.45, we can write $A^T = PL$, where P is orthogonal and L is lower triangular.
By Fact 4.3.9a, $A = (PL)^T = L^TP^T$.
Note that $R = L^T$ is upper triangular, and $Q = P^T$ is orthogonal (by Exercise 6).

30. **a.** If an $n \times n$ matrix A is orthogonal and upper triangular, then A^{-1} is both lower triangular (since $A^{-1} = A^T$) and upper triangular (being the inverse of an upper triangular matrix; compare with Exercise 2.3.35c). Therefore, $A^{-1} = A^T$ is a diagonal matrix, and so is A itself. Since A is orthogonal with positive diagonal entries, all the diagonal entries must be 1, so that $A = I_n$.

b. Using the terminology suggested in the hint, we observe that $Q_2^{-1}Q_1$ is orthogonal (by Fact 4.3.4) and $R_2R_1^{-1}$ is upper triangular with positive diagonal entries. By part a, the matrix $Q_2^{-1}Q_1 = R_2R_1^{-1}$ is I_n, so that $Q_1 = Q_2$ and $R_1 = R_2$, as claimed.

31. **a.** Using the terminology suggested in the hint, we observe that $I_m = Q_1^TQ_1 = (Q_2S)^TQ_2S = S^TQ_2^TQ_2S = S^TS$, so that S is orthogonal, by Fact 4.3.7.

b. Using the terminology suggested in the hint, we observe that $R_2R_1^{-1}$ is both orthogonal (let $S = R_2R_1^{-1}$ in part a) and upper triangular, with positive diagonal entries. By Exercise 30a, we have $R_2R_1^{-1} = I_m$, so that $R_1 = R_2$. Then $Q_1 = Q_2R_2R_1^{-1} = Q_2$, as claimed.

32. By Fact 4.3.9a, $A^T = \begin{bmatrix} 1 & 0 & 0 \\ 1 & 1 & 0 \\ -1 & 2 & 1 \end{bmatrix}\begin{bmatrix} 1 & 0 & 0 \\ 0 & -1 & 0 \\ 0 & 0 & 2 \end{bmatrix}\begin{bmatrix} 1 & 3 & 2 \\ 0 & 1 & 0 \\ 0 & 0 & 1 \end{bmatrix}$.

33. By Exercise 2.4.62b, the given LDU factorization of A is unique.
By Fact 4.3.9a, $A = A^T = (LDU)^T = U^TD^TL^T = U^TDL^T$ is another way to write the LDU factorization of A (since U^T is lower triangular and L^T is upper triangular). By the uniqueness of the LDU factorization, we have $U = L^T$ (and $L = U^T$), as claimed.

34. **a.** $\begin{bmatrix} A & -B^T \\ B & A^T \end{bmatrix} + \begin{bmatrix} C & -D^T \\ D & C^T \end{bmatrix} = \begin{bmatrix} A+C & -B^T - D^T \\ B+D & A^T + C^T \end{bmatrix} = \begin{bmatrix} A+C & -(B+D)^T \\ B+D & (A+C)^T \end{bmatrix}$ is of the required form.

b. $k\begin{bmatrix} A & -B^T \\ B & A^T \end{bmatrix} = \begin{bmatrix} kA & -kB^T \\ kB & kA^T \end{bmatrix} = \begin{bmatrix} kA & -(kB)^T \\ kB & (kA)^T \end{bmatrix}$ is of the required form.

c. $\begin{bmatrix} A & -B^T \\ B & A^T \end{bmatrix}\begin{bmatrix} C & -D^T \\ D & C^T \end{bmatrix} = \begin{bmatrix} AC - B^T D & -AD^T - B^T C^T \\ BC + A^T D & -BD^T + A^T C^T \end{bmatrix} = \begin{bmatrix} AC - B^T D & -(BC + A^T D)^T \\ BC + A^T D & (AC - B^T D)^T \end{bmatrix}$ is of the required form.

Note that A, B, C, D, and their transposes are rotation-dilation matrices, so that they all commute.

d. $\begin{bmatrix} A & -B^T \\ B & A^T \end{bmatrix}^T = \begin{bmatrix} A^T & B^T \\ -B & (A^T)^T \end{bmatrix}$ is of the required form.

e. Note that the columns $\vec{v}_1, \vec{v}_2, \vec{v}_3, \vec{v}_4$ or M are orthogonal, and they all have length $\sqrt{p^2+q^2+r^2+s^2}$.
Now $M^T M$ is the 4×4 matrix whose *ij*th entry is $\vec{v}_i \cdot \vec{v}_j$, so that $M^T M = (p^2+q^2+r^2+s^2)I_4$.

f. If $M \neq 0$, then $k = p^2+q^2+r^2+s^2 > 0$, and $\left(\frac{1}{k}M^T\right)M = I_4$, so that M is invertible, with

$$M^{-1} = \frac{1}{p^2+q^2+r^2+s^2}M^T.$$

By parts b and d, M^{-1} is in H as well.

g. No! $A = \begin{bmatrix} 0 & -1 & 0 & 0 \\ 1 & 0 & 0 & 0 \\ 0 & 0 & 0 & 1 \\ 0 & 0 & -1 & 0 \end{bmatrix}$ and $B = \begin{bmatrix} 0 & 0 & -1 & 0 \\ 0 & 0 & 0 & -1 \\ 1 & 0 & 0 & 0 \\ 0 & 1 & 0 & 0 \end{bmatrix}$ do not commute ($AB = -BA$).

4.4

1. A basis of $\ker(A^T)$ is $\begin{bmatrix} -3 \\ 2 \end{bmatrix}$.

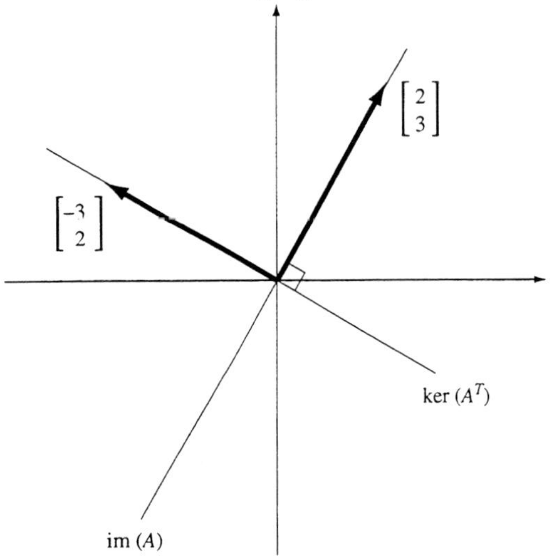

2. A basis of $\ker(A^T)$ is $\begin{bmatrix} 1 \\ -2 \\ 1 \end{bmatrix}$. im($A$) is the plane perpendicular to this line.

3. We will first show that the vectors $\vec{v}_1, \ldots, \vec{v}_p, \vec{w}_1, \ldots, \vec{w}_q$ span $\mathbb{R}^n$. Any vector $\vec{v}$ in $\mathbb{R}^n$ can be written as $\vec{v} = \text{proj}_V \vec{v} + (\vec{v} - \text{proj}_V \vec{v})$, where $\text{proj}_V \vec{v}$ is in V and $\vec{v} - \text{proj}_V \vec{v}$ is in $V^\perp$ (by definition of a projection, Fact 4.1.6).
Now $\text{proj}_V \vec{v}$ is a linear combination of $\vec{v}_1, \ldots, \vec{v}_p$, and $\vec{v} - \text{proj}_V \vec{v}$ is a linear combination of $\vec{w}_1, \ldots, \vec{w}_q$, showing that the vectors $\vec{v}_1, \ldots, \vec{v}_p, \vec{w}_1, \ldots, \vec{w}_q$ span $\mathbb{R}^n$.

Note that $p + q = n$, by Fact 4.4.2a; therefore, the vectors $\vec{v}_1, \ldots, \vec{v}_p, \vec{w}_1, \ldots, \vec{w}_q$ form a basis of $\mathbb{R}^n$, by Fact 3.3.4d.

4. By Fact 4.4.1, the equation $(\text{im } B)^\perp = \ker(B^T)$ holds for any matrix B. Now let $B = A^T$. Then $(\text{im}(A^T))^\perp = \ker(A)$. Taking transposes of both sides and using Fact 4.4.2b we obtain $\text{im}(A^T) = (\ker A)^\perp$, as claimed.

5. $V = \ker(A)$, where $A = \begin{bmatrix} 1 & 1 & 1 & 1 \\ 1 & 2 & 5 & 4 \end{bmatrix}$.

Then $V^\perp = (\ker A)^\perp = \text{im}(A^T)$, by Exercise 4.

The two columns of A^T form a basis of $V^\perp$:

$$\begin{bmatrix} 1 \\ 1 \\ 1 \\ 1 \end{bmatrix}, \begin{bmatrix} 1 \\ 2 \\ 5 \\ 4 \end{bmatrix}$$

6. Yes! For any matrix A,

$$\text{im}(A) \underset{\uparrow}{=} (\ker(A^T))^\perp \underset{\uparrow}{=} (\ker(AA^T))^\perp = (\ker(AA^T)^T)^\perp \underset{\uparrow}{=} \text{im}(AA^T).$$

Facts 4.4.1 and 4.4.2b | Fact 4.4.3a | Facts 4.4.1 and 4.4.2b

7. $\text{im}(A)$ and $\ker(A)$ are orthogonal complements by Fact 4.4.1:
$(\text{im } A)^\perp = \ker(A^T) = \ker(A)$

8. a. By Fact 4.4.7, $L^+(\vec{y}) = (A^T A)^{-1} A^T \vec{y}$.
The transformation L^+ is linear since it is "given by a matrix," by Definition 2.1.1.

b. If L (and therefore A) is invertible, then $L^+(\vec{y}) = A^{-1}(A^T)^{-1} A^T \vec{y} = A^{-1}\vec{y} = L^{-1}\vec{y}$, so that $L^+ = L^{-1}$.

c. $L^+(L(\vec{x})) = (\text{the unique least-squares solution } \vec{u} \text{ of } L(\vec{u}) = L(\vec{x})) = \vec{x}$.

d. $L(L^+(\vec{y})) = A(A^T A)^{-1} A^T \vec{y} = \text{proj}_V \vec{y}$, where $V = \text{im}(A)$, by Fact 4.4.8.

e. Here $A = \begin{bmatrix} 1 & 0 \\ 0 & 1 \\ 0 & 0 \end{bmatrix}$. Then $L^+(\vec{y}) = (A^T A)^{-1} A^T \vec{y} = \begin{bmatrix} 1 & 0 & 0 \\ 0 & 1 & 0 \end{bmatrix} \vec{y}$.

9.

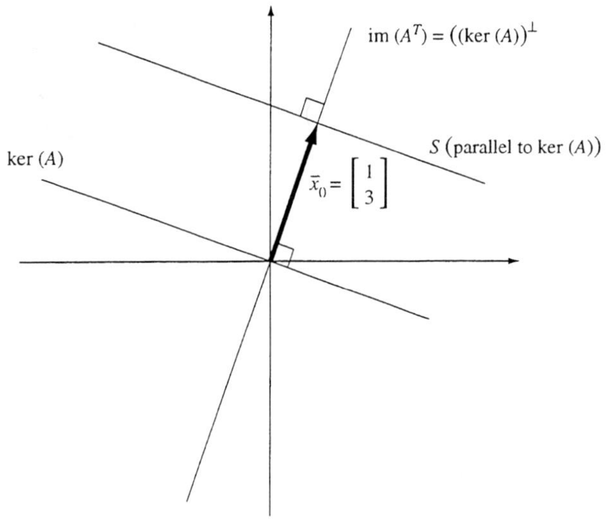

$\vec{x}_0$ is the shortest of all the vectors in S.

10. a. If $\vec{x}$ is an arbitrary solution of the system $A\vec{x} = \vec{b}$, let $\vec{x}_h = \text{proj}_V\vec{x}$, where $V = \ker(A)$, and $\vec{x}_0 = \vec{x} - \text{proj}_V\vec{x}$. Note that $\vec{b} = A\vec{x} = A(\vec{x}_h + \vec{x}_0) = A\vec{x}_h + A\vec{x}_0 = A\vec{x}_0$, since $\vec{x}_h$ is in ker(A).

b. If $\vec{x}_0$ and $\vec{x}_1$ are two solutions of the system $A\vec{x} = \vec{b}$, both from $(\ker A)^\perp$, then $\vec{x}_1 - \vec{x}_0$ is in the subspace $(\ker A)^\perp$ as well. Also, $A(\vec{x}_1 - \vec{x}_0) = A\vec{x}_1 - A\vec{x}_0 = \vec{b} - \vec{b} = \vec{0}$, so that $\vec{x}_1 - \vec{x}_0$ is in ker(A). By Fact 4.4.2c, it follows that $\vec{x}_1 - \vec{x}_0 = \vec{0}$, or $\vec{x}_1 = \vec{x}_0$, as claimed.

c. Write $\vec{x}_1 = \vec{x}_h + \vec{x}_0$ as in part a; note that $\vec{x}_h$ is orthogonal to $\vec{x}_0$. The claim now follows from the Pythagorean Theorem (Fact 4.1.8).

11. a. Note that $L^+(\vec{y}) = A^T(AA^T)^{-1}\vec{y}$; indeed, this vector is in $\text{im}(A^T) = (\ker A)^\perp$, and it is a solution of $L(\vec{x}) = A\vec{x} = \vec{y}$.

b. $L(L^+(\vec{y})) = \vec{y}$, by definition of L^+.

c. $L^+(L(\vec{x})) = A^T(AA^T)^{-1}A\vec{x} = \text{proj}_V\vec{x}$, where $V = \text{im}(A^T) = (\ker A)^\perp$, by Fact 4.4.8.

d. $\text{im}(L^+) = \text{im}(A^T)$, by part c, and $\ker(L^+) = \{\vec{0}\}$ (if $\vec{y}$ is in $\ker(L^+)$, then $\vec{y} = L(L^+(\vec{y})) = L(\vec{0}) = \vec{0}$, by part b).

e. Let $A = \begin{bmatrix} 1 & 0 & 0 \\ 0 & 1 & 0 \end{bmatrix}$; then the matrix of L^+ is $A^T(AA^T)^{-1} = A^T = \begin{bmatrix} 1 & 0 \\ 0 & 1 \\ 0 & 0 \end{bmatrix}$.

12. By Fact 4.4.6, the least-squares solutions of the linear system $A\vec{x} = \vec{b}$ are the exact solutions of the (consistent) system $A^TA\vec{x} = A^T\vec{b}$. The minimal solution of this normal equation (in the sense of Exercise 10) is called the minimal least-squares solution of the system $A\vec{x} = \vec{b}$.
Equivalently, the minimal least-squares solution of $A\vec{x} = \vec{b}$ can be defined as the minimal solution of the consistent system $A\vec{x} = \text{proj}_V\vec{b}$, where $V = \text{im}(A)$.

13. a. Suppose that $L^+(\vec{y}_1) = \vec{x}_1$ and $L^+(\vec{y}_2) = \vec{x}_2$; this means that $\vec{x}_1$ and $\vec{x}_2$ are both in $(\ker A)^\perp = \text{im}(A^T)$, $A^T A\vec{x}_1 = A^T\vec{y}_1$, and $A^T A\vec{x}_2 = A^T\vec{y}_2$. Then $\vec{x}_1 + \vec{x}_2$ is in $\text{im}(A^T)$ as well, and $A^T A(\vec{x}_1 + \vec{x}_2) = A^T(\vec{y}_1 + \vec{y}_2)$, so that $L^+(\vec{y}_1 + \vec{y}_2) = \vec{x}_1 + \vec{x}_2$.
The verification of the property $L^+(k\vec{y}) = kL^+(\vec{y})$ is analogous.

b. $L^+(L(\vec{x}))$ is the orthogonal projection of $\vec{x}$ onto $(\ker A)^\perp = \text{im}(A^T)$.

c. $L(L^+(\vec{y}))$ is the orthogonal projection of $\vec{y}$ onto $\text{im}(A) = (\ker(A^T))^\perp$.

d. $\text{im}(L^+) = \text{im}(A^T)$ and $\ker(L^+) = \ker(A^T)$, by parts b and c.

e. $L^+\begin{bmatrix} y_1 \\ y_2 \end{bmatrix} = \begin{bmatrix} \frac{y_1}{2} \\ 0 \\ 0 \end{bmatrix}$, so that the matrix of L^+ is $\begin{bmatrix} \frac{1}{2} & 0 \\ 0 & 0 \\ 0 & 0 \end{bmatrix}$.

14.

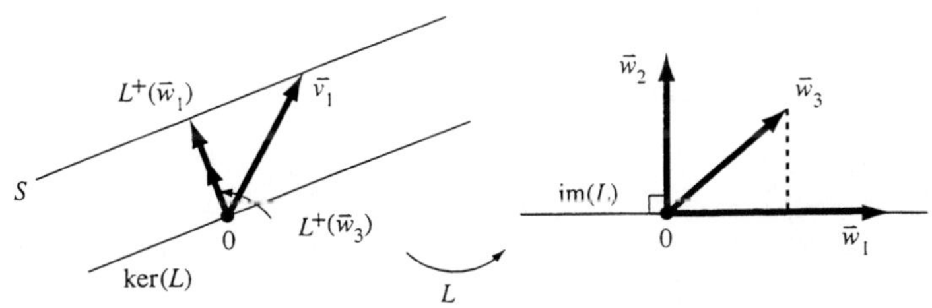

$L^+(\vec{w}_1)$ is the minimal solution of the system $L(\vec{x}) = \vec{w}_1$. The line *S* above shows all solutions of the system $L(\vec{x}) = \vec{w}_1$ (compare with Exercise 9). The minimal solution, $L^+(\vec{w}_1)$, is perpendicular to ker(*L*).
$L^+(\vec{w}_2) = L^+(\text{proj}_{\text{im}(L)}\vec{w}_2) = L^+(\vec{0}) = \vec{0}$
$L^+(\vec{w}_3) = L^+(\text{proj}_{\text{im}(L)}\vec{w}_3) \approx L^+(0.55\vec{w}_1) = 0.55L^+(\vec{w}_1)$

15. Note that $(A^TA)^{-1}A^TA = I_n$; let $B = (A^TA)^{-1}A^T$.

16. If *A* is an $m \times n$ matrix, then

$$\dim(\text{im } A)^\perp \underset{\text{Fact 4.4.2a}}{\uparrow{=}} m - \dim(\text{im } A) \underset{\text{Fact 3.3.8}}{\uparrow{=}} m - \text{rank}(A)$$

and

$$\dim(\ker(A^T)) \underset{\text{Fact 3.3.5}}{\uparrow{=}} m - \text{rank}(A^T)$$

It follows that $\text{rank}(A) = \text{rank}(A^T)$, as claimed.

17. Yes! By Fact 4.4.3, $\ker(A) = \ker(A^TA)$. Taking dimensions of both sides and using Fact 3.3.5, we find that $n - \text{rank}(A) = n - \text{rank}(A^TA)$; the claim follows.

18. Yes! By Exercise 17, $\text{rank}(A) = \text{rank}(A^TA)$. Substituting A^T for A in Exercise 17 and using Fact 4.3.9c, we find that $\text{rank}(A) = \text{rank}(A^T) = \text{rank}(AA^T)$. The claim follows.

19. $\vec{x}^* = (A^TA)^{-1}A^T\vec{b} = \begin{bmatrix}1\\1\end{bmatrix}$, by Fact 4.4.7.

20. Using Fact 4.4.7, we find $\vec{x}^* = \begin{bmatrix}2\\2\end{bmatrix}$ and $\vec{b} - A\vec{x}^* = \begin{bmatrix}-1\\1\\1\end{bmatrix}$.

Note that $\vec{b} - A\vec{x}^*$ is perpendicular to the two columns of A.

21. Using Fact 4.4.7, we find $\vec{x}^* = \begin{bmatrix}-1\\2\end{bmatrix}$ and $\vec{b} - A\vec{x}^* = \begin{bmatrix}-12\\36\\-18\end{bmatrix}$, so that $\left\|\vec{b} - A\vec{x}^*\right\| = 42$.

22. Using Fact 4.4.7, we find $\vec{x}^* = \begin{bmatrix}3\\-2\end{bmatrix}$ and $\vec{b} - A\vec{x}^* = \vec{0}$. This system is in fact consistent and $\vec{x}^*$ is the exact solution; the error $\left\|\vec{b} - A\vec{x}^*\right\|$ is 0.

23. Using Fact 4.4.7, we find $\vec{x}^* = \vec{0}$; here $\vec{b}$ is perpendicular to im(A).

24. Using Fact 4.4.7, we find $\vec{x}^* = [2]$.

25. In this case, the normal equation $A^TA\vec{x} = A^T\vec{b}$ is $\begin{bmatrix}5 & 15\\15 & 45\end{bmatrix}\begin{bmatrix}x_1\\x_2\end{bmatrix} = \begin{bmatrix}5\\15\end{bmatrix}$, which simplifies to $x_1 + 3x_2 = 1$, or $x_1 = 1 - 3x_2$. The solutions are of the form $\vec{x}^* = \begin{bmatrix}1-3t\\t\end{bmatrix}$, where t is an arbitrary constant.

26. Here, the normal equation $A^TA\vec{x} = A^T\vec{b}$ is $\begin{bmatrix}66 & 78 & 90\\78 & 93 & 108\\90 & 108 & 126\end{bmatrix}\begin{bmatrix}x_1\\x_2\\x_3\end{bmatrix} = \begin{bmatrix}1\\2\\3\end{bmatrix}$, with solutions $\vec{x}^* = \begin{bmatrix}t-\frac{7}{6}\\1-2t\\t\end{bmatrix}$, where t is an arbitrary constant.

27. The least-squares solutions of the system $SA\vec{x} = S\vec{b}$ are the exact solutions of the normal equation $(SA)^TSA\vec{x} = (SA)^TS\vec{b}$.
Note that $S^TS = I_n$, since S is orthogonal; therefore, the normal equation simplifies as follows:
$(SA)^TSA\vec{x} = A^TS^TSA\vec{x} = A^TA\vec{x}$ and $(SA)^TS\vec{b} = A^TS^TS\vec{b} = A^T\vec{b}$, so that the normal equation is $A^TA\vec{x} = A^T\vec{b}$, the same as the normal equation of the system $A\vec{x} = \vec{b}$. Therefore, the systems $A\vec{x} = \vec{b}$ and $SA\vec{x} = S\vec{b}$ have the same least-squares solution, $\vec{x}^* = \begin{bmatrix}7\\11\end{bmatrix}$.

28. The least-squares solutions of the system $A\vec{x} = \vec{v}_n$ are the exact solutions of $A\vec{x} = \text{proj}_{\text{im}(A)}\vec{v}_n$. Note that $\vec{v}_n$ is orthogonal to im(A), so that $\text{proj}_{\text{im}(A)}\vec{v}_n = \vec{0}$, and the unique least-squares solution is $\vec{x}^* = \vec{0}$.

29. By Fact 4.4.7, $\vec{x}^* = (A^TA)^{-1}A^T\vec{b} = \begin{bmatrix} 1+\varepsilon & 1 \\ 1 & 1+\varepsilon \end{bmatrix}^{-1} \begin{bmatrix} 1+\varepsilon \\ 1+\varepsilon \end{bmatrix} = \frac{1}{2+\varepsilon}\begin{bmatrix} 1+\varepsilon \\ 1+\varepsilon \end{bmatrix} \approx \begin{bmatrix} \frac{1}{2} \\ \frac{1}{2} \end{bmatrix}$, where $\varepsilon = 10^{-20}$.

If we use a hand-held calculator, due to roundoff errors we find the normal equation $\begin{bmatrix} 1 & 1 \\ 1 & 1 \end{bmatrix}\begin{bmatrix} x_1 \\ x_2 \end{bmatrix} = \begin{bmatrix} 1 \\ 1 \end{bmatrix}$, with infinitely many solutions.

30. We attempt to solve the system

$$\begin{aligned} c_0 + 0c_1 &= 0 \\ c_0 + 0c_1 &= 1 \\ c_0 + 1c_1 &= 1 \end{aligned}, \text{ or } \begin{bmatrix} 1 & 0 \\ 1 & 0 \\ 1 & 1 \end{bmatrix}\begin{bmatrix} c_0 \\ c_1 \end{bmatrix} = \begin{bmatrix} 0 \\ 1 \\ 1 \end{bmatrix}.$$

This system cannot be solved exactly; the least-squares solution is $\begin{bmatrix} c_0^* \\ c_1^* \end{bmatrix} = \begin{bmatrix} \frac{1}{2} \\ \frac{1}{2} \end{bmatrix}$. The line that fits the data points best is $f^*(t) = \frac{1}{2} + \frac{1}{2}t$.

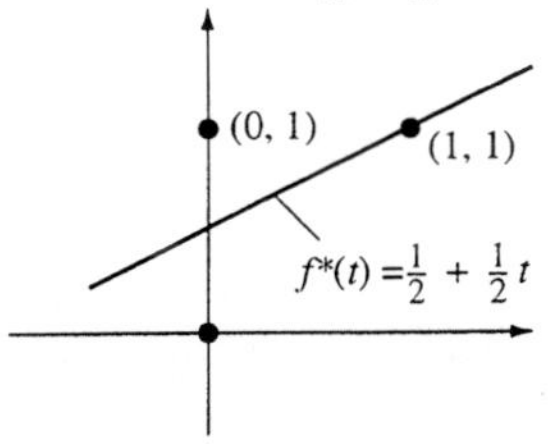

The line goes through the point (1, 1) and "splits the difference" between (0, 0) and (0, 1).

31. We want $\begin{bmatrix} c_0 \\ c_1 \end{bmatrix}$ such that

$$\begin{aligned} 3 &= c_0 + 0c_1 \\ 3 &= c_0 + 1c_1 \\ 6 &= c_0 + 1c_1 \end{aligned} \text{ or } \begin{bmatrix} 1 & 0 \\ 1 & 1 \\ 1 & 1 \end{bmatrix}\begin{bmatrix} c_1 \\ c_2 \end{bmatrix} = \begin{bmatrix} 3 \\ 3 \\ 6 \end{bmatrix}.$$

Since $\ker\begin{bmatrix} 1 & 0 \\ 1 & 1 \\ 1 & 1 \end{bmatrix} = \{\vec{0}\}$, $\begin{bmatrix} c_0 \\ c_1 \end{bmatrix}^* = \left(\begin{bmatrix} 1 & 1 & 1 \\ 0 & 1 & 1 \end{bmatrix}\begin{bmatrix} 1 & 0 \\ 1 & 1 \\ 1 & 1 \end{bmatrix}\right)^{-1}\begin{bmatrix} 1 & 1 & 1 \\ 0 & 1 & 1 \end{bmatrix}\begin{bmatrix} 3 \\ 3 \\ 6 \end{bmatrix} = \begin{bmatrix} 3 & 2 \\ 2 & 2 \end{bmatrix}^{-1}\begin{bmatrix} 12 \\ 9 \end{bmatrix} = \begin{bmatrix} 3 \\ \frac{3}{2} \end{bmatrix}$ so $f^*(t) = 3 + \frac{3}{2}t$.

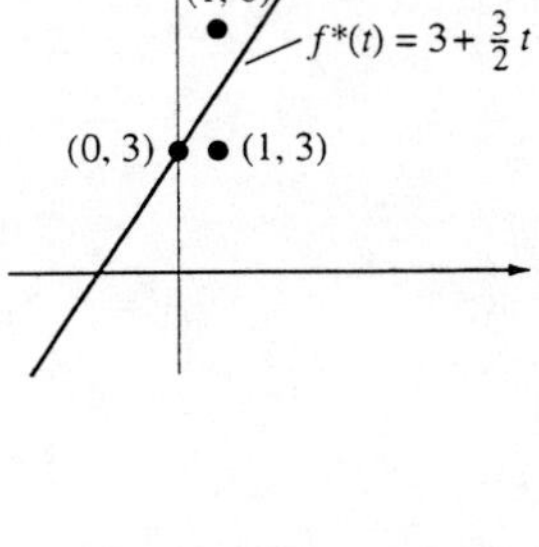

32. We want $\begin{bmatrix} c_0 \\ c_1 \\ c_2 \end{bmatrix}$ of $f(t) = c_0 + c_1 t + c_2 t^2$ such that

$$\begin{matrix} 27 = c_0 + 0c_1 + 0c_2 \\ 0 = c_0 + 1c_1 + 1c_2 \\ 0 = c_0 + 2c_1 + 4c_2 \\ 0 = c_0 + 3c_1 + 9c_2 \end{matrix} \text{ or } \begin{bmatrix} 1 & 0 & 0 \\ 1 & 1 & 1 \\ 1 & 2 & 4 \\ 1 & 3 & 9 \end{bmatrix} \begin{bmatrix} c_0 \\ c_1 \\ c_2 \end{bmatrix} = \begin{bmatrix} 27 \\ 0 \\ 0 \\ 0 \end{bmatrix}.$$

If we call the coefficient matrix A, we notice that $\ker(A) = \{\vec{0}\}$ so

$$\begin{bmatrix} c_0 \\ c_1 \\ c_2 \end{bmatrix}^* = (A^T A)^{-1} A^T \begin{bmatrix} 27 \\ 0 \\ 0 \\ 0 \end{bmatrix} = \begin{bmatrix} 25.65 \\ -28.35 \\ 6.75 \end{bmatrix} \text{ so } f^*(t) = 25.65 - 28.35t + 6.75t^2.$$

33. We want $\begin{bmatrix} c_0 \\ c_1 \\ c_2 \end{bmatrix}$ such that

$$\begin{matrix} 0 = c_0 + \sin(0)c_1 + \cos(0)c_2 \\ 1 = c_0 + \sin(1)c_1 + \cos(1)c_2 \\ 2 = c_0 + \sin(2)c_1 + \cos(2)c_2 \\ 3 = c_0 + \sin(3)c_1 + \cos(3)c_2 \end{matrix} \text{ or } \begin{bmatrix} 1 & 0 & 1 \\ 1 & \sin(1) & \cos(1) \\ 1 & \sin(2) & \cos(2) \\ 1 & \sin(3) & \cos(3) \end{bmatrix} \begin{bmatrix} c_1 \\ c_2 \\ c_3 \end{bmatrix} = \begin{bmatrix} 0 \\ 1 \\ 2 \\ 3 \end{bmatrix}.$$

Since the coefficient matrix has kernel $\{\vec{0}\}$, we compute $\begin{bmatrix} c_1 \\ c_2 \\ c_3 \end{bmatrix}^*$ using Fact 4.4.7, obtaining $\begin{bmatrix} c_0 \\ c_1 \\ c_2 \end{bmatrix}^* \approx \begin{bmatrix} 1.5 \\ 0.1 \\ -1.41 \end{bmatrix}$ so

$f^*(t) \approx 1.5 + 0.1\sin t - 1.41\cos t.$

34. We want $\begin{bmatrix} c_0 \\ c_1 \\ c_2 \\ c_3 \\ c_4 \end{bmatrix}$ such that

$$\begin{bmatrix} 1 & 0 & 1 & 0 & 1 \\ 1 & \sin(0.5) & \cos(0.5) & \sin(1) & \cos(1) \\ 1 & \sin(1) & \cos(1) & \sin(2) & \cos(2) \\ 1 & \sin(1.5) & \cos(1.5) & \sin(3) & \cos(3) \\ 1 & \sin(2) & \cos(2) & \sin(4) & \cos(4) \\ 1 & \sin(2.5) & \cos(2.5) & \sin(5) & \cos(5) \\ 1 & \sin(3) & \cos(3) & \sin(6) & \cos(6) \end{bmatrix} \begin{bmatrix} c_0 \\ c_1 \\ c_2 \\ c_3 \\ c_4 \end{bmatrix} = \begin{bmatrix} 0 \\ 0.5 \\ 1 \\ 1.5 \\ 2 \\ 2.5 \\ 3 \end{bmatrix}.$$

Since the columns of the coefficient matrix are linearly independent, its kernel is $\{\vec{0}\}$. We can use Fact 4.4.7 to

compute $\begin{bmatrix} c_0 \\ c_1 \\ c_2 \\ c_3 \\ c_4 \end{bmatrix}^* \approx \begin{bmatrix} 1.5 \\ 0.109 \\ -1.537 \\ 0.303 \\ 0.043 \end{bmatrix}$ so $f^*(t) \approx 1.5 + 0.109\sin(t) - 1.537\cos(t) + 0.303\sin(2t) + 0.043\cos(2t)$.

35. a. The ijth entry of $A_n^T A_n$ is the dot product of the ith row of A_n^T and the jth column of A_n, i.e.

$$A_n^T A_n = \begin{bmatrix} n & \sum_{i=1}^n \sin a_i & \sum_{i=1}^n \cos a_i \\ \sum_{i=1}^n \sin a_i & \sum_{i=1}^n \sin^2 a_i & \sum_{i=1}^n \sin a_i \cos a_i \\ \sum_{i=1}^n \cos a_i & \sum_{i=1}^n \sin a_i \cos a_i & \sum_{i=1}^n \cos^2 a_i \end{bmatrix} \text{ and } A_n^T \vec{b} = \begin{bmatrix} \sum_{i=1}^n g(a_i) \\ \sum_{i=1}^n g(a_i)\sin a_i \\ \sum_{i=1}^n g(a_i)\cos a_i \end{bmatrix}.$$

b. $$\lim_{n\to\infty} \frac{2\pi}{n} A_n^T A_n = \begin{bmatrix} 2\pi & \int_0^{2\pi} \sin t\,dt & \int_0^{2\pi} \cos t\,dt \\ \int_0^{2\pi} \sin t\,dt & \int_0^{2\pi} \sin^2 t\,dt & \int_0^{2\pi} \sin t \cos t\,dt \\ \int_0^{2\pi} \cos t\,dt & \int_0^{2\pi} \sin t \cos t\,dt & \int_0^{2\pi} \cos^2 t\,dt \end{bmatrix} = \begin{bmatrix} 2\pi & 0 & 0 \\ 0 & \pi & 0 \\ 0 & 0 & \pi \end{bmatrix}$$

and $$\lim_{n\to\infty} \frac{2\pi}{n} A_n^T \vec{b} = \begin{bmatrix} \int_0^{2\pi} g(t)dt \\ \int_0^{2\pi} g(t)\sin t\,dt \\ \int_0^{2\pi} g(t)\cos t\,dt \end{bmatrix}$$

(Here $\frac{2\pi}{n} = \Delta t$ so $\lim_{n\to\infty} \frac{2\pi}{n} \sum_{i=1}^n \cos(t_i) = \lim_{n\to\infty} \sum_{i=1}^n \cos(t_i)\Delta t = \int_0^{2\pi} \cos t\,dt$ for instance. All other limits are obtained similarly.)

c. $$\begin{bmatrix} c \\ p \\ q \end{bmatrix} = \lim_{n\to\infty} \begin{bmatrix} c_n \\ p_n \\ q_n \end{bmatrix} = \begin{bmatrix} 2\pi & 0 & 0 \\ 0 & \pi & 0 \\ 0 & 0 & \pi \end{bmatrix}^{-1} \begin{bmatrix} \int_0^{2\pi} g(t)\,dt \\ \int_0^{2\pi} g(t)\sin t\,dt \\ \int_0^{2\pi} g(t)\cos t\,dt \end{bmatrix} = \begin{bmatrix} \frac{1}{2\pi}\int_0^{2\pi} g(t)\,dt \\ \frac{1}{\pi}\int_0^{2\pi} g(t)\sin t\,dt \\ \frac{1}{\pi}\int_0^{2\pi} g(t)\cos t\,dt \end{bmatrix}$$

and $f(t) = c + p\sin t + q\cos t$, where c, p, q are given above.

36. We want $\begin{bmatrix} a \\ b \\ c \end{bmatrix}$ such that

$$a+b\sin\left(\frac{2\pi}{365}28\right)+c\cos\left(\frac{2\pi}{365}28\right)=10$$
$$a+b\sin\left(\frac{2\pi}{365}77\right)+c\cos\left(\frac{2\pi}{365}77\right)=12$$
$$a+b\sin\left(\frac{2\pi}{365}124\right)+c\cos\left(\frac{2\pi}{365}124\right)=14$$
$$a+b\sin\left(\frac{2\pi}{365}168\right)+c\cos\left(\frac{2\pi}{365}168\right)=15.$$

Using $A=\begin{bmatrix} 1 & \sin\left(\frac{2\pi}{365}28\right) & \cos\left(\frac{2\pi}{365}28\right) \\ 1 & \sin\left(\frac{2\pi}{365}77\right) & \cos\left(\frac{2\pi}{365}77\right) \\ 1 & \sin\left(\frac{2\pi}{365}124\right) & \cos\left(\frac{2\pi}{365}124\right) \\ 1 & \sin\left(\frac{2\pi}{365}168\right) & \cos\left(\frac{2\pi}{365}168\right) \end{bmatrix}$ and $\vec{b}=\begin{bmatrix} 10 \\ 12 \\ 14 \\ 15 \end{bmatrix}$ we compute $\begin{bmatrix} a \\ b \\ c \end{bmatrix}^* = (A^TA)^{-1}A^T\vec{b} \approx \begin{bmatrix} 12.25 \\ 0.394 \\ -2.726 \end{bmatrix}$ and

$$f^*(t)\approx 12.25+0.394\sin\left(\frac{2\pi}{365}t\right)-2.726\cos\left(\frac{2\pi}{365}t\right).$$

37. a. We want c_0, c_1 such that

$$\begin{aligned} c_0+c_1(35)&=\log 35 \\ c_0+c_1(46)&=\log 46 \\ c_0+c_1(59)&=\log 77 \\ c_0+c_1(69)&=\log 133 \end{aligned} \quad \text{or} \quad \underbrace{\begin{bmatrix} 1 & 35 \\ 1 & 46 \\ 1 & 59 \\ 1 & 69 \end{bmatrix}}_{A} \begin{bmatrix} c_0 \\ c_1 \end{bmatrix} = \underbrace{\begin{bmatrix} \log 35 \\ \log 46 \\ \log 77 \\ \log 133 \end{bmatrix}}_{\vec{b}}$$

so $\begin{bmatrix} c_0 \\ c_1 \end{bmatrix}^* = (A^TA)^{-1}A^T\vec{b} \approx \begin{bmatrix} 0.915 \\ 0.017 \end{bmatrix}$ so $\log(d)\approx 0.915+0.017t$.

b. $d\approx 10^{0.915}\cdot 10^{0.017t}\approx 8.22\cdot 10^{0.017t}$

c. If $t = 88$ then $d \approx 258$. Since the Airbus has only 93 displays, new technologies must have rendered the old trends obsolete.

38. We want $\begin{bmatrix} c_0 \\ c_1 \\ c_2 \end{bmatrix}$ such that

$$\begin{aligned} 110&=c_0+2c_1+c_2 \\ 180&=c_0+12c_1+0c_2 \\ 120&=c_0+5c_1+c_2 \\ 160&=c_0+11c_1+c_2 \\ 160&=c_0+6c_1+0c_2 \end{aligned} \quad \text{or} \quad \begin{bmatrix} 1 & 2 & 1 \\ 1 & 12 & 0 \\ 1 & 5 & 1 \\ 1 & 11 & 1 \\ 1 & 6 & 0 \end{bmatrix} \begin{bmatrix} c_0 \\ c_1 \\ c_2 \end{bmatrix} = \begin{bmatrix} 110 \\ 180 \\ 120 \\ 160 \\ 160 \end{bmatrix}.$$

The least-squares solution is $\begin{bmatrix} c_0 \\ c_1 \\ c_2 \end{bmatrix}^* = \begin{bmatrix} 125 \\ 5 \\ -25 \end{bmatrix}$, so that $w^* = 125 + 5h - 25g$.

For a general population, we expect c_0 and c_1 to be positive, since c_0 gives the weight of a 5' male, and increased height should contribute positively to the weight. We expect c_2 to be negative, since females tend to be lighter than males of equal height.

39. **a.** We want $\begin{bmatrix} c_0 \\ c_1 \end{bmatrix}$ such that

$$\begin{aligned} \log(250) &= c_0 + c_1 \log(600{,}000) \\ \log(60) &= c_0 + c_1 \log(200{,}000) \\ \log(25) &= c_0 + c_1 \log(60{,}000) \\ \log(12) &= c_0 + c_1 \log(10{,}000) \\ \log(5) &= c_0 + c_1 \log(2500) \end{aligned}.$$

The least-squares solution to the above system is $\begin{bmatrix} c_0 \\ c_1 \end{bmatrix}^* \approx \begin{bmatrix} -1.616 \\ 0.664 \end{bmatrix}$ so $\log z \approx -1.616 + 0.664 \log g$.

b. Exponentiating both sides of the answer to a, we get $z \approx 10^{-1.616} \cdot g^{0.664} \approx 0.0242 \cdot g^{0.664}$.

c. This model is close since $\sqrt{g} = g^{0.5}$.

40. First we look for $\begin{bmatrix} c_0 \\ c_1 \end{bmatrix}$ such that $\log D = c_0 + c_1 \log a$.

Proceeding as in Exercise 39, we get $\begin{bmatrix} c_0 \\ c_1 \end{bmatrix}^* \approx \begin{bmatrix} 0 \\ 1.5 \end{bmatrix}$, i.e. $\log D \approx 1.5 \log a$, hence $D \approx 10^{1.5 \log a} = a^{1.5}$.

Note that the formula $D = a^{1.5}$ is Kepler's third law of planetary motion.

41. **a.** We want $\begin{bmatrix} c_0 \\ c_1 \end{bmatrix}$ such that $\log D = c_0 + c_1 t$ (t in years since 1970), i.e.

$$\begin{aligned} \log 370 &= c_0 + c_1(0) \\ \log 533 &= c_0 + c_1(5) \\ \log 908 &= c_0 + c_1(10) \\ \log 1{,}823 &= c_0 + c_1(15) \\ \log 3{,}233 &= c_0 + c_1(20) \\ \log 4{,}871 &= c_0 + c_1(25) \end{aligned}.$$

The least-squares solution to the system is $\begin{bmatrix} c_0 \\ c_1 \end{bmatrix}^* \approx \begin{bmatrix} 2.53 \\ 0.047 \end{bmatrix}$, i.e. $\log D \approx 2.53 + 0.047t$

or $D \approx 10^{2.53} \cdot 10^{0.047t} \approx 339 \cdot 10^{0.047t}$.

b. In 2020, $t = 50$ and $D \approx 76{,}000$.
The formula predicts a debt of about 76 trillion dollars.

Chapter 5

5.1

1. $\det\begin{bmatrix}1 & 2\\ 3 & 4\end{bmatrix} = -2$

2. $\det\begin{bmatrix}7 & 11\\ 7 & 11\end{bmatrix} = 0$

3. $\det\begin{bmatrix}1 & 2 & 3\\ 0 & 2 & 3\\ 0 & 0 & 3\end{bmatrix} = 6$, by Fact 5.1.3.

4. $\det\begin{bmatrix}2 & 0 & 0\\ 3 & 2 & 0\\ 4 & 3 & 2\end{bmatrix} = 8$, by Fact 5.1.3.

5. $\det\begin{bmatrix}0 & 0 & 2\\ 0 & 3 & 0\\ 4 & 0 & 0\end{bmatrix} = -24$

6. $\det\begin{bmatrix}4 & 3 & 2\\ 4 & 3 & 0\\ 4 & 0 & 0\end{bmatrix} = -24$

7. $\det\begin{bmatrix}1 & 2 & 3\\ 4 & 5 & 6\\ 7 & 8 & 9\end{bmatrix} = 0$

8. $\det\begin{bmatrix}1 & 2 & 3\\ 4 & 5 & 6\\ 7 & 8 & 7\end{bmatrix} = 6$

9. $\det\begin{bmatrix}7 & 8 & 9\\ 0 & 0 & 0\\ 5 & 6 & 7\end{bmatrix} = 0$

10. $\det\begin{bmatrix}7 & 8 & 9\\ 7 & 8 & 9\\ 4 & 5 & 6\end{bmatrix} = 0$

11. $\det\begin{bmatrix}1 & 1 & 1\\ 1 & 1 & 1\\ 1 & 1 & 1\end{bmatrix} = 0$

12. $\det\begin{bmatrix} a & b & c \\ a & b & c \\ d & e & f \end{bmatrix} = 0$

13. $\det\begin{bmatrix} 2 & 3 & 4 & 5 \\ 0 & 6 & 7 & 8 \\ 0 & 0 & 3 & 2 \\ 0 & 0 & 0 & 1 \end{bmatrix} = 36,$ by Fact 5.1.3.

14. $\det\begin{bmatrix} 0 & 0 & 0 & 1 \\ 0 & 0 & 1 & 0 \\ 0 & 1 & 0 & 0 \\ 1 & 0 & 0 & 0 \end{bmatrix} = 1$

15. $\det\begin{bmatrix} 0 & 2 & 3 & 4 \\ 1 & 2 & 3 & 4 \\ 0 & 0 & 0 & 4 \\ 0 & 0 & 3 & 4 \end{bmatrix} = 24$

16. $\det\begin{bmatrix} 0 & 0 & 2 & 3 & 1 \\ 0 & 0 & 0 & 2 & 2 \\ 0 & 9 & 7 & 9 & 3 \\ 0 & 0 & 0 & 0 & 5 \\ 3 & 4 & 5 & 8 & 5 \end{bmatrix} = 540$

17. $\det\begin{bmatrix} 0 & 0 & 1 & 0 & 0 \\ 0 & 0 & 0 & 1 & 0 \\ 1 & 0 & 0 & 0 & 0 \\ 0 & 0 & 0 & 0 & 1 \\ 0 & 1 & 0 & 0 & 0 \end{bmatrix} = -1$

18. $\det\begin{bmatrix} a & b & c & d & e \\ 0 & 0 & 0 & 0 & 0 \\ f & g & h & i & j \\ k & l & m & n & p \\ q & r & s & t & u \end{bmatrix} = 0$

19. $\det\begin{bmatrix} a & b & c \\ 0 & p & q \\ 0 & r & s \end{bmatrix} = aps - aqr$

20. $\det\begin{bmatrix}1&1&1&1&1&1\\1&1&1&1&1&1\\1&1&1&1&1&1\\1&1&1&1&1&1\\1&1&1&1&1&1\\1&1&1&1&1&1\end{bmatrix}=0$

21. Invertible since its determinant is 2 (i.e. $\neq 0$).

22. Not invertible since its determinant is 0.

23. Invertible if $a \neq 0$, $b \neq 0$, $c \neq 0$ since its determinant is abc.

24. Not invertible since its determinant is 0.

25. Invertible since its determinant is 6 (i.e. $\neq 0$).

26. We will compute $\det(\lambda I_2 - A) = \det\begin{pmatrix}\lambda-1 & -3\\ 0 & \lambda-3\end{pmatrix} = (\lambda-1)(\lambda-3)$, so the matrix is not invertible when $(\lambda-1)(\lambda-3)=0$ i.e. when $\lambda=1$ or $\lambda=3$.

27. We will compute $\det(\lambda I_2 - A) = \det\begin{pmatrix}\lambda-4 & -2\\ -2 & \lambda-7\end{pmatrix} = \lambda^2-11\lambda+24=(\lambda-3)(\lambda-8)$, so the matrix is not invertible when $(\lambda-3)(\lambda-8)=0$ i.e. when $\lambda=3$ or $\lambda=8$.

28. We will compute $\det(\lambda I_2 - A) = \det\begin{pmatrix}\lambda-2 & 3\\ -3 & \lambda-2\end{pmatrix} = \lambda^2+4\lambda+13$. Since $\lambda^2-4\lambda+13=0$ has no real solutions, the matrix is invertible for all real values of λ.

29. We will compute $\det(\lambda I_2 - A) = \det\begin{pmatrix}\lambda-3 & -1\\ 4 & \lambda+1\end{pmatrix} = (\lambda+1)^2$, so the matrix is not invertible when $(\lambda-1)^2=0$, i.e. when $\lambda=1$.

30. We will compute $\det(\lambda I_3 - A)=(\lambda-1)(\lambda-2)(\lambda-3)$. The matrix is not invertible when $(\lambda-1)(\lambda-2)(\lambda-3)=0$, i.e. when $\lambda=1$ or $\lambda=2$ or $\lambda=3$.

31. We will compute $\det(\lambda I_3 - A) = \det\begin{bmatrix}\lambda & 0 & -1\\ 0 & \lambda & -1\\ -1 & -1 & \lambda-1\end{bmatrix} = \lambda(\lambda-2)(\lambda+1)$, so the matrix is not invertible when $\lambda(\lambda-2)(\lambda+1)=0$, i.e. when $\lambda=0$, $\lambda=2$, or $\lambda=-1$.

32. We will compute $\det(\lambda I_3 - A) = \det\begin{bmatrix}\lambda & -1 & 0\\ 0 & \lambda & -1\\ 0 & -8 & \lambda+2\end{bmatrix} = \lambda(\lambda+4)(\lambda-2)$, so the matrix is not invertible when $\lambda(\lambda+4)(\lambda-2)=0$, i.e. when $\lambda=0$, $\lambda=-4$, or $\lambda=2$.

33. Its determinant is zero, since each pattern will contain an entry from the row or column of zeros. Hence, each product associated with a pattern will be zero, and the determinant will be zero as well.

34. The possible values are ± 1 because there will be only one pattern which does not contain a zero, with entries 1; the sign will be determined by the number of inversions in the only nonzero pattern.

35. $\det(A) = 1$ since there is only one pattern which does not contain a 0, and since the number of inversions in that pattern is $1+2+3+\cdots+99 = \frac{99\cdot 100}{2} = 99\cdot 50$, which is even.

36. We need to consider only the pattern containing all the ones. As in Exercise 35, the number of inversions in this pattern is $1+2+\cdots+(n-2)+(n-1) = \frac{(n-1)n}{2} = S$. If n is divisible by 4 then $\frac{n}{2}$ is even, so that $S = (n-1)\frac{n}{2}$ is even as well. If dividing n by 4 leaves a remainder of 1 (that is $n = 5, 9, 13, \ldots$), then $\frac{n-1}{2}$ is even, and $S = \frac{n-1}{2}n$ is even as well.

If dividing n by 4 leaves a remainder of 2, then $n-1$ and $\frac{n}{2}$ are both odd, so that S is odd as well.

Finally, if the remainder is 3, then $\frac{n-1}{2}$, n, and S are all odd.

In summary:

$$\det(A) = \begin{cases} 1 \text{ if dividing } n \text{ by 4 leaves a remainder of 0 or 1, that is, } n = 4, 5, 8, 9, 12, 13, \ldots \\ -1 \text{ if dividing } n \text{ by 4 leaves a remainder of 2 or 3, that is, } n = 2, 3, 6, 7, 10, 11, \ldots \end{cases}$$

37. Let $A = \begin{bmatrix} a_1 & a_2 \\ a_3 & a_4 \end{bmatrix}$ and $C = \begin{bmatrix} c_1 & c_2 \\ c_3 & c_4 \end{bmatrix}$.

Then $\det(M) = a_1a_4c_1c_4 - a_2a_3c_1c_4 - a_1a_4c_2c_3 + a_2a_3c_2c_3 = \det(A)\det(C)$.

38. Not necessarily true; as a counter-example consider $A = \begin{bmatrix} 1 & 0 \\ 0 & 0 \end{bmatrix}$, $B = \begin{bmatrix} 0 & 0 \\ 1 & 0 \end{bmatrix}$, $C = \begin{bmatrix} 0 & 1 \\ 0 & 0 \end{bmatrix}$, $D = \begin{bmatrix} 0 & 0 \\ 0 & 1 \end{bmatrix}$

so $\det(A) = \det(B) = \det(C) = \det(D) = 0$ hence $\det(A)\det(D) - \det(B)\det(C) = 0$ but $\det(M) = -1$.

39. Pick any pattern other than the diagonal pattern. Let a_{ii} be the left-most diagonal entry of the matrix such that a_{ii} is *not* in the chosen pattern. Then the chosen pattern must have an entry in row i to the right of a_{ii} and in column i below a_{ii}. This is the case since the rows above a_{ii} and the columns to the left of a_{ii} were already represented in the pattern by diagonal entries.

40. $\det(A) = 2(\cos^2\alpha + \sin^2\alpha) = 2$ so A is invertible for all values of α.

41. $\det(A) = -x(x-1)^2$ so A is invertible when $x \neq 0$ and $x \neq 1$.

In Exercises 42 to 45, let $A = \begin{bmatrix} a_1 & a_2 \\ a_3 & a_4 \end{bmatrix}$.

42. $\det B = \det\begin{bmatrix} 3a_1 & 3a_2 \\ a_3 & a_4 \end{bmatrix} = 3a_1a_4 - 3a_2a_3 = 3\det A$ so $\det(B) = 3\det(A) = 3k$.

43. $\det C = \det\begin{bmatrix} a_3 & a_4 \\ a_1 & a_2 \end{bmatrix} = a_2a_3 - a_1a_4 = -\det(A)$ so $\det(C) = -\det(A) = -k$.

44. $\det(D)=\det\begin{bmatrix} a_1+6a_3 & a_2+6a_4 \\ a_3 & a_4 \end{bmatrix}=a_1a_4+6a_3a_4-a_2a_3-6a_3a_4=a_1a_4-a_2a_3=\det(A)$
so $\det(D)=\det(A)=k$.

45. $\det(A^T)=\det\begin{bmatrix} a_1 & a_3 \\ a_2 & a_4 \end{bmatrix}=a_1a_4-a_2a_3=\det(A)$ so $\det(A^T)=\det(A)=k$.

46. The work on page 240 in the text shows that $\det[\vec{u}\quad\vec{v}\quad\vec{w}]=\vec{u}\cdot(\vec{v}\times\vec{w})$, for any 3×3 matrix $[\vec{u}\quad\vec{v}\quad\vec{w}]$.
Therefore, $\det(A)=\det[\vec{v}\times\vec{w}\quad\vec{v}\quad\vec{w}]=(\vec{v}\times\vec{w})\cdot(\vec{v}\times\vec{w})=\|\vec{v}\times\vec{w}\|^2$.
The matrix A is invertible if $\vec{v}\times\vec{w}\neq 0$, that is, if $\vec{v}$ and $\vec{w}$ are not parallel.

47. Consider a pattern in A, with entries $a_1,\ldots,a_n$; the corresponding pattern in $-A$ will have the entries $-a_1,\ldots,-a_n$. The products associated with these patterns will be $a_1a_2\cdots a_n$ and $(-1)^n a_1a_2\cdots a_n$, respectively. Since these observations apply to all patterns, we can say that $\det(-A)=(-1)^n\det(A)$.

48. $\det(kA)=k^n\det(A)$
The argument is analogous to the one in Exercise 47.

49. Let $A=\begin{bmatrix} a_1 & a_2 \\ a_3 & a_4 \end{bmatrix}$. Then if $a_1a_4-a_2a_3\neq 0$, $A^{-1}=\dfrac{1}{\det(A)}\begin{bmatrix} a_4 & -a_2 \\ -a_3 & a_1 \end{bmatrix}$.

By Exercise 48, $\det(A^{-1})=\left(\dfrac{1}{\det(A)}\right)^2(a_1a_4-a_2a_3)=\left(\dfrac{1}{\det(A)}\right)^2\cdot\det(A)$ so $\det(A^{-1})=\dfrac{1}{\det(A)}$.

50. Using the language of Exercise 2.4.61 we list the principal submatrices of A, namely

$A^{(1)}=[7]$, $A^{(2)}=\begin{bmatrix} 7 & 4 \\ 5 & 3 \end{bmatrix}$, $A^{(3)}=\begin{bmatrix} 7 & 4 & 2 \\ 5 & 3 & 1 \\ 3 & 1 & 3 \end{bmatrix}$ so $\det(A^{(1)})=7$, $\det(A^{(2)})=1$, $\det(A^{(3)})=0$.

By Exercise 2.4.61, a matrix has an LU factorization if and only if all of its principal submatrices are invertible. Since $A^{(3)}$ is not invertible, A does not have an LU factorization.

51. Only one pattern makes a nonzero contribution. The number of inversions in this pattern is n^2: for each 1 in the lower I_n, there are n inversions corresponding to the n entries of the upper I_n, so

$$\det(A)=\begin{cases} 1 \text{ if } n \text{ (and } n^2\text{) is even} \\ -1 \text{ if } n \text{ (and } n^2\text{) is odd.} \end{cases}$$

52. The pattern containing all the 1000's has 4 inversions so it contributes $(1000)^5=10^{15}$ to the value of the determinant. There are $5!-1=119$ other patterns; the product associated with each of these patterns is less than $1000^3\cdot 9^2<10^{11}$. Therefore we can say that $\det(A)>0$.

5.2

1. $\det\begin{bmatrix} 1 & 2 & 3 \\ 4 & 5 & 6 \\ 7 & 8 & 10 \end{bmatrix}=-3$

2. $\det\begin{bmatrix}1&1&0&0\\2&9&1&0\\0&9&0&0\\1&9&9&5\end{bmatrix}=-45$

3. $\det\begin{bmatrix}1&-1&2&-2\\-1&2&1&6\\2&1&14&10\\-2&6&10&33\end{bmatrix}=9$

4. $\det\begin{bmatrix}0&2&1&0&1\\0&0&2&0&2\\0&5&3&9&9\\0&7&4&0&1\\3&9&5&4&8\end{bmatrix}=-324$

5. $\det\begin{bmatrix}1&1&1&1&1\\1&2&2&2&2\\1&1&3&3&3\\1&1&1&4&4\\1&1&1&1&5\end{bmatrix}=24$

6. There are many ways to do this problem; here is one possible approach:
Subtracting the second to last row from the last, we can make the last row into $[0\ \ 0\ \ \cdots\ \ 0\ \ 1]$.
Now expanding along the last row we see that $\det(M_n)=\det(M_{n-1})$.
Since $\det(M_1)=1$ we can conclude that $\det(M_n)=1$ for all n.

7. $\det(A)=1$

8. Since $\vec{v}_2, \ldots, \vec{v}_n$ are linearly independent, $T(\vec{x})=0$ only if $\vec{x}$ is a linear combination of the $\vec{v}_i$'s, (otherwise the matrix $[\vec{x}\ \ \vec{v}_2\ \ \cdots\ \ \vec{v}_n]$ is invertible, and $T(\vec{x})\neq 0$). Hence, the kernel of T is the span of $\vec{v}_2, \ldots, \vec{v}_n$, an $(n-1)$-dimensional subspace of $\mathbb{R}^n$. The image of T is the real line $\mathbb{R}$ (since it must be 1-dimensional).

9. By Fact 5.2.4a, the desired determinant is $(-9)(8)=-72$.

10. By Fact 5.2.4b, the desired determinant is -8.

11. By Fact 5.2.4b, applied twice, since there are two row swaps, the desired determinant is $(-1)(-1)(8)=8$.

12. By Fact 5.2.4c, the desired determinant is 8.

13. By Fact 5.2.4c, the desired determinant is 8.

14. This determinant is 0, since the first row is twice the last.

15. If a square matrix A has two equal columns then its columns are linearly dependent, hence A is not invertible, and $\det(A)=0$.

16. a. $f(t)=\det\begin{bmatrix}1&1&1\\t&a&b\\t^2&a^2&b^2\end{bmatrix}=(ab^2-a^2b)+(a^2-b^2)t+(b-a)t^2$ so $f(t)$ is a quadratic function of t. The coefficient of t^2 is $(b-a)$.

b. In the cases $t=a$ and $t=b$ the matrix has two identical columns; compare with Exercise 15. It follows that $f(t)=k(t-a)(t-b)$ with $k=$ coefficient of $t^2=(b-a)$.

c. The matrix is invertible for the values of t for which $f(t)\neq 0$, i.e. for $t\neq a, t\neq b$.

17. a. If $n=1$, $A=\begin{bmatrix}1&1\\a_0&a_1\end{bmatrix}$ so $\det(A)=a_1-a_0$ (and the product formula holds).

b. Expanding the given determinant about the right-most column, we see that the coefficient k of t^n is the $n-1$ Vandermonde determinant which we assume is $\prod_{n-1\geq i>j}(a_i-a_j)$.

$f(a_0)=f(a_1)=\cdots=f(a_{n-1})=0$ since in each case the given matrix has two identical columns, hence its determinant equals zero (see Exercise 15).

Therefore $f(t)=\left(\prod_{n-1\geq i>j}(a_i-a_j)\right)(t-a_0)(t-a_1)\cdots(t-a_{n-1})$ and $\det(A)=f(a_n)=\prod_{n\geq i>j}(a_i-a_j)$, as required.

18. By Exercise 17, we need to compute $\prod_{i>j}(a_i-a_j)$ where $a_0=1$, $a_1=2$, $a_2=3$, $a_3=4$, $a_4=5$ so

$\prod_{i>j}(a_i-a_j)=(2-1)(3-1)(3-2)(4-1)(4-2)(4-3)(5-1)(5-2)(5-3)(5-4)=288.$

19. Think of the ith column of the given matrix as $a_i\begin{bmatrix}1\\a_i\\a_i^2\\\vdots\\a_i^{n-1}\end{bmatrix}$ so by Fact 5.2.3b, the determinant can be written as

$$(a_1a_2\cdots a_n)\det\begin{bmatrix}1&1&\cdots&1\\a_1&a_2&\cdots&a_n\\a_1^2&a_2^2&\cdots&a_n^2\\\vdots&\vdots&&\vdots\\a_1^{n-1}&a_2^{n-1}&\cdots&a_n^{n-1}\end{bmatrix}.$$

The new determinant is a Vandermonde determinant (see Exercise 17), and we get $\prod_{i=1}^{n}a_i\prod_{i>j}(a_i-a_j)$.

20. a. Since A is invertible, $\text{rref}(A) = I_n$ so $\text{rref}[A \mid AB] = [I_n \mid M]$, for some matrix M. Using Fact 2.4.12, we compute $[A \mid AB]\begin{bmatrix} B\vec{e}_i \\ -\vec{e}_i \end{bmatrix} = AB\vec{e}_i - AB\vec{e}_i = \vec{0}$, so $\begin{bmatrix} B\vec{e}_i \\ -\vec{e}_i \end{bmatrix}$ is in the kernel of $[A \mid AB]$ for all i, hence in the kernel of $[I_n \mid M]$. In other words, $[I_n \mid M]\begin{bmatrix} B\vec{e}_i \\ -\vec{e}_i \end{bmatrix} = B\vec{e}_i - M\vec{e}_i = \vec{0}$, i.e. $B\vec{e}_i = M\vec{e}_i$ for all i. Thus B and M have identical columns, i.e. $B = M$, and $\text{rref}[A \mid AB] = [I_n \mid B]$ as claimed.

b. If $B = A^{-1}$ we get $\text{rref}[A \mid I_n] = [I_n \mid A^{-1}]$ which tells us how to compute A^{-1} (see Fact 2.3.5).

21. $\begin{bmatrix} x_1 \\ x_2 \end{bmatrix}$ must satisfy $\det\begin{bmatrix} 1 & 1 & 1 \\ x_1 & a_1 & b_1 \\ x_2 & a_2 & b_2 \end{bmatrix} = 0$, i.e. must satisfy the linear equation

$(a_1b_2 - a_2b_1) - x_1(b_2 - a_2) + x_2(b_1 - a_1) = 0.$

We can see that $\begin{bmatrix} x_1 \\ x_2 \end{bmatrix} = \begin{bmatrix} a_1 \\ a_2 \end{bmatrix}$ and $\begin{bmatrix} x_1 \\ x_2 \end{bmatrix} = \begin{bmatrix} b_1 \\ b_2 \end{bmatrix}$ satisfy this equation, since the matrix has two identical columns in these cases.

22. Expanding about the first column we see that the equation has the form $A - Bx_1 + Cx_2 - D(x_1^2 + x_2^2) = 0$.
If $D \neq 0$ this equation defines a circle; otherwise it is a line. From Exercise 21 we know that $D = 0$ if and only if the three given points $\begin{bmatrix} a_1 \\ a_2 \end{bmatrix}, \begin{bmatrix} b_1 \\ b_2 \end{bmatrix}, \begin{bmatrix} c_1 \\ c_2 \end{bmatrix}$ are collinear. Note that the circle or line runs through the three given points.

23. Applying Fact 5.2.7 to the equation $AA^{-1} = I_n$ we see that $\det(A)\det(A^{-1}) = 1$.
The only way the product of the two integers $\det(A)$ and $\det(A^{-1})$ can be 1 is that they are both 1 or both -1. Therefore, $\det(A) = 1$ or $\det(A) = -1$.

24. $\det(A^TA) = \det(A^T)\det(A) = [\det(A)]^2 = 9$

↑ Fact 5.2.7 (first equality) ↑ Fact 5.2.1 (second equality)

25. $\det(A^TA) = \det(A^T)\det(A) = [\det(A)]^2 > 0$

↑ Fact 5.2.7 (first equality) ↑ Fact 5.2.1 (second equality)

26. By Exercise 25, $\det(A^TA) = [\det(A)]^2$. Since A is orthogonal, $A^TA = I_n$ so that
$1 = \det(I_n) = \det(A^TA) = [\det(A)]^2$ and $\det(A) = \pm 1$.

27. a. $\begin{bmatrix} 0 & a \\ -a & 0 \end{bmatrix}$, $a \neq 0$

b. The diagonal entries a_{ii} must satisfy $a_{ii} = -a_{ii}$ since they remain on the diagonal under transposition. Hence, $a_{ii} = 0$ for all $1 \le i \le n$.

c. $\det(A) = \det(A^T) = \det(-A) = (-1)^n \det(A)$

↑ Fact 5.2.1 (first equality) ↑ Since A skew symmetric (second equality) ↑ By Fact 5.2.4 (third equality)

Since n is odd, $(-1)^n = -1$ so $\det(A) = -\det(A)$, hence $\det(A) = 0$ and A is not invertible.

28. $\det(A^TA) = \det((QR)^T QR) = \det(R^TQ^TQR) = \det(R^TI_nR) = \det(R^TR) = \det(R^T)\det(R)$

↑ Definition of A ↑ Since columns of Q orthonormal ↑ Fact 5.2.7

$$= [\det(R)]^2 = \left(\prod_{i=1}^{n} r_{ii}\right)^2 > 0.$$

↑ Fact 5.2.1 ↑ Since R is triangular.

29. $\det(A^TA) = \det\left(\begin{bmatrix}\vec{v}^T\\ \vec{w}^T\end{bmatrix}[\vec{v}\ \ \vec{w}]\right) = \det\begin{bmatrix}\vec{v}\cdot\vec{v} & \vec{v}\cdot\vec{w}\\ \vec{v}\cdot\vec{w} & \vec{w}\cdot\vec{w}\end{bmatrix} = \det\begin{bmatrix}\|\vec{v}\|^2 & \vec{v}\cdot\vec{w}\\ \vec{v}\cdot\vec{w} & \|\vec{w}\|^2\end{bmatrix} = \|\vec{v}\|^2\|\vec{w}\|^2 - (\vec{v}\cdot\vec{w})^2 \geq 0$

by the Cauchy-Schwarz inequality (Fact 4.1.10).

30. a. We claim that $\vec{v}_2 \times \vec{v}_3 \times \cdots \times \vec{v}_n \neq \vec{0}$ if and only if the vectors $\vec{v}_2, \ldots, \vec{v}_n$ are linearly independent. If the vectors $\vec{v}_2, \ldots, \vec{v}_n$ are linearly independent, then we can find a basis $\vec{x}, \vec{v}_2, \ldots, \vec{v}_n$ of $\mathbb{R}^n$ (any vector $\vec{x}$ that is not in span$(\vec{v}_2, \ldots, \vec{v}_n)$ will do). Then $\vec{x}\cdot(\vec{v}_2\ x \ldots x\,\vec{v}_n) = \det[\vec{x}\,\vec{v}_2\ \cdots\ \vec{v}_n] \neq 0$, so that $\vec{v}_2\ x\ldots x\ \vec{v}_n \neq \vec{0}$. Conversely, suppose that $\vec{v}_2 \times \vec{v}_3 \times \cdots \times \vec{v}_n \neq \vec{0}$; say the ith component of this vector is nonzero. Then $0 \neq \vec{e}_i \cdot (\vec{v}_2 \times \cdots \times \vec{v}_n) = \det[\vec{e}_i\ \ \vec{v}_2\ \ \cdots\ \ \vec{v}_n]$, so that the vectors $\vec{v}_2, \ldots, \vec{v}_n$ are linearly independent (being columns of an invertible matrix).

b. ith component of $\vec{e}_2 \times \vec{e}_3 \times \cdots \times \vec{e}_n = \det\begin{bmatrix}1 & 1 & & 1\\ \vec{e}_i & \vec{e}_2 & \cdots & \vec{e}_n\\ 1 & 1 & & 1\end{bmatrix} = \begin{cases}1 \text{ if } i=1\\ 0 \text{ if } i>1\end{cases}$

so $\vec{e}_2 \times \vec{e}_3 \times \cdots \times \vec{e}_n = \vec{e}_1$.

c. $\vec{v}_i \cdot (\vec{v}_2 \times \vec{v}_3 \times \cdots \times \vec{v}_n) = \det\begin{bmatrix}1 & 1 & 1 & & 1\\ \vec{v}_i & \vec{v}_2 & \vec{v}_3 & \cdots & \vec{v}_n\\ 1 & 1 & 1 & & 1\end{bmatrix} = 0$

for any $2 \leq i \leq n$ since the above matrix has two identical columns.

d. Compare the ith components of the two vectors:

$\det\begin{bmatrix}1 & 1 & 1 & & 1\\ \vec{e}_i & \vec{v}_2 & \vec{v}_3 & \cdots & \vec{v}_n\\ 1 & 1 & 1 & & 1\end{bmatrix}$ and $\det\begin{bmatrix}1 & 1 & 1 & & 1\\ \vec{e}_i & \vec{v}_3 & \vec{v}_2 & \cdots & \vec{v}_n\\ 1 & 1 & 1 & & 1\end{bmatrix}$.

The two determinants differ by a factor of -1 by Fact 5.2.4b, so that $\vec{v}_2 \times \vec{v}_3 \times \cdots \times \vec{v}_n = -\vec{v}_3 \times \vec{v}_2 \times \cdots \times \vec{v}_n$.

e. $\det[\vec{v}_2 \times \vec{v}_3 \times \cdots \times \vec{v}_n\ \ \vec{v}_2\ \ \vec{v}_3\ \ \cdots\ \ \vec{v}_n] = (\vec{v}_2 \times \vec{v}_3 \times \cdots \times \vec{v}_n)\cdot(\vec{v}_2 \times \vec{v}_3 \times \cdots \times \vec{v}_n) = \|\vec{v}_2 \times \cdots \times \vec{v}_n\|^2$

f. On page 240 of the text we saw that the "old" cross product satisfies the defining equation of the "new" cross product: $\vec{x}\cdot(\vec{v}_2 \times \vec{v}_3) = \det[\vec{x}\ \ \vec{v}_2\ \ \vec{v}_3]$.

31. $f(x)$ is a linear function so $f'(x)$ is the coefficient of x (the slope). But, by Fact 5.2.10, the coefficient of x is given by $-\det\begin{bmatrix}1&2&3&4\\0&2&3&4\\0&0&3&4\\0&0&0&4\end{bmatrix}=-24$ so $f'(x)=-24$.

32. **a.** $\det\begin{bmatrix}a&3&d\\b&3&e\\c&3&f\end{bmatrix}=3\det\begin{bmatrix}a&1&d\\b&1&e\\c&1&f\end{bmatrix}=21$

$\uparrow$ Facts 5.2.1 and 5.2.4a

b. $\det\begin{bmatrix}a&3&d\\b&5&e\\c&7&f\end{bmatrix}=\det\begin{bmatrix}a&2(1)+1&d\\b&2(2)+1&e\\c&2(3)+1&f\end{bmatrix}=\det\begin{bmatrix}a&2(1)&d\\b&2(2)&e\\c&2(3)&f\end{bmatrix}+\det\begin{bmatrix}a&1&d\\b&1&e\\c&1&f\end{bmatrix}$

$\uparrow$ Fact 5.2.3

$=2\det\begin{bmatrix}a&1&d\\b&2&e\\c&3&f\end{bmatrix}+\det\begin{bmatrix}a&1&d\\b&1&e\\c&1&f\end{bmatrix}=2\cdot 11+7=29$

$\uparrow$ Fact 5.2.4a

33. Yes! For example $T\begin{bmatrix}x&b\\y&d\end{bmatrix}=dx+by$ is given by the matrix $[d\ \ b]$, so that T is linear in the first column.

34. **a.** $D_1=2,\ D_2=3,\ D_3=4,\ D_4=5$

b. $D_n=2D_{n-1}-D_{n-2}$ (expand along the first column to see this)

c. $D_n=n+1$ (by induction)

35. $\det(Q_1)=\det(Q_2)=\det(Q_3)=1$
$\det(Q_n)=2\det(Q_{n-1})-\det(Q_{n-2})$ (expand along the first column), so that $\det(Q_n)=1$ for all n.

36. As suggested in the hint, let k be the number of pattern entries which are above and to the left of a_{ij}. Since there is one pattern entry in each of the $i-1$ rows above a_{ij}, there will be $i-1-k$ pattern entries above and to the right of a_{ij} (see the diagram below). Similarly, there will be $j-1-k$ pattern entries to the left and below a_{ij}. Therefore, there are $(i-1-k)+(j-1-k)=i+j-2-2k$ inversions involving a_{ij}. This number is even if $i+j$ is even and odd if $i+j$ is odd, as claimed.

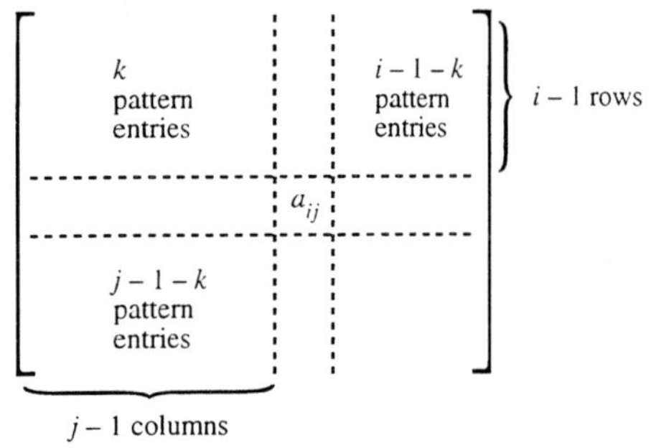

37. Expand along the first column, realizing that all but the first contribution is zero, since the other minors will have two equal rows. Therefore, $\det(P_n) = \det(P_{n-1})$.
Since $\det(P_1) = 1$ we can conclude that $\det(P_n) = 1$, for all n.

38. In Exercise 37 we have an example with $n - 1$ zeros. We will show that this is the smallest possible number: if a zero-one matrix of size $n \times n$ contains less than $n - 1$ zeros, then this matrix will have at least two rows without any zeros. These two rows will contain only ones, and will therefore be identical; such a matrix is not invertible.

39. a. See Exercise 23.

b. If $A = \begin{bmatrix} a & b \\ c & d \end{bmatrix}$, then $A^{-1} = \frac{1}{\det(A)}\begin{bmatrix} d & -b \\ -c & a \end{bmatrix}$ has integer entries.

40. $f(t) = (\det(A + tB))^2 - 1 = \left(\det\begin{bmatrix} a_1 + tb_1 & a_2 + tb_2 \\ a_3 + tb_3 & a_4 + tb_4 \end{bmatrix}\right)^2 - 1$ assuming $A = \begin{bmatrix} a_1 & a_2 \\ a_3 & a_4 \end{bmatrix}$, $B = \begin{bmatrix} b_1 & b_2 \\ b_3 & b_4 \end{bmatrix}$.
Then the determinant above is a polynomial of degree ≤ 2 so its square is a polynomial of degree ≤ 4. Hence $f(t)$ is a polynomial of degree ≤ 4.
Since A, $A + B$, $A + 2B$, $A + 3B$, $A + 4B$ are invertible and their inverses have integer entries, by Exercise 39a, it follows that their determinants are ± 1. Hence $f(0) = f(1) = f(2) = f(3) = f(4) = 0$.
Since f is a polynomial of degree ≤ 4 with at least 5 roots, it follows that $f(t) = 0$ for all t, in particular for $t = 5$, so $\det(A + 5B) = \pm 1$. Hence $A + 5B$ is an invertible 2×2 matrix whose inverse has integer entries by Exercise 39b.

41. Mimicking the proof of Fact 5.2.4c we can show:
If B is obtained from A by adding a multiple of a row of A to another row, then $D(B) = D(A)$. As in the proof of Algorithm 5.2.6 it follows that $D(A) = (-1)^s k_1 k_2 \ldots k_r D(\text{rref } A)$.
If A is not invertible, then $D(\text{rref } A) = 0$, by the linearity of D in the last row, so that $D(A) = \det(A) = 0$. If A is invertible, then $\text{rref}(A) = I_n$ and $D(\text{rref } A) = 1$, by property c of the function D. Therefore,
$D(A) = (-1)^s k_1 k_2 \cdots k_r = \det(A)$.
We have shown that $D(A) = \det(A)$ for all $n \times n$ matrices A, as claimed.

42. a. We show first that D is linear in the ith row.

$$\underset{A}{D\begin{bmatrix} \vec{v}_1 \\ \vdots \\ \vec{x} \\ \vdots \\ \vec{v}_n \end{bmatrix}} = \frac{1}{\det M}\det \underset{AM}{\begin{bmatrix} \vec{v}_1 M \\ \vdots \\ \vec{x}M \\ \vdots \\ \vec{v}_n M \end{bmatrix}}$$

The entries in the ith row of AM are linear combinations of the components x_i of the vector $\vec{x}$, while the other entries of AM are constants. Therefore, det(AM) is a linear combination of the x_i (think about the patterns). Since $\dfrac{1}{\det M}$ is a constant, we have $D\begin{bmatrix}\vec{v}_1\\ \vdots\\ \vec{x}\\ \vdots\\ \vec{v}_n\end{bmatrix} = c_1x_1 + c_2x_2 + \cdots + c_nx_n$ for some constants c_i, as claimed.

b. Secondly, we need to show that $D(B) = -D(A)$ if B is obtained from A by a row swap:

$$A = \begin{bmatrix}\vec{v}_1\\ \vdots\\ \vec{v}_i\\ \vdots\\ \vec{v}_j\\ \vdots\\ \vec{v}_n\end{bmatrix} \rightarrow B = \begin{bmatrix}\vec{v}_1\\ \vdots\\ \vec{v}_j\\ \vdots\\ \vec{v}_i\\ \vdots\\ \vec{v}_n\end{bmatrix}$$

$$D(B) = \frac{1}{\det M}\det\begin{bmatrix}\vec{v}_1M\\ \vdots\\ \vec{v}_jM\\ \vdots\\ \vec{v}_iM\\ \vdots\\ \vec{v}_nM\end{bmatrix} = -\frac{1}{\det M}\det\begin{bmatrix}\vec{v}_1M\\ \vdots\\ \vec{v}_iM\\ \vdots\\ \vec{v}_jM\\ \vdots\\ \vec{v}_nM\end{bmatrix} = -D(A).$$

c. The property $D(I_n) = 1$ is obvious.

It now follows from Exercise 41 that $\det(A) = D(A) = \dfrac{\det(AM)}{\det(M)}$ and therefore $\det(AM) = \det(A)\det(M)$.

43. $T\begin{bmatrix}\vec{x}\\ \vec{y}\end{bmatrix} = \vec{0}$ means that $[A_1 \quad A_2]\begin{bmatrix}\vec{x}\\ \vec{y}\end{bmatrix} = A_1\vec{x} + A_2\vec{y} = \vec{0}$ so for any $\vec{x}$ in $\mathbb{R}^n$ there is a unique $\vec{y}$ such that $T\begin{bmatrix}\vec{x}\\ \vec{y}\end{bmatrix} = \vec{0}$, namely, $\vec{y} = -A_2^{-1}A_1\vec{x}$. The transformation $\vec{x} \rightarrow \vec{y}$ is linear since it is given by $-A_2^{-1}A_1$.

5.3

1. By Fact 5.3.3 $\quad$ Area $= \left|\det\begin{bmatrix}3 & 8\\ 7 & 2\end{bmatrix}\right| = |-50| = 50.$

2. By Fact 5.3.3 $\quad$ Area $= \dfrac{1}{2}\left|\det\begin{bmatrix}3 & 8\\ 7 & 2\end{bmatrix}\right| = \dfrac{1}{2}|-50| = 25$

3. Area of triangle $= \frac{1}{2}\left|\det\begin{bmatrix} 6 & 1 \\ -2 & 4 \end{bmatrix}\right| = 13$

(See figure.)

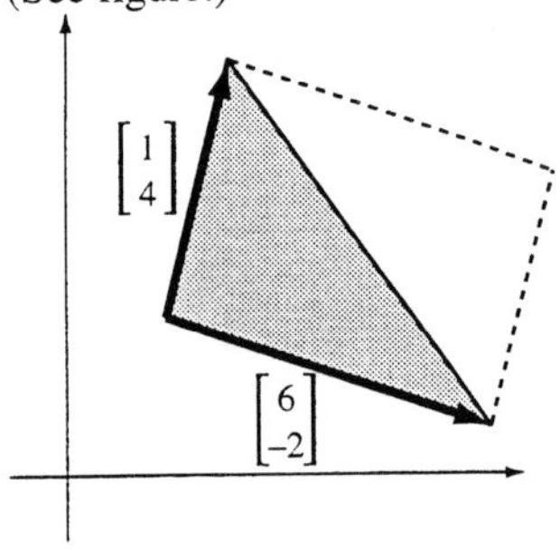

4. Note that area of triangle $= \frac{1}{2}\left|\det\begin{bmatrix} b_1 - a_1 & c_1 - a_1 \\ b_2 - a_2 & c_2 - a_2 \end{bmatrix}\right|$.

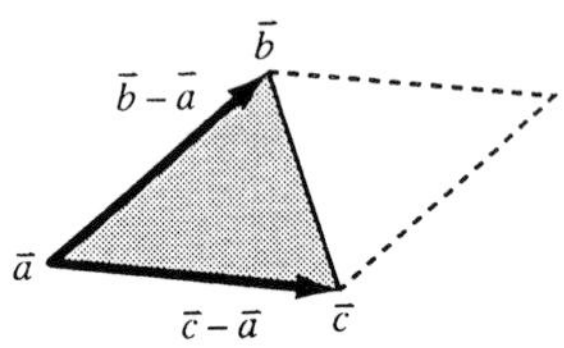

On the other hand, $\det\begin{bmatrix} a_1 & b_1 & c_1 \\ a_2 & b_2 & c_2 \\ 1 & 1 & 1 \end{bmatrix} = \det\begin{bmatrix} a_1 & b_1 - a_1 & c_1 - a_1 \\ a_2 & b_2 - a_2 & c_2 - a_2 \\ 1 & 0 & 0 \end{bmatrix} = \det\begin{bmatrix} b_1 - a_1 & c_1 - a_1 \\ b_2 - a_2 & c_2 - a_2 \end{bmatrix}$.

↑ subtract first column from second and third; ↑ expand down the first column

Therefore, area of triangle $= \frac{1}{2}\left|\det\begin{bmatrix} a_1 & b_1 & c_1 \\ a_2 & b_2 & c_2 \\ 1 & 1 & 1 \end{bmatrix}\right|$.

5. The volume of the tetrahedron T_0 defined by $\vec{e}_1, \vec{e}_2, \vec{e}_3$ is $\frac{1}{3}(\text{base})(\text{height}) = \frac{1}{6}$ (formula for the volume of a pyramid).

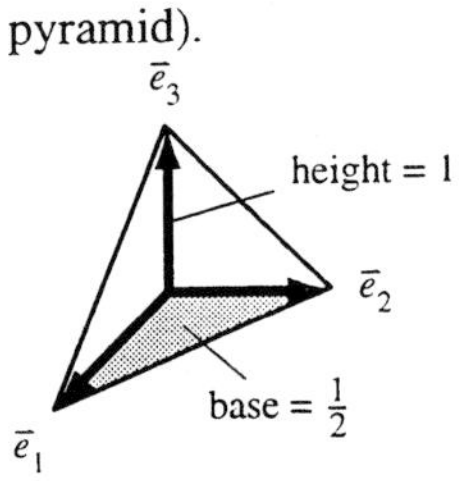

The tetrahedron T defined by $\vec{v}_1, \vec{v}_2, \vec{v}_3$ can be obtained by applying the linear transformation with matrix $[\vec{v}_1 \;\; \vec{v}_2 \;\; \vec{v}_3]$ to T_0.

Now we have $\text{vol}(T) = |\det[\vec{v}_1 \;\; \vec{v}_2 \;\; \vec{v}_3]|\text{vol}(T_0) = \frac{1}{6}|\det[\vec{v}_1 \;\; \vec{v}_2 \;\; \vec{v}_3]| = \frac{1}{6}V(\vec{v}_1, \vec{v}_2, \vec{v}_3)$.

↑ Fact 5.3.8 and page 280; ↑ Fact 5.3.5

6. From Exercise 5 we know that volume of tetrahedron $= \frac{1}{6}\left|\det\begin{bmatrix} a_1 & b_1 & c_1 \\ a_2 & b_2 & c_2 \\ 1 & 1 & 1 \end{bmatrix}\right|$, and Exercise 4 tells us that

area of triangle $= \frac{1}{2}\left|\det\begin{bmatrix} a_1 & b_1 & c_1 \\ a_2 & b_2 & c_2 \\ 1 & 1 & 1 \end{bmatrix}\right|$, so that area of tetrahedron $= \frac{1}{3}$(area of triangle).

We can see this result more directly if we think of the tetrahedron as an inverted pyramid whose base is the triangle and whose height is 1.

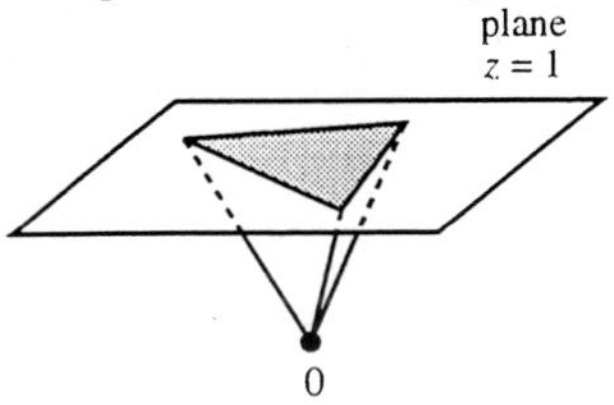

The three vertices of the shaded triangle are $\begin{bmatrix} a_1 \\ a_2 \\ 1 \end{bmatrix}, \begin{bmatrix} b_1 \\ b_2 \\ 1 \end{bmatrix}, \begin{bmatrix} c_1 \\ c_2 \\ 1 \end{bmatrix}$.

7. Area $= \frac{1}{2}\left|\det\begin{bmatrix} 10 & -2 \\ 11 & 13 \end{bmatrix}\right| + \frac{1}{2}\left|\det\begin{bmatrix} 8 & 10 \\ 2 & 11 \end{bmatrix}\right| = 110$

(See Figure.)

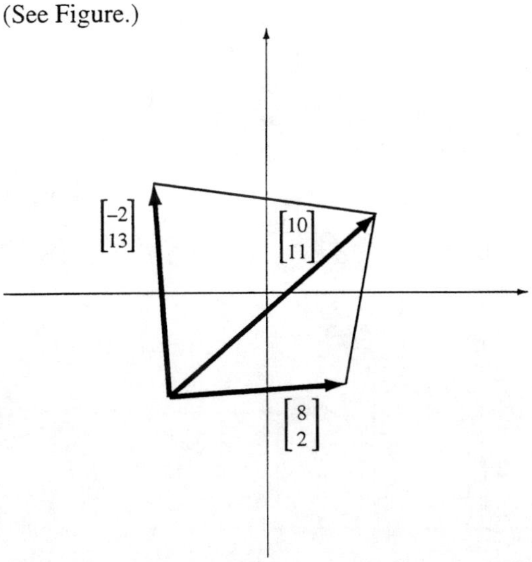

8. We need to show that both sides of the equation in Fact 5.3.4 give zero.
$|\det(A)| = 0$ since A is not invertible. On the other hand, since A is not invertible, the $\vec{v}_i$'s are linearly dependent, i.e., we can write $\vec{v}_i = c_1\vec{v}_1 + \cdots + c_{i-1}\vec{v}_{i-1}$ for some i (compare with Exercise 3.2.35). This implies that $\text{proj}_{V_{i-1}}\vec{v}_i = \vec{v}_i$ and therefore $\left\|\vec{v}_i - \text{proj}_{V_{i-1}}\vec{v}_i\right\| = 0$, so that the right hand side of the equation is 0, as claimed.

9. By Fact 5.3.3, $|\det[\vec{v}_1 \quad \vec{v}_2]| =$ area of the parallelogram defined by $\vec{v}_1$ and $\vec{v}_2$.
But $\|\vec{v}_1\|$ is the base of that parallelogram and $\|\vec{v}_2\|\sin\alpha$ is its height, hence $|\det[\vec{v}_1 \quad \vec{v}_2]| = \|\vec{v}_1\|\|\vec{v}_2\|\sin\alpha$.

10. $|\det(A)| \geq \|\vec{v}_1\|\|\vec{v}_2\|\cdots\|\vec{v}_n\|$ since $|\det(A)| = \|\vec{v}_1\|\|\vec{v}_2 - \text{proj}_{V_1}\vec{v}_2\|\cdots\|\vec{v}_n - \text{proj}_{V_{n-1}}\vec{v}_n\|$ and $\|\vec{v}_i\| \geq \|\vec{v}_i - \text{proj}_{V_{i-1}}\vec{v}_i\|$.
The equality holds if $\|\vec{v}_i\| = \|\vec{v}_i - \text{proj}_{V_{i-1}}\vec{v}_i\|$ for all i, that is, if the $\vec{v}_i$'s are mutually perpendicular.

11.

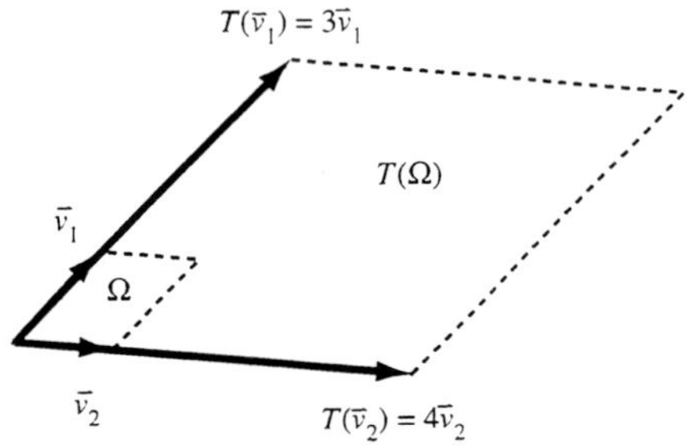

By Fact 5.3.8, $|\det A| = \dfrac{\text{area of } T(\Omega)}{\text{area of } \Omega} = 12.$

12. Denote the columns by $\vec{v}_1, \vec{v}_2, \vec{v}_3, \vec{v}_4$. From Fact 5.3.4 and Exercise 8 we know that $|\det(A)| \leq \|\vec{v}_1\|\|\vec{v}_2\|\|\vec{v}_3\|\|\vec{v}_4\|$; equality holds if the columns are orthogonal. Since the entries of the $\vec{v}_i$ are 0, 1, and −1, we have $\|\vec{v}_i\| \leq \sqrt{1+1+1+1} = 2$. Therefore, $|\det A| \leq 16$.
To build an example where $\det(A) = 16$ we want all 1's and −1's as entries, and the columns need to be orthogonal. A little experimentation produces $A = \begin{bmatrix} 1 & 1 & 1 & 1 \\ 1 & 1 & -1 & -1 \\ 1 & -1 & -1 & 1 \\ 1 & -1 & 1 & -1 \end{bmatrix}$ (there are other solutions).
Note that we need to *check* that $\det(A) = 16$ (and not −16).

13. By Fact 5.3.7, the desired 2-volume is $\sqrt{\det\left(\begin{bmatrix} 1 & 1 & 1 & 1 \\ 1 & 2 & 3 & 4 \end{bmatrix}\begin{bmatrix} 1 & 1 \\ 1 & 2 \\ 1 & 3 \\ 1 & 4 \end{bmatrix}\right)} = \sqrt{\det\begin{bmatrix} 4 & 10 \\ 10 & 30 \end{bmatrix}} = \sqrt{20}.$

14. By Fact 5.3.7, the desired 3-volume is $\sqrt{\det\left(\begin{bmatrix} 1 & 0 & 0 & 0 \\ 1 & 1 & 1 & 1 \\ 1 & 2 & 3 & 4 \end{bmatrix}\begin{bmatrix} 1 & 1 & 1 \\ 0 & 1 & 2 \\ 0 & 1 & 3 \\ 0 & 1 & 4 \end{bmatrix}\right)} = \sqrt{\det\begin{bmatrix} 1 & 1 & 1 \\ 1 & 4 & 10 \\ 1 & 10 & 30 \end{bmatrix}} = \sqrt{6}.$

15. If $\vec{v}_1, \vec{v}_2, \ldots, \vec{v}_k$ are linearly dependent and if $A = [\vec{v}_1 \ \cdots \ \vec{v}_k]$, then $\det(A^TA) = 0$ since A^TA and A have equal and nonzero kernels (by Fact 4.3.3), hence A^TA is not invertible.
On the other hand, since the $\vec{v}_i$'s are linearly dependent, $\vec{v}_i = a_1\vec{v}_1 + \cdots + a_{i-1}\vec{v}_{i-1}$ for some i.
Hence $\|\vec{v}_i - \text{proj}_{V_{i-1}}\vec{v}_i\| = 0$, so $V(\vec{v}_1, \ldots, \vec{v}_k) = 0$.
The above shows that both sides of the equation in Fact 5.3.7 are 0 if $\vec{v}_1, \ldots, \vec{v}_k$ are linearly dependent.

16. False

If T is given by $A = 2I_3$ then $|\det(A)| = 8$. But if Ω is the square defined by $\vec{e}_1, \vec{e}_2$ in $\mathbb{R}^3$ (of area 1), then $T(\Omega)$ is the square defined by $2\vec{e}_1, 2\vec{e}_2$ and the area of $T(\Omega)$ is 4.

17. a. Let $\vec{w} = \vec{v}_1 \times \vec{v}_2 \times \vec{v}_3$. Note that $\vec{w}$ is orthogonal to $\vec{v}_1, \vec{v}_2$, and $\vec{v}_3$, by Exercise 5.2.30c. Then

$$V(\vec{v}_1, \vec{v}_2, \vec{v}_3, \vec{w}) = V(\vec{v}_1, \vec{v}_2, \vec{v}_3)\|\vec{w} - \text{proj}_{V_3}\vec{w}\| = V(\vec{v}_1, \vec{v}_2, \vec{v}_3)\|w\|.$$

$\uparrow$ Definition 5.3.6 (first equality); $\uparrow$ $\text{proj}_{V_3}\vec{w} = \vec{0}$ (second equality)

b. By Exercise 5.2.30e,

$$V(\vec{v}_1, \vec{v}_2, \vec{v}_3, \vec{v}_1 \times \vec{v}_2 \times \vec{v}_3) = |\det[\vec{v}_1 \quad \vec{v}_2 \quad \vec{v}_3 \quad \vec{v}_1 \times \vec{v}_2 \times \vec{v}_3]| = |\det[\vec{v}_1 \times \vec{v}_2 \times \vec{v}_3 \quad \vec{v}_1 \quad \vec{v}_2 \quad \vec{v}_3]| = \|\vec{v}_1 \times \vec{v}_2 \times \vec{v}_3\|^2.$$

c. By parts a and b $V(\vec{v}_1, \vec{v}_2, \vec{v}_3) = \|\vec{v}_1 \times \vec{v}_2 \times \vec{v}_3\|$. If the vectors $\vec{v}_1, \vec{v}_2, \vec{v}_3$ are linearly dependent, then both sides of the equation are 0, by Exercise 15 and Exercise 5.2.30a.

18. a.

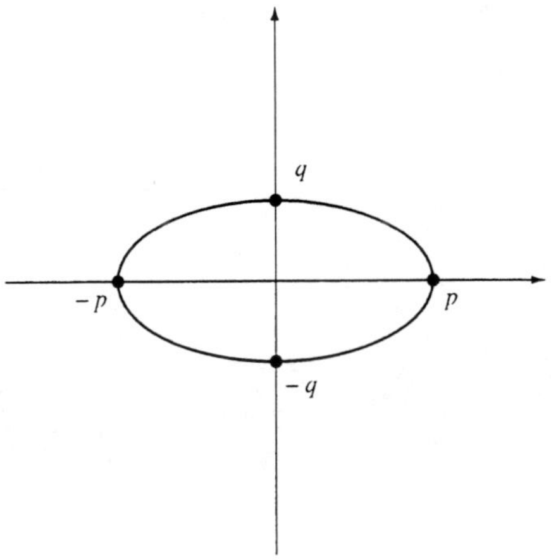

$$\begin{bmatrix} p & 0 \\ 0 & q \end{bmatrix}\begin{bmatrix} \cos(t) \\ \sin(t) \end{bmatrix} = \begin{bmatrix} p \cdot \cos(t) \\ q \cdot \sin(t) \end{bmatrix},$$ the ellipse with semi-axes $\pm\begin{bmatrix} p \\ 0 \end{bmatrix}$ and $\pm\begin{bmatrix} 0 \\ q \end{bmatrix}$.

(area of the ellipse) = $|\det(A)|$(area of the unit circle) = $pq\pi$

b. By Fact 5.3.8, $|\det(A)| = \dfrac{\text{area of the ellipse}}{\text{area of the unit circle}} = \dfrac{ab\pi}{\pi} = ab$ so $|\det(A)| = ab$.

c. The unit circle consists of all vectors of the form $\vec{x} = \cos(t)\dfrac{1}{\sqrt{2}}\begin{bmatrix} 1 \\ 1 \end{bmatrix} + \sin(t)\dfrac{1}{\sqrt{2}}\begin{bmatrix} 1 \\ -1 \end{bmatrix}$; its image is the ellipse consisting of all vectors $T(\vec{x}) = \cos(t)\underbrace{2\sqrt{2}\begin{bmatrix} 1 \\ 1 \end{bmatrix}}_{\text{semi-major axis}} + \sin(t)\underbrace{\sqrt{2}\begin{bmatrix} 1 \\ -1 \end{bmatrix}}_{\text{semi-minor axis}}$.

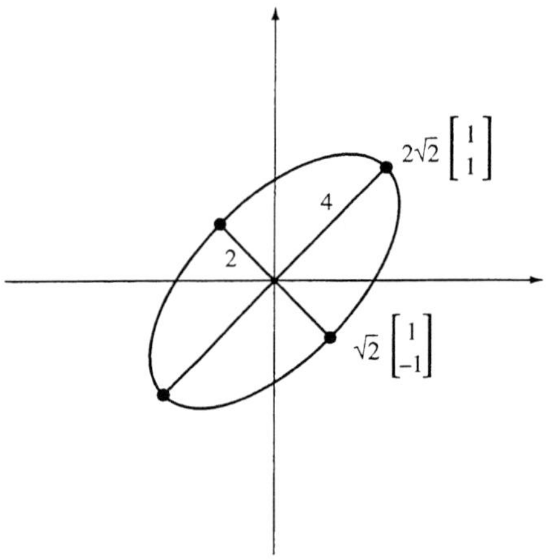

19. $\det[\vec{v}_1 \quad \vec{v}_2 \quad \vec{v}_3] = \vec{v}_1 \cdot (\vec{v}_2 \times \vec{v}_3) = \|\vec{v}_1\|\|\vec{v}_2 \times \vec{v}_3\| \cos\alpha$ where α is the angle between $\vec{v}_1$ and $\vec{v}_2 \times \vec{v}_3$ so $\det[\vec{v}_1 \quad \vec{v}_2 \quad \vec{v}_3] > 0$ if and only if $\cos\ \alpha > 0$, i.e., if and only if α is acute $\left(0 \le \alpha < \frac{\pi}{2}\right)$.

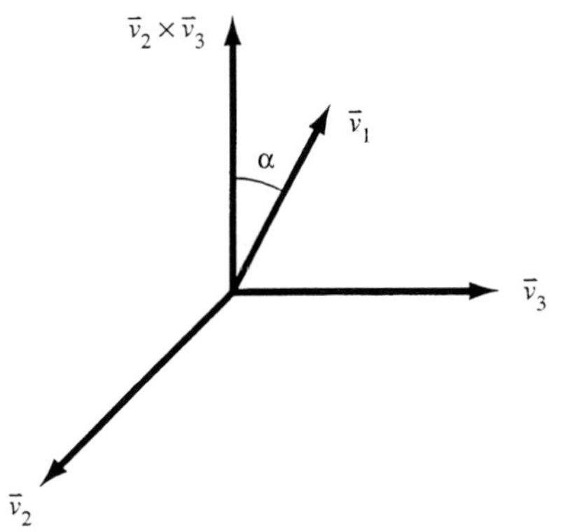

20. By Exercise 19, $\vec{v}_1, \vec{v}_2, \vec{v}_3$ constitute a positively oriented basis if and only if $\det[\vec{v}_1 \quad \vec{v}_2 \quad \vec{v}_3] > 0$. Assume that $\vec{v}_1, \vec{v}_2, \vec{v}_3$ is such a basis. We want to show that $A\vec{v}_1, A\vec{v}_2, A\vec{v}_3$ is positively oriented if and only if $\det(A) > 0$. We have $\det[A\vec{v}_1 \quad A\vec{v}_2 \quad A\vec{v}_3] = \det(A[\vec{v}_1 \quad \vec{v}_2 \quad \vec{v}_3]) = \det(A)\det[\vec{v}_1 \quad \vec{v}_2 \quad \vec{v}_3]$ so since $\det[\vec{v}_1 \quad \vec{v}_2 \quad \vec{v}_3] > 0$ by assumption, $\det[A\vec{v}_1 \quad A\vec{v}_2 \quad A\vec{v}_3] > 0$ if and only if $\det(A) > 0$. Hence A is orientation preserving if and only if $\det(A) > 0$.

21. **a.** Reverses
Consider $\vec{v}_2$ and $\vec{v}_3$ in the plane (not parallel), and let $\vec{v}_1 = \vec{v}_2 \times \vec{v}_3$; then $\vec{v}_1, \vec{v}_2, \vec{v}_3$ is a positively oriented basis, but $T(\vec{v}_1) = -\vec{v}_1, T(\vec{v}_2) = \vec{v}_2, T(\vec{v}_3) = \vec{v}_3$ is negatively oriented.

b. Preserves
Consider $\vec{v}_2$ and $\vec{v}_3$ orthogonal to the line (not parallel), and let $\vec{v}_1 = \vec{v}_2 \times \vec{v}_3$; then $\vec{v}_1, \vec{v}_2, \vec{v}_3$ is a positively oriented basis, and $T(\vec{v}_1) = \vec{v}_1, T(\vec{v}_2) = -\vec{v}_2, T(\vec{v}_3) = -\vec{v}_3$ is positively oriented as well.

c. Reverses
The standard basis $\vec{e}_1, \vec{e}_2, \vec{e}_3$ is positively oriented, but $T(\vec{e}_1) = -\vec{e}_1, T(\vec{e}_2) = -\vec{e}_2, T(\vec{e}_3) = -\vec{e}_3$ is negatively oriented.

22. Here $A=\begin{bmatrix}3 & 7\\ 4 & 11\end{bmatrix}$, $\det(A)=5$, $\vec{b}=\begin{bmatrix}1\\ 3\end{bmatrix}$, so by Fact 5.3.9

$$x=\frac{\det\begin{bmatrix}1 & 7\\ 3 & 11\end{bmatrix}}{5}=-2,\ y=\frac{\det\begin{bmatrix}3 & 1\\ 4 & 3\end{bmatrix}}{5}=1.$$

23. Here $A=\begin{bmatrix}5 & -3\\ -6 & 7\end{bmatrix}$, $\det(A)=17$, $\vec{b}=\begin{bmatrix}1\\ 0\end{bmatrix}$, so by Fact 5.3.9,

$$x_1=\frac{\det\begin{bmatrix}1 & -3\\ 0 & 7\end{bmatrix}}{17}=\frac{7}{17},\ x_2=\frac{\det\begin{bmatrix}5 & 1\\ -6 & 0\end{bmatrix}}{17}=\frac{6}{17}.$$

24. Here $A=\begin{bmatrix}2 & 3 & 0\\ 0 & 4 & 5\\ 6 & 0 & 7\end{bmatrix}$, $\det(A)=146$, $\vec{b}=\begin{bmatrix}8\\ 3\\ -1\end{bmatrix}$, so by Fact 5.3.9,

$$x=\frac{\det\begin{bmatrix}8 & 3 & 0\\ 3 & 4 & 5\\ -1 & 0 & 7\end{bmatrix}}{146}=1,\ y=\frac{\det\begin{bmatrix}2 & 8 & 0\\ 0 & 3 & 5\\ 6 & -1 & 7\end{bmatrix}}{146}=2,\ z=\frac{\det\begin{bmatrix}2 & 3 & 8\\ 0 & 4 & 3\\ 6 & 0 & -1\end{bmatrix}}{146}=-1.$$

25. By Fact 5.3.10, the *ij*th entry of adj(A) is given by $(-1)^{i+j}\det(A_{ji})$ so since $A=\begin{bmatrix}1 & 0 & 1\\ 0 & 1 & 0\\ 2 & 0 & 1\end{bmatrix}$ for $i=1, j=1$, we get

$(-1)^2\det\begin{bmatrix}1 & 0\\ 0 & 1\end{bmatrix}=1$, and for $i=1, j=2$ we get $(-1)^3\det\begin{bmatrix}0 & 1\\ 0 & 1\end{bmatrix}=0$, etc.

Completing this process gives $\text{adj}(A)=\begin{bmatrix}1 & 0 & -1\\ 0 & -1 & 0\\ -2 & 0 & 1\end{bmatrix}$, hence by Fact 5.3.10,

$$A^{-1}=\frac{1}{\det(A)}\text{adj}(A)=\frac{1}{-1}\begin{bmatrix}1 & 0 & -1\\ 0 & -1 & 0\\ -2 & 0 & 1\end{bmatrix}=\begin{bmatrix}-1 & 0 & 1\\ 0 & 1 & 0\\ 2 & 0 & -1\end{bmatrix}.$$

26. By Fact 5.3.10, $A^{-1}=\frac{1}{\det(A)}\text{adj}(A)$, so if det($A$) = 1, $A^{-1}=\text{adj}(A)$. If A has integer entries then $(-1)^{i+j}\det(A_{ji})$ will be an integer for all $1\le i,j\le n$, hence adj(A) will have integer entries. Therefore, A^{-1} will also have integer entries.

27. By Fact 5.3.9, using $A=\begin{bmatrix}a & -b\\ b & a\end{bmatrix}$, $\det(A)=a^2+b^2$, $\vec{b}=\begin{bmatrix}1\\ 0\end{bmatrix}$, we get

$x=\frac{\det\begin{bmatrix}1 & -b\\ 0 & a\end{bmatrix}}{a^2+b^2}=\frac{a}{a^2+b^2},\ y=\frac{\det\begin{bmatrix}a & 1\\ b & 0\end{bmatrix}}{a^2+b^2}=\frac{-b}{a^2+b^2}$, so x is positive, y is negative (since $a, b>0$), and as b increases, x decreases.

28. Here $A=\begin{bmatrix} s & a \\ m & -h \end{bmatrix}$, $\det(A)=-sh-ma$, $\vec{b}=\begin{bmatrix} I^\circ+G \\ M_s+M^\circ \end{bmatrix}$ so, by Fact 5.3.9,

$$Y=\frac{\det\begin{bmatrix} I^\circ+G & a \\ M_s-M^\circ & -h \end{bmatrix}}{-sh-ma}=\frac{-h(I^\circ+G)-a(M_s-M^\circ)}{-sh-ma}=\frac{h(I^\circ+G)+a(M_s-M^\circ)}{sh+ma},$$

$$r=\frac{\det\begin{bmatrix} s & I^\circ+G \\ m & M_s-M^\circ \end{bmatrix}}{-sh-ma}=\frac{s(M_s-M^\circ)-m(I^\circ+G)}{-sh-ma}=\frac{m(I^\circ+G)-s(M_s-M^\circ)}{sh+ma}.$$

29. By Fact 5.3.9,

$$dx_1=\frac{\det\begin{bmatrix} 0 & R_1 & -(1-\alpha) \\ 0 & 1-\alpha & -(1-\alpha)^2 \\ -R_2de_2 & -R_2 & -\frac{(1-\alpha)^2}{\alpha} \end{bmatrix}}{D}=\frac{R_1R_2(1-\alpha)^2de_2-R_2(1-\alpha)^2de_2}{D}$$

$$dy_1=\frac{\det\begin{bmatrix} -R_1 & 0 & -(1-\alpha) \\ \alpha & 0 & -(1-\alpha)^2 \\ R_2 & -R_2de_2 & -\frac{(1-\alpha)^2}{\alpha} \end{bmatrix}}{D}=\frac{R_2de_2(R_1(1-\alpha)^2+\alpha(1-\alpha))}{D}>0$$

$$dp=\frac{\det\begin{bmatrix} -R_1 & R_1 & 0 \\ \alpha & 1-\alpha & 0 \\ R_2 & -R_2 & -R_2de_2 \end{bmatrix}}{D}=\frac{R_1R_2de_2}{D}>0$$

30. In the simple case when $f(x, y)=1$ we have $\iint_{\Omega_2} f(x,y)dA=\iint_{\Omega_2} dA=\text{area of }\Omega_2=|\det M|$ and $\iint_{\Omega_1} g(u,v)dA=\iint_{\Omega_1} dA=\text{area of }\Omega_1=1$, so that $\iint_{\Omega_2} f(x,y)dA=|\det M|\cdot\iint_{\Omega_1} g(u,v)dA$.

This formula holds, in fact, for any continuous function $f(x, y)$; see an introductory text in multivariable calculus for a justification.

31. **a.** False; for example, $\det(I_2)=1$, $\det(2I_2)=4$.

b. True, by Facts 5.2.7 and 5.2.8

c. True, by Fact 5.3.8, since $\left|\det\begin{bmatrix} 1 & 2 \\ 3 & 5 \end{bmatrix}\right|=1$.

d. True, by Fact 5.2.7, since $\det(AA^T)=\det(A)\det(A^T)=\det(A^T)\det(A)=\det(A^TA)$.

e. True; the diagonal pattern makes an odd contribution to the determinant, while all other contributions are even. Thus, $\det(A)$ is odd. Since $\det(A)\neq 0$, the matrix A is invertible.

32. **a.** True; the pattern of large numbers prevents its determinant from being zero.

b. False; consider $A = \begin{bmatrix} 1 & 1 & -1 \\ 1 & 1 & 1 \\ 1 & -1 & 1 \end{bmatrix}$, with $\det(A) = 4$.

c. False; if $A = I_2$, $B = -I_2$ then $\det(A) + \det(B) = 2$ and $\det(A + B) = 0$.

d. True; $\det(B) = -\det(A)$ by Fact 5.2.4b so $\det(B) < \det(A)$ ensures $\det(A) \neq 0$.

e. False; counter-example $A = \begin{bmatrix} 2 & 1 & 1 \\ 1 & 2 & 1 \\ 1 & 5 & 2 \end{bmatrix}$

Chapter 6

6.1

1. If $\vec{v}$ is an eigenvector of A, *then* $A\vec{v} = \lambda\vec{v}$.
Hence $A^3\vec{v} = A^2(A\vec{v}) = A^2(\lambda\vec{v}) = A(A\lambda\vec{v}) = A(\lambda A\vec{v}) = A(\lambda^2\vec{v}) = \lambda^2 A\vec{v} = \lambda^3\vec{v}$, so $\vec{v}$ is an eigenvector of A^3 with eigenvalue λ^3.

2. We know $A\vec{v} = \lambda\vec{v}$ so $\vec{v} = A^{-1}A\vec{v} = A^{-1}\lambda\vec{v} = \lambda A^{-1}\vec{v}$, so $\vec{v} = \lambda A^{-1}\vec{v}$ or $A^{-1}\vec{v} = \frac{1}{\lambda}\vec{v}$.
Hence $\vec{v}$ is an eigenvector of A^{-1} with eigenvalue $\frac{1}{\lambda}$.

3. We know $A\vec{v} = \lambda\vec{v}$, so $(A + 2I_n)\vec{v} = A\vec{v} + 2I_n\vec{v} = \lambda\vec{v} + 2\vec{v} = (\lambda + 2)\vec{v}$, hence $\vec{v}$ is an eigenvector of $(A + 2I_n)$ with eigenvalue $\lambda + 2$.

4. We know $A\vec{v} = \lambda\vec{v}$, so $7A\vec{v} = 7\lambda\vec{v}$, hence $\vec{v}$ is an eigenvector of $7A$ with eigenvalue 7λ.

5. Assume $A\vec{v} = \lambda\vec{v}$ and $B\vec{v} = \beta\vec{v}$ for some eigenvalues λ, β. Then $(A + B)\vec{v} = A\vec{v} + B\vec{v} = \lambda\vec{v} + \beta\vec{v} = (\lambda + \beta)\vec{v}$ so $\vec{v}$ is an eigenvector of $A + B$ with eigenvalue $\lambda + \beta$.

6. True (See summary 6.1.4.)

7. We know $A\vec{v} = \lambda\vec{v}$ so $(\lambda I_n - A)\vec{v} = \lambda I_n\vec{v} - A\vec{v} = \lambda\vec{v} - \lambda\vec{v} = \vec{0}$ so a nonzero vector $\vec{v}$ is in the kernel of $(\lambda I_n - A)$ so $\ker(\lambda I_n - A) \neq \{\vec{0}\}$ and $\lambda I_n - A$ is not invertible.

8. We want all $\begin{bmatrix} a & b \\ c & d \end{bmatrix}$ such that $\begin{bmatrix} a & b \\ c & d \end{bmatrix}\begin{bmatrix} 1 \\ 0 \end{bmatrix} = 5\begin{bmatrix} 1 \\ 0 \end{bmatrix}$ hence $\begin{bmatrix} a \\ c \end{bmatrix} = \begin{bmatrix} 5 \\ 0 \end{bmatrix}$, i.e. the desired matrices must have the form $\begin{bmatrix} 5 & b \\ 0 & d \end{bmatrix}$.

9. We want $\begin{bmatrix} a & b \\ c & d \end{bmatrix}\begin{bmatrix} 1 \\ 0 \end{bmatrix} = \lambda\begin{bmatrix} 1 \\ 0 \end{bmatrix}$ for any λ. Hence $\begin{bmatrix} a \\ c \end{bmatrix} = \begin{bmatrix} \lambda \\ 0 \end{bmatrix}$, i.e. the desired matrices must have the form $\begin{bmatrix} \lambda & b \\ 0 & d \end{bmatrix}$, they must be upper triangular.

10. We want $\begin{bmatrix} a & b \\ c & d \end{bmatrix}\begin{bmatrix} 1 \\ 2 \end{bmatrix} = 5\begin{bmatrix} 1 \\ 2 \end{bmatrix}$, i.e. the desired matrices must have the form $\begin{bmatrix} 5-2b & b \\ 10-2d & d \end{bmatrix}$.

11. We want $\begin{bmatrix} a & b \\ c & d \end{bmatrix}\begin{bmatrix} 1 \\ 2 \end{bmatrix} = \lambda\begin{bmatrix} 1 \\ 2 \end{bmatrix}$ for some λ, so
$a + 2b = \lambda$
$c + 2d = 2\lambda$ i.e.

$c + 2d = 2(a + 2b)$ or $c = 2a + 4b - 2d$.
Hence the desired form is $\begin{bmatrix} a & b \\ 2a+4b-2d & d \end{bmatrix}$.

12. Solving $\begin{bmatrix} 2 & 0 \\ 3 & 4 \end{bmatrix}\begin{bmatrix} v_1 \\ v_2 \end{bmatrix} = 2\begin{bmatrix} v_1 \\ v_2 \end{bmatrix}$ we get $\begin{bmatrix} v_1 \\ v_2 \end{bmatrix} = \begin{bmatrix} t \\ -\frac{3}{2}t \end{bmatrix}$ and

solving $\begin{bmatrix} 2 & 0 \\ 3 & 4 \end{bmatrix}\begin{bmatrix} v_1 \\ v_2 \end{bmatrix} = 4\begin{bmatrix} v_1 \\ v_2 \end{bmatrix}$ we get $\begin{bmatrix} v_1 \\ v_2 \end{bmatrix} = \begin{bmatrix} 0 \\ t \end{bmatrix}$.

13. Solving $\begin{bmatrix} -6 & 6 \\ -15 & 13 \end{bmatrix}\begin{bmatrix} v_1 \\ v_2 \end{bmatrix} = 4\begin{bmatrix} v_1 \\ v_2 \end{bmatrix}$, we get $\begin{bmatrix} v_1 \\ v_2 \end{bmatrix} = \begin{bmatrix} \frac{3}{5}t \\ t \end{bmatrix}$.

14. We want a 4×4 matrix A such that $A\vec{e}_2 = \lambda\vec{e}_2$, i.e. the second column of A must be of the form $\begin{bmatrix} 0 \\ \lambda \\ 0 \\ 0 \end{bmatrix}$, so

$$A = \begin{bmatrix} a & 0 & c & d \\ e & \lambda & f & g \\ h & 0 & i & j \\ k & 0 & l & m \end{bmatrix}.$$

15. Any vector on L is unaffected by the reflection, so that a nonzero vector on L is an eigenvector with eigenvalue 1. Any vector on $L^\perp$ is flipped about L, so that a nonzero vector on $L^\perp$ is an eigenvector with eigenvalue -1. Picking a nonzero vector from L and one from $L^\perp$, we obtain an eigenbasis.

16. Rotation by 180° is a flip about the origin so every nonzero vector is an eigenvector with the eigenvalue -1. Any basis for $\mathbb{R}^2$ is an eigenbasis.

17. No (real) eigenvalues

18. Any nonzero vector in the plane is unchanged, hence is an eigenvector with the eigenvalue 1. Since any nonzero vector in $E^\perp$ is flipped about the origin, it is an eigenvector with eigenvalue -1. Pick any two non-collinear vectors from E and one from $E^\perp$ to form an eigenbasis.

19. Any nonzero vector in L is an eigenvector with eigenvalue 1, and any nonzero vector in the plane $L^\perp$ is an eigenvector with eigenvalue 0. Form an eigenbasis by picking any nonzero vector in L and any two nonzero, non-collinear vectors in $L^\perp$.

20. Any nonzero vector along the $\vec{e}_3$-axis is unchanged, hence is an eigenvector with eigenvalue 1. No other (real) eigenvalues can be found.

21. Any nonzero vector in $\mathbb{R}^3$ is an eigenvector with eigenvalue 5. Any basis for $\mathbb{R}^3$ is an eigenbasis.

22. Any nonzero scalar multiple of $\vec{v}$ is an eigenvector with eigenvalue 1.

23. **a.** Since $S = [\vec{v}_1 \ \cdots \ \vec{v}_n]$, $S^{-1}\vec{v}_i = S^{-1}(S\vec{e}_i) = \vec{e}_i$.

b. ith column of $S^{-1}AS$

$= S^{-1}AS\vec{e}_i$

$= S^{-1}A\vec{v}_i$ (by definition of S)

$= S^{-1}\lambda_i\vec{v}_i$ (since $\vec{v}_i$ is an eigenvector)

$= \lambda_i S^{-1}\vec{v}_i$

$= \lambda_i\vec{e}_i$ (by part a)

hence $S^{-1}AS = \begin{bmatrix} \lambda_1 & 0 & 0 & \cdots & 0 \\ 0 & \lambda_2 & 0 & \cdots & 0 \\ \vdots & & & & \\ 0 & 0 & 0 & \cdots & \lambda_n \end{bmatrix}$.

24.

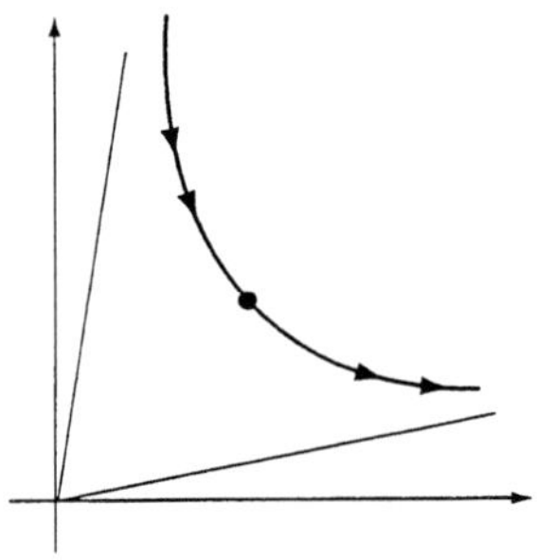

25.

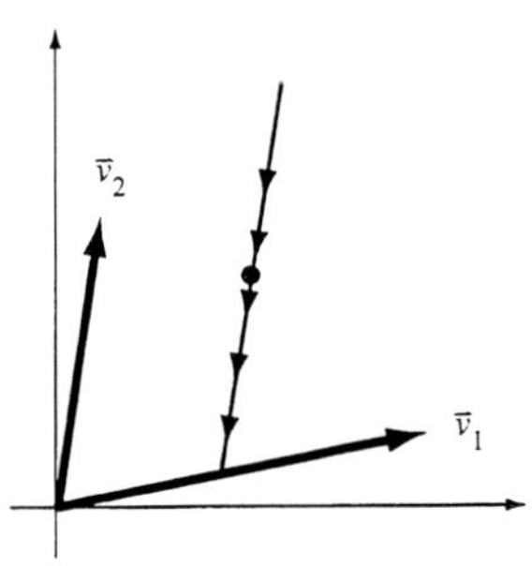

26.

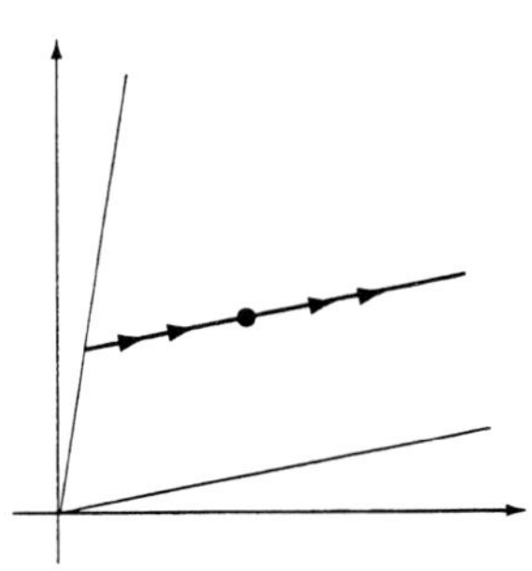

27.

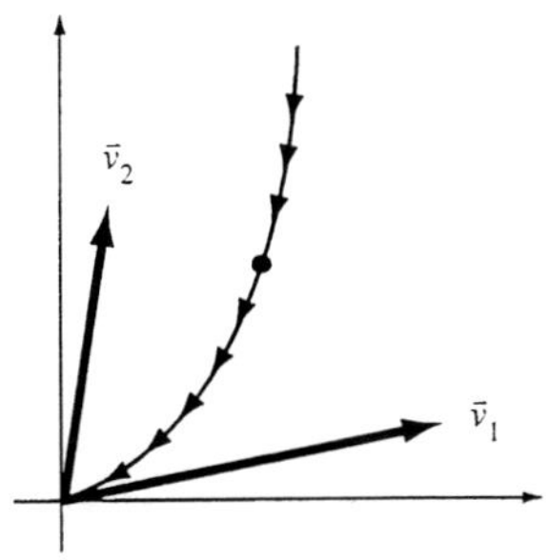

28.

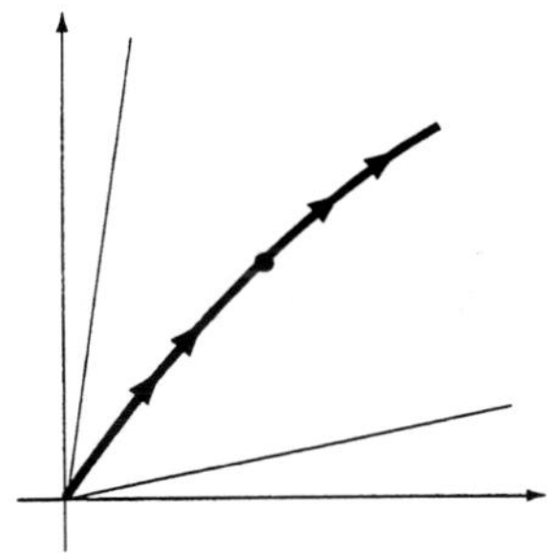

29.

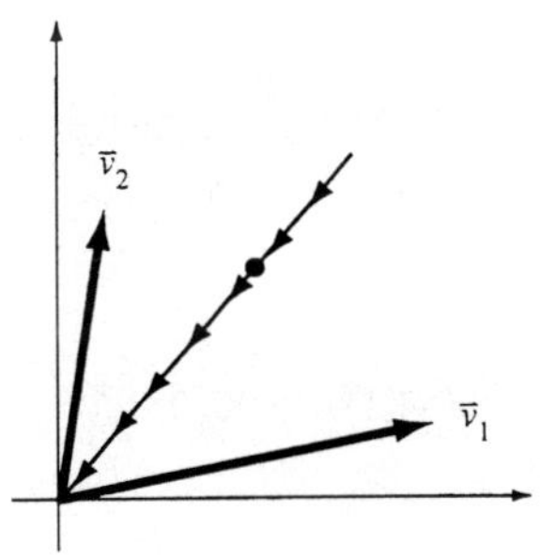

In Exercises 30–32, since the matrix is diagonal, $\vec{e}_1$ and $\vec{e}_2$ are eigenvectors.

30.

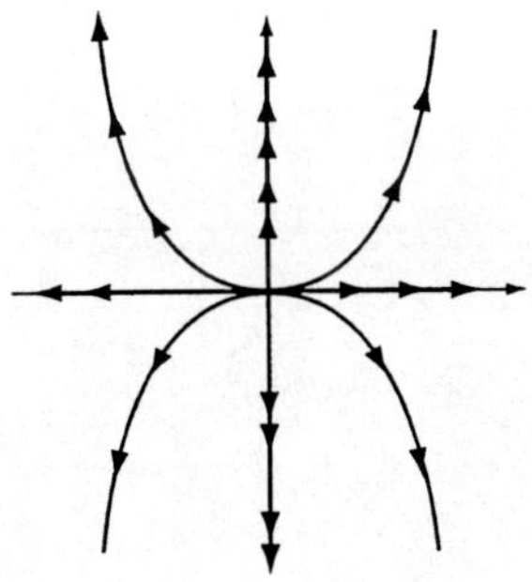

31. 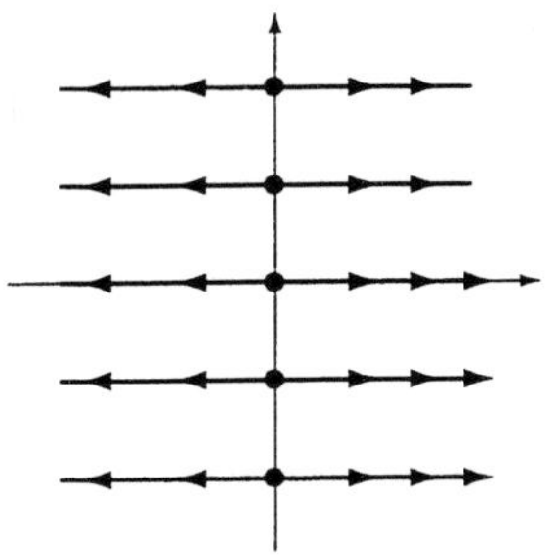

32. 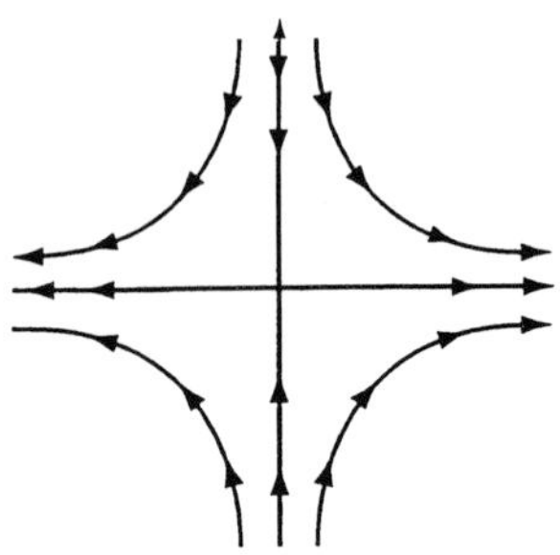

33. The eigenvalues of the system are $\lambda_1 = 1.1$, and $\lambda_2 = 0.9$ and the eigenvectors are $\vec{v}_1 = \begin{bmatrix} 100 \\ 300 \end{bmatrix}$ and $\vec{v}_2 = \begin{bmatrix} 200 \\ 100 \end{bmatrix}$ respectively. So if $\vec{x}_0 - \begin{bmatrix} 100 \\ 800 \end{bmatrix}$, we can see that $\vec{x}_0 = 3\vec{v}_1 - \vec{v}_2$. Therefore, by Fact 6.13

$\vec{x}(t) = 3(1.1)^t \begin{bmatrix} 100 \\ 300 \end{bmatrix} - (0.9)^t \begin{bmatrix} 200 \\ 100 \end{bmatrix}$, i.e. $c(t) = 300(1.1)^t - 200(0.9)^t$ and $r(t) = 900(1.1)^t - 100(0.9)^t$.

34. We want A such that $A\begin{bmatrix} 3 \\ 1 \end{bmatrix} = \begin{bmatrix} 15 \\ 5 \end{bmatrix}$ and $A\begin{bmatrix} 1 \\ 2 \end{bmatrix} = \begin{bmatrix} 10 \\ 20 \end{bmatrix}$, i.e. $A\begin{bmatrix} 3 & 1 \\ 1 & 2 \end{bmatrix} = \begin{bmatrix} 15 & 10 \\ 5 & 20 \end{bmatrix}$, so

$$A = \begin{bmatrix} 15 & 10 \\ 5 & 20 \end{bmatrix}\begin{bmatrix} 3 & 1 \\ 1 & 2 \end{bmatrix}^{-1} = \begin{bmatrix} 4 & 3 \\ -2 & 11 \end{bmatrix}.$$

35. We are given that $\vec{x}(t) = 2^t \begin{bmatrix} 1 \\ 1 \end{bmatrix} + 6^t \begin{bmatrix} -1 \\ 1 \end{bmatrix}$, hence we know that the eigenvalues are 2 and 6 with corresponding eigenvectors $\begin{bmatrix} 1 \\ 1 \end{bmatrix}$ and $\begin{bmatrix} -1 \\ 1 \end{bmatrix}$ respectively (see Fact 6.1.3), so we want a matrix A such that $A\begin{bmatrix} 1 & -1 \\ 1 & 1 \end{bmatrix} = \begin{bmatrix} 2 & -6 \\ 2 & 6 \end{bmatrix}$.

Multiplying on the right by $\begin{bmatrix} 1 & -1 \\ 1 & 1 \end{bmatrix}^{-1}$, we get $A = \begin{bmatrix} 4 & -2 \\ -2 & 4 \end{bmatrix}$.

36. **a.** $\begin{bmatrix} 0.978 & -0.006 \\ 0.004 & 0.992 \end{bmatrix}\begin{bmatrix} -1 \\ 2 \end{bmatrix} = \begin{bmatrix} -0.99 \\ 1.98 \end{bmatrix} = 0.99\begin{bmatrix} -1 \\ 2 \end{bmatrix}$, and $\begin{bmatrix} 0.978 & -0.006 \\ 0.004 & 0.992 \end{bmatrix}\begin{bmatrix} 3 \\ -1 \end{bmatrix} = \begin{bmatrix} 2.94 \\ -0.98 \end{bmatrix} = 0.98\begin{bmatrix} 3 \\ -1 \end{bmatrix}$.

The eigenvalues are $\lambda_1 = 0.99$ and $\lambda_2 = 0.98$.

b. $\vec{x}_0 = \begin{bmatrix} g_0 \\ l_0 \end{bmatrix} = \begin{bmatrix} 100 \\ 0 \end{bmatrix} = 20\begin{bmatrix} -1 \\ 2 \end{bmatrix} + 40\begin{bmatrix} 3 \\ -1 \end{bmatrix}$ so $\vec{x}(t) = 20(0.996)^t \begin{bmatrix} -1 \\ 2 \end{bmatrix} + 40(0.98)^t \begin{bmatrix} 3 \\ -1 \end{bmatrix}$, hence

$g(t) = -20(0.99)^t + 120(0.98)^t$ and $h(t) = 40(0.99)^t - 40(0.98)^t$.

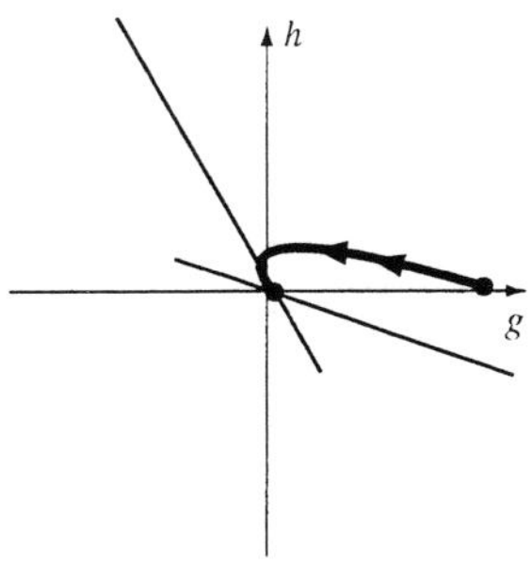

$h(t)$ first rises, then falls back to zero. $g(t)$ falls a little below zero, then goes back up to zero.

c. We set $g(t) = -20(0.99)^t + 120(0.98)^t = 0$.
Solving for t we get that $g(t) = 0$ for $t \approx 176$ minutes. (After $t = 176$, $g(t) < 0$).

37. a. $A = 5\begin{bmatrix} 0.6 & 0.8 \\ 0.8 & -0.6 \end{bmatrix}$ is a scalar multiple of an orthogonal matrix. By Fact 6.1.2, the possible eigenvalues of the orthogonal matrix are ± 1, so that the possible eigenvalues of A are ± 5. In part b we see that these are indeed eigenvalues.

b. Solve $A\vec{v} = \pm 5\vec{v}$ to get $\vec{v}_1 = \begin{bmatrix} 2 \\ 1 \end{bmatrix}$, $\vec{v}_2 = \begin{bmatrix} -1 \\ 2 \end{bmatrix}$, an eigenbasis for $\mathbb{R}^2$.

38. a. We are given that
$n(t+1) = 2a(t)$
$a(t+1) = n(t) + a(t)$,
so that the matrix is $A = \begin{bmatrix} 0 & 2 \\ 1 & 1 \end{bmatrix}$.

b. $A\begin{bmatrix} 1 \\ 1 \end{bmatrix} = \begin{bmatrix} 0 & 2 \\ 1 & 1 \end{bmatrix}\begin{bmatrix} 1 \\ 1 \end{bmatrix} = \begin{bmatrix} 2 \\ 2 \end{bmatrix} = 2\begin{bmatrix} 1 \\ 1 \end{bmatrix}$ and $A\begin{bmatrix} 2 \\ -1 \end{bmatrix} = \begin{bmatrix} 0 & 2 \\ 1 & 1 \end{bmatrix}\begin{bmatrix} 2 \\ -1 \end{bmatrix} = \begin{bmatrix} -2 \\ 1 \end{bmatrix} = -\begin{bmatrix} 2 \\ -1 \end{bmatrix}$, hence 2 and -1 are the eigenvalues associated with $\begin{bmatrix} 1 \\ 1 \end{bmatrix}$ and $\begin{bmatrix} 2 \\ -1 \end{bmatrix}$ respectively.

c. We are given $\vec{x}_0 = \begin{bmatrix} n_0 \\ a_0 \end{bmatrix} = \begin{bmatrix} 1 \\ 0 \end{bmatrix}$ so $\vec{x}_0 = \frac{1}{3}\begin{bmatrix} 1 \\ 1 \end{bmatrix} + \frac{1}{3}\begin{bmatrix} 2 \\ -1 \end{bmatrix}$, and $\vec{x}(t) = \frac{1}{3}2^t\begin{bmatrix} 1 \\ 1 \end{bmatrix} + \frac{1}{3}(-1)^t\begin{bmatrix} 2 \\ -1 \end{bmatrix}$ (by Fact 6.1.3),
hence $n(t) = \frac{1}{3}2^t + \frac{2}{3}(-1)^t$ and $a(t) = \frac{1}{3}2^t - \frac{1}{3}(-1)^t$.

39. Let λ be an eigenvalue of $S^{-1}AS$. Then for some nonzero vector $\vec{v}$, $S^{-1}AS\vec{v} = \lambda\vec{v}$, i.e., $AS\vec{v} = S\lambda\vec{v} = \lambda S\vec{v}$ so λ is an eigenvalue of A with eigenvector $S\vec{v}$.
Conversely, if α is an eigenvalue of A with eigenvector $\vec{w}$, then $A\vec{w} = \alpha\vec{w}$.
Therefore, $S^{-1}AS(S^{-1}\vec{w}) = S^{-1}A\vec{w} = S^{-1}\alpha\vec{w} = \alpha S^{-1}\vec{w}$, so $S^{-1}\vec{w}$ is an eigenvector of $S^{-1}AS$ with eigenvalue α

40. $(A^2 + 2A + 3I_n)\vec{v} = A^2\vec{v} + 2A\vec{v} + 3I_n\vec{v} = 4^2\vec{v} + 2\cdot 4\vec{v} + 3\vec{v} = (16+8+3)\vec{v} = 27\vec{v}$
so $\vec{v}$ is an eigenvector of $A^2 + 2A + 3I_n$ with eigenvalue 27.

41. a. True; if $A\vec{v} = \vec{0}$ for a nonzero $\vec{v}$, then A is not invertible so $\det(A) = 0$.

b. True; multiply to check.

c. True; if λ is an eigenvalue of A with eigenvector $\vec{v}$, then $A^4\vec{v} = \lambda^4\vec{v} = \vec{0}$ and $\lambda = 0$.

42. a. True; since $(\vec{v}\vec{v}^T)\vec{v} = \vec{v}(\vec{v}^T\vec{v}) = \vec{v}(\vec{v}\cdot\vec{v}) = (\vec{v}\cdot\vec{v})\vec{v} = \lambda\vec{v}$.

b. True; if $A\vec{v} = \lambda\vec{v}$ and $B\vec{v} = \alpha\vec{v}$ then $AB\vec{v} = A(B\vec{v}) = A(\alpha\vec{v}) = \alpha(A\vec{v}) = (\alpha\lambda)\vec{v}$.

c. True; suppose $A\vec{v} = \lambda\vec{v}$.
If $\lambda = 0$, then $A\vec{v} = \vec{0}$, so that $\vec{v}$ is in ker(A).
If $\lambda \neq 0$, then $\vec{v} = \frac{1}{\lambda}A\vec{v} = A\left(\frac{1}{\lambda}\vec{v}\right)$ is in im(A).

6.2

1. $\lambda_1 = 1$, $\lambda_2 = 3$ by Fact 6.2.2.

2. $\lambda_1 = 2$ (Algebraic multiplicity 2)
$\lambda_2 = 1$ (Algebraic multiplicity 2), by Fact 6.2.2.

3. $\det(\lambda I_2 - A) = \det\begin{bmatrix}\lambda-5 & 4\\ -2 & \lambda+1\end{bmatrix} = (\lambda-1)(\lambda-3) = 0$ so $\lambda_1 = 1$, $\lambda_2 = 3$.

4. $\det(\lambda I_2 - A) = \det\begin{bmatrix}\lambda & -4\\ 1 & \lambda-4\end{bmatrix} = \lambda(\lambda-4)+4 = (\lambda-2)^2 = 0$ so $\lambda = 2$ with algebraic multiplicity 2.

5. $\det(\lambda I_2 - A) = \det\begin{bmatrix}\lambda-11 & 15\\ -6 & \lambda+7\end{bmatrix} = \lambda^2 - 4\lambda + 13$ so $\det(\lambda I_2 - A) = 0$ for no real λ.

6. $\det(\lambda I_2 - A) = \det\begin{bmatrix}\lambda-1 & -2\\ -3 & \lambda-4\end{bmatrix} = \lambda^2 - 5\lambda - 2 = 0$ so $\lambda_{1,2} = \frac{5\pm\sqrt{33}}{2}$.

7. $\lambda = 1$ with algebraic multiplicity 3, by Fact 6.2.2.

8. $f_A(\lambda) = \lambda^2(\lambda+3)$ so
$\lambda_1 = 0$ (Algebraic multiplicity 2)
$\lambda_2 = -3$.

9. $f_A(\lambda) = (\lambda-2)^2(\lambda-1)$ so
$\lambda_1 = 2$ (Algebraic multiplicity 2)
$\lambda_2 = 1$

10. $f_A(\lambda) = (\lambda+1)^2(\lambda-1)$ so $\lambda_1 = -1$ (Algebraic multiplicity 2), $\lambda_2 = 1$.

11. $f_A(\lambda) = \lambda^3 + \lambda^2 + \lambda + 1 = (\lambda+1)(\lambda^2+1) = 0$
$\lambda = -1$ (Algebraic multiplicity 1)

12. $f_A(\lambda) = \lambda(\lambda+1)(\lambda-1)^2$ so $\lambda_1 = 0$, $\lambda_2 = -1$, $\lambda_3 = 1$ (Algebraic multiplicity 2).

13. $f_A(\lambda) = \lambda^3 - 1 = (\lambda-1)(\lambda^2+\lambda+1)$ so $\lambda = 1$ (Algebraic multiplicity 1).

14. $f_A(\lambda) = \det(\lambda I_2 - B)\det(\lambda I_2 - D)$ (see Exercise 5.1.37).
The eigenvalues of A are the eigenvalues of B and D. The eigenvalues of C are irrelevant.

15. $f_A(\lambda) = \lambda^2 - 2\lambda + (1-k) = 0$ if $\lambda_{1,2} = \dfrac{2 \pm \sqrt{4 - 4(1-k)}}{2} = 1 \pm \sqrt{k}$
The matrix has 2 distinct real eigenvalues when $k > 0$, no real eigenvalues when $k < 0$.

16. $f_A(\lambda) = \lambda^2 - (a+c)\lambda + (ac - b^2)$
The discriminant of this quadratic equation is $(a+c)^2 - 4(ac - b^2) = a^2 + 2ac + c^2 - 4ac + 4b^2 = (a-c)^2 + 4b^2$; this quantity is always positive (since $b \neq 0$). There will always be two distinct real eigenvalues.

17. $f_A(\lambda) = \lambda^2 - a^2 - b^2 = 0$ so $\lambda_{1,2} = \pm\sqrt{a^2 + b^2}$.
The matrix represents a reflection about a line followed by a dilation by $\sqrt{a^2 + b^2}$, hence the eigenvalues.

18. $f_A(\lambda) = \lambda^2 - 2a\lambda + a^2 - b^2$ so $\lambda_{1,2} = \dfrac{+2a \pm \sqrt{4a^2 - 4(a^2 - b^2)}}{2} = a \pm b$.
Hence the eigenvalues are $a \pm b$.

19. True, since $f_A(\lambda) = \lambda^2 - \text{tr}(A)\lambda + \det(A)$ and the discriminant $[\text{tr}(A)]^2 - 4\det(A)$ is positive if $\det(A)$ is negative.

20. By Fact 6.2.4, $f_A(\lambda) = \lambda^2 - \text{tr}(A)\lambda + \det(A)$; on the other hand,
$f_A(\lambda) = (\lambda - \lambda_1)(\lambda - \lambda_2) = \lambda^2 - (\lambda_1 + \lambda_2)\lambda + \lambda_1\lambda_2$.
Equating coefficients, we get $\text{tr}(A) = \lambda_1 + \lambda_2$ and $\det(A) = \lambda_1\lambda_2$.
If $A = \begin{bmatrix} 1 & 2 \\ 4 & 3 \end{bmatrix}$ then $\lambda_1 = 5$, $\lambda_2 = -1$, $\text{tr}(A) = 4$, $\det(A) = -5$.

21. By Fact 6.2.5, $f_A(\lambda) = \lambda^n - \text{tr}(A)\lambda^{n-1} + \cdots + (-1)^n \det(A)$.
On the other hand, since the λ_i are all distinct and there are n of them,
$f_A(\lambda) = (\lambda - \lambda_1)(\lambda - \lambda_2)\cdots(\lambda - \lambda_n) = \lambda^n - (\lambda_1 + \lambda_2 + \cdots + \lambda_n)\lambda^{n-1} + \cdots + (-1)^n \lambda_1\lambda_2\cdots\lambda_n$.
Equating the second and last coefficients, we get $\text{tr}(A) = \sum_{i=1}^{n} \lambda_i$, $\det(A) = (-1)^n \prod_{i=1}^{n} \lambda_i$.

22. By Fact 5.2.1, $f_A(\lambda) = \det(\lambda I_n - A) = \det(\lambda I_n - A)^T = \det(\lambda I_n - A^T) = f_{A^T}(\lambda)$. Since the characteristic polynomials of A and A^T are identical, the two matrices have the same eigenvalues, with the same algebraic multiplicities.

23. $f_B(\lambda) = \det(\lambda I_n - B) = \det(\lambda I_n - S^{-1}AS)$
$= \det(\lambda S^{-1} I_n S - S^{-1}AS)$
$= \det\left(S^{-1}(\lambda I_n - A)S\right) = \det(S^{-1})\det(\lambda I_n - A)\det(S)$
$= (\det(S))^{-1}\det(\lambda I_n - A)\det(S) = \det(\lambda I_n - A) = f_A(\lambda)$
Hence, since $f_A(\lambda) = f_B(\lambda)$, A and B have the same eigenvalues.

24. $\lambda_1 = 0.25$, $\lambda_2 = 1$

25. $A\begin{bmatrix} c \\ b \end{bmatrix} = \begin{bmatrix} ac+bc \\ bc+bd \end{bmatrix} = \begin{bmatrix} (a+b)c \\ (c+d)b \end{bmatrix} = \begin{bmatrix} c \\ b \end{bmatrix}$ since $a+b=c+d=1$; therefore, $\begin{bmatrix} c \\ b \end{bmatrix}$ is an eigenvector with eigenvalue $\lambda_1 = 1$.

Also, $A\begin{bmatrix} 1 \\ -1 \end{bmatrix} = \begin{bmatrix} a-c \\ b-d \end{bmatrix} = (a-c)\begin{bmatrix} 1 \\ -1 \end{bmatrix}$ since $a-c=-(b-d)$; therefore, $\begin{bmatrix} 1 \\ -1 \end{bmatrix}$ is an eigenvector with eigenvalue $\lambda_2 = a-c$. Notice that $|a-c|<1$ so a possible phase portrait is

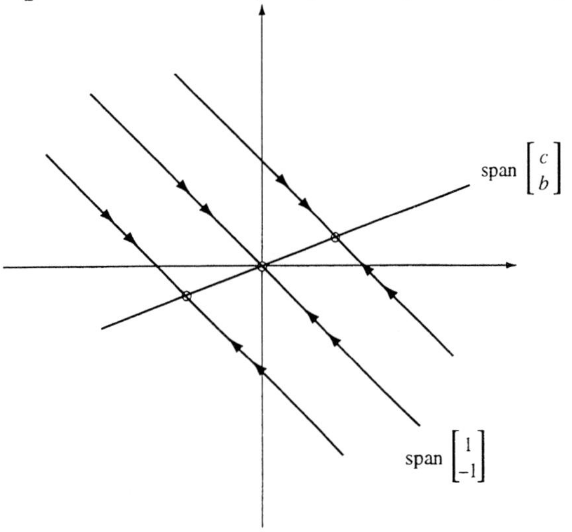

26. Here $\begin{bmatrix} c \\ b \end{bmatrix} = \begin{bmatrix} 0.25 \\ 0.5 \end{bmatrix}$ with $\lambda_1 = 1$ and $\begin{bmatrix} 1 \\ -1 \end{bmatrix}$ with $\lambda_2 = a-c = 0.25$.

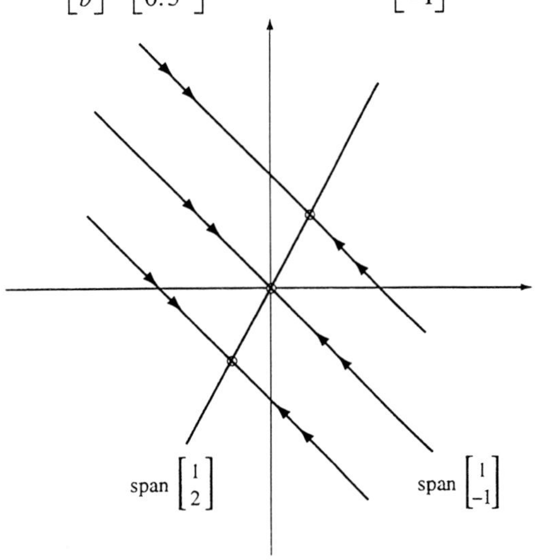

27. a. We know $\vec{v}_1 = \begin{bmatrix} 1 \\ 2 \end{bmatrix}$, $\lambda_1 = 1$ and $\vec{v}_2 = \begin{bmatrix} 1 \\ -1 \end{bmatrix}$, $\lambda_2 = \frac{1}{4}$. If $\vec{x}_0 = \begin{bmatrix} 1 \\ 0 \end{bmatrix}$ then $\vec{x}_0 = \frac{1}{3}\vec{v}_1 + \frac{2}{3}\vec{v}_2$, so by Fact 6.1.3,

$$x_1(t) = \frac{1}{3} + \frac{2}{3}\left(\frac{1}{4}\right)^t$$

$$x_2(t) = \frac{2}{3} - \frac{2}{3}\left(\frac{1}{4}\right)^t.$$

If $\vec{x}_0 = \begin{bmatrix} 0 \\ 1 \end{bmatrix}$ then $\vec{x}_0 = \frac{1}{3}\vec{v}_1 - \frac{1}{3}\vec{v}_2$, so by Fact 6.13,

$$x_1(t) = \frac{1}{3} - \frac{1}{3}\left(\frac{1}{4}\right)^t$$

$$x_2(t) = \frac{2}{3} + \frac{1}{3}\left(\frac{1}{4}\right)^t$$

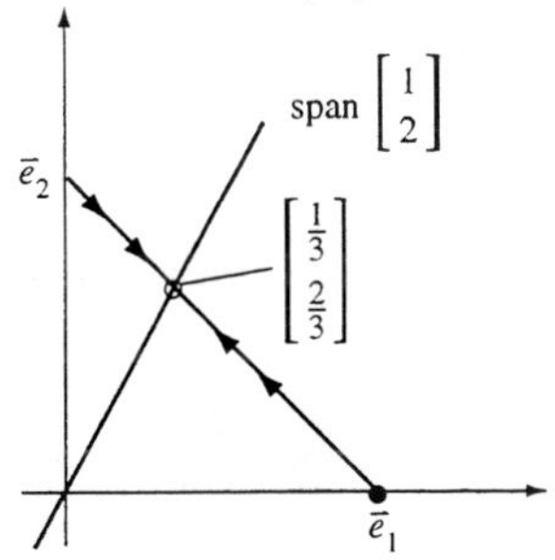

b. A^t approaches $\frac{1}{3}\begin{bmatrix} 1 & 1 \\ 2 & 2 \end{bmatrix}$, as $t \to \infty$.

c. Let us think about the first column of A^t, which is $A^t\vec{e}_1$. We can use Fact 6.1.3 to compute $A^t\vec{e}_1$.

Start by writing $\vec{e}_1 = c_1\begin{bmatrix} c \\ b \end{bmatrix} + c_2\begin{bmatrix} 1 \\ -1 \end{bmatrix}$; a straightforward computation shows that $c_1 = \frac{1}{c+b}$ and $c_2 = \frac{b}{c+b}$.

Now $A^t\vec{e}_1 = \frac{1}{c+b}\begin{bmatrix} c \\ b \end{bmatrix} + \frac{b}{c+b}(\lambda_2)^t\begin{bmatrix} 1 \\ -1 \end{bmatrix}$, where $\lambda_2 = a - c$.

Since $|\lambda_2| < 1$, the second summand goes to zero, so that $\lim_{t\to\infty}(A^t\vec{e}_1) = \frac{1}{c+b}\begin{bmatrix} c \\ b \end{bmatrix}$.

Likewise, $\lim_{t\to\infty}(A^t\vec{e}_2) = \frac{1}{c+b}\begin{bmatrix} c \\ b \end{bmatrix}$, so that $\lim_{t\to\infty} A^t = \frac{1}{c+b}\begin{bmatrix} c & c \\ b & b \end{bmatrix}$.

28. a. $w(t+1) = 0.8w(t) + 0.1m(t)$
$m(t+1) = 0.2w(t) + 0.9m(t)$

so $A = \begin{bmatrix} 0.8 & 0.1 \\ 0.2 & 0.9 \end{bmatrix}$ which is a regular transition matrix since its columns sum to 1 and its entries are positive.

b. The eigenvectors of A are $\begin{bmatrix} 0.1 \\ 0.2 \end{bmatrix}$ or $\begin{bmatrix} 1 \\ 2 \end{bmatrix}$ with $\lambda_1 = 1$, and $\begin{bmatrix} 1 \\ -1 \end{bmatrix}$ with $\lambda_2 = 0.7$.

$\vec{x}_0 = \begin{bmatrix} 1200 \\ 0 \end{bmatrix} = 400\begin{bmatrix} 1 \\ 2 \end{bmatrix} + 800\begin{bmatrix} 1 \\ -1 \end{bmatrix}$ so $\vec{x}(t) = 400\begin{bmatrix} 1 \\ 2 \end{bmatrix} + 800(0.7)^t\begin{bmatrix} 1 \\ -1 \end{bmatrix}$ or

$w(t) = 400 + 800(0.7)^t$

$m(t) = 800 - 800(0.7)^t$.

c. As $t \to \infty$, $w(t) \to 400$ so Wipfs won't have to close the store.

29. The ith entry of $A\vec{e}$ is $[a_{i1} \quad a_{i2} \quad \cdots \quad a_{in}]\vec{e} = \sum_{j=1}^{n} a_{ij} = 1$, so $A\vec{e} = \vec{e}$ and $\lambda = 1$ is an eigenvalue of A, corresponding to the eigenvector $\vec{e}$.

30. a. Let v_i be the largest component of the vector $\vec{v}$, that is, $v_i \ge v_j$ for $j = 1, \ldots, n$. Then the ith component of

$$A\vec{v} \text{ is } \lambda v_i = \sum_{j=1}^{n} a_{ij}v_j \underset{\substack{\uparrow \\ v_j \le v_i}}{\le} \sum_{j=1}^{n} a_{ij}v_i = \left(\sum_{j=1}^{n} a_{ij}\right)v_i \underset{\substack{\uparrow \\ \sum_{j=1}^{n} a_{ij} = 1}}{=} v_i$$

We can conclude that $\lambda v_i \le v_i$, and therefore $\lambda \le 1$, as claimed. Also note that if $\vec{v}$ is not a multiple of the eigenvector $\vec{e}$ discussed in Exercise 29, then $v_j < v_i$ for some index j, so that $\sum_{j=1}^{n} a_{ij}v_j < \sum_{j=1}^{n} a_{ij}v_i$ and therefore $\lambda < 1$.

b. Let v_i be the component of $\vec{v}$ with the largest absolute value, that is, $|v_i| \ge |v_j|$ for $j = 1, 2, \ldots, n$. Then the absolute value of the ith component of $A\vec{v}$ is $|\lambda||v_i| = \left|\sum_{j=1}^{n} a_{ij}v_j\right| \le \sum_{j=1}^{n} a_{ij}|v_j| \le \sum_{j=1}^{n} a_{ij}|v_i| = \left(\sum_{j=1}^{n} a_{ij}\right)|v_i| = |v_i|$ so that $|\lambda||v_i| \le |v_i|$ and $|\lambda| \le 1$, as claimed.

31. Since A and A^T have the same eigenvalues (by Exercise 22), Exercise 29 states that $\lambda = 1$ is an eigenvalue of A, and Exercise 30 says that $|\lambda| \le 1$ for all eigenvalues λ. $\vec{e}$ need not be an eigenvector of A; consider

$A = \begin{bmatrix} 0.9 & 0.9 \\ 0.1 & 0.1 \end{bmatrix}$.

32. $f_A(\lambda) = \lambda^3 - 3\lambda + k$

Setting $f_A'(\lambda) = 0$ we find that $f_A(\lambda)$ has a local minimum at $(1, k-2)$ and a local maximum at $(-1, k+2)$.

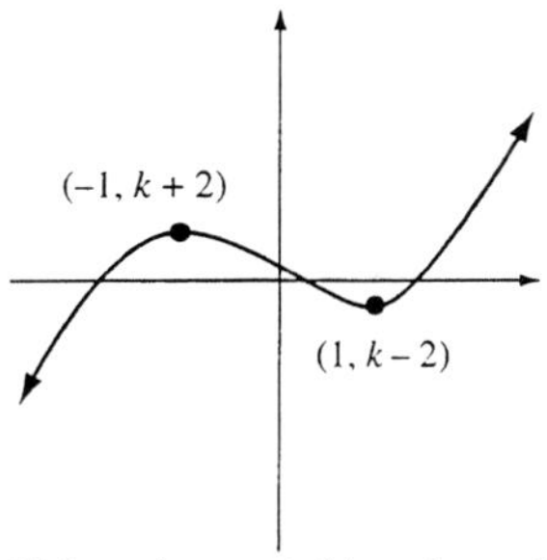

This polynomial has three distinct real roots if $k + 2$ is positive (that is, if $k > -2$) and $k - 2$ is negative (that is, $k < 2$). Both conditions are satisfied if $|k| < 2$. If $|k| = 2$, then there are two distinct roots; if $|k| > 2$, then there is only one real root.

33. a. $f_A(\lambda) = \det(\lambda I_3 - A) = \lambda^3 - c\lambda^2 - b\lambda - a$

b. By part a, $c = 17$, $b = -5$ and $a = \pi$, so $M = \begin{bmatrix} 0 & 1 & 0 \\ 0 & 0 & 1 \\ \pi & -5 & 17 \end{bmatrix}$.

34. Consider the possible graphs of $f_A(\lambda)$ assuming that it has 2 distinct real roots.

i)

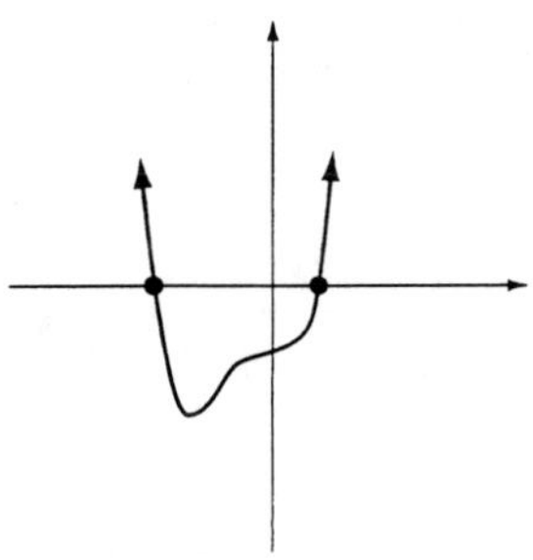

Algebraic multiplicity of each eigenvalue is 1. Example: $\begin{bmatrix} 1 & 0 & 0 & 0 \\ 0 & -2 & 0 & 0 \\ 0 & 0 & 0 & -1 \\ 0 & 0 & 1 & 0 \end{bmatrix}$

ii)

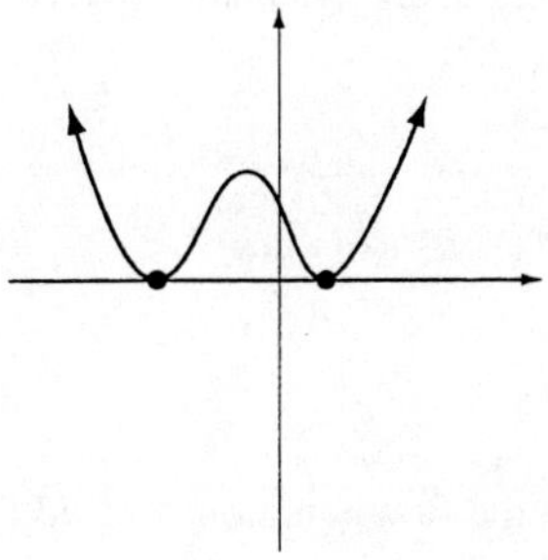

Algebraic multiplicity of each eigenvalue is 2. Example: $\begin{bmatrix} -2 & 0 & 0 & 0 \\ 0 & -2 & 0 & 0 \\ 0 & 0 & 1 & 0 \\ 0 & 0 & 0 & 1 \end{bmatrix}$

iii)

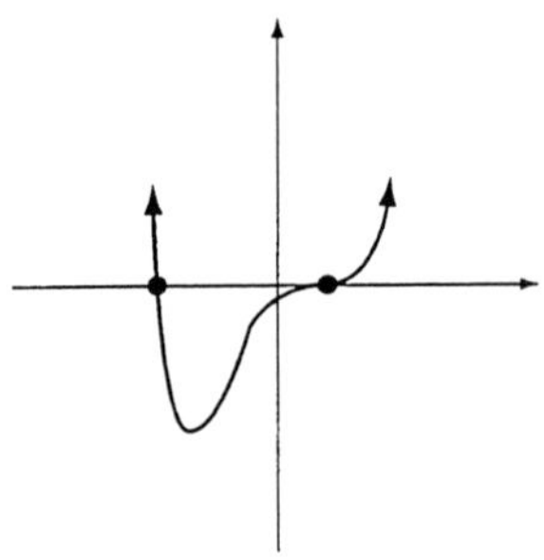

Algebraic multiplicity of λ_1 is 1, and of λ_2 is 3. Example: $\begin{bmatrix} -2 & 0 & 0 & 0 \\ 0 & 1 & 0 & 0 \\ 0 & 0 & 1 & 0 \\ 0 & 0 & 0 & 1 \end{bmatrix}$

35. $A = \begin{bmatrix} 0 & -1 & 0 & 0 \\ 1 & 0 & 0 & 0 \\ 0 & 0 & 0 & -1 \\ 0 & 0 & 1 & 0 \end{bmatrix}$, $f_A(\lambda) = (\lambda^2 + 1)^2$

36. Let $A = \begin{bmatrix} B & & & 0 \\ & B & & \\ & & \ddots & \\ 0 & & & B \end{bmatrix}$ where $B = \begin{bmatrix} 0 & -1 \\ 1 & 0 \end{bmatrix}$, $f_A(\lambda) = (\lambda^2 + 1)^n$.

37. We can write $f_A(\lambda) = (\lambda - \lambda_0)^2 g(\lambda)$, for some polynomial g. The product rule for derivatives tells us that $f'_A(\lambda) = 2(\lambda - \lambda_0)g(\lambda) + (\lambda - \lambda_0)^2 g'(\lambda)$, so that $f'_A(\lambda_0) = 0$, as claimed.

38. a. If $f(x) = x^3 + 6x - 20$ then $f'(x) = 3x^2 + 6$ so $f'(x) > 0$ for all x, i.e. f is always increasing, hence has only one real root.

b. If $v^3 - u^3 = 20$ and $vu = 2$ then
$(v-u)^3 + 6(v-u) = v^3 - 3v^2u + 3vu^2 - u^3 + 6(v-u) = v^3 - u^3 - 3vu(v-u) + 6(v-u)$
$= 20 - 6(v-u) + 6(v-u) = 20$
Hence $x = v - u$ satisfies the equation $x^3 + 6x = 20$.

c. The second equation tells us that $u = \frac{2}{v}$ or $u^3 = \frac{8}{v^3}$. Substituting into the first equation we find that

$v^3 - \frac{8}{v^3} = 20$, or, $(v^3)^2 - 8 = 20v^3$ or $(v^3)^2 - 20v^3 - 8 = 0$, with solutions

$v^3 = \frac{20 \pm \sqrt{400+32}}{2} = 10 \pm \sqrt{108} = 10 \pm 6\sqrt{3}$ and $v = \sqrt[3]{10 \pm \sqrt{108}}$.

Now $u^3 = v^3 - 20 = -10 \pm \sqrt{108}$ and $u = \sqrt[3]{-10 \pm \sqrt{108}}$.

d. Let $v = \sqrt[3]{\frac{q}{2} + \sqrt{\left(\frac{q}{2}\right)^2 + \left(\frac{p}{3}\right)^3}}$ and $u = \sqrt[3]{-\frac{q}{2} + \sqrt{\left(\frac{q}{2}\right)^2 + \left(\frac{p}{3}\right)^3}}$.

Then $v^3 - u^3 = q$ and $vu = \sqrt[3]{\left(\frac{q}{2}\right)^2 + \left(\frac{p}{3}\right)^3 - \left(\frac{q}{2}\right)^2} = \frac{p}{3}$.

Since $x = v - u$ we have

$x^3 + px = v^3 - 3v^2u + 3vu^2 - u^3 + p(v-u) = v^3 - u^3 - 3vu(v-u) + p(v-u)$

$= q - p(v-u) + p(v-u) = q$, as claimed.

If p is negative, the expression $\left(\frac{q}{2}\right)^2 + \left(\frac{p}{3}\right)^3$ may be negative. Also, the equation $x^3 + px = q$ may have more than one solution in this case.

e. Setting $x = t - \left(\frac{a}{3}\right)$ we get $\left(t - \frac{a}{3}\right)^3 + a\left(t - \frac{a}{3}\right)^2 + b\left(t - \frac{a}{3}\right) + c = 0$ or

$t^3 - at^2 + at^2 + (\text{linear and constant terms}) = 0$ or $t^3 + (\text{linear and constant terms}) = 0$, as claimed (bring the constant terms to the right-hand side).

6.3

1. $\lambda_1 = 7, \lambda_2 = 9, E_7 = \text{span}\begin{bmatrix}1\\0\end{bmatrix}, E_9 = \text{span}\begin{bmatrix}4\\1\end{bmatrix}$,

Eigenbasis: $\begin{bmatrix}1\\0\end{bmatrix}, \begin{bmatrix}4\\1\end{bmatrix}$

2. $\lambda_1 = 2, \lambda_2 = 0, E_2 = \text{span}\begin{bmatrix}1\\1\end{bmatrix}, E_0 = \text{span}\begin{bmatrix}-1\\1\end{bmatrix}$

Eigenbasis: $\begin{bmatrix}1\\1\end{bmatrix}, \begin{bmatrix}-1\\1\end{bmatrix}$

3. $\lambda_1 = 4, \lambda_2 = 9, E_4 = \text{span}\begin{bmatrix}3\\-2\end{bmatrix}, E_9 = \text{span}\begin{bmatrix}1\\1\end{bmatrix}$

Eigenbasis: $\begin{bmatrix}3\\-2\end{bmatrix}, \begin{bmatrix}1\\1\end{bmatrix}$

4. $\lambda_1 = \lambda_2 = 1, E_1 = \text{span}\begin{bmatrix}-1\\1\end{bmatrix}$

No eigenbasis

5. No real eigenvalues as $f_A(\lambda) = \lambda^2 - 2\lambda + 2$.

6. $\lambda_{1,2} = \dfrac{7 \pm \sqrt{57}}{2}$

Eigenbasis: $\begin{bmatrix} 3 \\ \lambda_1 - 2 \end{bmatrix} \approx \begin{bmatrix} 3 \\ 5.27 \end{bmatrix}, \begin{bmatrix} 3 \\ \lambda_2 - 2 \end{bmatrix} \approx \begin{bmatrix} 3 \\ -2.27 \end{bmatrix}$

7. $\lambda_1 = 1, \lambda_2 = 2, \lambda_3 = 3$, eigenbasis: $\vec{e}_1, \vec{e}_2, \vec{e}_3$

8. $\lambda_1 = 1, \lambda_2 = 2, \lambda_3 = 3$, eigenbasis: $\begin{bmatrix} 1 \\ 0 \\ 0 \end{bmatrix}, \begin{bmatrix} 1 \\ 1 \\ 0 \end{bmatrix}, \begin{bmatrix} 1 \\ 2 \\ 1 \end{bmatrix}$

9. $\lambda_1 = \lambda_2 = 1, \lambda_3 = 0$, eigenbasis: $\begin{bmatrix} 1 \\ 0 \\ 0 \end{bmatrix}, \begin{bmatrix} 0 \\ 1 \\ 0 \end{bmatrix}, \begin{bmatrix} -1 \\ 0 \\ 1 \end{bmatrix}$

10. $\lambda_1 = \lambda_2 = 1, \lambda_3 = 0, E_1 = \text{span}\begin{bmatrix} 1 \\ 0 \\ 0 \end{bmatrix}, E_0 = \text{span}\begin{bmatrix} 0 \\ 0 \\ 1 \end{bmatrix}$

No eigenbasis

11. $\lambda_1 = \lambda_2 = 0, \lambda_3 = 3$, eigenbasis: $\begin{bmatrix} 1 \\ -1 \\ 0 \end{bmatrix}, \begin{bmatrix} 1 \\ 0 \\ -1 \end{bmatrix}, \begin{bmatrix} 1 \\ 1 \\ 1 \end{bmatrix}$

12. $\lambda_1 = \lambda_2 = \lambda_3 = 1, \ E_1 = \text{span}\begin{bmatrix} 1 \\ 0 \\ 0 \end{bmatrix}$, no eigenbasis

13. $\lambda_1 = 0, \lambda_2 = 1, \lambda_3 = -1$, eigenbasis: $\begin{bmatrix} 0 \\ 1 \\ 0 \end{bmatrix}, \begin{bmatrix} 1 \\ -3 \\ 1 \end{bmatrix}, \begin{bmatrix} 1 \\ -1 \\ 2 \end{bmatrix}$

14. $\lambda_1 = 0, \lambda_1 = \lambda_3 = 1$, eigenbasis: $\begin{bmatrix} 0 \\ 1 \\ 0 \end{bmatrix}, \begin{bmatrix} 1 \\ -5 \\ 0 \end{bmatrix}, \begin{bmatrix} 0 \\ 2 \\ 1 \end{bmatrix}$

15. $\lambda_1 = 0, \lambda_2 = \lambda_3 = 1, E_0 = \text{span}\begin{bmatrix} 0 \\ 1 \\ 0 \end{bmatrix}, E_1 = \text{span}\begin{bmatrix} 1 \\ -1 \\ 2 \end{bmatrix}$

No eigenbasis

16. $\lambda_1 = 0$ (no other real eigenvalues), with eigenvector $\begin{bmatrix} 1 \\ -1 \\ 1 \end{bmatrix}$.

No real eigenbasis

17. $\lambda_1 = \lambda_2 = 0,\ \lambda_3 = \lambda_4 = 1$

With eigenbasis $\begin{bmatrix}1\\0\\0\\0\end{bmatrix}, \begin{bmatrix}0\\-1\\1\\0\end{bmatrix}, \begin{bmatrix}0\\1\\0\\0\end{bmatrix}, \begin{bmatrix}0\\0\\0\\1\end{bmatrix}$

18. $\lambda_1 = \lambda_2 = 0,\ \lambda_3 = \lambda_4 = 1,\ E_0 = \text{span}(\vec{e}_1, \vec{e}_3),\ E_1 = \text{span}(\vec{e}_2)$
No eigenbasis

19. We find the eigenspace $E_1 = \ker(I_3 - A) = \ker\begin{bmatrix}0 & -a & -b\\0 & 0 & -c\\0 & 0 & 0\end{bmatrix} = \ker\begin{bmatrix}0 & a & b\\0 & 0 & c\\0 & 0 & 0\end{bmatrix}$.

If $a = b = c = 0$ then E_1 is 3-dimensional with eigenbasis $\vec{e}_1, \vec{e}_2, \vec{e}_3$.
If $a \neq 0$ and $c \neq 0$ then E_1 is 1-dimensional and otherwise E_1 is 2-dimensional. The geometric multiplicity of the eigenvalue 1 is $\dim(E_1)$.

20. For $\lambda_1 = 1$, $E_1 = \ker\begin{bmatrix}0 & a & b\\0 & 0 & c\\0 & 0 & 1\end{bmatrix} = \ker\begin{bmatrix}0 & a & 0\\0 & 0 & 1\\0 & 0 & 0\end{bmatrix}$ so if $a = 0$ then E_1 is 2-dimensional, otherwise it is 1 dimensional.

For $\lambda_2 = 2$, $E_2 = \ker\begin{bmatrix}1 & -a & -b\\0 & 1 & -c\\0 & 0 & 0\end{bmatrix}$ so E_2 is 1-dimensional.

Hence, there is an eigenbasis if $a = 0$.

21. We want A such that $A\begin{bmatrix}1\\2\end{bmatrix} = \begin{bmatrix}1\\2\end{bmatrix}$ and $A\begin{bmatrix}2\\3\end{bmatrix} = 2\begin{bmatrix}2\\3\end{bmatrix} = \begin{bmatrix}4\\6\end{bmatrix}$, i.e. $A\begin{bmatrix}1 & 2\\2 & 3\end{bmatrix} = \begin{bmatrix}1 & 4\\2 & 6\end{bmatrix}$ so

$A = \begin{bmatrix}1 & 4\\2 & 6\end{bmatrix}\begin{bmatrix}1 & 2\\2 & 3\end{bmatrix}^{-1} = \begin{bmatrix}5 & -2\\6 & -2\end{bmatrix}$.

The answer is unique.

22. We want A such that $A\vec{e}_1 = 7\vec{e}_1$ and $A\vec{e}_2 = 7\vec{e}_2$ hence $A = \begin{bmatrix}7 & 0\\0 & 7\end{bmatrix}$.

23. $\lambda_1 = \lambda_2 = 1$ and $E_1 = \text{span}(\vec{e}_1)$, hence there is no eigenbasis. The matrix represents a shear parallel to the x-axis.

24. Any *shear* parallel to the line spanned by $\begin{bmatrix}2\\1\end{bmatrix}$ will do; for example, we can find the matrix A such that

$A\begin{bmatrix}2\\1\end{bmatrix} = \begin{bmatrix}2\\1\end{bmatrix}$ and $A\begin{bmatrix}0\\1\end{bmatrix} = \begin{bmatrix}0\\1\end{bmatrix} + \begin{bmatrix}2\\1\end{bmatrix} = \begin{bmatrix}2\\2\end{bmatrix}$.

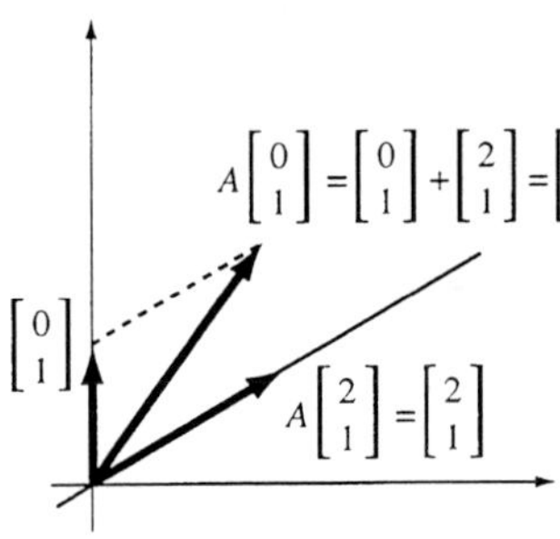

We want $A\begin{bmatrix}2 & 0\\1 & 1\end{bmatrix}=\begin{bmatrix}2 & 2\\1 & 2\end{bmatrix}$, so that $A=\begin{bmatrix}2 & 2\\1 & 2\end{bmatrix}\begin{bmatrix}2 & 0\\1 & 1\end{bmatrix}^{-1}=\begin{bmatrix}0 & 2\\-\frac{1}{2} & 2\end{bmatrix}$. There are other solutions; they are of the form $\begin{bmatrix}1+2c & -4c\\c & 1-2c\end{bmatrix}$, where c is an arbitrary constant $\left(\text{in our example, } c=-\frac{1}{2}\right)$.

25. If λ is an eigenvalue of A, then $E_\lambda=\ker(\lambda I_3-A)=\ker\begin{bmatrix}\lambda & -1 & 0\\0 & \lambda & -1\\-a & -b & \lambda-c\end{bmatrix}$.

The row-reduced echelon form of the above matrix will contain two leading 1's hence E_λ is always 1-dimensional, i.e. the geometric multiplicity of λ is 1.

26. True:
Note that $f_A(0)=\det(0I_6-A)=\det(-A)=(-1)^6\det(A)=\det(A)$ is negative. Since $\lim_{\lambda\to\infty} f_A(\lambda)=\infty$, there must be a positive root, by the Intermediate Value Theorem (see Exercise 2.2.47c). Therefore, the matrix A has a positive eigenvalue.

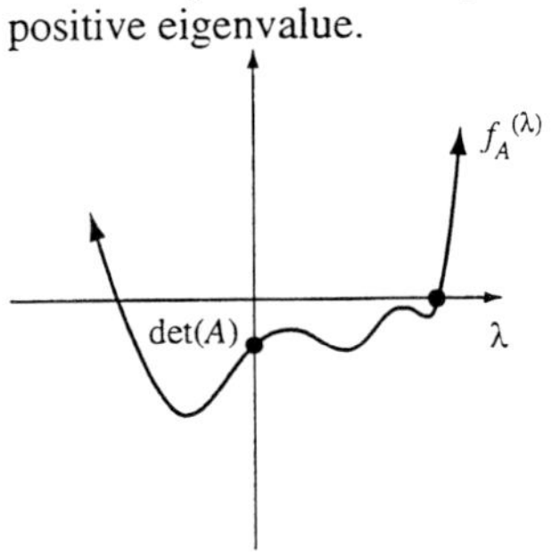

27. By Fact 6.2.4, $f_A(\lambda)=\lambda^2-5\lambda+6=(\lambda-3)(\lambda-2)$ so $\lambda_1=2$, $\lambda_2=3$.

28. Since $J_n(\lambda)$ is triangular, its eigenvalues are its diagonal entries, hence its only eigenvalue is λ. Moreover,

$$E_\lambda=\ker(\lambda I_n-J_n(\lambda))=\ker\begin{bmatrix}0 & 1 & 0 & \cdots & 0\\0 & 0 & 1 & & \vdots\\\vdots & \vdots & 0 & & \vdots\\\vdots & \vdots & \vdots & & 1\\0 & 0 & 0 & & 0\end{bmatrix}.$$

Since only the first variable is free, the geometric multiplicity of λ is 1 while its algebraic multiplicity is n.

29. Note that r is the number of nonzero diagonal entries of A, since the nonzero columns of A form a basis of im(A). Therefore, there are $n-r$ zeros on the diagonal, so that the algebraic multiplicity of the eigenvalue 0 is $n-r$. It is true for any $n\times n$ matrix A that the geometric multiplicity of the eigenvalue 0 is dim(ker(A)) $= n-\text{rank}(A)=n-r$.

30. Since A is triangular, $f_\lambda(A) = (\lambda - a_{11})(\lambda - a_{22})\cdots(\lambda - a_{mm})(\lambda - 0)^{n-m}$.
Hence the algebraic multiplicity of $\lambda = 0$ is $(n-m)$.
Also note that the rank of A is at least m, since the first m columns of A are linearly independent. Therefore, the geometric multiplicity of the eigenvalue 0 is $\dim(\ker(A)) = n - \text{rank}(A) \le n - m$.

31. They must be the same. For if they are not, by Fact 6.3.3, the geometric multiplicities would not add up to n.

32. Recall that a matrix and its transpose have the same rank (Fact 4.3.9c). The geometric multiplicity of λ as an eigenvalue of A is $\dim\left(\ker(\lambda I_n - A)\right) = n - \text{rank}(\lambda I_n - A)$. The geometric multiplicity of λ as an eigenvalue of A^T is $\dim\left(\ker(\lambda I_n - A^T)\right) = \dim\left(\ker(\lambda I_n - A)^T\right) = n - \text{rank}(\lambda I_n - A)^T = n - \text{rank}(\lambda I_n - A)$.
We can see that the two multiplicities are the same.

33. a. $A\vec{v}\cdot\vec{w} = (A\vec{v})^T\vec{w} = (\vec{v}^T A^T)\vec{w} = (\vec{v}^T A)\vec{w} = v^T(A\vec{w}) = \vec{v}\cdot A\vec{w}$

$\uparrow$ (under the third equals sign) *A* symmetric

b. Assume $A\vec{v} = \lambda\vec{v}$ and $A\vec{w} = \alpha\vec{w}$ for $\lambda \ne \alpha$, then $(A\vec{v})\cdot\vec{w} = (\lambda\vec{v})\cdot\vec{w} = \lambda(\vec{v}\cdot\vec{w})$, and $\vec{v}\cdot A\vec{w} = \vec{v}\cdot\alpha\vec{w} = \alpha(\vec{v}\cdot\vec{w})$.
By part a, $\lambda(\vec{v}\cdot\vec{w}) = \alpha(\vec{v}\cdot\vec{w})$ i.e. $(\lambda - \alpha)(\vec{v}\cdot\vec{w}) = 0$.
Since $\lambda \ne \alpha$, it must be that $\vec{v}\cdot\vec{w} = 0$, i.e. $\vec{v}$ and $\vec{w}$ are perpendicular.

34. Note that $f_A(0) = \det(0I_3 - A) = \det(-A) = (-1)^3\det(A) = -1$.
Since $\lim_{\lambda\to\infty} f_A(\lambda) = \infty$, the polynomial $f_A(\lambda)$ must have a positive root λ_0, by the Intermediate Value Theorem. In other words, the matrix A will have a positive eigenvalue λ_0. Since A is orthogonal, this eigenvalue λ_0 will be 1, by Fact 6.1.2. This means that there is a nonzero vector $\vec{v}$ in $\mathbb{R}^3$ such that $A\vec{v} = 1\vec{v} = \vec{v}$, as claimed.

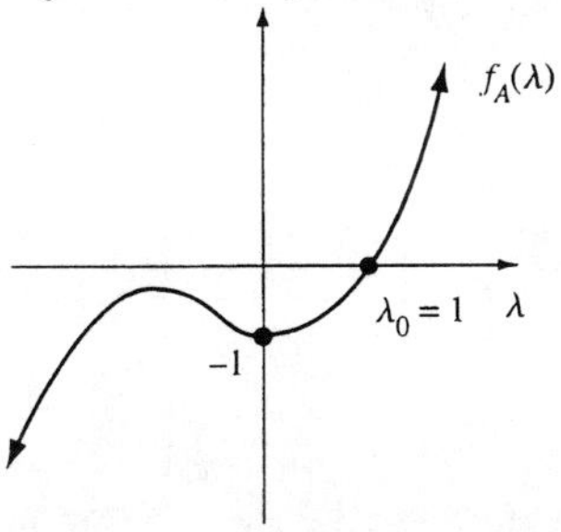

35. a. There are two eigenvalues, $\lambda_1 = 1$ (with $E_1 = V$) and $\lambda_2 = 0$ (with $E_0 = V^\perp$).
Now geometric multiplicity(1) = $\dim(E_1) = \dim(V) = m$, and
geometric multiplicity(0) = $\dim(E_0) = \dim(V^\perp) = n - \dim(V) = n - m$.
Since geometric multiplicity(λ) $\le$ algebraic multiplicity(λ), by Fact 6.3.3, and the algebraic multiplicities cannot add up to more than n, the geometric and algebraic multiplicities of the eigenvalues are the same here.

b. Analogous to part a: $E_1 = V$ and $E_{-1} = V^\perp$.
geometric multiplicity(1) = algebraic multiplicity(1) = $\dim(V) = m$, and
geometric multiplicity(-1) = algebraic multiplicity(-1) = $\dim(V^\perp) = n - m$.

36. The matrix of the dynamical system is $A=\begin{bmatrix}a & b\\ b & a\end{bmatrix}$ so $f_A(\lambda)=(\lambda-a)^2-b^2$.

Hence, $\lambda_{1,2}=a\pm b$, and the respective eigenvectors are $\begin{bmatrix}1\\1\end{bmatrix}$ and $\begin{bmatrix}-1\\1\end{bmatrix}$.

Since $\vec{x}(0)=\begin{bmatrix}3\\ \frac{1}{2}\end{bmatrix}=\frac{7}{4}\begin{bmatrix}1\\1\end{bmatrix}-\frac{5}{4}\begin{bmatrix}-1\\1\end{bmatrix}$, by Fact 6.1.3, $\vec{x}(t)=\frac{7}{4}(a+b)^t\begin{bmatrix}1\\1\end{bmatrix}+\frac{5}{4}(a-b)^t\begin{bmatrix}1\\-1\end{bmatrix}$.

Note that $a-b$ is between 0 and 1, so that the second summand in the formula above goes to $\vec{0}$ as t goes to infinity. Qualitatively different outcomes occur depending on whether $a+b$ exceeds 1, equals 1, or is less than 1:

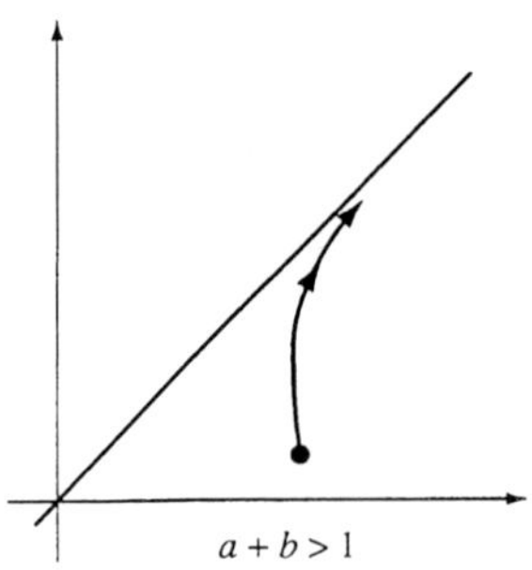

The spending of both nations grows: an arms race

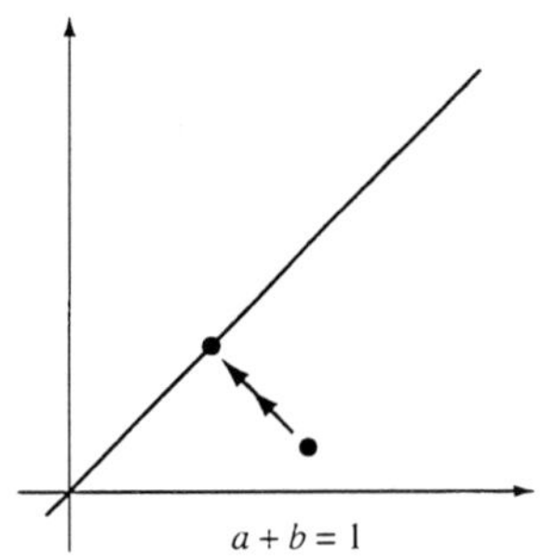

The spending of both nations stabilizes at an equal level of $b 1.75.

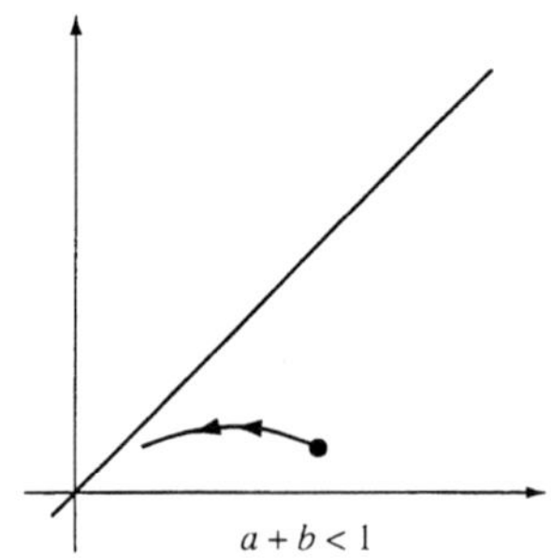

The spending of both nations decreases to zero: disarmament

37. The eigenvalues of A are 1.2, –0.8, –0.4 with eigenvectors $\begin{bmatrix}9\\6\\2\end{bmatrix}, \begin{bmatrix}2\\-2\\1\end{bmatrix}, \begin{bmatrix}1\\-2\\2\end{bmatrix}$.

Since $\vec{x}_0=50\begin{bmatrix}9\\6\\2\end{bmatrix}+50\begin{bmatrix}2\\-2\\1\end{bmatrix}+50\begin{bmatrix}1\\-2\\2\end{bmatrix}$ we have $\vec{x}(t)=50(1.2)^t\begin{bmatrix}9\\6\\2\end{bmatrix}+50(-0.8)^t\begin{bmatrix}2\\-2\\1\end{bmatrix}+50(-0.4)^t\begin{bmatrix}1\\-2\\2\end{bmatrix}$ so as t goes to infinity $j(t):n(t):a(t)$ approaches the proportion 9 : 6 : 2.

38. $C(t+1)=0.8C(t)+10$ so if $A\begin{bmatrix}C(t)\\1\end{bmatrix}=\begin{bmatrix}C(t+1)\\1\end{bmatrix}$, $A=\begin{bmatrix}0.8 & 10\\0 & 1\end{bmatrix}$. A has eigenvectors $\begin{bmatrix}1\\0\end{bmatrix}, \begin{bmatrix}50\\1\end{bmatrix}$ corresponding to $\lambda_1=0.8$ and $\lambda_2=1$. Since $\begin{bmatrix}C(0)\\1\end{bmatrix}=\begin{bmatrix}0\\1\end{bmatrix}$, and $\begin{bmatrix}0\\1\end{bmatrix}=-50\begin{bmatrix}1\\0\end{bmatrix}+\begin{bmatrix}50\\1\end{bmatrix}$, we have $C(t)=-50(0.9)^t+50$, hence in the long run, there will be 50 spectators. The graph of $C(t)$ look like:

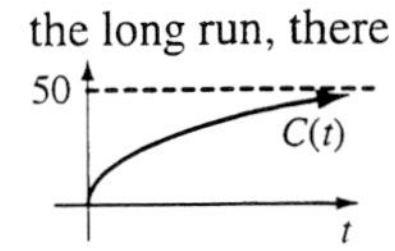

39. a. $A=\frac{1}{2}\begin{bmatrix}0 & 1 & 1\\1 & 0 & 1\\1 & 1 & 0\end{bmatrix}$

b. After 10 rounds, we have $A^{10}\begin{bmatrix}7\\11\\5\end{bmatrix}\approx\begin{bmatrix}7.6660156\\7.6699219\\7.6640625\end{bmatrix}$.

After 50 rounds, we have $A^{50}\begin{bmatrix}7\\11\\5\end{bmatrix}\approx\begin{bmatrix}7.66666666667\\7.66666666667\\7.66666666667\end{bmatrix}$.

c. The eigenvalues of A are 1 and $-\frac{1}{2}$ with $E_1=\text{span}\begin{bmatrix}1\\1\\1\end{bmatrix}$ and $E_{-\frac{1}{2}}=\text{span}\left(\begin{bmatrix}0\\1\\-1\end{bmatrix},\begin{bmatrix}-1\\-1\\2\end{bmatrix}\right)$ so

$$\vec{x}(t)=\left(1+\frac{c_0}{3}\right)\begin{bmatrix}1\\1\\1\end{bmatrix}+\left(-\frac{1}{2}\right)^t\begin{bmatrix}0\\1\\-1\end{bmatrix}+\left(-\frac{1}{2}\right)^t\frac{c_0}{3}\begin{bmatrix}-1\\-1\\2\end{bmatrix}.$$

After 1001 rounds, Alberich will be ahead of Brunnhilde $\left(\text{by }\left(\frac{1}{2}\right)^{1001}\right)$, so that Carl needs to beat Alberich to win the game. A straightforward computation shows that $c(1001)-a(1001)=\left(\frac{1}{2}\right)^{1001}(1-c_0)$; Carl wins if this quantity is positive, which is the case if c_0 is less than 1.

Alternatively, observe that the ranking of the players is reversed in each round: Whoever is first will be last after the next round. Since the total number of rounds is odd (1001), Carl wants to be last initially to win the game; he wants to choose a smaller number than both Alberich and Brunnhilde.

40. a. $a_{11}=0.7$ means that only 70% of the pollutant present in Lake Silvaplana at a given time is still there a week later; some is carried down to Lake Sils by the river Inn, and some is absorbed or evaporates.
The other diagonal entries can be interpreted analogously. $a_{21}=0.1$ means that 10% of the pollutant present in Lake Silvaplana at any given time can be found in Lake Sils a week later, carried down by the river Inn.
The significance of the coefficient $a_{32}=0.2$ is analogous; $a_{31}=0$ means that no pollutant is carried down from Lake Silvaplana to Lake St. Moritz in just one week.
The matrix is lower triangular since no pollutant is carried from Lake Sils to Lake Silvaplana, for example (the river Inn flows the other way).

b. The eigenvalues of A are 0.8, 0.6, 0.7 with corresponding eigenvectors $\begin{bmatrix}0\\0\\1\end{bmatrix},\begin{bmatrix}0\\1\\-1\end{bmatrix},\begin{bmatrix}1\\1\\-2\end{bmatrix}$.

$\vec{x}(0)=\begin{bmatrix}100\\0\\0\end{bmatrix}=100\begin{bmatrix}0\\0\\1\end{bmatrix}-100\begin{bmatrix}0\\1\\-1\end{bmatrix}+100\begin{bmatrix}1\\1\\-2\end{bmatrix}$ so $\vec{x}(t)=100(0.8)^t\begin{bmatrix}0\\0\\1\end{bmatrix}-100(0.6)^t\begin{bmatrix}0\\1\\-1\end{bmatrix}+100(0.7)^t\begin{bmatrix}1\\1\\-2\end{bmatrix}$ or

$x_1(t)=100(0.7)^t$

$x_2(t)=100(0.7)^t-100(0.6)^t$

$x_3(t)=100(0.8)^t+100(0.6)^t-200(0.7)^t$

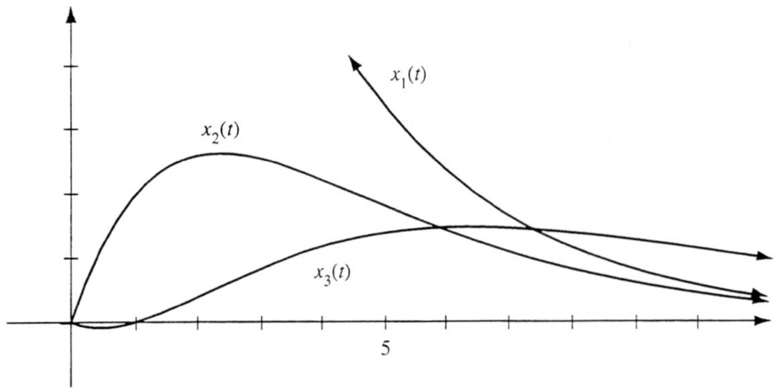

Using calculus, we find that the function $x_2(t) = 100(0.7)^t - 100(0.6)^t$ reaches its maximum at $t \approx 2.33$. Keep in mind, however, that our model holds for integer t only.

41. a. $A = \begin{bmatrix} 0.1 & 0.2 \\ 0.4 & 0.3 \end{bmatrix}, \vec{b} = \begin{bmatrix} 1 \\ 2 \end{bmatrix}$

b. $B = \begin{bmatrix} A & \vec{b} \\ 0 & 1 \end{bmatrix}$

c. The eigenvalues of A are 0.5 and –0.1 with associated eigenvectors $\begin{bmatrix} 1 \\ 2 \end{bmatrix}$ and $\begin{bmatrix} 1 \\ -1 \end{bmatrix}$.

The eigenvalues of B are 0.5, –0.1, and 1. If $A\vec{v} = \lambda\vec{v}$ then $B\begin{bmatrix} \vec{v} \\ 0 \end{bmatrix} = \lambda\begin{bmatrix} \vec{w} \\ 0 \end{bmatrix}$ so $\begin{bmatrix} \vec{v} \\ 0 \end{bmatrix}$ is an eigenvector of B.

Furthermore, $\begin{bmatrix} 2 \\ 4 \\ 1 \end{bmatrix}$ is an eigenvector of B corresponding to the eigenvalue 1. Note that this vector is

$\begin{bmatrix} -(A - I_2)^{-1}\vec{b} \\ 1 \end{bmatrix}$.

d. Write $\vec{y}(0) = \begin{bmatrix} x_1(0) \\ x_2(0) \\ 1 \end{bmatrix} = c_1\begin{bmatrix} 1 \\ 2 \\ 0 \end{bmatrix} + c_2\begin{bmatrix} 1 \\ -1 \\ 0 \end{bmatrix} + c_3\begin{bmatrix} 2 \\ 4 \\ 1 \end{bmatrix}$.

Note that $c_3 = 1$.

Now $\vec{y}(t) = c_1(0.5)^t\begin{bmatrix} 1 \\ 2 \\ 0 \end{bmatrix} + c_2(-0.1)^t\begin{bmatrix} 1 \\ -1 \\ 0 \end{bmatrix} + \begin{bmatrix} 2 \\ 4 \\ 1 \end{bmatrix} \xrightarrow{t\to\infty} \begin{bmatrix} 2 \\ 4 \\ 1 \end{bmatrix}$ so that $\vec{x}(t) \xrightarrow{t\to\infty} \begin{bmatrix} 2 \\ 4 \end{bmatrix}$.

42. a. $T_1(t+1) = 0.6T_1(t) + 0.1T_2(t) + 20$
$T_2(t+1) = 0.1T_1(t) + 0.6T_2(t) + 0.1T_3(t) + 20$
$T_3(t+1) = 0.1T_2(t) + 0.6T_3(t) + 40$

so $A = \begin{bmatrix} 0.6 & 0.1 & 0 \\ 0.1 & 0.6 & 0.1 \\ 0 & 0.1 & 0.6 \end{bmatrix}, \vec{b} = \begin{bmatrix} 20 \\ 20 \\ 40 \end{bmatrix}$

b. $B = \begin{bmatrix} A & \vec{b} \\ 0 & 1 \end{bmatrix}$

c. $\vec{y}(10) = B^{10}\begin{bmatrix} 0 \\ 0 \\ 0 \\ 1 \end{bmatrix} \approx \begin{bmatrix} 70.86 \\ 93.95 \\ 120.56 \\ 1 \end{bmatrix}$

$\vec{y}(30) = B^{30}\begin{bmatrix} 0 \\ 0 \\ 0 \\ 1 \end{bmatrix} \approx \begin{bmatrix} 74.989 \\ 99.985 \\ 124.989 \\ 1 \end{bmatrix}$

Seems to approach $\begin{bmatrix} 75 \\ 100 \\ 125 \\ 1 \end{bmatrix}$

d. The eigenvalues of A are $\lambda_1 \approx 0.45858$, $\lambda_2 = 0.6$, $\lambda_3 \approx 0.74142$ so the eigenvalues of B are $\lambda_1 \approx 0.45858$, $\lambda_2 = 0.6$, $\lambda_3 \approx 0.74142$, $\lambda_4 = 1$.

If $\vec{v}_1, \vec{v}_2, \vec{v}_3$ are eigenvectors of A (with $A\vec{v}_i = \lambda_i\vec{v}_i$), then $\begin{bmatrix} \vec{v}_1 \\ 0 \end{bmatrix}, \begin{bmatrix} \vec{v}_2 \\ 0 \end{bmatrix}, \begin{bmatrix} \vec{v}_3 \\ 0 \end{bmatrix}$ are corresponding eigenvectors of B. Furthermore, $\begin{bmatrix} 75 \\ 100 \\ 125 \\ 1 \end{bmatrix}$ is an eigenvector if B with eigenvalue 1. Since $\lambda_1, \lambda_2, \lambda_3$ are all less than 1, $\lim_{t\to\infty} \vec{x}(t) = \begin{bmatrix} 75 \\ 100 \\ 125 \end{bmatrix}$, as in Exercise 41.

43. a. If $\vec{x}(t) = \begin{bmatrix} r(t) \\ p(t) \\ w(t) \end{bmatrix}$, then $\vec{x}(t+1) = A\vec{x}(t)$ with $A = \begin{bmatrix} \frac{1}{2} & \frac{1}{4} & 0 \\ \frac{1}{2} & \frac{1}{2} & \frac{1}{2} \\ 0 & \frac{1}{4} & \frac{1}{2} \end{bmatrix}$.

The eigenvalues of A are $0, \frac{1}{2}, 1$ with eigenvectors $\begin{bmatrix} 1 \\ -2 \\ 1 \end{bmatrix}, \begin{bmatrix} 1 \\ 0 \\ -1 \end{bmatrix}, \begin{bmatrix} 1 \\ 2 \\ 1 \end{bmatrix}$.

Since $\vec{x}(0) = \begin{bmatrix} 1 \\ 0 \\ 0 \end{bmatrix} = \frac{1}{4}\begin{bmatrix} 1 \\ -2 \\ 1 \end{bmatrix} + \frac{1}{2}\begin{bmatrix} 1 \\ 0 \\ -1 \end{bmatrix} + \frac{1}{4}\begin{bmatrix} 1 \\ 2 \\ 1 \end{bmatrix}$, $\vec{x}(t) = \frac{1}{2}\left(\frac{1}{2}\right)^t \begin{bmatrix} 1 \\ 0 \\ -1 \end{bmatrix} + \frac{1}{4}\begin{bmatrix} 1 \\ 2 \\ 1 \end{bmatrix}$ for $t > 0$.

b. As $t \to \infty$ the ratio is 1 : 2 : 1 (since the first term of $\vec{x}(t)$ drops out).

44. a. We are told that
$a(t+1) = a(t) + j(t)$
$j(t+1) = a(t)$, so that $A = \begin{bmatrix} 1 & 1 \\ 1 & 0 \end{bmatrix}$.

b. $f_A(\lambda) = \lambda(\lambda - 1) - 1 = \lambda^2 - \lambda - 1$ so $\lambda_{1,2} = \frac{1 \pm \sqrt{5}}{2}$ with eigenvectors $\begin{bmatrix} \lambda_1 \\ 1 \end{bmatrix}$ and $\begin{bmatrix} \lambda_2 \\ 1 \end{bmatrix}$.

Since $\vec{x}(0) = \begin{bmatrix} 1 \\ 0 \end{bmatrix} = \frac{1}{\sqrt{5}}\begin{bmatrix} \lambda_1 \\ 1 \end{bmatrix} - \frac{1}{\sqrt{5}}\begin{bmatrix} \lambda_2 \\ 1 \end{bmatrix}$ we have $\vec{x}(t) = \frac{1}{\sqrt{5}}(\lambda_1)^t \begin{bmatrix} \lambda_1 \\ 1 \end{bmatrix} - \frac{1}{\sqrt{5}}(\lambda_2)^t \begin{bmatrix} \lambda_2 \\ 1 \end{bmatrix}$, i.e.

$a(t) = \frac{1}{\sqrt{5}}\left((\lambda_1)^{t+1} - (\lambda_2)^{t+1}\right)$

$j(t) = \frac{1}{\sqrt{5}}\left((\lambda_1)^t - (\lambda_2)^t\right)$.

c. As $t \to \infty$, $\frac{a(t)}{j(t)} \to \lambda_1 = \frac{1+\sqrt{5}}{2}$, since $|\lambda_2| < 1$.

45. This "random" matrix $A = [\vec{0} \;\; \vec{v}_2 \;\; \cdots \;\; \vec{v}_n]$ is unlikely to have any zeros above the diagonal. In this case, the columns $\vec{v}_2, \ldots, \vec{v}_n$ will be linearly independent (none of them is a linear combination of the previous ones), so that rank(A) = $n - 1$ and geometric multiplicity(0) = dim(ker(A)) = n − rank(A) = 1. Alternatively, you can argue in terms of rref(A).

46. a. $f_A(\lambda) = (\lambda - 2)^2 - 3$ so $\lambda_{1,2} = 2 \pm \sqrt{3}$ (or approximately 3.73 and 0.27) with eigenvectors $\begin{bmatrix} 1 \\ \sqrt{3} \end{bmatrix}$ and $\begin{bmatrix} -1 \\ \sqrt{3} \end{bmatrix}$.

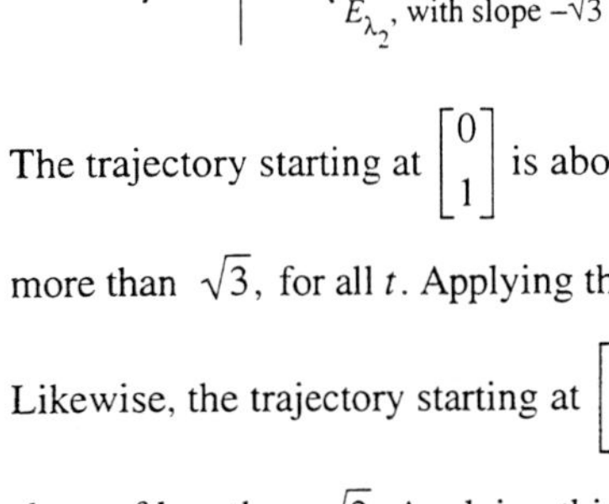

b. The trajectory starting at $\begin{bmatrix} 0 \\ 1 \end{bmatrix}$ is above the line E_{λ_1}, so that $A^t\begin{bmatrix} 0 \\ 1 \end{bmatrix}$ = (second column of A^t) has a slope of more than $\sqrt{3}$, for all t. Applying this to $t = 6$ gives the estimate $\sqrt{3} < \frac{1351}{780}$.

Likewise, the trajectory starting at $\begin{bmatrix} 1 \\ 1 \end{bmatrix}$ is below E_{λ_1}, so that $A^t\begin{bmatrix} 1 \\ 1 \end{bmatrix}$ = (sum of the two columns of A^t) has a slope of less than $\sqrt{3}$. Applying this to $t = 4$ gives $\frac{265}{153} < \sqrt{3}$.

c. $\det(A^6) = (\det A)^6 = 1$ and $\det(A^6) = 1351^2 - 780 \cdot 2340$, so that $1351^2 - 780 \cdot 2340 = 1$.

Dividing both sides by $1351 \cdot 780$ we obtain $\frac{1351}{780} - \frac{2340}{1351} = \frac{1}{780 \cdot 1351} < 10^{-6}$.

Now note that $\frac{2340}{1351}$ is the slope of $A^6\begin{bmatrix}1\\0\end{bmatrix}$, which is less than $\sqrt{3}$.

Therefore $\frac{1351}{780} - \sqrt{3} < \frac{1351}{780} - \frac{2340}{1351} < 10^{-6}$, as claimed.

d. The slope of $A^6\begin{bmatrix}1\\1\end{bmatrix} = \begin{bmatrix}2131\\3691\end{bmatrix}$ is less than $\sqrt{3}$, i.e. $\frac{3691}{2131} < \sqrt{3}$.

6.4

1. $z = 3 - 3i$ so $|z| = \sqrt{3^2 + (-3)^2} = \sqrt{18}$ and $\arg(z) = -\frac{\pi}{4}$, so $z = \sqrt{18}\left(\cos\left(-\frac{\pi}{4}\right) + i\sin\left(-\frac{\pi}{4}\right)\right)$.

2. If $z = r(\cos\phi + i\sin\phi)$ then $z^4 = r^4(\cos 4\phi + i\sin 4\phi)$.

$z^4 = 1$ if $r = 1$, $\cos 4\phi = 1$ and $\sin 4\phi = 0$ so $4\phi = 2k\pi$ for an integer k, and $\phi = \frac{k\pi}{2}$,

i.e. $z = \cos\left(\frac{k\pi}{2}\right) + i\sin\left(\frac{k\pi}{2}\right)$, $k = 0, 1, 2, 3$.

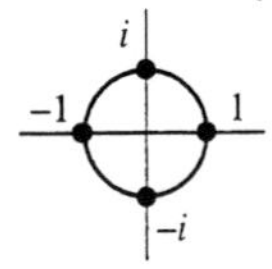

3. If $z = r(\cos\phi + i\sin\phi)$, then $z^n = r^n(\cos(n\phi) + i\sin(n\phi))$.

$z^n = 1$ if $r = 1$, $\cos(n\phi) = 1$, $\sin(n\phi) = 0$ so $n\phi = 2k\pi$ for an integer k, and $\phi = \frac{2k\pi}{n}$,

i.e. $z = \cos\left(\frac{2k\pi}{n}\right) + i\sin\left(\frac{2k\pi}{n}\right)$, $k = 0, 1, 2, \ldots, n-1$.

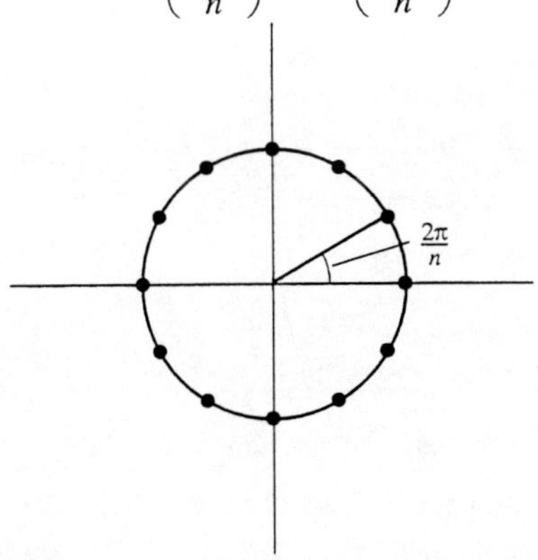

4. Let $z = r(\cos\phi + i\sin\phi)$ then $w = \sqrt{r}\left(\cos\left(\frac{\phi + 2\pi k}{2}\right) + i\sin\left(\frac{\phi + 2\pi k}{2}\right)\right)$, $k = 0, 1$.

5. Let $z = r(\cos\phi + i\sin\phi)$ then $w = \sqrt[n]{r}\left(\cos\left(\frac{\phi + 2\pi k}{n}\right) + i\sin\left(\frac{\phi + 2\pi k}{n}\right)\right)$, $k = 0, 1, 2, \ldots, n-1$.

6. If we have $z = r(\cos\phi + i\sin\phi)$ then $\frac{1}{z}$ must have the property that $z \cdot \frac{1}{z} = 1 = \cos 0 + i \sin 0$

i.e. $|z| \cdot \left|\frac{1}{z}\right| = 1$ and $\arg\left(z \cdot \frac{1}{z}\right) = \arg(z) + \arg\left(\frac{1}{z}\right) = 0$ so $\frac{1}{z} = \frac{1}{r}(\cos(-\phi) + i\sin(-\phi)) = \frac{1}{r}(\cos\phi - i\sin\phi)$ (since cosine is even, sine odd). Hence $\frac{1}{z}$ is a real scalar multiple of $\bar{z}$.

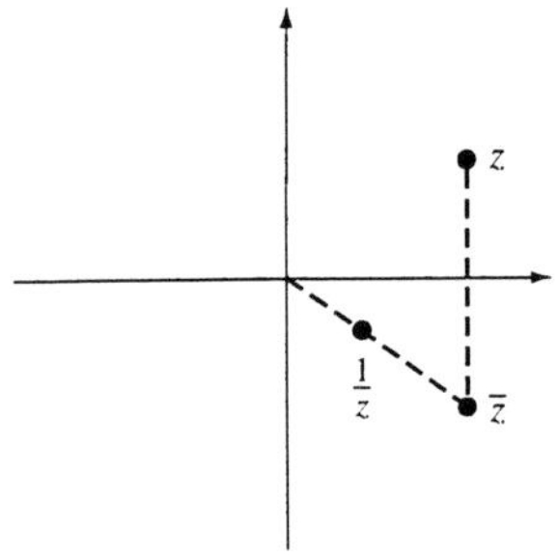

7. $|T(z)| = |z|\sqrt{2}$ and $\arg(T(z)) = \arg(1 - i) + \arg(z) = -\frac{\pi}{4} + \arg(z)$ so T is a clockwise rotation by $\frac{\pi}{4}$ followed by a dilation by $\sqrt{2}$.

8. By Fact 6.4.1, $(\cos 3\phi + i\sin 3\phi) = (\cos\phi + i\sin\phi)^3$, i.e.
$\cos 3\phi + i\sin 3\phi = \cos^3\phi + 3i\cos^2\phi\sin\phi - 3\cos\phi\sin^2\phi - i\sin^3\phi$
$= (\cos^3\phi - 3\cos\phi\sin^2\phi) + i(3\cos^2\phi\sin\phi - \sin^3\phi)$.
Equating real and imaginary parts, we get
$\cos 3\phi = \cos^3\phi - 3\cos\phi\sin^2\phi$
$\sin 3\phi = 3\cos^2\phi\sin\phi - \sin^3\phi$.

9. $|z| = \sqrt{0.8^2 + 0.7^2} = \sqrt{1.15}$, $\arg(z) = \arctan\left(-\frac{0.7}{0.8}\right) \approx -0.72$

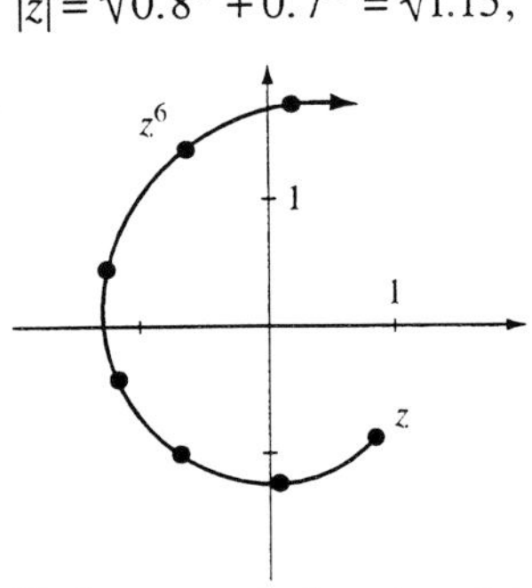

Trajectory spirals outward, in the clockwise direction.

10. Let $p(x) = ax^3 + bx^2 + cx + d$, where $a \neq 0$. Since p must have a real root, say λ_1, we can write $p(x) = a(x - \lambda_1)g(x)$ where $g(x)$ is of the form $g(x) = x^2 + px + q$. On page 347 we see that $g(x) = (x - \lambda_2)(x - \lambda_3)$, so that $p(x) = a(x - \lambda_1)(x - \lambda_2)(x - \lambda_3)$, as claimed.

11. Notice that $f(1) = 0$ so $\lambda = 1$ is a root of f. Hence $f(\lambda) = (\lambda - 1)g(\lambda)$, where $g(\lambda) = \frac{f(\lambda)}{\lambda - 1} = \lambda^2 - 2\lambda + 5$. Setting $g(\lambda) = 0$ we get $\lambda = 1 \pm 2i$ so that $f(\lambda) = (\lambda - 1)(\lambda - 1 - 2i)(\lambda - 1 + 2i)$.

12. We will use the facts:
i) $\overline{z+w} = \overline{z} + \overline{w}$ and
ii) $\left(\overline{z^n}\right) = \overline{z}^n$

These are easy to check. Assume λ_0 is a complex root of $f(\lambda) = a_n\lambda^n + \cdots + a_1\lambda + a_0$ where the coefficients a_i are real. Since λ_0 is a root of *f*, we have $a_n\lambda_0^n + a_{n-1}\lambda_0^{n-1} + \cdots + a_1\lambda_0 + a_0 = 0$.
Taking the conjugate of both sides we get $\overline{a_n\lambda_0^n + a_{n-1}\lambda_0^{n-1} + \cdots + a_1\lambda_0 + a_0} = \overline{0}$ so by fact i), and factoring the real constants we get $a_n\overline{\lambda_0^n} + a_{n-1}\overline{\lambda_0^{n-1}} + \cdots + a_1\overline{\lambda_0} + a_0 = 0$.
Now, by fact ii), $a_n(\overline{\lambda}_0)^n + a_{n-1}(\overline{\lambda}_0)^{n-1} + \cdots + a_1\overline{\lambda}_0 + a_0 = 0$, i.e. $\overline{\lambda}_0$ is also a root of *f*, as claimed.

13. Yes, $\mathbb{Q}$ is a field. Check the axioms on p 348.

14. No, $\mathbb{Z}$ is not a field since multiplicative inverses do not exist, i.e. division within $\mathbb{Z}$ is not possible (Axiom 8 does not hold).

15. Yes, check the axioms on p 348. (additive identity 0 and multiplicative identity 1)

16. Yes, check the axioms on p 348.
(additive identity $\begin{bmatrix} 0 & 0 \\ 0 & 0 \end{bmatrix}$, multiplicative identity $\begin{bmatrix} 1 & 0 \\ 0 & 1 \end{bmatrix}$,
also notice that rotations - dilations commute when multiplied.)

17. No, since multiplication is not commutative; Axiom 5 does not hold.

18. Let $\vec{v}_1, \vec{v}_2$ be two eigenvectors of *A*. Then they define a parallelogram of area $S = |\det[\vec{v}_1 \ \ \vec{v}_2]|$.
Now $A\vec{v}_1 = \lambda_1\vec{v}_1$ and $A\vec{v}_2 = \lambda_2\vec{v}_2$ define a parallelogram of area $S_1 = |\det[\lambda_1\vec{v}_1 \ \ \lambda_2\vec{v}_2]| = |\lambda_1\lambda_2 \det[\vec{v}_1 \ \ \vec{v}_2]|$ so $\frac{S_1}{S} = |\lambda_1\lambda_2| = |\det(A)|$ by Fact 5.3.8. Hence $|\det(A)| = |\lambda_1\lambda_2|$, as claimed. In $\mathbb{R}^3$, a similar argument holds if we replace areas by volumes.

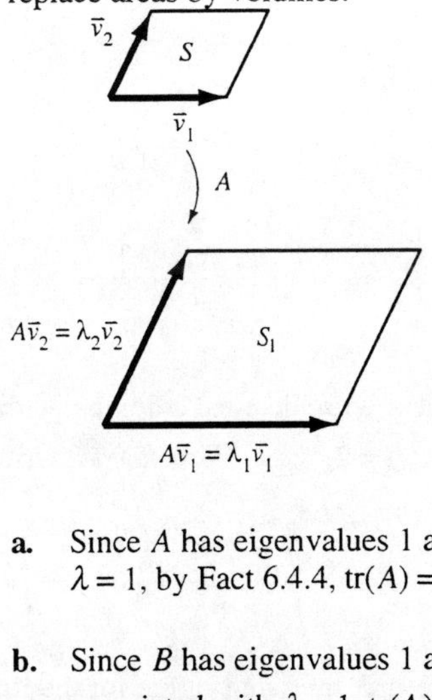

19. a. Since *A* has eigenvalues 1 and 0 associated with *V* and $V^{\perp}$ respectively and since *V* is the eigenspace of $\lambda = 1$, by Fact 6.4.4, tr(A) = m, det(A) = 0.

b. Since *B* has eigenvalues 1 and −1 associated with *V* and $V^{\perp}$ respectively and since *V* is the eigenspace associated with $\lambda = 1$, $\text{tr}(A) = m - (n-m) = 2m - n$, $\det B = (-1)^{n-m}$.

20. $f_A(\lambda) = (\lambda - 3)(\lambda + 3) + 10 = \lambda^2 + 1$ so $\lambda_{1,2} = \pm i$.

21. $f_A(\lambda) = (\lambda - 11)(\lambda + 7) + 90 = \lambda^2 - 4\lambda + 13$ so $\lambda_{1,2} = 2 \pm 3i$.

22. $f_A(\lambda) = (\lambda - 1)(\lambda - 10) + 12 = \lambda^2 - 11\lambda + 22$ so $\lambda_{1,2} = \dfrac{11 \pm \sqrt{33}}{2}$.

23. $f_A(\lambda) = \lambda^3 + 1 = (\lambda - 1)(\lambda^2 + \lambda + 1)$ so $\lambda_1 = 1$, $\lambda_{2,3} = \dfrac{-1 \pm \sqrt{3}i}{2}$.

24. $f_A(\lambda) = \lambda^3 - 3\lambda^2 + 7\lambda - 5$ so $\lambda_1 = 1$, $\lambda_{2,3} = 1 \pm 2i$. (See Exercise 11.)

25. $f_A(\lambda) = \lambda^4 - 1 = (\lambda^2 - 1)(\lambda^2 + 1) = (\lambda - 1)(\lambda + 1)(\lambda - i)(\lambda + i)$ so $\lambda_{1,2} = \pm 1$ and $\lambda_{3,4} = \pm i$

26. $f_A(\lambda) = (\lambda^2 - 2\lambda + 2)(\lambda^2 - 2\lambda) = (\lambda^2 - 2\lambda + 2)(\lambda - 2)\lambda = 0$, so $\lambda_{1,2} = 1 \pm i$, $\lambda_3 = 2$, $\lambda_4 = 0$.

27. By Fact 6.4.4 $\operatorname{tr}(A) = \lambda_1 + \lambda_2 + \lambda_3$, $\det(A) = \lambda_1\lambda_2\lambda_3$ but $\lambda_1 = \lambda_2 \neq \lambda_3$ by assumption, so $\operatorname{tr}(A) = 1 = 2\lambda_2 + \lambda_3$ and $\det(A) = 3 = \lambda_2^2\lambda_3$.
Solving for λ_2, λ_3 we get $-1, 3$ hence $\lambda_1 = \lambda_2 = -1$ and $\lambda_3 = 3$. (Note that the eigenvalues must be real; why?)

28. By Fact 6.4.4,
$2z\bar{z} = 50$
$2 + z + \bar{z} = 8$
If $z = a + bi$ we get $(a + bi) + (a - bi) = 6$ so $a = 3$.
$(a + bi)(a - bi) = 25$, i.e. $a^2 + b^2 = 25$ or $9 + b^2 = 25$ so $b = \pm 4$.
Hence $\lambda_{2,3} = 3 \pm 4i$.

29. $\operatorname{tr}(A) = 0$ so $\lambda_1 + \lambda_2 + \lambda_3 = 0$.
Also, we can compute $\det(A) = bcd > 0$ since $b, c, d > 0$. Therefore, $\lambda_1\lambda_2\lambda_3 > 0$.
Hence two of the eigenvalues must be negative, and the largest one (in absolute value) must be positive.

30. a. The ith entry of $A\vec{x}$ is $\sum_{k=1}^{n} a_{ik}x_k$, so that the sum of all the entries of $A\vec{x}$ is

$$\sum_{i=1}^{n}\sum_{k=1}^{n} a_{ik}x_k = \sum_{k=1}^{n}\sum_{i=1}^{n} a_{ik}x_k = \sum_{k=1}^{n}\Bigg(\underbrace{\sum_{i=1}^{n} a_{ik}}_{\uparrow\ 1}\Bigg)x_k = \sum_{k=1}^{n} x_k = 1.$$

b. As we do some computer experiments, A^t appears to approach a matrix with identical columns, with column sum 1. Let $\vec{v}_1, \vec{v}_2, \ldots, \vec{v}_n$ be an eigenbasis with $\lambda_1 = 1$ and $|\lambda_j| < 1$ for $j = 2, \ldots, n$. For a *fixed* i, write

$\vec{e}_i = c_1\vec{v}_1 + c_2\vec{v}_2 + \cdots + c_n\vec{v}_n$, so that ($i$th column of A^t) $= A^t\vec{e}_i = c_1\vec{v}_1 + [c_2\lambda_2^t\vec{v}_2 + \cdots + c_n\lambda_n^t\vec{v}_n]$.
(The term in *square* brackets goes to zero as t goes to infinity.)
Therefore, $\lim_{t\to\infty}$ (ith column of A^t) $= \lim_{t\to\infty}(A^t\vec{e}_i) = c_1\vec{v}_1$.

Furthermore, the entries of $A^t\vec{e}_i$ add up to 1, for all t, by part a. Therefore, the same is true for the limit (since the limit of a sum is the sum of the limits).
It follows that $\lim_{t\to\infty}(A^t)$ exists and has identical columns, with column sum 1, as claimed.

31. No matter how we choose A, $\frac{1}{15}A$ is a regular transition matrix, so that $\lim_{t\to\infty}\left(\frac{1}{15}A\right)^t$ is a matrix with identical columns by Exercise 30. Therefore, the columns of A^t "become more and more alike" as t approaches infinity, in the sense that $\lim_{t\to\infty}\frac{ij\text{th entry of } A^t}{ik\text{th entry of } A^t} = 1$ for all i, j, k.

32. a. $\vec{x}(t) = \begin{bmatrix} a(t) \\ m(t) \\ s(t) \end{bmatrix} = \begin{bmatrix} 0.6a(t)+0.1m(t)+0.5s(t) \\ 0.2a(t)+0.7m(t)+0.1s(t) \\ 0.2a(t)+0.2m(t)+0.4s(t) \end{bmatrix}$ so $A = \begin{bmatrix} 0.6 & 0.1 & 0.5 \\ 0.2 & 0.7 & 0.1 \\ 0.2 & 0.2 & 0.4 \end{bmatrix}$.

Note that A is a regular transition matrix.

b. By Exercise 30, $\lim_{t\to\infty}(A^t) = [\vec{v} \ \ \vec{v} \ \ \vec{v}]$, where $\vec{v}$ is the unique eigenvector of A with eigenvalue 1 and column sum 1. We find that $\vec{v} = \begin{bmatrix} 0.4 \\ 0.35 \\ 0.25 \end{bmatrix}$.

Now $\lim_{t\to\infty}\vec{x}(t) = \lim_{t\to\infty}(A^t\vec{x}_0) = \left(\lim_{t\to\infty} A^t\right)\vec{x}_0 = [\vec{v} \ \ \vec{v} \ \ \vec{v}]\vec{x}_0 = \vec{v}$, since the components of $\vec{x}_0$ add up to 1.

The market shares approach 40%, 35%, and 25%, respectively, regardless of the initial shares.

33. a. C is obtained from B by dividing each column of B by its first component. Thus, the first row of C will consist of 1's.

b. We observe that the columns of C are almost identical, so that the columns of B are "almost parallel" (that is, almost scalar multiples of each other).

c. Let $\lambda_1, \lambda_2, \ldots, \lambda_5$ be the eigenvalues. Assume λ_1 real and positive and $\lambda_1 > |\lambda_j|$ for $2 \le j \le 5$.

Let $\vec{v}_1, \ldots, \vec{v}_5$ be corresponding eigenvectors. For a fixed i, write $\vec{e}_i = \sum_{j=1}^{5} c_j\vec{v}_j$; then

(ith column of A^t) $= A^t\vec{e}_i = c_1\lambda_1^t\vec{v}_1 + \cdots + c_5\lambda_5^t\vec{v}_5$.

But in the last expression, for large t, the first term is dominant, so the ith column of A^t is almost parallel to $\vec{v}_1$, the eigenvector corresponding to the dominant eigenvalue.

d. By part c, the columns of B and C are almost eigenvectors of A associated with the largest eigenvalue, λ_1. Since the first row of C consists of 1's, the entries in the first row of AC will be close to λ_1.

34. a. The eigenvalues of $\lambda I_n - A$ are $\lambda - \lambda_1, \lambda - \lambda_2, \ldots, \lambda - \lambda_n$, and we were told that $|\lambda - \lambda_1| < |\lambda - \lambda_i|$ for $i = 2, \ldots, n$. We may assume that $\lambda \ne \lambda_1$ (otherwise we are done).

The eigenvalues of $(\lambda I_n - A)^{-1}$ are $(\lambda - \lambda_1)^{-1}, (\lambda - \lambda_2)^{-1}, \ldots, (\lambda - \lambda_n)^{-1}$, and $(\lambda - \lambda_1)^{-1}$ has the largest modulus. The matrices A, $\lambda I_n - A$, and $(\lambda I_n - A)^{-1}$ have the same eigenvectors.

For large t, the columns of the tth power of $(\lambda I_n - A)^{-1}$ will be almost eigenvectors of A. If $\vec{v}$ is such a column, compare $\vec{v}$ and $A\vec{v}$ to find an approximation of λ_1.

b.

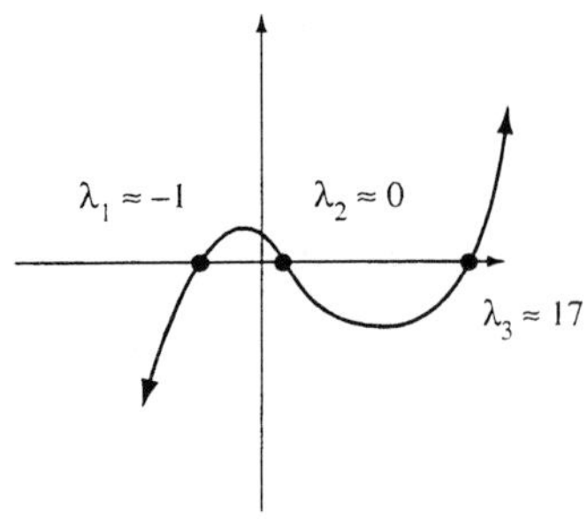

Let $\lambda = -1$.

$$\lambda I_3 - A = \begin{bmatrix} -2 & -2 & -3 \\ -4 & -6 & -6 \\ -7 & -8 & -11 \end{bmatrix}, N = (\lambda I_3 - A)^{-1} = \begin{bmatrix} -9 & -1 & 3 \\ 1 & -0.5 & 0 \\ 5 & 1 & -2 \end{bmatrix}, \text{ and } B = N^{20}.$$

Obtain C from B as in Exercise 33:

$$C \approx \begin{bmatrix} 1 & 1 & 1 \\ -0.98922005729 & -0.98922005729 & -0.098922005729 \\ -0.569298722688 & -0.569298722688 & -0.569298722688 \end{bmatrix}$$

$$AC \approx \begin{bmatrix} -0.905740179522 & -0.905740179522 & -0.905740179522 \\ * & * & * \\ * & * & * \end{bmatrix}$$

The entries in the first row of AC give us a good approximation for λ_1, and the columns of C give us a good approximation for a corresponding eigenvector.

35. If $f_A(\lambda)$ is the characteristic polynomial of A, then, by Fact 6.4.2,

$f_A(\lambda) = (\lambda - \lambda_1)(\lambda - \lambda_2)\cdots(\lambda - \lambda_n) = \lambda^n - (\lambda_1 + \lambda_2 + \lambda_3 + \cdots + \lambda_n)\lambda^{n-1} + \cdots + (-1)^n \lambda_1 \cdots \lambda_n$.

But, by Fact 6.2.5, the coefficient of λ^{n-1} is $-\text{tr}(A)$, hence $\text{tr}(A) = \lambda_1 + \cdots + \lambda_n$.

36. a. The entries in the first row are age-specific birth rates and the entries just below the diagonal are age-specific survival rates. For example, the entry 1.6 in the first row tells us that during the next 15 years the people who are 15–30 years old today will *on average* have 1.6 children (3.2 per couple) who will survive to the next census. The entry 0.53 tells us that 53% of those in the age group 45–60 today will still be alive in 15 years (they will then be in the age group 60–75).

b. Using techonology, we find the largest eigenvalue $\lambda_1 = 1.908$ with associated eigenvector $\vec{v}_1 \approx \begin{bmatrix} 0.574 \\ 0.247 \\ 0.115 \\ 0.047 \\ 0.014 \\ 0.002 \end{bmatrix}$.

The components of $\vec{v}_1$ give the distribution of the population among the age groups in the long run, assuming that current trends continue. λ_1 gives the factor by which the population will grow in the long run in a period of 15 years; this translates to an annual growth factor of $\sqrt[15]{1.908} \approx 1.044$, or an annual growth of about 4.4%.

37. a. Use that $\overline{w}+\overline{z}=\overline{w+z}$ and $\overline{w}\,\overline{z}=\overline{wz}$.

$$\begin{bmatrix} w_1 & -\overline{z}_1 \\ z_1 & \overline{w}_1 \end{bmatrix}+\begin{bmatrix} w_2 & -\overline{z}_2 \\ z_2 & \overline{w}_2 \end{bmatrix}=\begin{bmatrix} w_1+w_2 & -(\overline{z_1+z_2}) \\ z_1+z_2 & \overline{w_1+w_2} \end{bmatrix} \text{ is in } \mathbb{H}.$$

$$\begin{bmatrix} w_1 & -\overline{z}_1 \\ z_1 & \overline{w}_1 \end{bmatrix}\begin{bmatrix} w_2 & -\overline{z}_2 \\ z_2 & \overline{w}_2 \end{bmatrix}=\begin{bmatrix} w_1w_2-\overline{z}_1z_2 & -(\overline{z_1w_2+\overline{w}_1z_2}) \\ z_1w_2+\overline{w}_1z_2 & \overline{w_1w_2-\overline{z}_1z_2} \end{bmatrix} \text{ is in } \mathbb{H}.$$

b. If A in $\mathbb{H}$ is nonzero, then $\det(A)=w\overline{w}+z\overline{z}=|w|^2+|z|^2>0$, so that A is invertible.

c. Yes; if $A=\begin{bmatrix} w & -\overline{z} \\ z & \overline{w} \end{bmatrix}$, then $A^{-1}=\dfrac{1}{|w|^2+|z|^2}\begin{bmatrix} \overline{w} & \overline{z} \\ -z & w \end{bmatrix}$ is in $\mathbb{H}$.

d. For example, if $A=\begin{bmatrix} i & 0 \\ 0 & -i \end{bmatrix}$ and $B=\begin{bmatrix} 0 & -1 \\ 1 & 0 \end{bmatrix}$, then $AB=\begin{bmatrix} 0 & -i \\ -i & 0 \end{bmatrix}$ and $BA=\begin{bmatrix} 0 & i \\ i & 0 \end{bmatrix}$.

38. a. $C_4^2=\begin{bmatrix} 0&0&1&0 \\ 0&0&0&1 \\ 1&0&0&0 \\ 0&1&0&0 \end{bmatrix}$, $C_4^3=\begin{bmatrix} 0&1&0&0 \\ 0&0&1&0 \\ 0&0&0&1 \\ 1&0&0&0 \end{bmatrix}$, $C_4^4=I_4$, then $C_4^{4+k}=C_4^k$.

The diagram below illustrates how C_4 acts on the basis vectors $\vec{e}_i$:

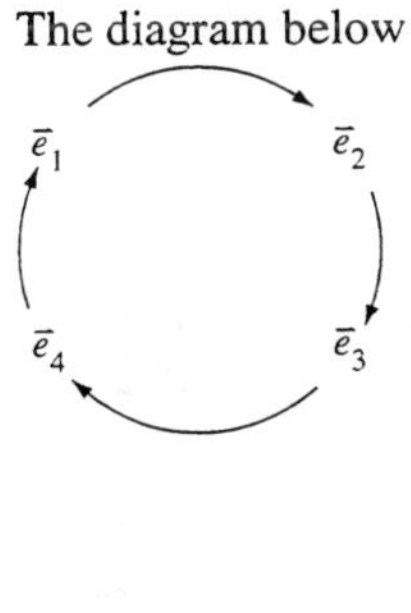

b. The eigenvalues are $\lambda_1=1$, $\lambda_2=-1$, $\lambda_3=i$, and $\lambda_4=-i$, and for each eigenvalue λ_k, $\vec{v}_k=\begin{bmatrix} \lambda_k^3 \\ \lambda_k^2 \\ \lambda_k \\ 1 \end{bmatrix}$ is an associated eigenvector.

c. $M=aI_4+bC_4+cC_4^2+dC_4^3$

If $\vec{v}$ is an eigenvector of C_4 with eigenvalue λ, then $M\vec{v}=a\vec{v}+b\lambda\vec{v}+c\lambda^2\vec{v}+d\lambda^3\vec{v}=(a+b\lambda+c\lambda^2+d\lambda^3)\vec{v}$, so that $\vec{v}$ is an eigenvector of M as well, with eigenvalue $a+b\lambda+c\lambda^2+d\lambda^3$.
The eigenbasis for C_4 we found in part b is an eigenbasis for all circulant 4×4 matrices.

39. The diagram below illustrates how C_n acts on the basis $\vec{e}_1, \vec{e}_2, \ldots, \vec{e}_n$ of $\mathbb{R}^n$:

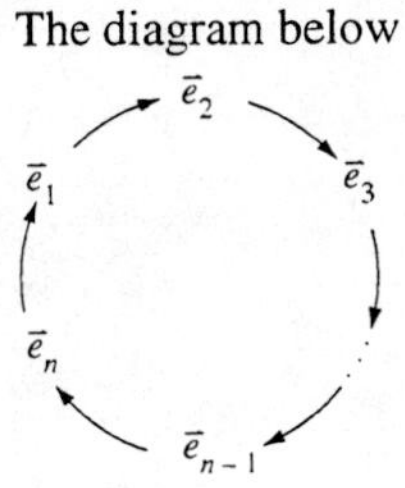

a. Based on the diagram above we see that C_n^k takes $\vec{e}_i$ to $\vec{e}_{i+k}$ "modulo n," that is, if $i + k$ exceeds n then C_n^k takes $\vec{e}_i$ to $\vec{e}_{i+k-n}$ (for $k = 1, \ldots, n-1$).

To put it differently: C_n^k is the matrix whose ith column is $\vec{e}_{i+k}$ if $i + k \leq n$ and $\vec{e}_{i+k-n}$ if $i + k > n$ (for $k = 1, \ldots, n-1$).

b. The characteristic polynomial is $\lambda^n - 1$, so that the eigenvalues are the n distinct solutions of the equation $\lambda^n = 1$ (the so-called nth roots of unity), equally spaced points along the unit circle, $\lambda_k = \cos\left(\frac{2\pi k}{n}\right) + i\sin\left(\frac{2\pi k}{n}\right)$, for $k = 0, 1, \ldots, n-1$ (compare with Exercise 5). For each eigenvalue λ_k,

$$\vec{v}_k = \begin{bmatrix} \lambda_k^{n-1} \\ \vdots \\ \lambda_k^2 \\ \lambda_k \\ 1 \end{bmatrix} \text{ is an associated eigenvector.}$$

c. The eigenbasis $\vec{v}_0, \vec{v}_1, \ldots, \vec{v}_{n-1}$ for C_n we found in part b is in fact an eigenbasis for all circulant $n \times n$ matrices.

40. In Exercise 6.2.38 we derived the formula $x = \sqrt[3]{\frac{q}{2} + \sqrt{\left(\frac{q}{2}\right)^2 + \left(\frac{p}{3}\right)^3}} + \sqrt[3]{\frac{q}{2} - \sqrt{\left(\frac{q}{2}\right)^2 + \left(\frac{p}{3}\right)^3}}$

for the solution of the equation $x^3 + px - q$. Here $\left(\frac{q}{2}\right)^2 + \left(\frac{p}{3}\right)^3$ is negative, and we can write

$$x = \sqrt[3]{\frac{q}{2} + i\sqrt{-\left(\frac{q}{2}\right)^2 + \left(\frac{-p}{3}\right)^3}} + \sqrt[3]{\frac{q}{2} - i\sqrt{-\left(\frac{q}{2}\right)^2 + \left(\frac{-p}{3}\right)^3}}.$$

Let us write this solution in polar coordinates:

$$x = \sqrt[3]{\left(-\frac{p}{3}\right)^{3/2}(\cos\alpha + i\sin\alpha)} + \sqrt[3]{\left(-\frac{p}{3}\right)^{3/2}(\cos\alpha - i\sin\alpha)}$$

$$= \sqrt{-\frac{p}{3}}\left(\cos\frac{\alpha + 2\pi k}{3} + i\sin\frac{\alpha + 2\pi k}{3}\right) + \sqrt{\frac{-p}{3}}\left(\cos\frac{\alpha + 2\pi k}{3} - i\sin\frac{\alpha + 2\pi k}{3}\right)$$

$$= 2\sqrt{-\frac{p}{3}}\cos\left(\frac{\alpha + 2\pi k}{3}\right), \quad k = 0, 1, 2$$

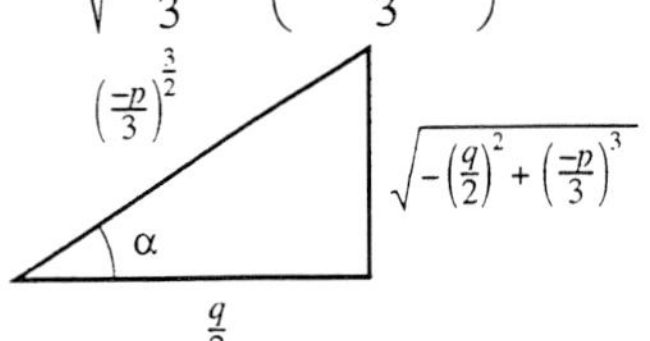

$$\cos\alpha = \frac{\frac{q}{2}}{\left(\frac{-p}{3}\right)^{\frac{3}{2}}}$$

Answer:

$$x_{1,2,3} = 2\sqrt{\frac{-p}{3}}\cos\left(\frac{\alpha + 2\pi k}{3}\right), \quad k = 0, 1, 2, \text{ where } \alpha = \arccos\left(\frac{\frac{q}{2}}{\left(-\frac{p}{3}\right)^{3/2}}\right).$$

Note that x is on the interval $\left(\sqrt{\frac{-p}{3}}, 2\sqrt{\frac{-p}{3}}\right)$ when $k = 0$, on $\left(-2\sqrt{\frac{-p}{3}}, -\sqrt{\frac{-p}{3}}\right)$ when $k = 1$ and on $\left(-\sqrt{\frac{-p}{3}}, \sqrt{\frac{-p}{3}}\right)$ when $k = -1$ (Think about it!).

41. Substitute $\rho = \frac{1}{x}$ into $14\rho^2 + 12\rho^3 - 1 = 0$;

$\frac{14}{x^2} + \frac{12}{x^3} - 1 = 0$

$14x + 12 - x^3 = 0$

$\underline{x^3 - 14x = 12}$

Now use the formula derived in Exercise 40 to find x, with $p = -14$ and $q = 12$. There is only one positive solution, $x \approx 4.114$, so that $\rho = \frac{1}{x} \approx 0.243$.

6.5

1. $\lambda_1 = 0.9$, $\lambda_2 = 0.8$, so by Fact 6.5.2, $\vec{0}$ is a stable equilibrium.

2. $\lambda_1 = -1.1$, $\lambda_2 = 0.9$, so by Fact 6.5.2, $\vec{0}$ is not a stable equilibrium. $(|\lambda_1| > 1)$

3. $\lambda_{1,2} = 0.8 \pm (0.7)i$ so $|\lambda_1| = |\lambda_2| = \sqrt{0.64 + 0.49} > 1$ so $\vec{0}$ is not a stable equilibrium.

4. $\lambda_{1,2} = -0.9 \pm (0.4)i$ so $|\lambda_1| = |\lambda_2| = \sqrt{0.81 + 0.16} < 1$ so $\vec{0}$ is a stable equilibrium.

5. $\lambda_1 = 0.8$, $\lambda_2 = 1.1$ so $\vec{0}$ is not a stable equilibrium.

6. $\lambda_{1,2} = 0.8 \pm (0.6)i$ so $|\lambda_1| = |\lambda_2| = \sqrt{0.64 + 0.36} = 1$ and $\vec{0}$ is not a stable equilibrium.

7. $\lambda_{1,2} = 0.9 \pm (0.5)i$ so $|\lambda_1| = |\lambda_2| = \sqrt{0.81 + 0.25} > 1$ and $\vec{0}$ is not a stable equilibrium.

8. $\lambda_1 = 0.9$, $\lambda_2 = 0.8$ so $\vec{0}$ is a stable equilibrium.

9. $\lambda_{1,2} = 0.8 \pm (0.6)i$, $\lambda_3 = 0.7$, so since $|\lambda_1| = |\lambda_2| = 1$ and $\vec{0}$ is not a stable equilibrium.

10. $\lambda_{1,2} = 0$, $\lambda_3 = 0.9$ so $\vec{0}$ is a stable equilibrium.

11. $\lambda_1 = k$, $\lambda_2 = 0.9$ so $\vec{0}$ is a stable equilibrium if $|k| < 1$.

12. $\lambda_{1,2} = 0.6 \pm ik$ so $\vec{0}$ is a stable equilibrium if $|\lambda_1| = |\lambda_2| = \sqrt{0.36 + k^2} < 1$ i.e. if $k^2 < 0.64$ or $|k| < 0.8$.

13. Since $\lambda_1 = 0.7$, $\lambda_2 = -0.9$, $\vec{0}$ is a stable equilibrium regardless of the value of k.

14. $\lambda_1 = 0$, $\lambda_2 = 2k$ so $\vec{0}$ is a stable equilibrium if $|2k| < 1$ or $|k| < \frac{1}{2}$.

15. $\lambda_{1,2} = 1 \pm \frac{1}{10}\sqrt{k}$

If $k \geq 0$ then $\lambda_1 = 1 + \frac{1}{10}\sqrt{k} \geq 1$. If $k < 0$ then $|\lambda_1| = |\lambda_2| > 1$. Thus, the zero state isn't a stable equilibrium for any real k.

16. $\lambda_{1,2} = \frac{2 \pm \sqrt{1+30k}}{10}$ so $|2 \pm \sqrt{1+30k}|$ must be less than 10. $\lambda_{1,2}$ are *real* if $k \geq -\frac{1}{30}$. In this case it is required that $2 + \sqrt{1+30k} < 10$ and $-10 < 2 - \sqrt{1+30k}$, which means that $\sqrt{1+30k} < 8$ or $k < \frac{21}{10}$.

$\lambda_{1,2}$ are *complex* if $k < -\frac{1}{30}$. Here it is required that $4 + (-1 - 30k) < 100$ or $k > -\frac{97}{30}$. Overall, $\vec{0}$ is a stable equilibrium if $-\frac{97}{30} < k < \frac{21}{10}$.

17. $\lambda_{1,2} = 0.6 \pm (0.8)i = 1(\cos\phi \pm i \cdot \sin\phi)$, where $\phi = \arctan\left(\frac{0.8}{0.6}\right) = \arctan\left(\frac{4}{3}\right) \approx 0.927$.

$E_{\lambda_1} = \ker\begin{bmatrix} 0.8i & 0.8 \\ -0.8 & 0.8i \end{bmatrix} = \text{span}\begin{bmatrix} -1 \\ i \end{bmatrix}$ so $\vec{w} = \begin{bmatrix} 0 \\ 1 \end{bmatrix}$, $\vec{v} = \begin{bmatrix} -1 \\ 0 \end{bmatrix}$.

$\vec{x}_0 = \begin{bmatrix} 0 \\ 1 \end{bmatrix} = 1\vec{w} + 0\vec{v}$, so $a = 1$ and $b = 0$. Now we use Fact 6.5.3:

$\vec{x}(t) = \begin{bmatrix} 0 & -1 \\ 1 & 0 \end{bmatrix}\begin{bmatrix} \cos(\phi t) & -\sin(\phi t) \\ \sin(\phi t) & \cos\phi t \end{bmatrix}\begin{bmatrix} 1 \\ 0 \end{bmatrix} = \begin{bmatrix} 0 & -1 \\ 1 & 0 \end{bmatrix}\begin{bmatrix} \cos\phi t \\ \sin\phi t \end{bmatrix} = \begin{bmatrix} -\sin\phi t \\ \cos\phi t \end{bmatrix}$, where $\phi = \arctan\left(\frac{4}{3}\right) \approx 0.927$.

The trajectory is a circle:

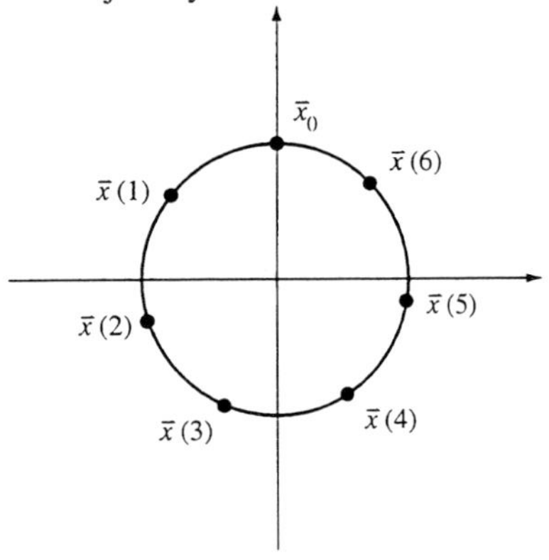

18. $\lambda_{1,2} = \frac{-4 \pm 2\sqrt{3}i}{5} = r(\cos\phi \pm i\sin\phi)$, where $r \approx 1.058$ and $\phi = \pi - \arctan\left(\frac{\frac{2\sqrt{3}}{5}}{\frac{4}{5}}\right) \approx 2.428$ (second quadrant).

$E_{\lambda_1} = \text{span}\begin{bmatrix} \frac{\sqrt{3}}{2} \\ i \end{bmatrix}$, so $\vec{w} = \begin{bmatrix} 0 \\ 1 \end{bmatrix}$, $\vec{v} = \begin{bmatrix} \frac{\sqrt{3}}{2} \\ 0 \end{bmatrix}$.

$\vec{x}_0 = 1\vec{w} + 0\vec{v}$, so $a = 1$, $b = 0$.

$\vec{x}(t) = r^t[\vec{w} \quad \vec{v}]\begin{bmatrix} \cos(\phi t) & -\sin(\phi t) \\ \sin(\phi t) & \cos(\phi t) \end{bmatrix}\begin{bmatrix} 1 \\ 0 \end{bmatrix} \approx (1.058)^t\begin{bmatrix} 0 & 0.866 \\ 1 & 0 \end{bmatrix}\begin{bmatrix} \cos(\phi t) \\ \sin(\phi t) \end{bmatrix} \approx (1.058)^t\begin{bmatrix} 0.866 \cdot \sin(2.428t) \\ \cos(2.428t) \end{bmatrix}$

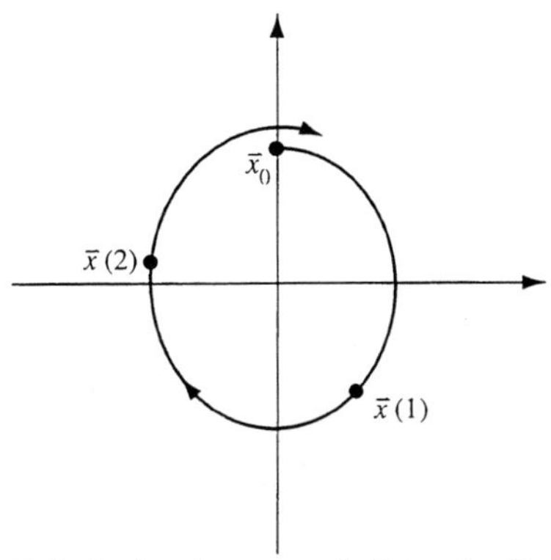

Spirals slowly outwards (plot the first few points).

19. $\lambda_{1,2} = 2 \pm 3i$, $r = \sqrt{13}$, and $\phi = \arctan\left(\frac{3}{2}\right) \approx 0.98$, so

$\lambda_1 \approx \sqrt{13}(\cos(0.98) + i\sin(0.98))$, $[\vec{w}\ \ \vec{v}] = \begin{bmatrix} 0 & -1 \\ 1 & 0 \end{bmatrix}$, $\begin{bmatrix} a \\ b \end{bmatrix} = \begin{bmatrix} 1 \\ 0 \end{bmatrix}$ and $\vec{x}(t) \approx \sqrt{13}^t \begin{bmatrix} -\sin(0.98t) \\ \cos(0.98t) \end{bmatrix}$.

Spirals outwards (rotation-dilation).

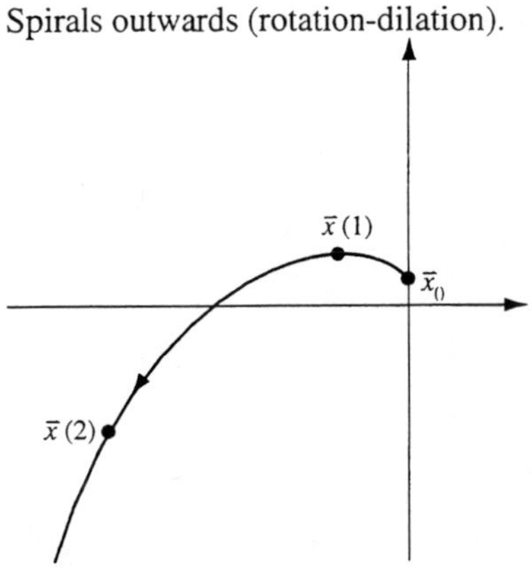

20. $\lambda_{1,2} = 4 \pm 3i$, $r = 5$, $\phi = \arctan\left(\frac{3}{4}\right) \approx 0.64$, so $\lambda_1 \approx 5(\cos(0.64) + i\sin(0.64))$, $[\vec{w}\ \ \vec{v}] = \begin{bmatrix} 0 & 1 \\ 1 & 0 \end{bmatrix}$, $\begin{bmatrix} a \\ b \end{bmatrix} = \begin{bmatrix} 1 \\ 0 \end{bmatrix}$ and

$\vec{x}(t) \approx 5^t \begin{bmatrix} \sin(0.64t) \\ \cos(0.69t) \end{bmatrix}$.

Spirals outwards (rotation-dilation).

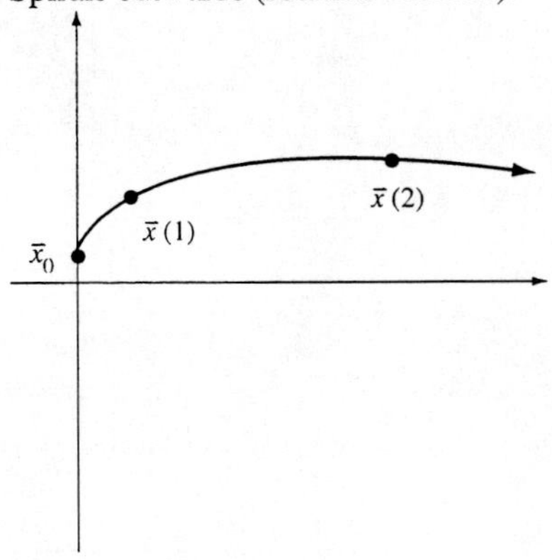

21. $\lambda_{1,2} = 4 \pm i$, $r = \sqrt{17}$, $\phi = \arctan\left(\frac{1}{4}\right) \approx 0.245$ so

$\lambda_1 \approx \sqrt{17}(\cos(0.245) + i\sin(0.245))$, $[\vec{w} \quad \vec{v}] = \begin{bmatrix} 0 & 5 \\ 1 & 3 \end{bmatrix}$, $\begin{bmatrix} a \\ b \end{bmatrix} = \begin{bmatrix} 1 \\ 0 \end{bmatrix}$ and $\vec{x}(t) \approx \sqrt{17}^t \begin{bmatrix} 5\sin(0.245t) \\ \cos(0.245t) + 3\sin(0.245t) \end{bmatrix}$

Spirals outwards.

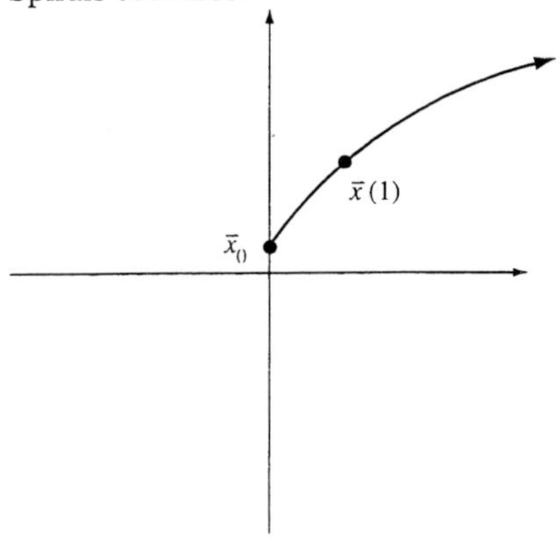

22. $\lambda_{1,2} = -2 \pm 3i$, $r = \sqrt{13}$, $\phi \approx 2.16$ (in second quadrant)

$[\vec{w} \quad \vec{v}] = \begin{bmatrix} 0 & -5 \\ 1 & -3 \end{bmatrix}$, $\begin{bmatrix} a \\ b \end{bmatrix} = \begin{bmatrix} 1 \\ 0 \end{bmatrix}$ so $\vec{x}(t) = \sqrt{13}^t \begin{bmatrix} -5\sin(\phi t) \\ \cos(\phi t) - 3\sin(\phi t) \end{bmatrix}$, where $\phi \approx 2.16$.

Spirals outwards.

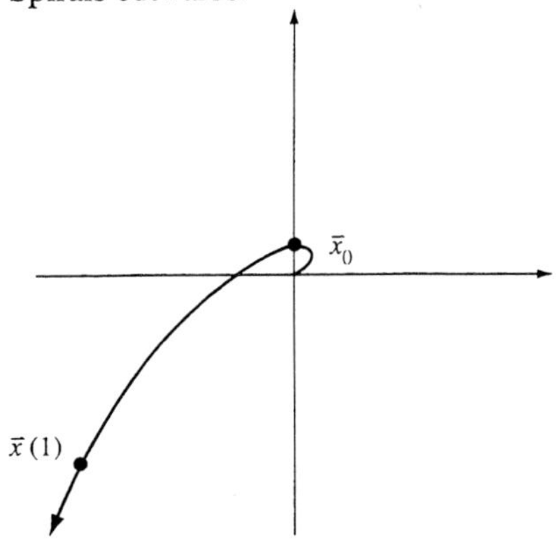

23. $\lambda_{1,2} = 0.4 \pm 0.3i$, $r = \frac{1}{2}$, $\phi = \arctan\left(\frac{0.3}{0.4}\right) \approx 0.643$

$[\vec{w} \quad \vec{v}] = \begin{bmatrix} 0 & 5 \\ 1 & 3 \end{bmatrix}$, $\begin{bmatrix} a \\ b \end{bmatrix} = \begin{bmatrix} 1 \\ 0 \end{bmatrix}$ so $\vec{x}(t) = \left(\frac{1}{2}\right)^t \begin{bmatrix} 5\sin(\phi t) \\ \cos(\phi t) + 3\sin\phi(t) \end{bmatrix}$.

Spirals inwards.

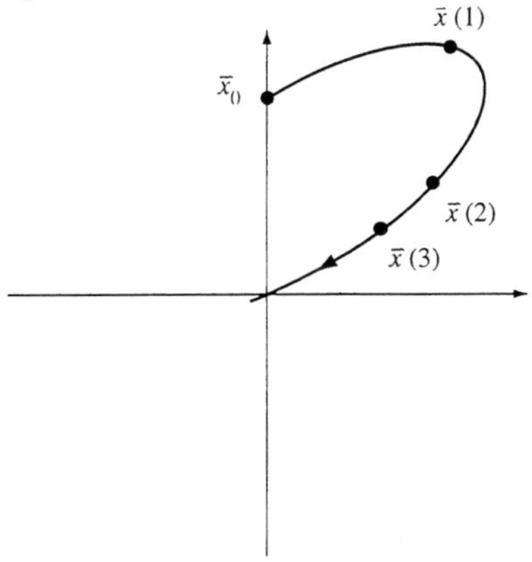

24. $\lambda_{1,2} = -0.8 \pm 0.6i,\ r = 1,\ \phi = \pi - \arctan\left(\frac{.6}{.8}\right) \approx 2.5$ (second quadrant)

$[\vec{w}\ \ \vec{v}] = \begin{bmatrix} 0 & -5 \\ 1 & -3 \end{bmatrix}, \begin{bmatrix} a \\ b \end{bmatrix} = \begin{bmatrix} 1 \\ 0 \end{bmatrix}$ so $\vec{x}(t) = \begin{bmatrix} -5\sin(\phi t) \\ \cos(\phi t) - 3\sin(\phi t) \end{bmatrix}$, an ellipse.

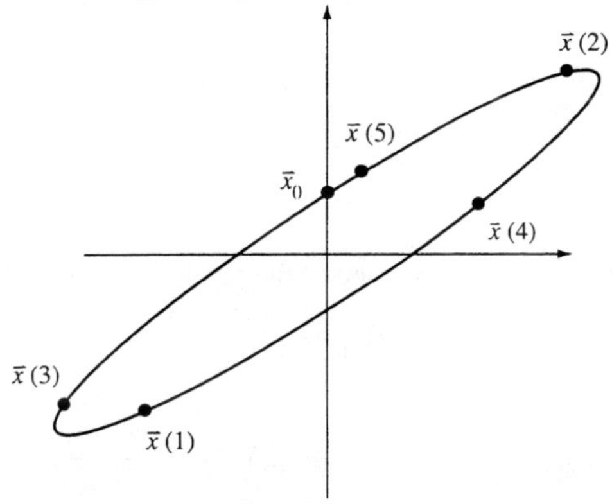

25. Not stable since if λ is an eigenvalue of A, then $\frac{1}{\lambda}$ is an eigenvalue of A^{-1} and $\left|\frac{1}{\lambda}\right| = \frac{1}{|\lambda|} > 1$.

26. Stable since A and A^T have the same eigenvalues.

27. Stable since if λ is an eigenvalue of $-A$, then $-\lambda$ is an eigenvalue of $-A$ and $|-\lambda| = |\lambda|$.

28. Not stable, since if λ is an eigenvalue of A, then $(\lambda - 2)$ is an eigenvalue of $(A - 2I_n)$ and $|\lambda - 2| > 1$.

29. Cannot tell; for example, if $A = \begin{bmatrix} \frac{1}{2} & 0 \\ 0 & \frac{1}{2} \end{bmatrix}$, then $A + I_2$ is $\begin{bmatrix} \frac{3}{2} & 0 \\ 0 & \frac{3}{2} \end{bmatrix}$ and the zero state is not stable, but if

$A = \begin{bmatrix} -\frac{1}{2} & 0 \\ 0 & -\frac{1}{2} \end{bmatrix}$ then $A + I_2 = \begin{bmatrix} \frac{1}{2} & 0 \\ 0 & \frac{1}{2} \end{bmatrix}$ and the zero state is stable.

30. We need to determine for which values of det(A) and tr(A) the modulus of both eigenvalues is less than 1. We will first think about the border line case and examine when one of the moduli is exactly 1:
If one of the eigenvalues is 1 and the other is λ, then tr(A) = $\lambda + 1$ and det(A) = λ, so that det(A) = tr(A) − 1.
If one of the eigenvalues is −1 and the other is λ, then tr(A) = $\lambda - 1$ and det(A) = $-\lambda$, so that det(A) = −tr(A) − 1.
If the eigenvalues are complex conjugates with modulus 1, then det(A) = 1 and $|\text{tr}(A)| < 2$ (think about it!).
It is convenient to represent these conditions in the tr-det plane, where each 2×2 matrix A is represented by the point (tr$A$, det$A$):

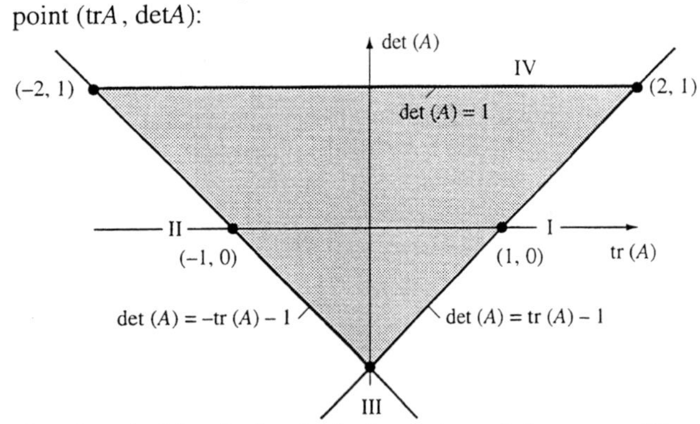

If tr(A) = det(A) = 0, then both eigenvalues of A are zero. We can conclude that throughout the shaded triangle above the modulus of both eigenvalues will be less than 1, since the modulus of the eigenvalues changes continuously with tr(A) and det(A) (consider the quadratic formula!). Conversely, we can choose sample points to show that in all the other four regions above the modulus of at least one of the eigenvalues exceeds one; consider the matrices $\begin{bmatrix} 2 & 0 \\ 0 & 0 \end{bmatrix}$ (in (I)), $\begin{bmatrix} -2 & 0 \\ 0 & 0 \end{bmatrix}$ (in (II)), $\begin{bmatrix} 2 & 0 \\ 0 & -2 \end{bmatrix}$ (in (III)), and $\begin{bmatrix} 0 & -2 \\ 2 & 0 \end{bmatrix}$ (in (IV)).
It follows that throughout these four regions, (I), (II), (III), and (IV), at least one of the eigenvalues will have a modulus exceeding one.
The point (tr$A$, det$A$) is in the shaded triangle if det(A) < 1, det(A) > tr(A) − 1 and det(A) > −tr(A) − 1. This means that $|\text{tr}\,A| - 1 < \det(A) < 1$, as claimed.

31. **a.** We will use the fact that for any two complex numbers z and w, $\overline{z+w} = \bar{z} + \bar{w}$ and $\overline{zw} = \bar{z}\,\bar{w}$.

The ijth entry of $\overline{AB}$ is $\overline{\sum_{k=1}^{n} a_{ik}b_{kj}} = \sum_{k=1}^{n} \overline{a_{ik}b_{kj}} = \sum_{k=1}^{n} \overline{a_{ik}}\,\overline{b_{kj}}$, which is the ijth entry of $\bar{A}\,\bar{B}$, as claimed.

b. Use part a, where B is the $n \times 1$ matrix $\vec{v} + i\vec{w}$. We are told that $AB = \lambda B$, where $\lambda = p + iq$. Then $A\bar{B} = \bar{A}\,\bar{B} = \overline{AB} = \overline{\lambda B} = \bar{\lambda}\bar{B}$, or $A(\vec{v} - i\vec{w}) = (p - iq)(\vec{v} - i\vec{w})$.

32. Take conjugates of both sides of the equation $\vec{x}_0 = c_1(\vec{v} + i\vec{w}) + c_2(\vec{v} - i\vec{w})$:
$\vec{x}_0 = \bar{\vec{x}}_0 = \overline{c_1(\vec{v} + i\vec{w}) + c_2(\vec{v} - i\vec{w})} = \bar{c}_1(\vec{v} - i\vec{w}) + \bar{c}_2(\vec{v} + i\vec{w}) = \bar{c}_2(\vec{v} + i\vec{w}) + \bar{c}_1(\vec{v} - i\vec{w})$.
The claim that $c_2 = \bar{c}_1$ now follows from the fact that the representation of $\vec{x}_0$ as a linear combination of the linearly independent vectors $\vec{v} + i\vec{w}$ and $\vec{v} - i\vec{w}$ is unique.

33. $\lambda_{1,2} = 0.99 \pm (0.01)i$ with eigenvector $\begin{bmatrix} 0 \\ 1 \end{bmatrix} + i\begin{bmatrix} 1 \\ 0 \end{bmatrix}$ for λ_1, so $\vec{w} = \begin{bmatrix} 1 \\ 0 \end{bmatrix}$, $\vec{v} = \begin{bmatrix} 0 \\ 1 \end{bmatrix}$, $\phi = \arctan\left(\frac{0.01}{0.99}\right) \approx 0.01$,

$r \approx 0.99$. Hence $\vec{x}(0) = \begin{bmatrix} 100 \\ 0 \end{bmatrix} = 100\vec{w} + 0\vec{v}$ and

$\vec{x}(t) \approx 0.99^t \begin{bmatrix} 1 & 0 \\ 0 & 1 \end{bmatrix}\begin{bmatrix} \cos(0.01t) & -\sin(0.01t) \\ \sin(0.01t) & \cos(0.01t) \end{bmatrix}\begin{bmatrix} 100 \\ 0 \end{bmatrix} = 0.99^t \cdot 100\begin{bmatrix} \cos(0.01t) \\ \sin(0.01t) \end{bmatrix}$ i.e. $\vec{x}(t)$ spirals in.

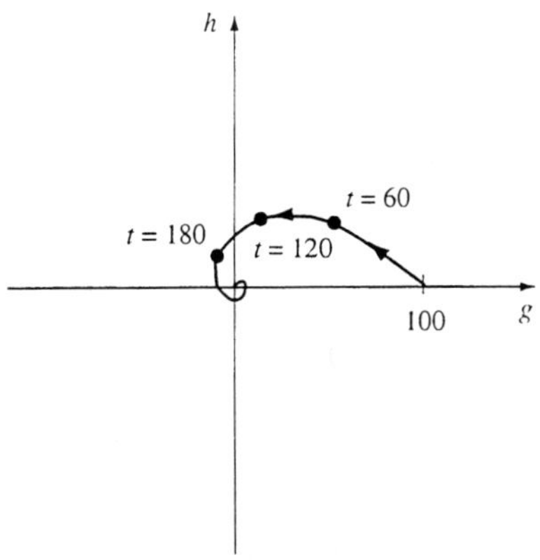

Hence, both glucose and excess insulin oscillate with damped oscillations until all of the excess sugar has been eliminated.

34. a. If $|\det A| = |\lambda_1\lambda_2\cdots\lambda_n| = |\lambda_1||\lambda_2|\cdots|\lambda_n| > 1$ then at least one eigenvalue is greater than one in modulus and the zero state fails to be stable.

b. If $|\det A| = |\lambda_1||\lambda_2|\cdots|\lambda_n| < 1$ we cannot conclude anything about the stability of $\vec{0}$.
$|2||0.1| < 1$ and $|0.2||0.1| < 1$ but in the first case we would not have stability, in the second case we would.

35. a. Let $\vec{v}_1, \ldots, \vec{v}_n$ be an eigenbasis for A. Then $\vec{x}(t) = \sum_{i=1}^{n} c_i\lambda_i^t\vec{v}_i$ and

$$\|\vec{x}(t)\| = \left\|\sum_{i=1}^{n} c_i\lambda_i^t\vec{v}_i\right\| \le \sum_{i=1}^{n}\left\|c_i\lambda_i^t\vec{v}_i\right\| = \sum_{i=1}^{n}\underbrace{|\lambda_i|^t}_{\le 1}\|c_i\vec{v}_i\| \le \sum_{i=1}^{n}\|c_i\vec{v}_i\|.$$

The last quantity, $\sum_{i=1}^{n}\|c_i\vec{v}_i\|$, gives the desired bound M.

b. $A = \begin{bmatrix} 1 & 1 \\ 0 & 1 \end{bmatrix}$ represents a shear parallel to the x-axis, with $A\begin{bmatrix} k \\ 1 \end{bmatrix} = \begin{bmatrix} k+1 \\ 1 \end{bmatrix}$, so that $\vec{x}(t) = A^t\begin{bmatrix} 0 \\ 1 \end{bmatrix} = \begin{bmatrix} t \\ 1 \end{bmatrix}$ is not bounded. This does not contradict part a, since there is no eigenbasis for A.

36. If the zero state is stable, then $\lim_{t\to\infty}$ (ith column of A^t) $= \lim_{t\to\infty}(A^t\vec{e}_i) = \vec{0}$, so that all columns and therefore all entries of A^t approach 0.
Conversely, if $\lim_{t\to\infty} A^t = 0$, then $\lim_{t\to\infty}(A^t\vec{x}_0) = \left(\lim_{t\to\infty} A^t\right)\vec{x}_0 = \vec{0}$ for all $\vec{x}_0$ (check the details).

37. a. Write $Y(t+1) = Y(t) = Y$, $C(t+1) = C(t) = C$, $I(t+1) = I(t) = I$.

$$\left|\begin{aligned} Y &= C + I + G_0 \\ C &= \gamma Y \\ I &= 0 \end{aligned}\right| \to \begin{aligned} Y &= \gamma Y + G_0 \\ Y &= \frac{G_0}{1-\gamma} \end{aligned}$$

$$Y = \frac{G_0}{1-\gamma},\ C = \frac{\gamma G_0}{1-\gamma},\ I = 0$$

b. $y(t) = Y(t) - \frac{G_0}{1-\gamma}$, $c(t) = C(t) - \frac{\gamma G_0}{1-\gamma}$, $i(t) = I(t)$

Substitute to verify the equations.

$$\begin{bmatrix} C(t+1) \\ i(t+1) \end{bmatrix} = \begin{bmatrix} \gamma & \gamma \\ \alpha\gamma - \alpha & \alpha\gamma \end{bmatrix} \begin{bmatrix} c(t) \\ i(t) \end{bmatrix}$$

c. $A = \begin{bmatrix} 0.2 & 0.2 \\ -4 & 1 \end{bmatrix}$ eigenvalues $0.6 \pm 0.8i$

not stable

d. $A = \begin{bmatrix} \gamma & \gamma \\ \gamma - 1 & \gamma \end{bmatrix}$ $\begin{matrix} \text{tr}A = 2\gamma \\ \det A = \gamma \end{matrix}$ stable (use Exercise 30)

e. $A = \begin{bmatrix} \gamma & \gamma \\ \alpha\gamma - \alpha & \alpha\gamma \end{bmatrix}$ $\begin{matrix} \text{tr}A = \gamma(1+\alpha) > 0 \\ \det A = \alpha\gamma \end{matrix}$

Use Exercise 30; stable if $\det(A) = \alpha\gamma < 1$ and $\text{tr}\,A - 1 = \alpha\gamma + \gamma - 1 < \alpha\gamma$.

The second condition is satisfied since $\gamma < 1$.

Stable if $\gamma < \frac{1}{\alpha}$

$\left(\text{eigenvalues are real if } \gamma \ge \frac{4\alpha}{(1+\alpha)^2}\right)$

38. a. $T(\vec{v}) = A\vec{v} + \vec{b} = \vec{v}$ if $\vec{v} - A\vec{v} = \vec{b}$ or $(I_n - A)\vec{v} = \vec{b}$

$I_n - A$ is invertible since 1 is not an eigenvalue of A. Therefore, $\vec{v} = (I_n - A)^{-1}\vec{b}$ is the only solution.

b. Let $\vec{y}(t) = \vec{x}(t) - \vec{v}$ be the deviation of $\vec{x}(t)$ from the equilibrium $\vec{v}$.

Then $\vec{y}(t+1) = \vec{x}(t+1) - \vec{v} = A\vec{x}(t) + \vec{b} - \vec{v} = A(\vec{y}(t) + \vec{v}) + \vec{b} - \vec{v} = A\vec{y}(t) + A\vec{v} + \vec{b} - \vec{v} = A\vec{y}(t)$, so that $\vec{y}(t) = A^t\vec{y}(0)$, or $\vec{x}(t) = \vec{v} + A^t(\vec{x}_0 - \vec{v})$.

$\lim_{t\to\infty} \vec{x}(t) = \vec{v}$ for all $\vec{x}_0$ if $\lim_{t\to\infty} A^t(\vec{x}_0 - \vec{v}) = \vec{0}$. This is the case if the modulus if all the eigenvalues of A is less than 1.

39. Use Exercise 38: $\vec{v} = (I_2 - A)^{-1}\vec{b} = \begin{bmatrix} 0.9 & -0.2 \\ -0.4 & 0.7 \end{bmatrix}^{-1} \begin{bmatrix} 1 \\ 2 \end{bmatrix} = \begin{bmatrix} 2 \\ 4 \end{bmatrix}$.

$\begin{bmatrix} 2 \\ 4 \end{bmatrix}$ is a stable equilibrium since the eigenvalues of A are 0.5 and -0.1.

40. Note that A can be partitioned as $A = \begin{bmatrix} B & -C^T \\ C & B^T \end{bmatrix}$, where B and C are rotation-dilation matrices. Also note that $BC = CB$, $B^TB = (p^2+q^2)I_2$, and $C^TC = (r^2+s^2)I_2$.

a. $A^TA = \begin{bmatrix} B^T & C^T \\ -C & B \end{bmatrix} \begin{bmatrix} B & -C^T \\ C & B^T \end{bmatrix} = (p^2+q^2+r^2+s^2)I_4$

b. By part a, $A^{-1} = \frac{1}{p^2+q^2+r^2+s^2}A^T$ if $A \neq 0$.

c. $(\det A)^2 = (p^2+q^2+r^2+s^2)^4$, by part a, so that $\det A = \pm(p^2+q^2+r^2+s^2)^2$.
The diagonal pattern makes the contribution $+p^4$, so that $\det A = (p^2+q^2+r^2+s^2)^2$.

d. Consider $\det(\lambda I_4 - A)$. Note that the matrix $\lambda I_4 - A$ has the same "format" as A, with p replaced by $\lambda - p$ and q, r, s, replaced by $-q, -r, -s$, respectively. By part c, $\det(\lambda I_4 - A) = \left((\lambda-p)^2+q^2+r^2+s^2\right)^2 = 0$ when

$$(\lambda-p)^2 = -q^2-r^2-s^2$$
$$\lambda - p = \pm i\sqrt{q^2+r^2+s^2}$$
$$\lambda_{1,2} = p \pm i\sqrt{q^2+r^2+s^2}.$$

Each of these eigenvalues has algebraic multiplicity 2 (if $q=r=s=0$ then $\lambda = p$ has algebraic multiplicity 4).

e. By part a we can write $A = \sqrt{p^2+q^2+r^2+s^2}\underbrace{\left(\frac{1}{\sqrt{p^2+q^2+r^2+s^2}}A\right)}_{S}$, where S is orthogonal.

Therefore, $\|A\vec{x}\| = \left\|\sqrt{p^2+q^2+r^2+s^2}\,(S\vec{x})\right\| = \sqrt{p^2+q^2+r^2+s^2}\,\|\vec{x}\|$.

f. Let $A = \begin{bmatrix} 3 & -3 & -4 & -5 \\ 3 & 3 & 5 & -4 \\ 4 & -5 & 3 & 3 \\ 5 & 4 & -3 & 3 \end{bmatrix}$ and $\vec{x} = \begin{bmatrix} 1 \\ 2 \\ 4 \\ 4 \end{bmatrix}$; then $A\vec{x} = \begin{bmatrix} -39 \\ 13 \\ 18 \\ 13 \end{bmatrix}$.

By part e, $\|A\vec{x}\|^2 = (3^2+3^2+4^2+5^2)\|\vec{x}\|^2$, or
$39^2+13^2+18^2+13^2 = (3^2+3^2+4^2+5^2)(1^2+2^2+4^2+4^2)$, as desired.

g. Any positive integer m can be written as $m = p_1p_2\cdots p_n$. Using part f repeatedly we see that the numbers $p_1, p_1p_2, p_1p_2p_3, \ldots, p_1p_2p_3\cdots p_{n-1}$, and finally $m = p_1\cdots p_n$ can be expressed as the sums of four squares.

41. Find the 2 × 2 matrix A that transforms $\begin{bmatrix} 8 \\ 6 \end{bmatrix}$ into $\begin{bmatrix} -3 \\ 4 \end{bmatrix}$ and $\begin{bmatrix} -3 \\ 4 \end{bmatrix}$ into $\begin{bmatrix} -8 \\ -6 \end{bmatrix}$:

$A\begin{bmatrix} 8 & -3 \\ 6 & 4 \end{bmatrix} = \begin{bmatrix} -3 & -8 \\ 4 & -6 \end{bmatrix}$ and $A = \begin{bmatrix} -3 & -8 \\ 4 & -6 \end{bmatrix}\begin{bmatrix} 8 & -3 \\ 6 & 4 \end{bmatrix}^{-1} = \frac{1}{50}\begin{bmatrix} 36 & -73 \\ 52 & -36 \end{bmatrix}$.

There are many other correct answers.

42. **a.** $x(t+1) = x(t) - ky(t)$

$y(t+1) = kx(t) + y(t) = kx(t) + (1-k^2)y(t)$ so $\begin{bmatrix} x(t+1) \\ y(t+1) \end{bmatrix} = \begin{bmatrix} 1 & -k \\ k & 1-k^2 \end{bmatrix}\begin{bmatrix} x(t) \\ y(t) \end{bmatrix}$.

b. $f_A(\lambda) = \lambda^2 - (2-k^2)\lambda + 1 = 0$

The discriminant is $(2-k^2)^2 - 4 = -4k^2 + k^4 = k^2(k^2-4)$, which is negative if k is a small positive number ($k < 2$). Therefore, the eigenvalues are complex. By Fact 6.5.4, the trajectory will be an ellipse, since $\det(A) = 1$.

Chapter 7

7.1

1. The equation $\begin{bmatrix} 7 \\ 16 \end{bmatrix} = c_1 \begin{bmatrix} 2 \\ 5 \end{bmatrix} + c_2 \begin{bmatrix} 5 \\ 12 \end{bmatrix}$ has the solution $c_1 = -4$, $c_2 = 3$, so that the desired coordinate vector is $\begin{bmatrix} -4 \\ 3 \end{bmatrix}$.

2. Solving a system of four equations with four unknowns we find the coordinate vector $\begin{bmatrix} 2 \\ -1 \\ -1 \\ 1 \end{bmatrix}$.

3. $A = \begin{bmatrix} 1 & 2 \\ 3 & 4 \end{bmatrix}$, $S = \begin{bmatrix} 1 & 1 \\ 1 & 2 \end{bmatrix}$, $S^{-1} = \begin{bmatrix} 2 & -1 \\ -1 & 1 \end{bmatrix}$

By Fact 7.1.4 the new matrix of T, namely B, is given by $B = S^{-1}AS = \begin{bmatrix} -1 & -1 \\ 4 & 6 \end{bmatrix}$.

4. $A = \begin{bmatrix} 7 & -1 \\ -6 & 8 \end{bmatrix}$, $S = \begin{bmatrix} 1 & -1 \\ 2 & 3 \end{bmatrix}$, $S^{-1} = \frac{1}{5}\begin{bmatrix} 3 & 1 \\ -2 & 1 \end{bmatrix}$

By Fact 7.1.4, the new matrix of T, namely B, is given by $B = S^{-1}AS = \begin{bmatrix} 5 & 0 \\ 0 & 10 \end{bmatrix}$.

5. a. Let B be the matrix of T with respect to $\mathcal{B} = \left(\begin{bmatrix} 3 \\ 1 \end{bmatrix}, \begin{bmatrix} -1 \\ 3 \end{bmatrix} \right)$.

By Fact 7.1.5, $B = \left[[T(\vec{v}_1)]_{\mathcal{B}} [T(\vec{v}_2)]_{\mathcal{B}} \right] = \left[[\vec{v}_1]_{\mathcal{B}} [\vec{0}]_{\mathcal{B}} \right] = \begin{bmatrix} 1 & 0 \\ 0 & 0 \end{bmatrix}$.

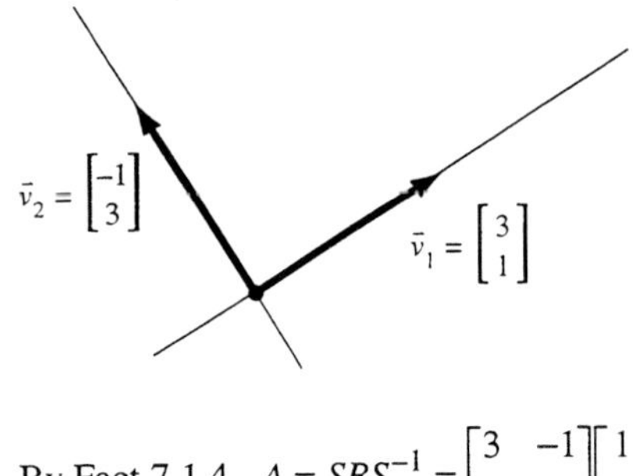

b. By Fact 7.1.4, $A = SBS^{-1} = \begin{bmatrix} 3 & -1 \\ 1 & 3 \end{bmatrix} \begin{bmatrix} 1 & 0 \\ 0 & 0 \end{bmatrix} \begin{bmatrix} 3 & -1 \\ 1 & 3 \end{bmatrix}^{-1} = \begin{bmatrix} 0.9 & 0.3 \\ 0.3 & 0.1 \end{bmatrix}$.

6. a. $\begin{bmatrix}1\\1\\-1\end{bmatrix}$ and $\begin{bmatrix}-1\\2\\-1\end{bmatrix}$ are on the plane, and $\begin{bmatrix}1\\2\\3\end{bmatrix}$ is perpendicular to it.

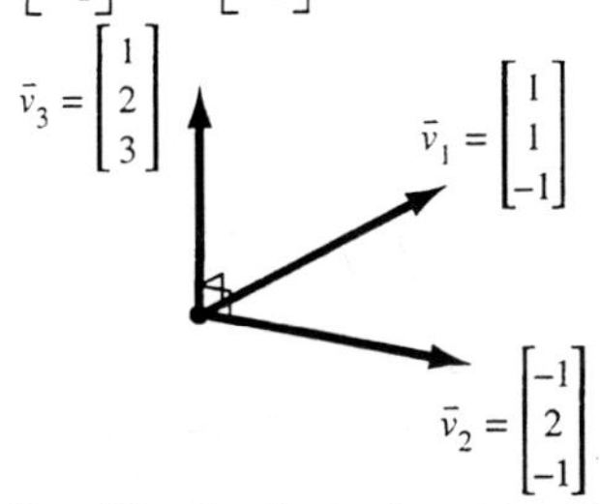

Let B be the desired matrix. By Fact 7.1.5,

$$B=\left[[T(\vec{v}_1)]_{\mathcal{B}}\ \ [T(\vec{v}_2)]_{\mathcal{B}}\ \ [T(\vec{v}_3)]_{\mathcal{B}}\right]=\left[[\vec{v}_1]_{\mathcal{B}}\ \ [\vec{v}_2]_{\mathcal{B}}\ \ [\vec{0}]_{\mathcal{B}}\right]=\begin{bmatrix}1&0&0\\0&1&0\\0&0&0\end{bmatrix}.$$

b. By Fact 7.1.4, $A=SBS^{-1}=\begin{bmatrix}1&-1&1\\1&2&2\\-1&-1&3\end{bmatrix}\begin{bmatrix}1&0&0\\0&1&0\\0&0&0\end{bmatrix}\begin{bmatrix}1&-1&1\\1&2&2\\-1&-1&3\end{bmatrix}^{-1}=\frac{1}{14}\begin{bmatrix}13&-2&-3\\-2&10&-6\\-3&-6&5\end{bmatrix}.$

7. By Fact 7.1.4, $B=S^{-1}AS=\begin{bmatrix}3&5\\5&8\end{bmatrix}^{-1}\begin{bmatrix}1&9\\9&4\end{bmatrix}\begin{bmatrix}3&5\\5&8\end{bmatrix}=\begin{bmatrix}-149&-231\\99&154\end{bmatrix}.$

8. By Fact 7.1.4, $A=SBS^{-1}=\begin{bmatrix}3&5\\5&8\end{bmatrix}\begin{bmatrix}1&9\\9&7\end{bmatrix}\begin{bmatrix}3&5\\5&8\end{bmatrix}^{-1}=\begin{bmatrix}-74&54\\-111&82\end{bmatrix}.$

9. By Fact 7.1.4, $A=SBS^{-1}=\begin{bmatrix}0&1\\1&0\end{bmatrix}\begin{bmatrix}a&b\\c&d\end{bmatrix}\begin{bmatrix}0&1\\1&0\end{bmatrix}^{-1}=\begin{bmatrix}d&c\\b&a\end{bmatrix}.$

10. By Fact 7.1.2, $\begin{bmatrix}5\\3\end{bmatrix}_{\mathcal{B}}=\begin{bmatrix}3&-1\\1&1\end{bmatrix}^{-1}\begin{bmatrix}5\\3\end{bmatrix}=\begin{bmatrix}2\\1\end{bmatrix}.$

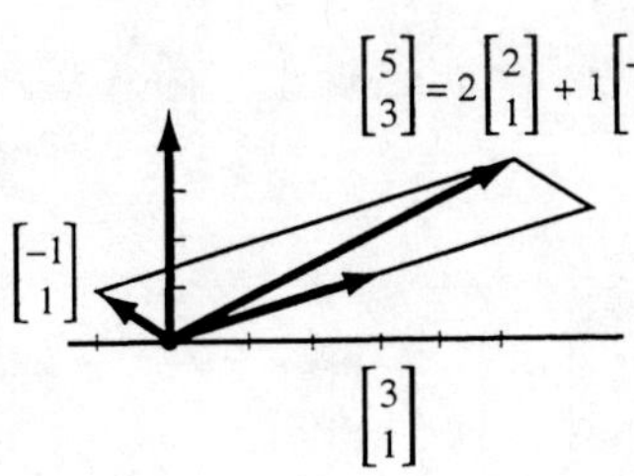

11. By Fact 7.1.2, $\begin{bmatrix} 2 \\ 0 \end{bmatrix}_{\mathcal{B}} = \begin{bmatrix} 3 & -1 \\ 1 & 1 \end{bmatrix}^{-1} \begin{bmatrix} 2 \\ 0 \end{bmatrix} = \begin{bmatrix} \frac{1}{2} \\ -\frac{1}{2} \end{bmatrix}$.

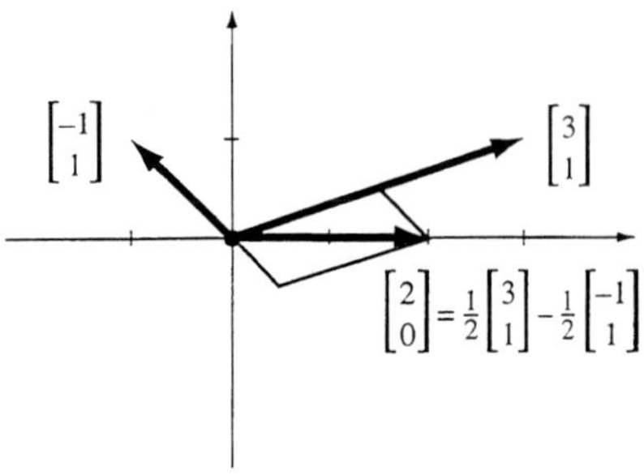

12. $[\vec{x}]_{\mathcal{B}} = \begin{bmatrix} -1 \\ 2 \end{bmatrix}$ means that $\vec{x} = -\vec{v} + 2\vec{w}$.

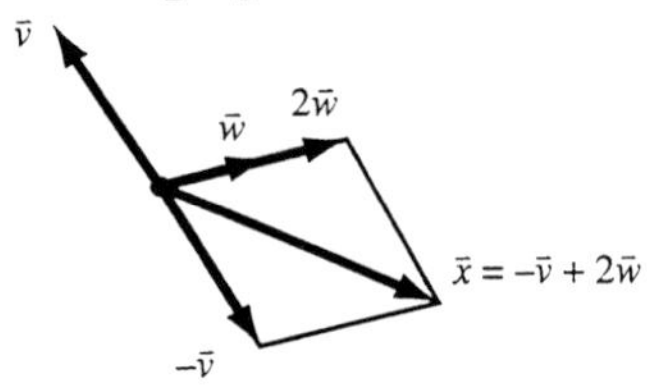

13. From the picture, $\vec{u} + \vec{v} = -\vec{w}$ so that $\vec{w} = -\vec{u} - \vec{v}$, i.e. $[\vec{w}]_{\mathcal{B}} = \begin{bmatrix} -1 \\ -1 \end{bmatrix}$.

14. a. $\overrightarrow{OP} = \vec{w} + 2\vec{v}$, so that $[\overrightarrow{OP}]_{\mathcal{B}} = \begin{bmatrix} 2 \\ 1 \end{bmatrix}$. $\overrightarrow{OQ} = \vec{v} + 2\vec{w}$, so that $[\overrightarrow{OQ}]_{\mathcal{B}} = \begin{bmatrix} 1 \\ 2 \end{bmatrix}$.

b. $\overrightarrow{OR} = 3\vec{v} + 2\vec{w}$

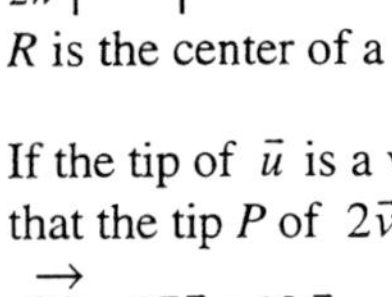

R is the center of a tile.

c. If the tip of $\vec{u}$ is a vertex, then so is the tip of $\vec{u} + 3\vec{v}$ and also the tip of $\vec{u} + 3\vec{w}$ (draw a sketch!). We know that the tip P of $2\vec{v} + \vec{w}$ is a vertex (see part a). Therefore, the tip S of $\overrightarrow{OS} = 17\vec{v} + 13\vec{w} = (2\vec{v} + \vec{w}) + 5(3\vec{v}) + 4(3\vec{w})$ is a vertex as well.

15. Solving the system $\begin{bmatrix} 1 \\ 0 \\ 0 \end{bmatrix} = c_1 \begin{bmatrix} 1 \\ 2 \\ 3 \end{bmatrix} + c_2 \begin{bmatrix} 0 \\ 1 \\ 2 \end{bmatrix} + c_3 \begin{bmatrix} 0 \\ 0 \\ 1 \end{bmatrix}$ gives the coordinate vector $\begin{bmatrix} c_1 \\ c_2 \\ c_3 \end{bmatrix} = \begin{bmatrix} 1 \\ -2 \\ 1 \end{bmatrix}$.

16. Yes; $T(\vec{x}) = [\vec{x}]_{\mathcal{B}} = S^{-1}\vec{x}$, so that T is "given by a matrix." (See Definition 2.1.1.)

17. By Fact 7.1.2, $\vec{x} = S[\vec{x}]_{\mathcal{B}} = \begin{bmatrix} 1 & 3 \\ 2 & 4 \end{bmatrix} \begin{bmatrix} 7 \\ 11 \end{bmatrix} = \begin{bmatrix} 40 \\ 58 \end{bmatrix}$.

18. Let Q be the matrix whose columns are the vectors of the basis $\mathcal{T}$. Then $\left[[\vec{v}_1]_{\mathcal{T}} \ \cdots \ [\vec{v}_n]_{\mathcal{T}}\right] = [Q^{-1}\vec{v}_1 \ \cdots \ Q^{-1}\vec{v}_n] = Q^{-1}[\vec{v}_1 \ \cdots \ \vec{v}_n]$ is an invertible matrix, so that the vectors $[\vec{v}_1]_{\mathcal{T}} \cdots [\vec{v}_n]_{\mathcal{T}}$ form a basis of $\mathbb{R}^n$.

19. By Fact 7.1.2, $\vec{x} = \begin{bmatrix} 1 & 1 \\ 1 & 2 \end{bmatrix}[\vec{x}]_{\mathcal{B}}$ and $\vec{x} = \begin{bmatrix} 1 & 3 \\ 2 & 4 \end{bmatrix}[\vec{x}]_{\mathcal{R}}$, so that $\begin{bmatrix} 1 & 1 \\ 1 & 2 \end{bmatrix}[\vec{x}]_{\mathcal{B}} = \begin{bmatrix} 1 & 3 \\ 2 & 4 \end{bmatrix}[\vec{x}]_{\mathcal{R}}$ and

$$[\vec{x}]_{\mathcal{R}} = \underbrace{\begin{bmatrix} 1 & 3 \\ 2 & 4 \end{bmatrix}^{-1}\begin{bmatrix} 1 & 1 \\ 1 & 2 \end{bmatrix}}_{P}[\vec{x}]_{\mathcal{B}}, \text{ i.e. } P = \begin{bmatrix} -\frac{1}{2} & 1 \\ \frac{1}{2} & 0 \end{bmatrix}.$$

20. Let $S = [\vec{v}_1 \ \vec{v}_2]$ where $\vec{v}_1, \vec{v}_2$ is the desired basis. Then by Fact 7.1.2, $\begin{bmatrix} 1 \\ 2 \end{bmatrix} = S\begin{bmatrix} 3 \\ 5 \end{bmatrix}$ and $\begin{bmatrix} 3 \\ 4 \end{bmatrix} = S\begin{bmatrix} 2 \\ 3 \end{bmatrix}$, i.e.

$S\begin{bmatrix} 3 & 2 \\ 5 & 3 \end{bmatrix} = \begin{bmatrix} 1 & 3 \\ 2 & 4 \end{bmatrix}$. Hence $S = \begin{bmatrix} 1 & 3 \\ 2 & 4 \end{bmatrix}\begin{bmatrix} 3 & 2 \\ 5 & 3 \end{bmatrix}^{-1} = \begin{bmatrix} 12 & -7 \\ 14 & -8 \end{bmatrix}$. The desired basis is $\begin{bmatrix} 12 \\ 14 \end{bmatrix}, \begin{bmatrix} -7 \\ -8 \end{bmatrix}$.

21. $T(\vec{v}_1) = \vec{v}_1 \times \vec{v}_2 = \vec{v}_3$ so $[T(\vec{v}_1)]_{\mathcal{B}} = \begin{bmatrix} 0 \\ 0 \\ 1 \end{bmatrix}$.

$T(\vec{v}_2) = \vec{v}_2 \times \vec{v}_2 = \vec{0}$ so $[T(\vec{v}_2)]_{\mathcal{B}} = \begin{bmatrix} 0 \\ 0 \\ 0 \end{bmatrix}$.

$T(\vec{v}_3) = \vec{v}_3 \times \vec{v}_2 = -\vec{v}_1$ so $[T(\vec{v}_3)]_{\mathcal{B}} = \begin{bmatrix} -1 \\ 0 \\ 0 \end{bmatrix}$.

Hence, by Fact 7.1.5, the desired matrix is $\begin{bmatrix} 0 & 0 & -1 \\ 0 & 0 & 0 \\ 1 & 0 & 0 \end{bmatrix}$.

22. a. Consider a linear relation $c_1A^2\vec{v} + c_2A\vec{v} + c_3\vec{v} = \vec{0}$.

Multiplying A^2 with the vectors on both sides and using that $A^3\vec{v} = \vec{0}$ and $A^4\vec{v} = \vec{0}$, we find that $c_3A^2\vec{v} = \vec{0}$ and therefore $c_3 = 0$, since $A^2\vec{v} \neq \vec{0}$.

Therefore, our relation simplifies to $c_1A^2\vec{v} + c_2A\vec{v} = \vec{0}$.

Multiplying A with the vectors on both sides we find that $c_2A^2\vec{v} = \vec{0}$ and therefore $c_2 = 0$. Then $c_1 = 0$ as well. We have shown that there is only the trivial relation among the vectors $A^2\vec{v}$, $A\vec{v}$, and $\vec{v}$, so that these three vectors form a basis of $\mathbb{R}^3$, as claimed.

b. $T(A^2\vec{v}) = A^3\vec{v} = \vec{0}$ so $[T(A^2\vec{v})]_{\mathcal{B}} = \begin{bmatrix}0\\0\\0\end{bmatrix}$.

$T(A\vec{v}) = A^2\vec{v}$ so $[T(A\vec{v})]_{\mathcal{B}} = \begin{bmatrix}1\\0\\0\end{bmatrix}$.

$T(\vec{v}) = A\vec{v}$ so $[T(\vec{v})]_{\mathcal{B}} = \begin{bmatrix}0\\1\\0\end{bmatrix}$.

Hence, by Fact 7.1.5, the desired matrix is $\begin{bmatrix}0&1&0\\0&0&1\\0&0&0\end{bmatrix}$.

23. a. By inspection, we can find an orthonormal basis $\vec{v}_1 = \vec{v}$, $\vec{v}_2$, $\vec{v}_3$ of $\mathbb{R}^3$:

$\vec{v}_1 = \vec{v} = \begin{bmatrix}0.6\\0.8\\0\end{bmatrix}$, $\vec{v}_2 = \begin{bmatrix}0\\0\\1\end{bmatrix}$, $\vec{v}_3 = \begin{bmatrix}0.8\\-0.6\\0\end{bmatrix}$

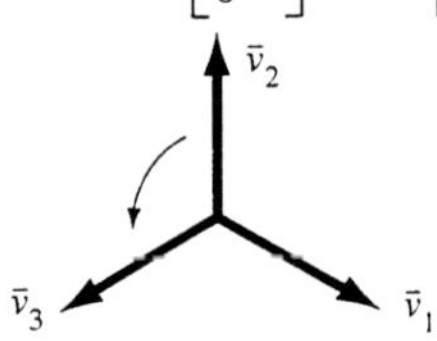

b. Now $T(\vec{v}_1) = \vec{v}_1$, $T(\vec{v}_2) = \vec{v}_3$ and $T(\vec{v}_3) = -\vec{v}_2$, so that the matrix B of T with respect to the basis $\vec{v}_1, \vec{v}_2, \vec{v}_3$ is

$B = \begin{bmatrix}1&0&0\\0&0&-1\\0&1&0\end{bmatrix}$. Then $A = SBS^{-1} = \begin{bmatrix}0.36&0.48&0.8\\0.48&0.64&-0.6\\-0.8&0.6&0\end{bmatrix}$.

24. a. $\vec{v}_0 + \vec{v}_1 + \vec{v}_2 + \vec{v}_3 = \begin{bmatrix}1\\1\\1\end{bmatrix} + \begin{bmatrix}1\\-1\\-1\end{bmatrix} + \begin{bmatrix}-1\\1\\-1\end{bmatrix} + \begin{bmatrix}-1\\-1\\1\end{bmatrix} = \begin{bmatrix}0\\0\\0\end{bmatrix}$

b. If $\mathcal{B}$ is the basis $\vec{v}_1, \vec{v}_2, \vec{v}_3$, then $\vec{v}_0 + \vec{v}_1 + \vec{v}_2 + \vec{v}_3 = \vec{0}$ (by part a) so $\vec{v}_0 = -\vec{v}_1 - \vec{v}_2 - \vec{v}_3$, i.e. $[\vec{v}_0]_{\mathcal{B}} = \begin{bmatrix}-1\\-1\\-1\end{bmatrix}$.

c. $T(\vec{v}_2) = T(-\vec{v}_0 - \vec{v}_1 - \vec{v}_3) = -T(\vec{v}_0) - T(\vec{v}_1) - T(\vec{v}_3) = -\vec{v}_3 - \vec{v}_0 - \vec{v}_1 = \vec{v}_2$

Hence, T is a rotation by 120° about the line spanned by $\vec{v}_2$. Its matrix, B, is given by $\left[[T(\vec{v}_1)]_{\mathcal{B}} \;\; [T(\vec{v}_2)]_{\mathcal{B}} \;\; [T(\vec{v}_3)]_{\mathcal{B}}\right]$ where

$T(\vec{v}_1) = \vec{v}_0 = -\vec{v}_1 - \vec{v}_2 - \vec{v}_3$ so $[T(\vec{v}_1)]_{\mathcal{B}} = \begin{bmatrix}-1\\-1\\-1\end{bmatrix}$

$T(\vec{v}_2) = \vec{v}_2$ so $T(\vec{v}_2)]_{\mathcal{B}} = \begin{bmatrix} 0 \\ 1 \\ 0 \end{bmatrix}$

$T(\vec{v}_3) = \vec{v}_1$ so $[T(\vec{v}_3)]_{\mathcal{B}} = \begin{bmatrix} 1 \\ 0 \\ 0 \end{bmatrix}$

and $B = \begin{bmatrix} -1 & 0 & 1 \\ -1 & 1 & 0 \\ -1 & 0 & 0 \end{bmatrix}$.

$f_B(\lambda) = \lambda^3 - 1$, so that $\lambda_1 = 1$, $\lambda_{2,3} = -\frac{1}{2} \pm \frac{\sqrt{3}}{2} i = \cos(120°) \pm i\sin(120°)$.

$B^3 = I_3$ since if the tetrahedron rotates by 120° three times, it returns to the original position.

25. a. Let S be the matrix whose columns are the vectors of the orthonormal basis $\mathcal{B}$; note that S is an orthogonal matrix. Then $B = S^{-1}AS$ is orthogonal as well, since inverses and products of orthogonal matrices are orthogonal. Also, $\det(B) = \det(A) = 1$ (see Example 6 of Section 5.2), so that B is a rotation matrix.

b. The first column of B is $[T(\vec{v}_1)]_{\mathcal{B}} = [\vec{v}_1]_{\mathcal{B}} = \begin{bmatrix} 1 \\ 0 \\ 0 \end{bmatrix}$, that is, $B = \begin{bmatrix} 1 & * & * \\ 0 & * & * \\ 0 & * & * \end{bmatrix}$.

Since the first column of B is orthogonal to the two others, the first row of B is $[1 \quad 0 \quad 0]$, so that $B = \begin{bmatrix} 1 & 0 \quad 0 \\ 0 & \\ 0 & B_{11} \end{bmatrix}$, for some 2×2 matrix B_{11}. Since the three columns of B are orthonormal, the two columns of B_{11} are orthonormal as well, so that B_{11} is an orthogonal matrix. Also, $1 = \det(B) = 1 \cdot \det(B_{11})$, so that $\det(B_{11}) = 1$, and B_{11} is a rotation matrix:

$B = \begin{bmatrix} 1 & 0 & 0 \\ 0 & \cos\phi & -\sin\phi \\ 0 & \sin\phi & \cos\phi \end{bmatrix}$ for some ϕ.

This shows that B_{11} defines a rotation through a certain angle ϕ in the $\vec{v}_2 - \vec{v}_3$ plane, perpendicular to the line spanned by $\vec{v}_1$. The matrix B defines a rotation about the line spanned by $\vec{v}_1$.

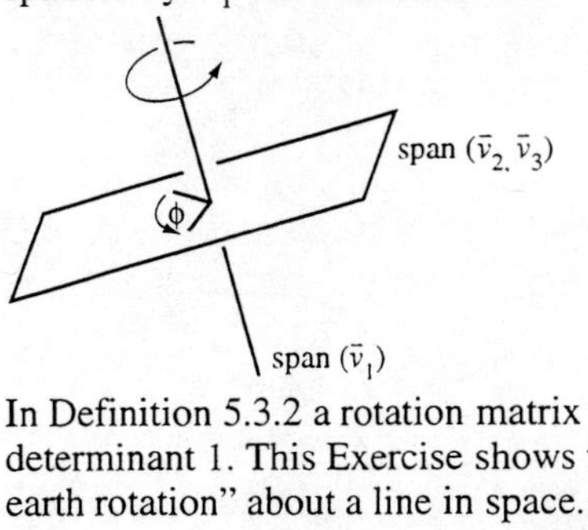

In Definition 5.3.2 a rotation matrix was defined somewhat abstractly as an orthogonal matrix with determinant 1. This Exercise shows that a 3×3 rotation matrix in this abstract sense defines a "down-to-earth rotation" about a line in space.

26. Let S be the $n \times n$ matrix whose columns are $-\vec{e}_1, \vec{e}_2, -\vec{e}_3, \ldots, (-1)^n \vec{e}_n$. Note that S is a diagonal matrix with $s_{ii} = (-1)^i$. Also, $S^{-1} = S$.

Now $B = S^{-1}AS$. Since S is diagonal, the entries of B are $b_{ij} = s_{ii}a_{ij}s_{jj} = (-1)^i a_{ij}(-1)^j = (-1)^{i+j} a_{ij}$.

Answer: $b_{ij} = (-1)^{i+j} a_{ij}$

27. Let S be the $n \times n$ matrix whose columns are $\vec{e}_n, \vec{e}_{n-1}, \ldots, \vec{e}_1$. Note that S has all 1's on "the other diagonal" and 0's elsewhere:

$$s_{ij} = \begin{cases} 1 & \text{if } i+j = n+1 \\ 0 & \text{otherwise} \end{cases}$$

Also, $S^{-1} = S$.

Now $B = S^{-1}AS = SAS$; the entries of B are $b_{ij} = s_{i,\,n+1-i}a_{n+1-i,\,n+1-j}s_{n+1-j,\,j} = a_{n+1-i,\,n+1-j}$.

Answer: $b_{ij} = a_{n+1-i,\,n+1-j}$

B is obtained from A by reversing the order of the rows and of the columns.

28. Note first that the diagonal entry s_{ii} of S gives the unit price of good i.

If a_{ij} tells us how many dollars' worth of good i are required to produce one dollar's worth of good j, then $a_{ij}s_{jj}$ tells us how many dollars' worth of good i are required to produce one unit of good j, and $s_{ii}^{-1}a_{ij}s_{jj}$ is the number of units of good i required to produce one unit of good j. Thus $b_{ij} = s_{ii}^{-1}a_{ij}s_{jj}$, and $B = S^{-1}AS$.

29. Note that $\det\begin{bmatrix} i & -i \\ 1 & 1 \end{bmatrix} = 2i \neq 0$, so that the matrix $\begin{bmatrix} i & -i \\ 1 & 1 \end{bmatrix}$ is invertible. Also, $[\vec{v} + i\vec{w} \quad \vec{v} - i\vec{w}]$ is invertible since the columns are eigenvectors with distinct eigenvalues (see Fact 6.3.5). Therefore,

$$[\vec{w} \quad \vec{v}] = [\vec{v} + i\vec{w} \quad \vec{v} - i\vec{w}]\begin{bmatrix} i & -i \\ 1 & 1 \end{bmatrix}^{-1}$$ is invertible as well, as claimed.

30. By Fact 7.1.5, $B = \left[[T(\vec{v})]_{\mathcal{B}} \quad [T(\vec{w})_{\mathcal{B}}]\right]$. By Example 5, we have

$$\begin{matrix} A\vec{v} = p\vec{v} - q\vec{w} \\ A\vec{w} = q\vec{v} + p\vec{w} \end{matrix} \text{ so}$$

$T(\vec{v}) = A\vec{v} = p\vec{v} - q\vec{w}$, i.e. $[T(\vec{v})]_{\mathcal{B}} = \begin{bmatrix} p \\ -q \end{bmatrix}$

$T(\vec{w}) = A\vec{w} = q\vec{v} + p\vec{w}$, i.e. $[T(\vec{w})]_{\mathcal{B}} = \begin{bmatrix} q \\ p \end{bmatrix}$

so $B = \begin{bmatrix} p & q \\ -q & p \end{bmatrix}$.

7.2

1. Diagonalizable; $S = I_2$, $D = A$.

2. Diagonalizable since $\lambda_{1,2} = 3, 2$ and $\vec{v}_1 = \begin{bmatrix} 2 \\ 1 \end{bmatrix}$, $\vec{v}_2 = \begin{bmatrix} 1 \\ 1 \end{bmatrix}$, $S = \begin{bmatrix} 2 & 1 \\ 1 & 1 \end{bmatrix}$, $D = \begin{bmatrix} 3 & 0 \\ 0 & 2 \end{bmatrix}$.

3. Diagonalizable; eigenvectors $\begin{bmatrix} 2 \\ 1 \end{bmatrix}, \begin{bmatrix} 1 \\ 1 \end{bmatrix}$ correspond to eigenvalues $1, -1$ so $S = \begin{bmatrix} 2 & 1 \\ 1 & 1 \end{bmatrix}$, $D = \begin{bmatrix} 1 & 0 \\ 0 & -1 \end{bmatrix}$.

4. Not diagonalizable since $\lambda_1 = \lambda_2 = 1$ and $A \neq I_2$ (compare with Example 3).

5. Not diagonalizable over $\mathbb{R}$ since $\lambda_{1,2} = 2 \pm i$ so no real eigenbasis exists.

6. Diagonalizable since $\lambda_1 = \lambda_3 = 1$, $\lambda_2 = -1$ with an eigenbasis $\begin{bmatrix}2\\1\\0\end{bmatrix}, \begin{bmatrix}1\\1\\0\end{bmatrix}, \begin{bmatrix}0\\0\\1\end{bmatrix}$ so $S = \begin{bmatrix}2 & 1 & 0\\1 & 1 & 0\\0 & 0 & 1\end{bmatrix}$, $D = \begin{bmatrix}1 & 0 & 0\\0 & -1 & 0\\0 & 0 & 1\end{bmatrix}$.

7. Diagonalizable since $\lambda_1 = 3$, $\lambda_2 = 2$, $\lambda_3 = 1$, with an eigenbasis $\begin{bmatrix}2\\0\\1\end{bmatrix}, \begin{bmatrix}1\\0\\1\end{bmatrix}, \begin{bmatrix}0\\1\\0\end{bmatrix}$ so

$$S = \begin{bmatrix}2 & 1 & 0\\0 & 0 & 1\\1 & 1 & 0\end{bmatrix},\ D = \begin{bmatrix}3 & 0 & 0\\0 & 2 & 0\\0 & 0 & 1\end{bmatrix}.$$

8. Diagonalizable since $\lambda_1 = \lambda_3 = 0$, $\lambda_2 = 3$, with an eigenbasis $\begin{bmatrix}-1\\1\\0\end{bmatrix}, \begin{bmatrix}1\\1\\1\end{bmatrix}, \begin{bmatrix}-1\\0\\1\end{bmatrix}$ so

$$S = \begin{bmatrix}-1 & 1 & -1\\1 & 1 & 0\\0 & 1 & 1\end{bmatrix},\ D = \begin{bmatrix}0 & 0 & 0\\0 & 3 & 0\\0 & 0 & 0\end{bmatrix}.$$

9. Not diagonalizable since the eigenspace corresponding to $\lambda_1 = \lambda_2 = 1$ is 1-dimensional, so no eigenbasis exists.

10. Diagonalizable since $\lambda_1 = 1$, $\lambda_2 = 2$, $\lambda_3 = 3$ with an eigenbasis $\begin{bmatrix}1\\0\\0\end{bmatrix}, \begin{bmatrix}2\\1\\0\end{bmatrix}, \begin{bmatrix}9\\6\\2\end{bmatrix}$ so $S = \begin{bmatrix}1 & 2 & 9\\0 & 1 & 6\\0 & 0 & 2\end{bmatrix}$, $D = \begin{bmatrix}1 & 0 & 0\\0 & 2 & 0\\0 & 0 & 3\end{bmatrix}$.

11. Diagonalizable since $\lambda_1 = 1$, $\lambda_2 = 0$, $\lambda_3 = -1$ with an eigenbasis $\begin{bmatrix}1\\1\\1\end{bmatrix}, \begin{bmatrix}1\\2\\3\end{bmatrix}, \begin{bmatrix}1\\3\\6\end{bmatrix}$ so

$$S = \begin{bmatrix}1 & 1 & 1\\1 & 2 & 3\\1 & 3 & 6\end{bmatrix},\ D = \begin{bmatrix}1 & 0 & 0\\0 & 0 & 0\\0 & 0 & -1\end{bmatrix}.$$

12. Not diagonalizable since $\lambda_{1,2,3,4} = 1$ and E_1 is 1-dimensional, so no eigenbasis exists.

13. The eigenvalues of A are $p \pm iq$. If $q \neq 0$ then $E_{p+iq} = \ker\begin{bmatrix}iq & -q\\q & iq\end{bmatrix} = \ker\begin{bmatrix}-1 & -i\\1 & i\end{bmatrix} = \text{span}\begin{bmatrix}-i\\1\end{bmatrix}$.

Similarly, $E_{p-iq} = \text{span}\begin{bmatrix}i\\1\end{bmatrix}$, so $S = \begin{bmatrix}-i & i\\1 & 1\end{bmatrix}$, $D = \begin{bmatrix}p+iq & 0\\0 & p-iq\end{bmatrix}$ and $S^{-1}\begin{bmatrix}p & q\\-q & p\end{bmatrix}S = D$.

This solution works for $q = 0$ as well.

14. The eigenvalues of A are $2 \pm 2i$.

$E_{2+2i} = \text{span}\begin{bmatrix} 3-i \\ 2 \end{bmatrix}$, and $E_{2-2i} = \text{span}\begin{bmatrix} 3+i \\ 2 \end{bmatrix}$ so $S = \begin{bmatrix} 3-i & 3+i \\ 2 & 2 \end{bmatrix}$, $D = \begin{bmatrix} 2+2i & 0 \\ 0 & 2-2i \end{bmatrix}$ and

$S^{-1}\begin{bmatrix} -4 & 10 \\ -4 & 8 \end{bmatrix}S = D.$

15. Use the ideas presented in Examples 6, 7, and 8.

$f_A(\lambda) = (\lambda + 1)^2$, so that $M = -A$ represents a shear. Modifying the approach in Example 6 slightly, we observe that if $\vec{w}$ is not an eigenvector of M then the matrix of $T(\vec{x}) = M\vec{x}$ with respect to the basis $\vec{w} - M\vec{w}$, $\vec{w}$ will be $\begin{bmatrix} 1 & -1 \\ 0 & 1 \end{bmatrix}$.

If we let $\vec{w} = \begin{bmatrix} 0 \\ 1 \end{bmatrix}$ then $\vec{w} - M\vec{w} = \begin{bmatrix} 8 \\ -4 \end{bmatrix}$ and $S^{-1}MS = \begin{bmatrix} 1 & -1 \\ 0 & 1 \end{bmatrix}$ where $S = \begin{bmatrix} 8 & 0 \\ -4 & 1 \end{bmatrix}$.

Then $S^{-1}AS = S^{-1}(-M)S = -\begin{bmatrix} 1 & -1 \\ 0 & 1 \end{bmatrix} = \begin{bmatrix} -1 & 1 \\ 0 & -1 \end{bmatrix}$, as required.

16. Use the ideas presented in Example 6.

$f_A(\lambda) = (\lambda - 1)^2$, so that A represents a shear, by Example 6. If $\vec{w} \neq \vec{0}$ is not an eigenvector of A then the matrix of $T(\vec{x}) = A\vec{x}$ with respect to the basis $A\vec{w} - \vec{w}$, $\vec{w}$ is $\begin{bmatrix} 1 & 1 \\ 0 & 1 \end{bmatrix}$.

If we let $\vec{w} = \begin{bmatrix} 0 \\ 1 \end{bmatrix}$ then $A\vec{w} - \vec{w} = \begin{bmatrix} -1 \\ -1 \end{bmatrix}$ and $S^{-1}AS = \begin{bmatrix} 1 & 1 \\ 0 & 1 \end{bmatrix}$ where $S = \begin{bmatrix} -1 & 0 \\ -1 & 1 \end{bmatrix}$.

17. Let D be the matrix of T with respect to $\mathcal{B}$. Denote the vectors of the basis $\mathcal{B}$ by $\vec{v}_1, \ldots, \vec{v}_n$.

The ith column of D is $[T(\vec{v}_i)]_{\mathcal{B}} = \begin{bmatrix} 0 \\ 0 \\ \vdots \\ d_{ii} \\ \vdots \\ 0 \end{bmatrix}$, which means that $T(\vec{v}_i) = d_{ii}\vec{v}_i$, which in turn means that $\vec{v}_i$ is an eigenvector of T, as claimed.

18. The eigenvalues of A are $\lambda_1 = 3$, $\lambda_2 = 2$, and corresponding eigenvectors are $\begin{bmatrix} 2 \\ 1 \end{bmatrix}$ and $\begin{bmatrix} 1 \\ 1 \end{bmatrix}$, so by Fact 7.2.2, for

$S = \begin{bmatrix} 2 & 1 \\ 1 & 1 \end{bmatrix}$, $\begin{bmatrix} 3 & 0 \\ 0 & 2 \end{bmatrix} = S^{-1}AS$, so $A = S\begin{bmatrix} 3 & 0 \\ 0 & 2 \end{bmatrix}S^{-1}$ and $A^t = S\begin{bmatrix} 3^t & 0 \\ 0 & 2^t \end{bmatrix}S^{-1} = \begin{bmatrix} 2\cdot 3^t - 2^t & -2\cdot 3^t + 2\cdot 2^t \\ 3^t - 2^t & -3^t + 2\cdot 2^t \end{bmatrix}$.

Hence, as $t \to \infty$, $a(t)$:$b(t)$:$c(t)$:$d(t)$ approaches the proportion 2:(–2):1:(–1).

19. $\lambda_1 = \lambda_2 = 2$ so the matrix $M = \frac{1}{2}A$ has eigenvalues 1, and by Example 6, M is a shear so A is a shear followed by a dilation.
By Example 7,
$M^t\vec{x}_0 = \vec{x}_0 + t(M\vec{x}_0 - \vec{x}_0) = \vec{x}_0 + tM\vec{x}_0 - t\vec{x}_0 = (I_2 + tM - tI_2)\vec{x}_0$ so $M^t = I_2 + tM - tI_2 = \begin{bmatrix} 1-t & t \\ -t & 1+t \end{bmatrix}$ hence
$A^t = 2^t M^t = 2^t \begin{bmatrix} 1-t & t \\ -t & 1+t \end{bmatrix}$.

20. **i.** A is similar to itself since $A = I_n^{-1} A I_n$.

ii. Assume $B = S^{-1}AS$. Then $A = SBS^{-1}$, so setting $S^{-1} = Q$ we get $A = Q^{-1}BQ$ so B is similar to A.

iii. Assume $B = S^{-1}AS$ and $C = Q^{-1}BQ$ (i.e. A is similar to B and B is similar to C).
Then $C = Q^{-1}BQ = Q^{-1}S^{-1}ASQ = (SQ)^{-1}A(SQ) = P^{-1}AP$, where $P = SQ$, so that A is similar to C, as claimed.

21. Yes; let $S = \begin{bmatrix} 0 & 1 \\ 1 & 0 \end{bmatrix}$ then $\begin{bmatrix} 2 & 0 \\ 0 & 3 \end{bmatrix} = S^{-1}\begin{bmatrix} 3 & 0 \\ 0 & 2 \end{bmatrix}S$.

22. Yes, since $\begin{bmatrix} -1 & 6 \\ -2 & 6 \end{bmatrix}$ has two distinct eigenvalues $\lambda_1 = 2$, $\lambda_2 = 3$ an eigenbasis $\vec{v}_1$, $\vec{v}_2$ exists.
Then $S^{-1}\begin{bmatrix} -1 & 6 \\ -2 & 6 \end{bmatrix}S = \begin{bmatrix} 2 & 0 \\ 0 & 3 \end{bmatrix}$ where $S = [\vec{v}_1 \;\; \vec{v}_2]$.

23. Yes; by Example 6, both shears are similar to $\begin{bmatrix} 1 & 1 \\ 0 & 1 \end{bmatrix}$, so by Exercise 20, they are similar to one another.

24. A has eigenvalues $\lambda_1 = 2$, $\lambda_2 = 3$ with eigenvectors $\vec{v}_1 = \begin{bmatrix} 1 \\ 1 \end{bmatrix}$, $\vec{v}_2 = \begin{bmatrix} 2 \\ 1 \end{bmatrix}$ so $Q^{-1}AQ = \begin{bmatrix} 2 & 0 \\ 0 & 3 \end{bmatrix}$ for $Q = \begin{bmatrix} 1 & 2 \\ 1 & 1 \end{bmatrix}$.
B has eigenvalues $\lambda_1 = 2$, $\lambda_2 = 3$ with eigenvectors $\vec{w}_1 = \begin{bmatrix} 2 \\ 1 \end{bmatrix}$ and $\vec{w}_2 = \begin{bmatrix} 5 \\ 3 \end{bmatrix}$, so $P^{-1}BP = \begin{bmatrix} 2 & 0 \\ 0 & 3 \end{bmatrix}$ for $P = \begin{bmatrix} 2 & 5 \\ 1 & 3 \end{bmatrix}$.
Therefore, $P^{-1}BP = Q^{-1}AQ$ and $B = PQ^{-1}AQP^{-1} = S^{-1}AS$ where $S = QP^{-1} = \begin{bmatrix} 1 & 2 \\ 1 & 1 \end{bmatrix}\begin{bmatrix} 2 & 5 \\ 1 & 3 \end{bmatrix}^{-1} = \begin{bmatrix} 1 & -1 \\ 2 & -3 \end{bmatrix}$.

25. No; consider $\begin{bmatrix} 1 & 0 \\ 0 & 1 \end{bmatrix}$ and $\begin{bmatrix} 1 & 1 \\ 0 & 1 \end{bmatrix}$. By Example 3, I_2 is similar only to I_2.

26. Yes; by Fact 6.2.4, $f_A(\lambda) = f_B(\lambda) = \lambda^2 - 7\lambda + 7$ so both A and B have two distinct eigenvalues $\lambda_{1,2}$.
Hence each has an eigenbasis, so each is similar to $\begin{bmatrix} \lambda_1 & 0 \\ 0 & \lambda_2 \end{bmatrix}$. By Exercise 20, A is similar to B.

27. **a.** Note that $AS = SB$.
If $\vec{x}$ is in the kernel of B, then $A(S\vec{x}) = AS\vec{x} = SB\vec{x} = S\vec{0} = \vec{0}$, so that $S\vec{x}$ is in the kernel of A, as claimed.

b. Use part a: if $\vec{v}_1, \ldots, \vec{v}_p$ is a basis of the kernel of B, then $S\vec{v}_1, \ldots, S\vec{v}_p$ will be linearly independent vectors in ker(A), since S is invertible. Therefore, dim(ker B) $\leq$ dim(ker A). Reversing the roles of A and B we see that in fact dim(ker B) = dim(ker A). Then rank(A) = n − dim(ker A) = n − dim(ker B) = rank(B), as claimed.

28. Note that $AS = SB$. If $B\vec{x} = \lambda\vec{x}$, then $A(S\vec{x}) = AS\vec{x} = SB\vec{x} = S\lambda\vec{x} = \lambda(S\vec{x})$, as claimed.
Let $E_{A,\lambda}$ and $E_{B,\lambda}$ be the eigenspaces for A and B associated with the eigenvalue λ. If $\vec{v}_1, \ldots, \vec{v}_p$ is a basis of $E_{B,\lambda}$, then $S\vec{v}_1, \ldots, S\vec{v}_p$ will be linearly independent vectors in $E_{A,\lambda}$.
Therefore, $\dim(E_{B,\lambda}) \leq \dim(E_{A,\lambda})$. Reversing the roles of A and B we see that in fact $\dim(E_{A,\lambda}) = \dim(E_{B,\lambda})$. This means that the geometric multiplicities of λ as eigenvalues of A and B are the same.

29. A represents a shear, so by Example 7, $\vec{x}(t) = \vec{x}_0 + t(A\vec{x}_0 - \vec{x}_0) = \begin{bmatrix} 0 \\ 1 \end{bmatrix} + t\left(\begin{bmatrix} 2 \\ 1 \end{bmatrix} - \begin{bmatrix} 0 \\ 1 \end{bmatrix}\right) = \begin{bmatrix} 2t \\ 1 \end{bmatrix}$.

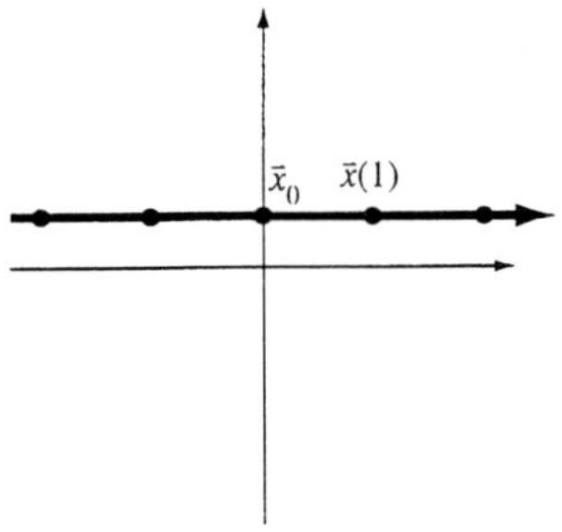

30. A represents a shear, so by Example 7, $\vec{x}(t) = \vec{x}_0 + t(A\vec{x}_0 - \vec{x}_0) = \begin{bmatrix} 0 \\ 1 \end{bmatrix} + t\left(\begin{bmatrix} -1 \\ 1 \end{bmatrix} - \begin{bmatrix} 0 \\ 1 \end{bmatrix}\right) = \begin{bmatrix} -t \\ 1 \end{bmatrix}$.

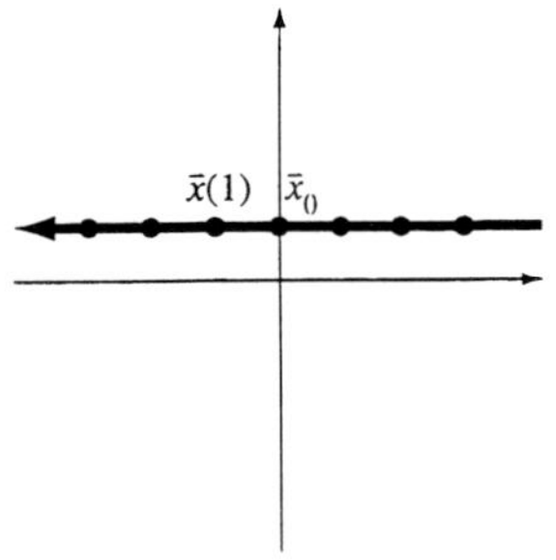

31. A represents a shear which leaves $\vec{x}_0 = \begin{bmatrix} 0 \\ 1 \end{bmatrix}$ fixed so $\vec{x}(t) = \begin{bmatrix} 0 \\ 1 \end{bmatrix}$.

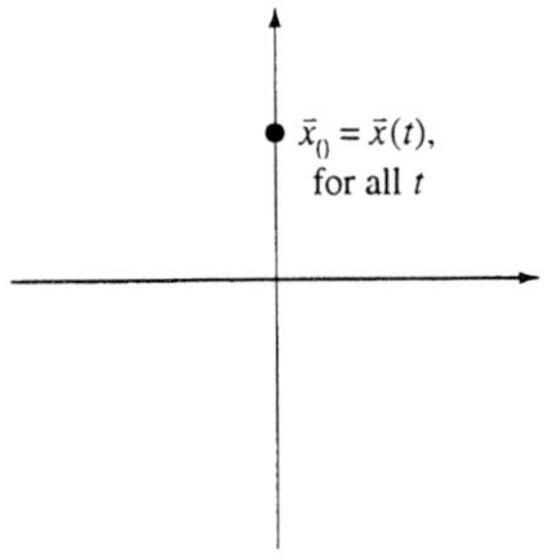

32. $\frac{1}{2}A = M$ represents a shear (since $f_M(\lambda) = (\lambda - 1)^2$), so, by Example 7,

$$M^t\vec{x}_0 = \vec{x}_0 + t(M\vec{x}_0 - \vec{x}_0) = \begin{bmatrix} 0 \\ 1 \end{bmatrix} + t\left(\begin{bmatrix} \frac{1}{2} \\ 1 \end{bmatrix} - \begin{bmatrix} 0 \\ 1 \end{bmatrix}\right) = \begin{bmatrix} \frac{t}{2} \\ 1 \end{bmatrix}.$$

Hence, $\vec{x}(t) = A^t\vec{x}_0 = 2^t M^t\vec{x}_0 = 2^t\begin{bmatrix} \frac{t}{2} \\ 1 \end{bmatrix}$.

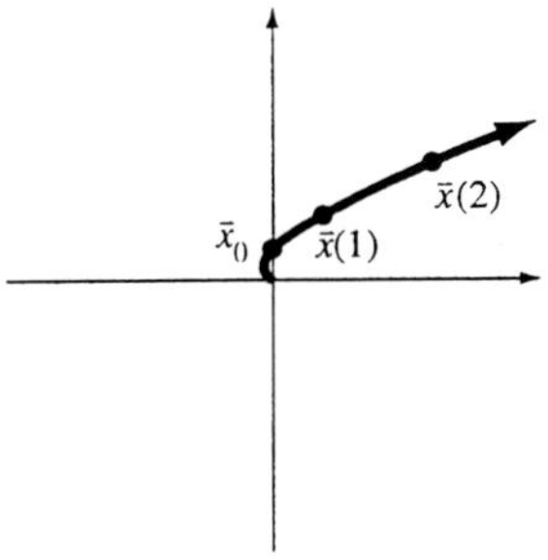

33. A represents a shear (since $f_A(\lambda) = (\lambda - 1)^2$), so by Example 7,

$$\vec{x}(t) = \vec{x}_0 + t(A\vec{x}_0 - \vec{x}_0) = \begin{bmatrix} 0 \\ 1 \end{bmatrix} + t\left(\begin{bmatrix} -1 \\ 0 \end{bmatrix} - \begin{bmatrix} 0 \\ 1 \end{bmatrix}\right) = \begin{bmatrix} -t \\ 1-t \end{bmatrix}.$$

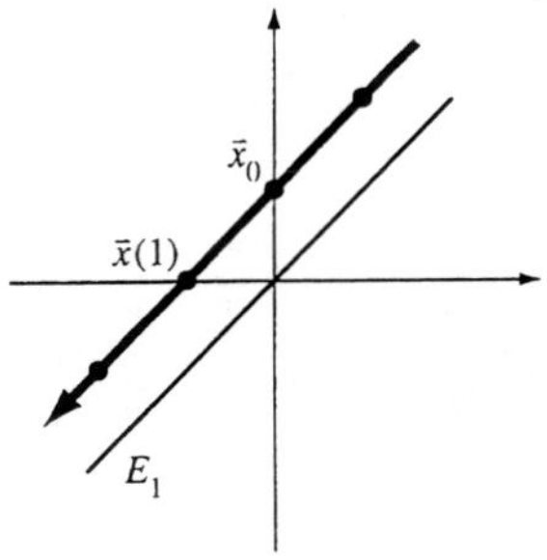

34. A represents a shear (since $f_A(\lambda) = (\lambda - 1)^2$), so by Example 7,

$$\vec{x}(t) = \vec{x}_0 + t(A\vec{x}_0 - \vec{x}_0) = \begin{bmatrix} 0 \\ 1 \end{bmatrix} + t\left(\begin{bmatrix} 1 \\ 3 \end{bmatrix} - \begin{bmatrix} 0 \\ 1 \end{bmatrix}\right) = \begin{bmatrix} t \\ 1+2t \end{bmatrix}.$$

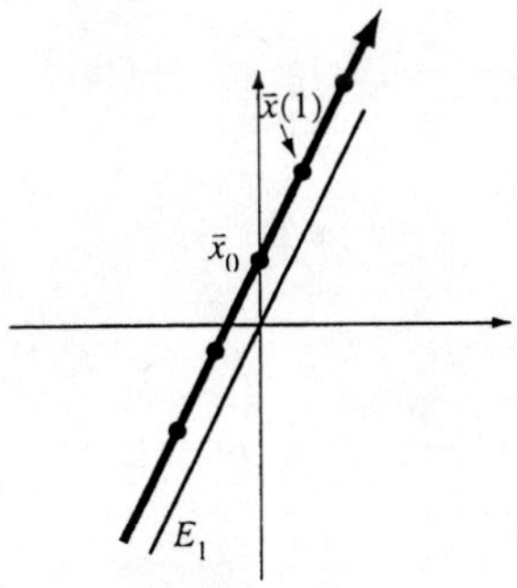

35. $M = \frac{1}{2}A$ represents a shear, so by Example 7, $M^t\vec{x}_0 = \vec{x}_0 + t(M\vec{x}_0 - \vec{x}_0) = \begin{bmatrix}0\\1\end{bmatrix} + t\left(\begin{bmatrix}-\frac{1}{2}\\\frac{1}{2}\end{bmatrix} - \begin{bmatrix}0\\1\end{bmatrix}\right) = \begin{bmatrix}-\frac{t}{2}\\1-\frac{t}{2}\end{bmatrix}$.

$\vec{x}(t) = A^t\vec{x}_0 = 2^t M^t\vec{x}_0 = 2^t\begin{bmatrix}-\frac{t}{2}\\1-\frac{t}{2}\end{bmatrix} = 2^{t-1}\begin{bmatrix}-t\\2-t\end{bmatrix}$

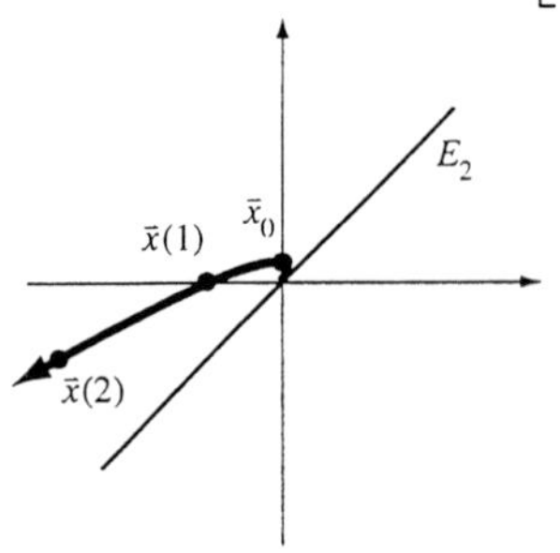

36. $M = \frac{1}{2}A$ represents a shear, so by Example 7, $M^t\vec{x}_0 = \vec{x}_0 + t(M\vec{x}_0 - \vec{x}_0) = \begin{bmatrix}0\\1\end{bmatrix} + t\left(\begin{bmatrix}\frac{1}{2}\\2\end{bmatrix} - \begin{bmatrix}0\\1\end{bmatrix}\right) = \begin{bmatrix}\frac{t}{2}\\1+t\end{bmatrix}$ so

$\vec{x}(t) = A^t\vec{x}_0 = 2^t M^t\vec{x}_0 = 2^t\begin{bmatrix}\frac{t}{2}\\1+t\end{bmatrix}$.

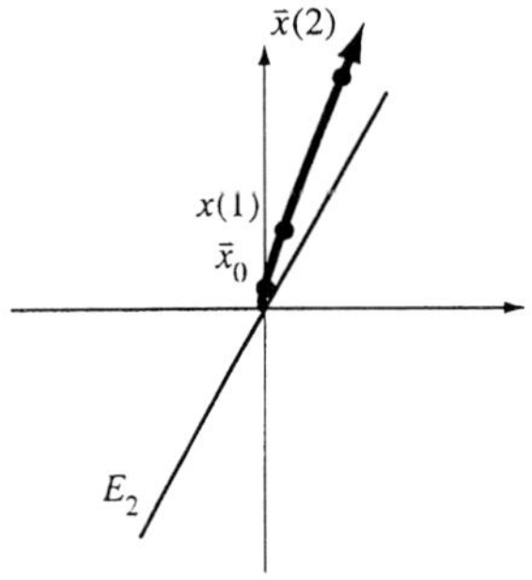

37. A direct computation shows that $A^2 = 0$ and $B^2 \neq 0$. We will argue indirectly, assuming that *A* is similar to *B*: $B = S^{-1}AS$ for some invertible *S*. Then $B^2 = (S^{-1}AS)(S^{-1}AS) = S^{-1}A^2S = 0$, a contradiction. Thus *A* and *B* are not similar.

38. False; consider *A* and *B* from Exercise 37. Zero is the eigenvalue of both matrices, with algebraic multiplicity 4, and, in both instances, the geometric multiplicity of $\lambda = 0$ is 2. And yet, *A* and *B* are not similar.

39. Yes; if $B = S^{-1}AS$ then $B^{-1} = (S^{-1}AS)^{-1} = S^{-1}(S^{-1}A)^{-1} = S^{-1}A^{-1}S$ so A^{-1} is similar to B^{-1}.

40. Yes; $B = S^{-1}AS$ is the product of invertible matrices.

41. Yes; if $B = S^{-1}AS$ then $B^T = (S^{-1}AS)^T = (AS)^T(S^{-1})^T = S^TA^T(S^{-1})^T = S^TA^T(S^T)^{-1}$ so $B^T = Q^{-1}A^TQ$ if $Q = (S^T)^{-1}$.

42. Yes; if *A* is similar to a diagonal matrix *D*, then A^T is similar to $D^T = D$, by Exercise 41, so that *A* is similar to A^T, by Exercise 20.

43. Yes; if $D = S^{-1}AS$ where *D* is diagonal then $D^2 = (S^{-1}AS)(S^{-1}AS) = S^{-1}A^2S$. But D^2 is diagonal so A^2 is diagonalizable.

44. Yes; if $D = S^{-1}AS$ where D is diagonal then $D^t = (S^{-1}AS)^t = S^{-1}A^tS$. But D^t is diagonal so A^t is diagonalizable.

45. Yes; if $D = S^{-1}AS$ then $D^{-1} = (S^{-1}AS)^{-1} = S^{-1}A^{-1}S$ but D^{-1} is diagonal so A^{-1} is diagonalizable.

46. No; if $A = \begin{bmatrix} 1 & 1 \\ 0 & 1 \end{bmatrix}$ then A is invertible, but it does not have an eigenbasis, so it is not diagonalizable.

Conversely, $B = \begin{bmatrix} 1 & 0 \\ 0 & 0 \end{bmatrix}$ is diagonalizable (as it is already diagonal), but B is not invertible.

47. No; if $A = \begin{bmatrix} 1 & 0 \\ 0 & 0 \end{bmatrix}$, $B = \begin{bmatrix} 0 & 1 \\ 0 & 1 \end{bmatrix}$ then A and B are diagonalizable since they have distinct eigenvalues, but $A + B = \begin{bmatrix} 1 & 1 \\ 0 & 1 \end{bmatrix}$ is not since it does not have an eigenbasis.

48. No; consider the counter example $\begin{bmatrix} 1 & 0 \\ 0 & 0 \end{bmatrix}\begin{bmatrix} 0 & 1 \\ 0 & 1 \end{bmatrix} = \begin{bmatrix} 0 & 1 \\ 0 & 0 \end{bmatrix}$.

49. No; if $A = \begin{bmatrix} 0 & 1 \\ 0 & 0 \end{bmatrix}$ then A is not diagonalizable, but $A^2 = \begin{bmatrix} 0 & 0 \\ 0 & 0 \end{bmatrix}$ is.

50. $BA = I_nBA = (A^{-1}A)BA = A^{-1}(AB)A$ so AB is similar to BA provided that A^{-1} exists.

51. Let $A = \begin{bmatrix} 1 & 0 \\ 0 & 0 \end{bmatrix}$, $B = \begin{bmatrix} 0 & 1 \\ 0 & 0 \end{bmatrix}$. Then $AB = \begin{bmatrix} 0 & 1 \\ 0 & 0 \end{bmatrix}$ and $BA = \begin{bmatrix} 0 & 0 \\ 0 & 0 \end{bmatrix}$, and $\begin{bmatrix} 0 & 0 \\ 0 & 0 \end{bmatrix}$ is similar only to itself.

52. a. By Exercise 50, AB is similar to BA since A is invertible. So, by Fact 7.2.6, AB and BA have the same characteristic polynomial.

b. Consider $f(x) = \det(\lambda I_n - (A - xI_n)B) - \det(\lambda I_n - B(A - xI_n))$.
Then f is a polynomial in x since it consists of two determinants, each of which is a polynomial in x. Its degree is $\le n$ since we are taking the determinants of two matrices whose entries are of the form $px + q$ for some constants p and q. If x is not an eigenvalue of A then $A - xI_n$ is invertible and $B(A - xI_n)$ and $(A - xI_n)B$ have the same characteristic polynomial (by part a) i.e.
$\det(\lambda I_n - (A - xI_n)B) = \det(\lambda I_n - B(A - xI_n))$, so $f(x) = 0$.
Since f is a polynomial of degree $\le n$, it can have at most n roots unless it is identically zero. But $f(x) = 0$ for all x which are not eigenvalues of A, therefore, $f(x) = 0$ for *all* x.
In particular $f(0) = 0$, meaning that $\det(\lambda I_n - AB) = \det(\lambda I_n - BA)$, so that AB and BA have the same characteristic polynomial and the same eigenvalues.
(For a different proof of the fact that characteristic polynomials of AB and of BA are equal see *Mathematics Magazine*, vol. 61, no. 5(Dec. 1988), p. 294.)

53. $\begin{bmatrix}1 & a\\0 & 1\end{bmatrix}^2=\begin{bmatrix}1 & a\\0 & 1\end{bmatrix}\begin{bmatrix}1 & a\\0 & 1\end{bmatrix}=\begin{bmatrix}1 & 2a\\0 & 1\end{bmatrix}$

$\begin{bmatrix}1 & a\\0 & 1\end{bmatrix}^3=\begin{bmatrix}1 & 2a\\0 & 1\end{bmatrix}\begin{bmatrix}1 & a\\0 & 1\end{bmatrix}=\begin{bmatrix}1 & 3a\\0 & 1\end{bmatrix}$

$\vdots$

$\begin{bmatrix}1 & a\\0 & 1\end{bmatrix}^t=\begin{bmatrix}1 & ta\\0 & 1\end{bmatrix}$

For a formal proof use induction: If $\begin{bmatrix}1 & a\\0 & 1\end{bmatrix}^{t-1}=\begin{bmatrix}1 & (t-1)a\\0 & 1\end{bmatrix}$ then

$\begin{bmatrix}1 & a\\0 & 1\end{bmatrix}^t=\begin{bmatrix}1 & a\\0 & 1\end{bmatrix}^{t-1}\begin{bmatrix}1 & a\\0 & 1\end{bmatrix}=\begin{bmatrix}1 & (t-1)a\\0 & 1\end{bmatrix}\begin{bmatrix}1 & a\\0 & 1\end{bmatrix}=\begin{bmatrix}1 & ta\\0 & 1\end{bmatrix}$, as claimed.

54. If $\lambda \neq 0$, then $\begin{bmatrix}\lambda & 1\\0 & \lambda\end{bmatrix}^t=\lambda^t\begin{bmatrix}1 & \frac{1}{\lambda}\\0 & 1\end{bmatrix}^t=\lambda^t\begin{bmatrix}1 & \frac{t}{\lambda}\\0 & 1\end{bmatrix}=\begin{bmatrix}\lambda^t & t\lambda^{t-1}\\0 & \lambda^t\end{bmatrix}$.

$\uparrow$

Exercise 53

In the case $\lambda = 0$ the formula is easy to check.

55. Yes; by Fact 6.2.4, $f_A(\lambda)=\lambda^2-tr(A)\lambda+\det(A)=\lambda^2-2\lambda+1=(\lambda-1)^2$ so by Example 6, A represents a shear.

56. The proof is analogous to Example 6: Consider a basis $\vec{v}$, $\vec{w}$ of $\mathbb{R}^2$ where $\vec{v}$ is an eigenvector of A with eigenvalue 0. Then the matrix of $T(\vec{x})=A\vec{x}$ with respect to this basis is of the form $B=\begin{bmatrix}0 & a\\0 & b\end{bmatrix}$, and $b=0$ since tr(B) = tr(A) = 0. Therefore, $B=\begin{bmatrix}0 & a\\0 & 0\end{bmatrix}$ for some nonzero a, that is, $A\vec{w}=a\vec{v}$. Finally, the matrix of T with respect to the basis $a\vec{v}$, $\vec{w}$ is $\begin{bmatrix}0 & 1\\0 & 0\end{bmatrix}$, proving our claim. If $S^{-1}AS=\begin{bmatrix}0 & 1\\0 & 0\end{bmatrix}$, then $A=S\begin{bmatrix}0 & 1\\0 & 0\end{bmatrix}S^{-1}$ and

$A^2=S\begin{bmatrix}0 & 1\\0 & 0\end{bmatrix}^2S^{-1}=\begin{bmatrix}0 & 0\\0 & 0\end{bmatrix}$.

57. The case when A is diagonalizable was discussed in Exercise 42.
If A is not diagonalizable, then both A and A^T have an eigenvalue λ with algebraic multiplicity two, so that both A and A^T are similar to $\begin{bmatrix}\lambda & 1\\0 & \lambda\end{bmatrix}$ (by Examples 6 and 8), and thus similar to each other.

58. Consider a complex 3×3 matrix A. Let $\mathcal{B}$ be a basis of $\mathbb{C}^3$ whose first vector is an eigenvector for A with eigenvalue λ. Then the matrix of $T(\vec{x})=A\vec{x}$ with respect to $\mathcal{B}$ will be of the form $B=\begin{bmatrix}\lambda & * & *\\0 & * & *\\0 & * & *\end{bmatrix}=\left[\begin{array}{c|c}\lambda & \vec{v}\\\hline 0 & M\end{array}\right]$.

Now M is similar to an upper triangular 2×2 matrix N (think about it!): $S^{-1}MS=N$ for some invertible 2×2 matrix S. Then $\left[\begin{array}{c|c}1 & 0\\\hline 0 & S\end{array}\right]^{-1}B\left[\begin{array}{c|c}1 & 0\\\hline 0 & S\end{array}\right]=\left[\begin{array}{c|c}1 & 0\\\hline 0 & S^{-1}\end{array}\right]\left[\begin{array}{c|c}\lambda & \vec{v}\\\hline 0 & M\end{array}\right]\left[\begin{array}{c|c}1 & 0\\\hline 0 & S\end{array}\right]=\left[\begin{array}{c|c}\lambda & \vec{v}S\\\hline 0 & N\end{array}\right]=C$ is upper triangular.

Since A is similar to B and B is similar to C, A will be similar to the upper triangular matrix C.

59. Note that $f(x)$ is not the zero polynomial, since $f(i) = \det(S_1 + iS_2) = \det(S) \neq 0$, as S is invertible. A nonzero polynomial has only finitely many zeros, so that there is a real number x_0 such that $f(x_0) = \det(S_1 + x_0S_2) \neq 0$, that is, $S_1 + x_0S_2$ is invertible. Now $SB = AS$ or $(S_1 + iS_2)B = A(S_1 + iS_2)$. Considering the real and the imaginary part, we can conclude that $S_1B = AS_1$ and $S_2B = AS_2$ and therefore $(S_1 + x_0S_2)B = A(S_1 + x_0S_2)$. Since $S_1 + x_0S_2$ is invertible, we have $B = (S_1 + x_0S_2)^{-1}A(S_1 + x_0S_2)$, as claimed.

60. a. We are told that the dimension of the eigenspace E_{λ_0} is d. We can choose a basis $\vec{v}_1, \ldots, \vec{v}_d$ of E_{λ_0} and a basis $\vec{v}_{d+1}, \ldots, \vec{v}_n$ of $(E_{\lambda_0})^{\perp}$; then $\vec{v}_1, \ldots, \vec{v}_n$ will be a basis of $\mathbb{R}^n$ of the desired form.

b. For $i = 1, \ldots, d$, (ith component of B) $= [T(\vec{v}_i)]_{\mathcal{B}} = [\lambda_0\vec{v}_i]_{\mathcal{B}} = \begin{bmatrix} 0 \\ 0 \\ \vdots \\ \lambda_0 \\ 0 \\ 0 \end{bmatrix} \leftarrow i\text{th component},$

so that $B = \left[\begin{array}{c|c} \lambda_0 I_d & * \\ \hline 0 & * \end{array}\right]$.

c. $f_B(\lambda) = \det\left[\begin{array}{c|c} (\lambda - \lambda_0)I_d & * \\ \hline 0 & * \end{array}\right] = (\lambda - \lambda_0)^d g(\lambda)$, so that

algebraic multiplicity of λ_0 as an eigenvalue of A = algebraic multiplicity of λ_0 as an eigenvalue of B
$\geq d$
= geometric multiplicity of λ_0 as an eigenvalue of A.

61. The key here is to recall that A and B have the same characteristic polynomial (Fact 7.2.6a). Therefore

a. A and B have the same eigenvalues with the same algebraic multiplicities (by definition), and

b, c. $\text{tr}(A) = \text{tr}(B)$ and $\det(A) = \det(B)$ by Fact 6.2.5: trace and determinant are coefficients of the characteristic polynomial (up to the sign).

62. a. B is diagonalizable since it has three distinct eigenvalues, so that $S^{-1}BS$ is diagonal for some invertible S. But $S^{-1}AS = S^{-1}I_3S = I_3$ is diagonal as well, proving our claim.

b. There is an invertible S such that $S^{-1}AS = D_1$ and $S^{-1}BS = D_2$ are both diagonal. Then $A = SD_1S^{-1}$ and $B = SD_2S^{-1}$, so that $AB = (SD_1S^{-1})(SD_2S^{-1}) = SD_1D_2S^{-1}$ and $BA = (SD_2S^{-1})(SD_1S^{-1}) = SD_2D_1S^{-1}$. These two results agree, since $D_1D_2 = D_2D_1$ for the diagonal matrices D_1 and D_2.

c. Let A be I_n and B a nondiagonalizable $n \times n$ matrix, for example, $A = \begin{bmatrix} 1 & 0 \\ 0 & 1 \end{bmatrix}$ and $B = \begin{bmatrix} 1 & 1 \\ 0 & 1 \end{bmatrix}$.

d. Suppose $BD = DB$ for a diagonal D with distinct diagonal entries. The ijth entry of the matrix $BD = DB$ is $b_{ij}d_{jj} = d_{ii}b_{ij}$. For $i \neq j$ this implies that $b_{ij} = 0$, as claimed.

e. Since A has n distinct eigenvalues, A is diagonalizable, that is, there is an invertible S such that $S^{-1}AS = D$ is a diagonal matrix with n distinct diagonal entries. We claim that $S^{-1}BS$ is diagonal as well; by part d it suffices to show that $S^{-1}BS$ commutes with $D = S^{-1}AS$. This is easy to verify:
$(S^{-1}BS)D = (S^{-1}BS)(S^{-1}AS) = S^{-1}BAS = S^{-1}ABS = (S^{-1}AS)(S^{-1}BS) = D(S^{-1}BS)$.

63. Let $B_i = A - \lambda_i I_n$; note that B_i and B_j commute for any two indices i and j. If $\vec{v}$ is an eigenvector of A with eigenvalue λ_i, then $B_i\vec{v} = \vec{0}$ and $B_1B_2\cdots B_i\cdots B_m\vec{v} = B_1\cdots B_{i-1}B_{i+1}\cdots B_m B_i\vec{v} = \vec{0}$.
Since A is diagonalizable, any vector $\vec{x}$ in $\mathbb{R}^n$ can be written as a linear combination of eigenvectors, so that $B_1B_2\cdots B_m\vec{x} = \vec{0}$ and therefore $B_1B_2\cdots B_m = 0$, as claimed.

64. If $\vec{v}$ is an eigenvector with eigenvalue λ, then

$$\begin{aligned} f_A(A)\vec{v} &= (A^n + a_{n-1}A^{n-1} + \cdots + a_1A + a_0I_n)\vec{v} \\ &= \lambda^n\vec{v} + a_{n-1}\lambda^{n-1}\vec{v} + \cdots + a_1\lambda\vec{v} + a_0\vec{v} \\ &= (\lambda^n + a_{n-1}\lambda^{n-1} + \cdots + a_1\lambda + a_0)\vec{v} \\ &= f_A(\lambda)\vec{v} = 0\vec{v} = \vec{0}. \end{aligned}$$

Since A is diagonalizable, any vector $\vec{x}$ in $\mathbb{R}^n$ can be written as a linear combination of eigenvectors, so that $f_A(A)\vec{x} = \vec{0}$. Since this equation holds for all $\vec{x}$ in $\mathbb{R}^n$, we have $f_A(A) = 0$, as claimed.

65. Example 6 tells us that A represents a shear, so that $A\vec{w} - \vec{w}$ is in E_1 for all $\vec{w}$ in $\mathbb{R}^2$.

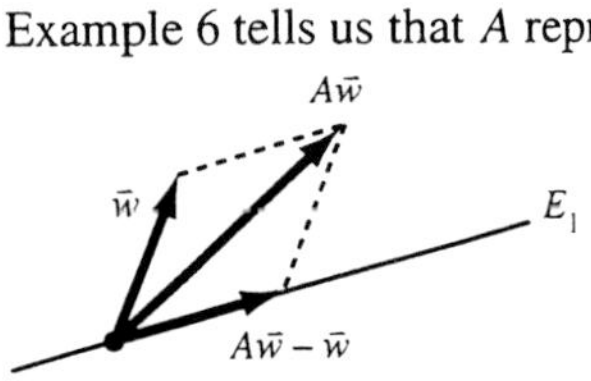

This means that $A(A\vec{w} - \vec{w}) = A\vec{w} - \vec{w}$ or $(A - I_2)(A\vec{w} - \vec{w}) = \vec{0}$ or $(A - I_2)^2\vec{w} = \vec{0}$.
Since this equation holds for all $\vec{w}$, we have $f_A(A) = (A - I_2)^2 = 0$.

66. Since $36 = \det(A^TA) = (\det A)^2$, it is required that $\det(A) = \lambda_1\lambda_2\lambda_3 = \pm 6$.
Also, A cannot have 3 distinct eigenvalues (a 3 × 3 matrix with 3 distinct eigenvalues is diagonalizable).
One possible choice is $A = \begin{bmatrix} 1 & 1 & 0 \\ 0 & 1 & 0 \\ 0 & 0 & 6 \end{bmatrix}$ (note that $\dim(E_1) = 1$, so that A is not diagonalizable).

67. *True*; A has three distinct eigenvalues, namely, a real eigenvalue and the conjugate complex eigenvalues $2 \pm 3i$.

68. We have to show that $\lim_{t\to\infty}(A^t) = 0$ if and only if the modulus of both eigenvalues of A is less than 1. In Section 6.5 we dealt with the case when A is diagonalizable over $\mathbb{C}$. If A is not diagonalizable over $\mathbb{C}$, then there is an invertible S such that $S^{-1}AS = \begin{bmatrix} \lambda & 1 \\ 0 & \lambda \end{bmatrix}$ for some real number λ (see Examples 6 and 8).

Now $A = S\begin{bmatrix} \lambda & 1 \\ 0 & \lambda \end{bmatrix}S^{-1}$ and $A^t = S\begin{bmatrix} \lambda^t & t\lambda^{t-1} \\ 0 & \lambda^t \end{bmatrix}S^{-1}$ (see Exercise 54).

Then $\lim_{t\to\infty}(\lambda^t) = 0$ if and only if $|\lambda| < 1$ and likewise $\lim_{t\to\infty}(t\lambda^{t-1}) = 0$ if and only if $|\lambda| < 1$.
The claim follows.

7.3

1. $\vec{e}_1, \vec{e}_2$ is an orthonormal eigenbasis.

2. $\frac{1}{\sqrt{2}}\begin{bmatrix}1\\1\end{bmatrix}, \frac{1}{\sqrt{2}}\begin{bmatrix}1\\-1\end{bmatrix}$ is an orthonormal eigenbasis.

3. $\frac{1}{\sqrt{5}}\begin{bmatrix}2\\1\end{bmatrix}, \frac{1}{\sqrt{5}}\begin{bmatrix}-1\\2\end{bmatrix}$ is an orthonormal eigenbasis.

4. $\frac{1}{\sqrt{3}}\begin{bmatrix}1\\1\\-1\end{bmatrix}, \frac{1}{\sqrt{2}}\begin{bmatrix}1\\-1\\0\end{bmatrix}, \frac{1}{\sqrt{6}}\begin{bmatrix}1\\1\\2\end{bmatrix}$ is an orthonormal eigenbasis.

5. Eigenvalues $-1, -1, 2$

   Choose $\vec{v}_1 = \frac{1}{\sqrt{2}}\begin{bmatrix}-1\\1\\0\end{bmatrix}$ in E_{-1} and $\vec{v}_2 = \frac{1}{\sqrt{3}}\begin{bmatrix}1\\1\\1\end{bmatrix}$ in E_2 and let $\vec{v}_3 = \vec{v}_1 \times \vec{v}_2 = \frac{1}{\sqrt{6}}\begin{bmatrix}1\\1\\-2\end{bmatrix}$.

6. $\frac{1}{3}\begin{bmatrix}2\\2\\1\end{bmatrix}, \frac{1}{3}\begin{bmatrix}2\\-1\\-2\end{bmatrix}, \frac{1}{3}\begin{bmatrix}1\\-2\\2\end{bmatrix}$ is an orthonormal eigenbasis.

7. $\frac{1}{\sqrt{2}}\begin{bmatrix}1\\1\end{bmatrix}, \frac{1}{\sqrt{2}}\begin{bmatrix}1\\-1\end{bmatrix}$ is an orthonormal eigenbasis, so $S = \frac{1}{\sqrt{2}}\begin{bmatrix}1 & 1\\1 & -1\end{bmatrix}$ and $D = \begin{bmatrix}5 & 0\\0 & 1\end{bmatrix}$.

8. $\frac{1}{\sqrt{10}}\begin{bmatrix}3\\1\end{bmatrix}, \frac{1}{\sqrt{10}}\begin{bmatrix}-1\\3\end{bmatrix}$ is an orthonormal eigenbasis, with $\lambda_1 = 4$ and $\lambda_2 = -6$, so $S = \frac{1}{\sqrt{10}}\begin{bmatrix}3 & -1\\1 & 3\end{bmatrix}$ and $D = \begin{bmatrix}4 & 0\\0 & -6\end{bmatrix}$.

9. $\frac{1}{\sqrt{2}}\begin{bmatrix}1\\0\\1\end{bmatrix}, \frac{1}{\sqrt{2}}\begin{bmatrix}-1\\0\\1\end{bmatrix}, \begin{bmatrix}0\\1\\0\end{bmatrix}$ is an orthonormal eigenbasis, with $\lambda_1 = 3$, $\lambda_2 = -3$, and $\lambda_3 = 2$, so $S = \frac{1}{\sqrt{2}}\begin{bmatrix}1 & -1 & 0\\0 & 0 & \sqrt{2}\\1 & 1 & 0\end{bmatrix}$ and $D = \begin{bmatrix}3 & 0 & 0\\0 & -3 & 0\\0 & 0 & 2\end{bmatrix}$.

10. $\lambda_1 = \lambda_2 = 0$ and $\lambda_3 = 9$.

$\vec{v}_1 = \frac{1}{\sqrt{5}}\begin{bmatrix}2\\1\\0\end{bmatrix}$ is in E_0 and $\vec{v}_2 = \frac{1}{3}\begin{bmatrix}1\\-2\\2\end{bmatrix}$ is in E_9.

Let $\vec{v}_3 = \vec{v}_1 \times \vec{v}_2 = \frac{1}{3\sqrt{5}}\begin{bmatrix}2\\-4\\-5\end{bmatrix}$; then $\vec{v}_1, \vec{v}_2, \vec{v}_3$ is an orthonormal eigenbasis.

$$S = \begin{bmatrix}\frac{2}{\sqrt{5}} & \frac{1}{3} & \frac{2}{3\sqrt{5}}\\ \frac{1}{\sqrt{5}} & -\frac{2}{3} & -\frac{4}{3\sqrt{5}}\\ 0 & \frac{2}{3} & -\frac{\sqrt{5}}{3}\end{bmatrix} \text{ and } D = \begin{bmatrix}0&0&0\\0&9&0\\0&0&0\end{bmatrix}.$$

11. $\frac{1}{\sqrt{2}}\begin{bmatrix}1\\0\\1\end{bmatrix}, \frac{1}{\sqrt{2}}\begin{bmatrix}-1\\0\\1\end{bmatrix}, \begin{bmatrix}0\\1\\0\end{bmatrix}$ is an orthonormal eigenbasis, with $\lambda_1 = 2$, $\lambda_2 = 0$, and $\lambda_3 = 1$, so $S = \frac{1}{\sqrt{2}}\begin{bmatrix}1&-1&0\\0&0&\sqrt{2}\\1&1&0\end{bmatrix}$

and $D = \begin{bmatrix}2&0&0\\0&0&0\\0&0&1\end{bmatrix}$.

12. **a.** $E_1 = \text{span}\begin{bmatrix}1\\0\\2\end{bmatrix}$ and $E_{-1} = (E_1)^{\perp}$. An orthonormal eigenbasis is $\frac{1}{\sqrt{5}}\begin{bmatrix}1\\0\\2\end{bmatrix}, \begin{bmatrix}0\\1\\0\end{bmatrix}, \frac{1}{\sqrt{5}}\begin{bmatrix}2\\0\\-1\end{bmatrix}$.

b. Use Fact 7.2.1: $B = \begin{bmatrix}1&0&0\\0&-1&0\\0&0&-1\end{bmatrix}$.

c. $A = SBS^{-1} = \begin{bmatrix}-0.6&0&0.8\\0&-1&0\\0.8&0&0.6\end{bmatrix}$, where $S = \begin{bmatrix}\frac{1}{\sqrt{5}}&0&\frac{2}{\sqrt{5}}\\0&1&0\\\frac{2}{\sqrt{5}}&0&-\frac{1}{\sqrt{5}}\end{bmatrix}$.

13. Yes; if $\vec{v}$ is an eigenvector of A with eigenvalue λ, then $\vec{v} = I_3\vec{v} = A^2\vec{v} = \lambda^2\vec{v}$, so that $\lambda^2 = 1$ and $\lambda = 1$ or $\lambda = -1$. Since A is symmetric, E_1 and E_{-1} will be orthogonal complements, so that A represents the reflection in E_1.

14. Let S be as in Exercise 3. Then $S^{-1}AS = \begin{bmatrix}0&0&0\\0&0&0\\0&0&3\end{bmatrix}$.

a. This matrix is $2A$ so that $S^{-1}(2A)S = \begin{bmatrix}0&0&0\\0&0&0\\0&0&6\end{bmatrix}$.

b. This is $A-3I_3$, so that $S^{-1}(A-3I_3)S=S^{-1}AS-3I_3=\begin{bmatrix}-3&0&0\\0&-3&0\\0&0&0\end{bmatrix}$.

c. This is $\frac{1}{2}(A-I_3)$, so that $S^{-1}\left(\frac{1}{2}(A-I_3)\right)S=\frac{1}{2}(S^{-1}AS-I_3)=\begin{bmatrix}-\frac{1}{2}&0&0\\0&-\frac{1}{2}&0\\0&0&1\end{bmatrix}$.

15. Yes, if $A\vec{v}=\lambda\vec{v}$, then $A^{-1}\vec{v}=\frac{1}{\lambda}\vec{v}$, so that an orthonormal eigenbasis for A is also an orthonormal eigenbasis for A^{-1} (with reciprocal eigenvalues).

16. a. $\ker(A)$ is four-dimensional, so that the eigenvalue 0 has multiplicity 4, and the remaining eigenvalue is $\text{tr}(A)=5$.

b. $B=A+2I_5$, so that the eigenvalues are 2, 2, 2, 2, 7.

c. $\det(B)=2^4\cdot 7=112$ (product of eigenvalues)

17. If A is the $n\times n$ matrix with all 1's, then the eigenvalues of A are 0 (with multiplicity $n-1$) and n. Now $B=qA+(p-q)I_n$, so that the eigenvalues of B are $p-q$ (with multiplicity $n-1$) and $qn+p-q$. Thus $\det(B)=(p-q)^{n-1}(qn+p-q)$.

18. By Fact 5.3.7, the volume is $|\det A|=\sqrt{\det(A^TA)}$. Now $\vec{v}_i\cdot\vec{v}_j=\|\vec{v}_i\|\|\vec{v}_j\|\cdot\cos(\alpha)=\frac{1}{2}$, so that A^TA has all 1's on the diagonal and $\frac{1}{2}$'s outside. By Exercise 17 $\left(\text{with } p=1 \text{ and } q=\frac{1}{2}\right)$,

$\det(A^TA)=\left(\frac{1}{2}\right)^{n-1}\left(\frac{1}{2}n+\frac{1}{2}\right)=\left(\frac{1}{2}\right)^n(n+1)$, so that the volume is $\sqrt{\det(A^TA)}=\left(\frac{1}{2}\right)^{n/2}\sqrt{n+1}$.

19. Let $L(\vec{x})=A\vec{x}$. Then A^TA is symmetric, since $(A^TA)^T=A^T(A^T)^T=A^TA$, so that there is an orthonormal eigenbasis $\vec{v}_1,\dots,\vec{v}_n$ for A^TA. Then the vectors $A\vec{v}_1,\dots,A\vec{v}_n$ are orthogonal, since $A\vec{v}_i\cdot A\vec{v}_j=(A\vec{v}_i)^TA\vec{v}_j=\vec{v}_i^TA^TA\vec{v}_j=\vec{v}_i\cdot(A^TA\vec{v}_j)=\vec{v}_i\cdot(\lambda_j\vec{v}_j)=\lambda_j(\vec{v}_i\cdot\vec{v}_j)=0$ if $i\neq j$.

20. By Exercise 19, there is an orthonormal basis $\vec{v}_1,\dots,\vec{v}_n$ of $\mathbb{R}^n$ such that $T(\vec{v}_1),\dots,T(\vec{v}_n)$ are orthogonal. Suppose that $T(\vec{v}_1),\dots,T(\vec{v}_r)$ are nonzero and $T(\vec{v}_{r+1}),\dots,T(\vec{v}_n)$ are zero. Then let $\vec{w}_i=\frac{1}{\|T(\vec{v}_i)\|}T(\vec{v}_i)$ for $i=1,\dots,r$ and choose an orthonormal basis $\vec{w}_{r+1},\dots,\vec{w}_m$ of $[\text{span}(\vec{w}_1,\dots,\vec{w}_r)]^\perp$. Then $\vec{w}_1,\dots,\vec{w}_m$ does the job.

21. For each eigenvalue there are two unit eigenvectors: $\pm\vec{v}_1$, $\pm\vec{v}_2$, and $\pm\vec{v}_3$. We have 6 choices for the first column of S, 4 choices remaining for the second column, and 2 for the third.
Answer: $6\cdot 4\cdot 2=48$.

22. a. If we let $k=2$ then A is symmetric and therefore (orthogonally) diagonalizable.

b. If we let $k=0$ then 0 is the only eigenvalue (but $A\neq 0$), so that A is not diagonalizable.

23. The eigenvalues are real (by Fact 7.3.3), so that the only possible eigenvalues are ± 1 (by Fact 6.1.2). Since A is symmetric, E_1 and E_{-1} are orthogonal complements. Thus A represents a *reflection* in E_1.

24. Note that A is symmetric and orthogonal, so that the eigenvalues are 1 and -1 (see Exercise 23).

$E_1 = \text{span}\left(\begin{bmatrix}1\\0\\0\\1\end{bmatrix}, \begin{bmatrix}0\\1\\1\\0\end{bmatrix}\right)$ and $E_{-1} = \text{span}\left(\begin{bmatrix}1\\0\\0\\-1\end{bmatrix}, \begin{bmatrix}0\\1\\-1\\0\end{bmatrix}\right)$, so that $\frac{1}{\sqrt{2}}\begin{bmatrix}1\\0\\0\\1\end{bmatrix}, \frac{1}{\sqrt{2}}\begin{bmatrix}0\\1\\1\\0\end{bmatrix}, \frac{1}{\sqrt{2}}\begin{bmatrix}1\\0\\0\\-1\end{bmatrix}, \frac{1}{\sqrt{2}}\begin{bmatrix}0\\1\\-1\\0\end{bmatrix}$ is an orthonormal eigenbasis.

25. Note that A is symmetric an orthogonal, so that the eigenvalues of A are 1 and -1.

$E_1 = \text{span}\left(\begin{bmatrix}1\\0\\0\\0\\1\end{bmatrix}, \begin{bmatrix}0\\1\\0\\1\\0\end{bmatrix}, \begin{bmatrix}0\\0\\1\\0\\0\end{bmatrix}\right)$, $E_{-1} = \text{span}\left(\begin{bmatrix}1\\0\\0\\0\\-1\end{bmatrix}, \begin{bmatrix}0\\1\\0\\-1\\0\end{bmatrix}\right)$

The columns of S must form an eigenbasis for A: $S = \begin{bmatrix}1 & 0 & 0 & 1 & 0\\ 0 & 1 & 0 & 0 & 1\\ 0 & 0 & 1 & 0 & 0\\ 0 & 1 & 0 & 0 & -1\\ 1 & 0 & 0 & -1 & 0\end{bmatrix}$ is one possible choice.

26. Since J_n is both orthogonal and symmetric, the eigenvalues are 1 and -1. If n is even, then both have multiplicity $\frac{n}{2}$ (as in Exercise 24). If n is odd, then the multiplicities are $\frac{n+1}{2}$ for 1 and $\frac{n-1}{2}$ for -1 (as in Exercise 25). One way to see this is to observe that $\text{tr}(J_n)$ is 0 for even n, and 1 for odd n (recall that the trace is the sum of the eigenvalues).

27. If n is even, then this matrix is $J_n + I_n$, for the J_n introduced in Exercise 26, so that the eigenvalues are 0 and 2, with multiplicity $\frac{n}{2}$ each. E_2 is the span of all $e_i + e_{n+1-i}$, for $i = 1, \ldots, \frac{n}{2}$, and E_0 is spanned by all $e_i - e_{n+1-i}$.

If n is odd, then E_2 is spanned by all $\vec{e}_i + \vec{e}_{n+1-i}$, for $i = 1, \ldots, \frac{n-1}{2}$; E_0 is spanned by all $\vec{e}_i - \vec{e}_{n+1-i}$, for $i = 1, \ldots, \frac{n-1}{2}$, and E_1 is spanned by $\vec{e}_{\frac{n+1}{2}}$.

28. For $\lambda \neq 0$,

$$F_A(\lambda) = \det\left[\begin{array}{cccc|c} \lambda & & & 0 & -1\\ & \lambda & & & -1\\ & & \ddots & & \vdots\\ 0 & & & \lambda & -1\\ \hline -1 & -1 & \cdots & -1 & \lambda - 1\end{array}\right] = \frac{1}{\lambda}\det\left[\begin{array}{cccc|c} \lambda & & & 0 & -1\\ & \lambda & & & -1\\ & & \ddots & & \vdots\\ 0 & & & \lambda & -1\\ \hline -\lambda & -\lambda & \cdots & -\lambda & \lambda^2 - \lambda\end{array}\right]$$

$$= \frac{1}{\lambda}\det\left[\begin{array}{cccc|c} \lambda & & & 0 & -1 \\ & \lambda & & & -1 \\ & & \ddots & & \vdots \\ 0 & & & \lambda & -1 \\ \hline 0 & 0 & \cdots & 0 & \lambda^2-\lambda-12 \end{array}\right] = \lambda^{11}(\lambda^2-\lambda-12) = \lambda^{11}(\lambda-4)(\lambda+3) = 0$$

↑
Add first 12 rows to last

Eigenvalues are 0 (with multiplicity 11), 4 and –3.
Eigenvalues for 0 are $\vec{e}_1 - \vec{e}_i$ ($i = 2, \ldots, 12$),

$$E_4 = \text{span}\begin{bmatrix} 1 \\ 1 \\ \vdots \\ 1 \\ 4 \end{bmatrix} \text{ (12 ones)}, \quad E_{-3} = \text{span}\begin{bmatrix} 1 \\ 1 \\ \vdots \\ 1 \\ -3 \end{bmatrix} \text{ (12 ones)}$$

so

$$S = \begin{bmatrix} 1 & 1 & 1 & 1 & 1 & 1 & 1 & 1 & 1 & 1 & 1 & 1 & 1 \\ -1 & 0 & 0 & 0 & 0 & 0 & 0 & 0 & 0 & 0 & 0 & 1 & 1 \\ 0 & -1 & 0 & 0 & 0 & 0 & 0 & 0 & 0 & 0 & 0 & 1 & 1 \\ 0 & 0 & -1 & 0 & 0 & 0 & 0 & 0 & 0 & 0 & 0 & 1 & 1 \\ 0 & 0 & 0 & -1 & 0 & 0 & 0 & 0 & 0 & 0 & 0 & 1 & 1 \\ 0 & 0 & 0 & 0 & -1 & 0 & 0 & 0 & 0 & 0 & 0 & 1 & 1 \\ 0 & 0 & 0 & 0 & 0 & -1 & 0 & 0 & 0 & 0 & 0 & 1 & 1 \\ 0 & 0 & 0 & 0 & 0 & 0 & -1 & 0 & 0 & 0 & 0 & 1 & 1 \\ 0 & 0 & 0 & 0 & 0 & 0 & 0 & -1 & 0 & 0 & 0 & 1 & 1 \\ 0 & 0 & 0 & 0 & 0 & 0 & 0 & 0 & -1 & 0 & 0 & 1 & 1 \\ 0 & 0 & 0 & 0 & 0 & 0 & 0 & 0 & 0 & -1 & 0 & 1 & 1 \\ 0 & 0 & 0 & 0 & 0 & 0 & 0 & 0 & 0 & 0 & -1 & 1 & 1 \\ 0 & 0 & 0 & 0 & 0 & 0 & 0 & 0 & 0 & 0 & 0 & 4 & -3 \end{bmatrix}$$

diagonalizes A, and $D = S^{-1}AS$ will have all zeros as entries except $d_{12,\,12} = 4$ and $d_{13,\,13} = -3$.

29. By Fact 4.4.1 $(\text{im } A)^{\perp} = \ker(A^T) = \ker(A)$, so that $\vec{v}$ is orthogonal to $\vec{w}$.

30. The columns $\vec{v}, \vec{v}_2, \ldots, \vec{v}_n$ of R form an orthogonal eigenbasis for $A = \vec{v}\vec{v}^T$, with eigenvalues 1, 0, 0, …, 0 ($n - 1$ zeros), since
$A\vec{v} = \vec{v}v^T\vec{v} = \vec{v}(\vec{v}\cdot\vec{v}) = \vec{v}$, (since $\vec{v}\cdot\vec{v} = 1$) and $A\vec{v}_i = \vec{v}v^T\vec{v}_i = \vec{v}(\vec{v}\cdot\vec{v}_i) = \vec{0}$ (since $\vec{v}\cdot\vec{v}_i = 0$).

Therefore we can let $S = R$, and $D = \begin{bmatrix} 1 & 0 & \cdots & 0 \\ 0 & 0 & \cdots & 0 \\ \vdots & & & \\ 0 & 0 & \cdots & 0 \end{bmatrix}$.

31. True; A is diagonalizable, that is, A is similar to a diagonal matrix D; then A^2 is similar to D^2.
Now $\text{rank}(D) = \text{rank}(D^2)$ is the number of nonzero entries on the diagonal of D (and D^2). Since similar matrices have the same rank (by Fact 7.2.6e) we can conclude that $\text{rank}(A) = \text{rank}(D) = \text{rank}(D^2) = \text{rank}(A^2)$.

32. By Exercise 17, $\det(A) = (1-q)^{n-1}(qn+1-q)$. A is invertible if $\det(A) \neq 0$, that is, if $q \neq 1$ and $q \neq \dfrac{1}{1-n}$.

33. The angles must add up to 2π, so $\alpha = \dfrac{2\pi}{3} = 120°$.

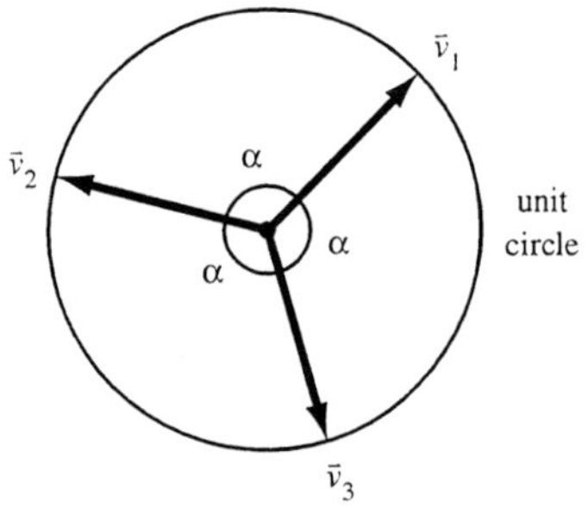

Algebraically, we can see this as follows: let $A = [\vec{v}_1 \quad \vec{v}_2 \quad \vec{v}_3]$, a 2×3 matrix.

Then $A^T A = \begin{bmatrix} 1 & \cos\alpha & \cos\alpha \\ \cos\alpha & 1 & \cos\alpha \\ \cos\alpha & \cos\alpha & 1 \end{bmatrix}$ is a noninvertible 3×3 matrix, so that $\cos\alpha = \dfrac{1}{1-3} = -\dfrac{1}{2}$, by Exercise 32,

and $\alpha = \dfrac{2\pi}{3} = 120°$.

34. Let $\vec{v}_1, \vec{v}_2, \vec{v}_3, \vec{v}_4$ be such vectors. Form $A = [\vec{v}_1 \quad \vec{v}_2 \quad \vec{v}_3 \quad \vec{v}_4]$, a 3×4 matrix.

Then $A^T A = \begin{bmatrix} 1 & \cos\alpha & \cos\alpha & \cos\alpha \\ \cos\alpha & 1 & \cos\alpha & \cos\alpha \\ \cos\alpha & \cos\alpha & 1 & \cos\alpha \\ \cos\alpha & \cos\alpha & \cos\alpha & 1 \end{bmatrix}$ is noninvertible, so that $\cos\alpha = \dfrac{1}{1-4} = -\dfrac{1}{3}$, by Exercise 32, and

$\alpha = \arccos\left(-\dfrac{1}{3}\right) \approx 109.5°$.

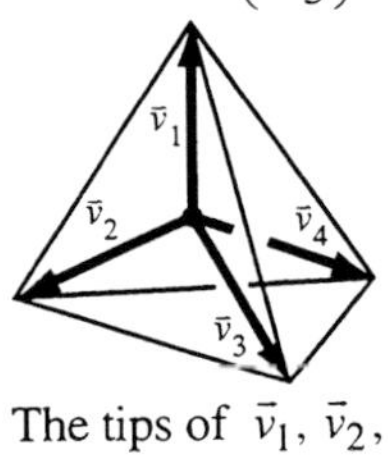

The tips of $\vec{v}_1, \vec{v}_2, \vec{v}_3, \vec{v}_4$ form a regular tetrahedron.

35. Let $\vec{v}_1, \ldots, \vec{v}_{n+1}$ be these vectors. Form $A = [\vec{v}_1 \quad \cdots \quad \vec{v}_{n+1}]$, an $n \times (n+1)$ matrix.

Then $A^T A = \begin{bmatrix} 1 & \cos\alpha & \cdots & \cos\alpha \\ \cos\alpha & 1 & \cdots & \cos\alpha \\ \vdots & & \ddots & \\ \cos\alpha & & \cdots & 1 \end{bmatrix}$ is a noninvertible $(n+1) \times (n+1)$ matrix with 1's on the diagonal and

$\cos\alpha$ outside, so that $\cos\alpha = \dfrac{1}{1-n}$, by Exercise 32, and $\alpha = \arccos\left(\dfrac{1}{1-n}\right)$.

36. If $\vec{v}$ is an eigenvector with eigenvalue λ, then $\lambda\vec{v} = A\vec{v} = A^2\vec{v} = \lambda^2\vec{v}$, so that $\lambda = \lambda^2$ and therefore $\lambda = 0$ or $\lambda = 1$. Since A is symmetric, E_0 and E_1 are orthogonal complements, so that A represents the orthogonal projection onto E_1.

37. a. If $S^{-1}AS$ is upper triangular then the first column of S is an eigenvector of A. Therefore, any matrix without real eigenvectors fails to be triangulizable over $\mathbb{R}$, for example, $\begin{bmatrix} 0 & -1 \\ 1 & 0 \end{bmatrix}$.

b. Proof by induction on n: For an $n \times n$ matrix A we can choose a complex invertible $n \times n$ matrix P whose first column is an eigenvector for A. Then $P^{-1}AP = \begin{bmatrix} \lambda & \vec{v} \\ 0 & B \end{bmatrix}$.

By induction, B is triangulizable, that is, there is an invertible $(n-1) \times (n-1)$ matrix Q such that $Q^{-1}BQ = T$ is upper triangular.

Now let $R = \begin{bmatrix} 1 & 0 \\ 0 & Q \end{bmatrix}$. Then $R^{-1}\begin{bmatrix} \lambda & \vec{v} \\ 0 & B \end{bmatrix}R = \begin{bmatrix} 1 & 0 \\ 0 & Q^{-1} \end{bmatrix}\begin{bmatrix} \lambda & \vec{v} \\ 0 & B \end{bmatrix}\begin{bmatrix} 1 & 0 \\ 0 & Q \end{bmatrix} = \begin{bmatrix} \lambda & \vec{v}Q \\ 0 & T \end{bmatrix}$ is upper triangular.

$R^{-1}\begin{bmatrix} \lambda & \vec{v} \\ 0 & B \end{bmatrix}R = R^{-1}P^{-1}APR = S^{-1}AS$, where $S = PR$, proving our claim.

38. a. By definition of an upper triangular matrix, $\vec{e}_1$ is in ker U, $\vec{e}_2$ is in $\ker(U^2)$, ..., $\vec{e}_n$ is in $\ker(U^n)$, so that all $\vec{x}$ in $\mathbb{C}^n$ are in $\ker(U^n)$, that is, $U^n = 0$.

b. By Exercise 37b, there is an invertible S such that $S^{-1}AS = U$ is upper triangular. The diagonal entries of U are all zero, since A and U have the same eigenvalues; therefore $U^n = 0$ by part a. Now $A = SUS^{-1}$ and $A^n = SU^nS^{-1} = 0$, as claimed.

39. a. For all i, j, $\left|\sum_{k=1}^{n} a_{ik}b_{kj}\right| \le \sum_{k=1}^{n} |a_{ik}b_{kj}| = \sum_{k=1}^{n} |a_{ik}||b_{kj}|$

↑ triangle inequality (applies to the $\le$)

b. By induction on t: $|A^t| = |A^{t-1}A| \le |A^{t-1}||A| \le |A|^{t-1}|A| = |A|^t$

↑ part a (first $\le$), ↑ by induction (second $\le$)

40. If $t \ge n-1$ then $(I_n + U)^t = I_n + \binom{t}{1}U + \binom{t}{2}U^2 + \cdots + \binom{t}{n-1}U^{n-1}$, since $U^n = 0$. Now $\binom{t}{k} \le t^n$ for $k = 1, \ldots, n-1$, so that $(I_n + U)^t \le t^n(I_n + U + \cdots + U^{n-1})$, as claimed. Check that the formula holds for $t < n-1$ as well.

41. Let λ be the largest $|r_{ii}|$; note that $\lambda < 1$. Then $|R| = \begin{bmatrix} |r_{11}| & & * \\ & \ddots & \\ 0 & & |r_{nn}| \end{bmatrix} \le \begin{bmatrix} \lambda & & * \\ & \ddots & \\ 0 & & \lambda \end{bmatrix} = \lambda \begin{bmatrix} 1 & & * \\ & \ddots & \\ 0 & & 1 \end{bmatrix} = \lambda(I_n + U)$,

and $|R^t| \le |R|^t \le \lambda^t(I_n + U)^t \le \lambda^t t^n(I_n + U + \cdots + U^{n-1})$.

We learn in Calculus that $\lim_{t\to\infty}(\lambda^t t^n) = 0$, so that $\lim_{t\to\infty}(R^t) = 0$.

42. a. From Exercise 37b we know that there is an invertible S and an upper triangular R such that $S^{-1}AS = R$, and $|r_{ii}| < 1$ for all i, since the diagonal entries of R are the eigenvalues of A. Now $\lim_{t\to\infty} R^t = 0$ by Exercise 41.

Note that $A = SRS^{-1}$ and $A^t = SR^tS^{-1}$, so that $\lim_{t\to\infty} A^t = 0$, as claimed.

b. See the remark after Definition 6.5.1.

7.4

1. $A = \begin{bmatrix} 6 & -\frac{7}{2} \\ -\frac{7}{2} & 8 \end{bmatrix}$

2. $A = \begin{bmatrix} 0 & \frac{1}{2} \\ \frac{1}{2} & 0 \end{bmatrix}$

3. $A = \begin{bmatrix} 3 & 0 & 3 \\ 0 & 4 & \frac{7}{2} \\ 3 & \frac{7}{2} & 5 \end{bmatrix}$

4. $A = \begin{bmatrix} 6 & 2 \\ 2 & 3 \end{bmatrix}$, positive definite

5. $A = \begin{bmatrix} 1 & 2 \\ 2 & 1 \end{bmatrix}$, indefinite (since det($A$) < 0)

6. $A = \begin{bmatrix} 2 & 3 \\ 3 & 4 \end{bmatrix}$, indefinite (since det($A$) < 0)

7. $A = \begin{bmatrix} 0 & 0 & 2 \\ 0 & 3 & 0 \\ 2 & 0 & 0 \end{bmatrix}$, indefinite (eigenvalues 2, –2, 3)

8. If $S^{-1}AS = D$ is diagonal, then $S^{-1}A^2S = D^2$, so that all eigenvalues of A^2 are ≥0. So A^2 is a positive semi-definite; it is positive definite if and only if A is invertible.

9. a. $(A^2)^T = (A^T)^2 = (-A)^2 = A^2$, so that A^2 is symmetric.

b. $q(\vec{x}) = \vec{x}^T A^2 \vec{x} = \vec{x}^T AA\vec{x} = -\vec{x}^T A^T A\vec{x} = -(A\vec{x})\cdot(A\vec{x}) = -\|A\vec{x}\|^2 \le 0$ for all $\vec{x}$, so that A^2 is negative semi-definite.

c. If $\vec{v}$ is a complex eigenvector of A with eigenvalue λ, then $A^2\vec{v} = \lambda^2\vec{v}$, and $\lambda^2 \le 0$, by part b. Therefore, λ is *imaginary*, that is, $\lambda = bi$ for a real b. Thus, the zero matrix is the only skew-symmetric matrix that is diagonalizable over $\mathbb{R}$.

10. $L(\vec{x}) = (\vec{x}+\vec{v})^T A(\vec{x}+\vec{v}) - \vec{x}^T A\vec{x} - \vec{v}^T A\vec{v} = \vec{x}^T A\vec{x} + \vec{x}^T A\vec{v} + \vec{v}^T A\vec{x} + \vec{v}^T A\vec{v} - \vec{x}^T A\vec{x} + \vec{v}^T A\vec{v}$
$= \vec{x}^T A\vec{v} + \vec{v}^T A\vec{x} = \vec{v}^T A\vec{x} + \vec{v}^T A\vec{x} = (2\vec{v}^T A)\vec{x},$
$\uparrow$
note that $\vec{x}^T A\vec{v}$ is a scalar so that $\vec{x}^T A\vec{v} = (\vec{x}^T A\vec{v})^T = \vec{v}^T A^T\vec{x} = \vec{v}^T A\vec{x}$ if A is symmetric.

So L is linear with matrix $2\vec{v}^T A$.

11. The eigenvalues of A^{-1} are the reciprocals of those of A, so that A and A^{-1} have the same definiteness.

12. det(A) is the product of the two (real) eigenvalues. q is indefinite if an only if those have different signs, that is, their product is negative.

13. $q(\vec{e}_i) = \vec{e}_i \cdot A\vec{e}_i = a_{ii} > 0$

14. If det(A) is positive then both eigenvalues have the same sign, so that A is positive definite or negative definite. Since $\vec{e}_1 \cdot A\vec{e}_1 = a > 0$, A is in fact positive definite.

15. $A = \begin{bmatrix} 6 & 2 \\ 2 & 3 \end{bmatrix}$; eigenvalues $\lambda_1 = 7$ and $\lambda_2 = 2$

orthonormal eigenbasis $\vec{v}_1 = \frac{1}{\sqrt{5}}\begin{bmatrix} 2 \\ 1 \end{bmatrix}, \vec{v}_2 = \frac{1}{\sqrt{5}}\begin{bmatrix} -1 \\ 2 \end{bmatrix}$

$\lambda_1 c_1^2 + \lambda_2 c_2^2 = 1$ or $7c_1^2 + 2c_2^2 = 1$

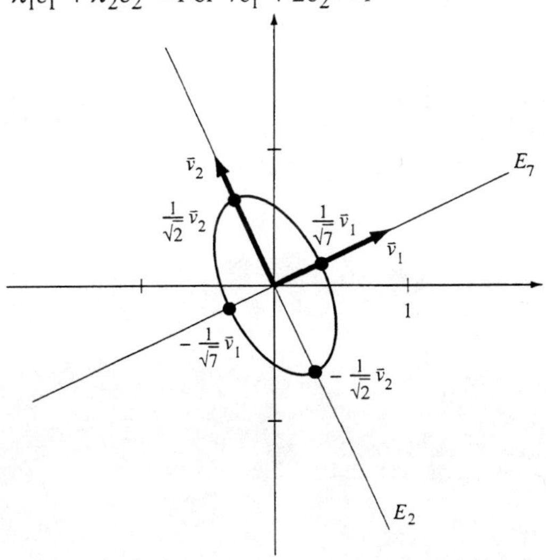

16. $A=\begin{bmatrix}0 & \frac{1}{2}\\ \frac{1}{2} & 0\end{bmatrix}$; eigenvalues $\lambda_1=\frac{1}{2}$, and $\lambda_2=-\frac{1}{2}$

orthonormal eigenbasis $\vec{v}_1=\frac{1}{\sqrt{2}}\begin{bmatrix}1\\1\end{bmatrix}$ and $\vec{v}_2=\frac{1}{\sqrt{2}}\begin{bmatrix}1\\-1\end{bmatrix}$

$\frac{1}{2}c_1^2-\frac{1}{2}c_2^2=1$

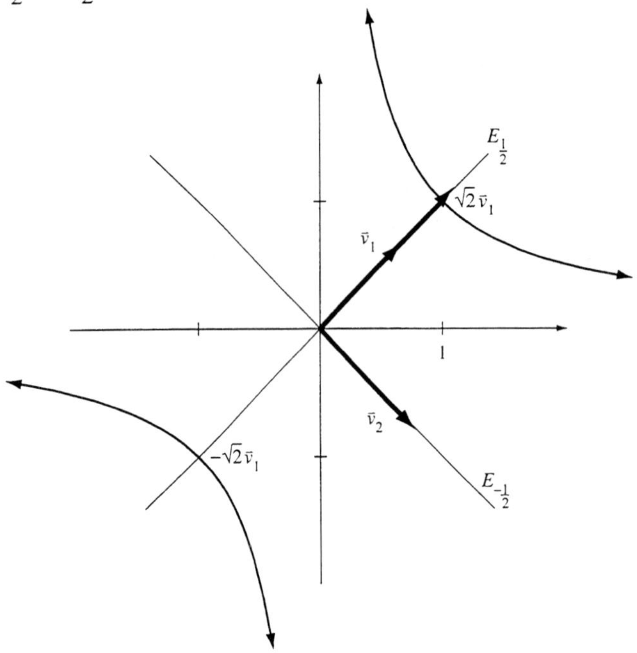

17. $A=\begin{bmatrix}3 & 2\\ 2 & 0\end{bmatrix}$, eigenvalues $\lambda_1=4$, $\lambda_2=-1$

orthonormal eigenbasis $\vec{v}_1=\frac{1}{\sqrt{5}}\begin{bmatrix}2\\1\end{bmatrix}$, $\vec{v}_2=\frac{1}{\sqrt{5}}\begin{bmatrix}-1\\2\end{bmatrix}$

$4c_1^2-c_2^2=1$ (hyperbola)

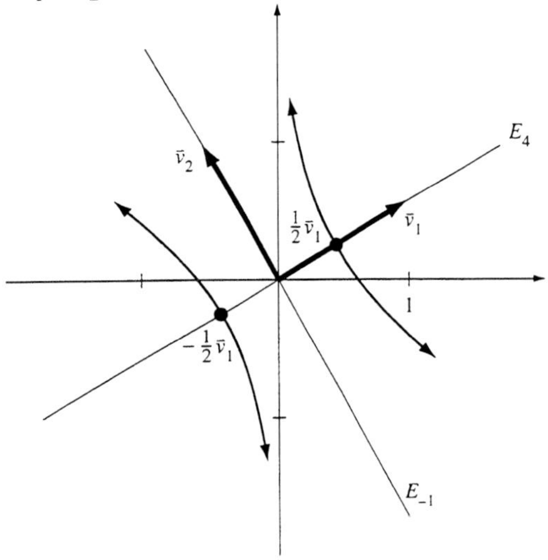

18. $A=\begin{bmatrix} 9 & -2 \\ -2 & 6 \end{bmatrix}$, eigenvalues $\lambda_1=10$, $\lambda_2=5$

orthonormal eigenbasis $\vec{v}_1=\frac{1}{\sqrt{5}}\begin{bmatrix} 2 \\ 1 \end{bmatrix}$, $\vec{v}_2=\frac{1}{\sqrt{5}}\begin{bmatrix} -1 \\ 2 \end{bmatrix}$

$10c_1^2+5c_2^2=1$ (ellipse)

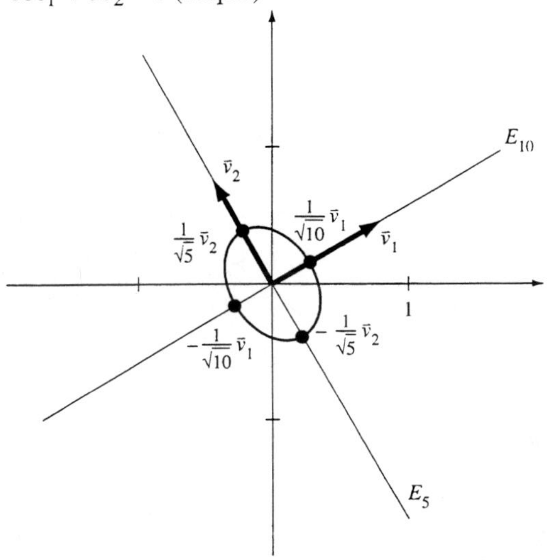

19. $A=\begin{bmatrix} 1 & 2 \\ 2 & 4 \end{bmatrix}$; eigenvalues $\lambda_1=5$, $\lambda_2=0$

eigenvectors $\vec{v}_1=\frac{1}{\sqrt{5}}\begin{bmatrix} 1 \\ 2 \end{bmatrix}$, $\vec{v}_2=\frac{1}{\sqrt{5}}\begin{bmatrix} -2 \\ 1 \end{bmatrix}$

$5c_1^2=1$ (a pair of lines)

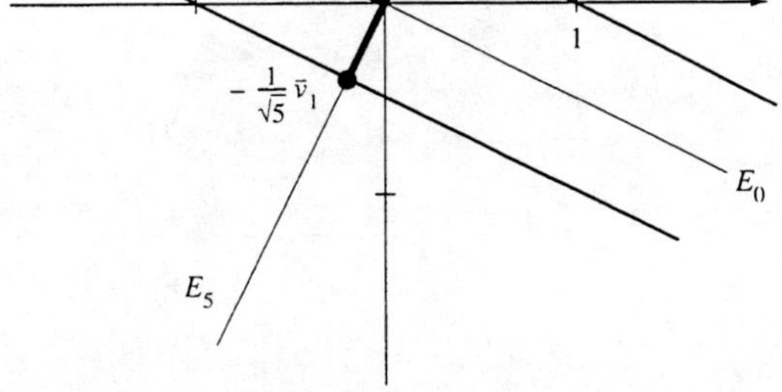

Note that $(x_1^2+4x_1x_2+4x_2^2)=(x_1+2x_2)^2=1$, so that $x_1+2x_2=\pm 1$, and the two lines are $x_2=\frac{1-x_1}{2}$ and $x_2=\frac{-1-x_1}{2}$.

20. $A = \begin{bmatrix} -3 & 3 \\ 3 & 5 \end{bmatrix}$; eigenvalues $\lambda_1 = 6$ and $\lambda_2 = -4$

orthonormal eigenbasis $\vec{v}_1 = \frac{1}{\sqrt{10}}\begin{bmatrix} 1 \\ 3 \end{bmatrix}$, $\vec{v}_2 = \frac{1}{\sqrt{10}}\begin{bmatrix} -3 \\ 1 \end{bmatrix}$

$6c_1^2 - 4c_2^2 = 1$ (hyperbola)

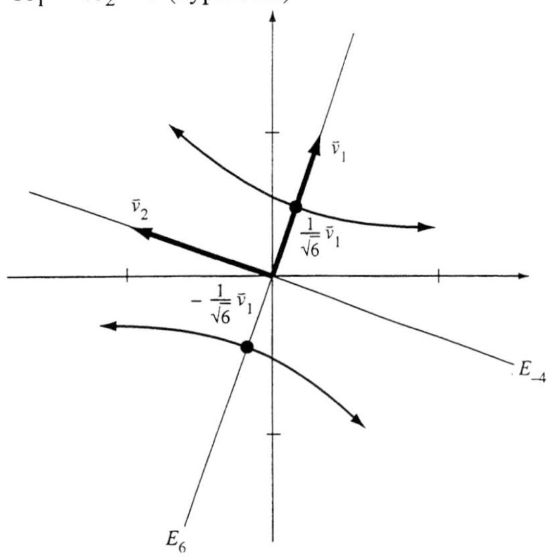

21. a. In each case, it is informative to think about the intersections with the three coordinate planes: $x_1 - x_2$, $x_1 - x_3$, and $x_2 - x_3$.

For the surface $x_1^2 + 4x_2^2 + 9x_3^2 = 1$, all these intersections are *ellipses*, and the surface itself is an *ellipsoid*.

This surface is connected and bounded; the points closest to the origin are $\pm\begin{bmatrix} 0 \\ 0 \\ \frac{1}{3} \end{bmatrix}$, and those farthest $\pm\begin{bmatrix} 1 \\ 0 \\ 0 \end{bmatrix}$.

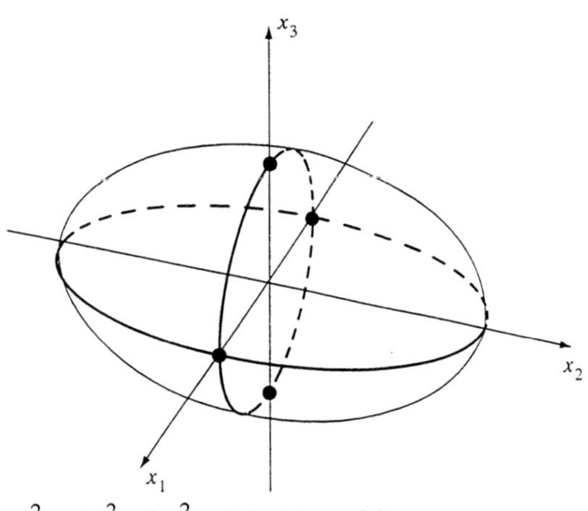

$x_1^2 + 4x_2^2 + 9x_3^2 = 1$ (not to scale)
an *ellipsoid*

In the case of $x_1^2 + 4x_2^2 - 9x_3^2 = 1$, the intersection with the $x_1 - x_2$ plane is an ellipse, and the two other intersections are hyperbolas. The surface is connected and not bounded; the points closest to the origin are $\pm\begin{bmatrix} 0 \\ \frac{1}{2} \\ 0 \end{bmatrix}$.

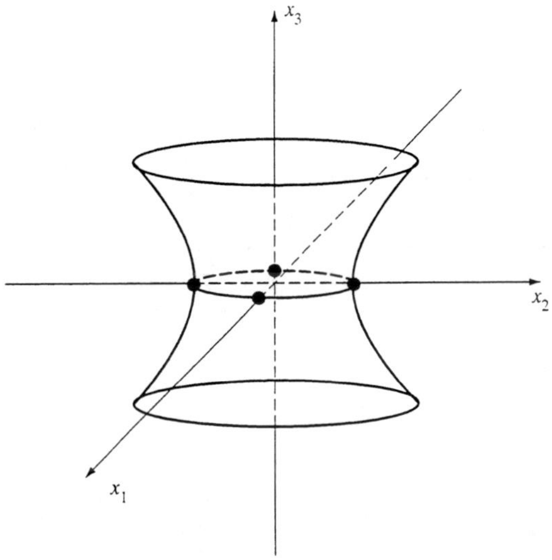

$x_1^2 + 4x_2^2 - 9x_3^2 = 1$ (not to scale)
a *hyperboloid of one sheet*

In the case $-x_1^2 - 4x_2^2 + 9x_3^2 = 1$, the intersection with the $x_1 - x_2$ plane is empty, and the two other intersections are hyperbolas. The surface consists of two pieces and is unbounded. The points closest to the origin are $\pm\begin{bmatrix} 0 \\ 0 \\ 113 \end{bmatrix}$.

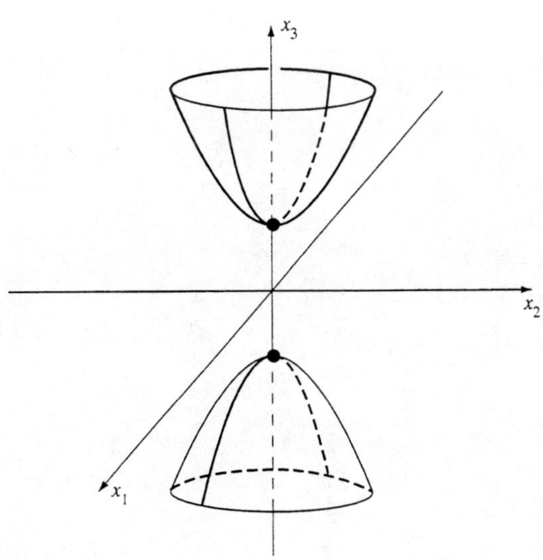

$-x_1^2 - 4x_2^2 + 9x_3^2 = 1$ (not to scale)
a *hyperboloid of two sheets*

b. $A = \begin{bmatrix} 1 & \frac{1}{2} & 1 \\ \frac{1}{2} & 2 & \frac{3}{2} \\ 1 & \frac{3}{2} & 3 \end{bmatrix}$ is positive definite, with three positive eigenvalues $\lambda_1, \lambda_2, \lambda_3$.

Surface is given by $\lambda_1 c_1^2 + \lambda_2 c_2^2 + \lambda_3 c_3^2 = 1$ with respect to principal axes, an *ellipsoid*
To find points closest to and farthest from origin, use technology to find eigenvalues and eigenvectors:
eigenvalues: $\lambda_1 \approx 0.56$, $\lambda_2 \approx 4.44$, $\lambda_3 = 1$

unit eigenvectors: $\vec{v}_1 \approx \begin{bmatrix} 0.86 \\ 0.19 \\ -0.47 \end{bmatrix}$, $\vec{v}_2 \approx \begin{bmatrix} 0.31 \\ 0.54 \\ 0.78 \end{bmatrix}$, $\vec{v}_3 = \frac{1}{\sqrt{6}} \begin{bmatrix} 1 \\ -2 \\ 1 \end{bmatrix}$

Equation: $0.56c_1^2 + 4.44c_2^2 + c_3^2 = 1$

Farthest points when $c_1 = \pm \frac{1}{\sqrt{0.56}}$ and $c_2 = c_3 = 0$

Closest points when $c_2 = \pm \frac{1}{\sqrt{4.44}}$ and $c_1 = c_3 = 0$

Farthest points $\approx \pm \frac{1}{\sqrt{0.56}} \begin{bmatrix} 0.86 \\ 0.19 \\ -0.47 \end{bmatrix} \approx \pm \begin{bmatrix} 1.15 \\ 0.26 \\ -0.63 \end{bmatrix}$

Closest points $\approx \pm \frac{1}{\sqrt{4.44}} \begin{bmatrix} 0.31 \\ 0.54 \\ 0.78 \end{bmatrix} \approx \pm \begin{bmatrix} 0.15 \\ 0.26 \\ 0.37 \end{bmatrix}$

22. $A = \begin{bmatrix} -1 & 0 & 5 \\ 0 & 1 & 0 \\ 5 & 0 & -1 \end{bmatrix}$; eigenvalues $\lambda_1 = 4$, $\lambda_2 = -6$, $\lambda_3 = 1$

Equation with respect to principal axes: $4c_1^2 - 6c_2^2 + c_3^2 = 1$, a hyperboloid of one sheet (see solutions to 21a).

Closest to origin when $c_1 = \pm \frac{1}{2}$, $c_2 = c_3 = 0$.

A unit eigenvector for eigenvalue 4 is $\vec{v} = \frac{1}{\sqrt{2}} \begin{bmatrix} 1 \\ 0 \\ 1 \end{bmatrix}$, so that the desired points are $\pm \frac{1}{2} \frac{1}{\sqrt{2}} \begin{bmatrix} 1 \\ 0 \\ 1 \end{bmatrix} \approx \pm \begin{bmatrix} 0.35 \\ 0 \\ 0.35 \end{bmatrix}$.

23. Yes; $M = \frac{1}{2}(A + A^T)$ is symmetric, and

$$\vec{x}^T M \vec{x} = \frac{1}{2}\vec{x}^T A \vec{x} + \frac{1}{2}\vec{x}^T A^T \vec{x} \underset{\uparrow}{=} \frac{1}{2}\vec{x}^T A \vec{x} + \frac{1}{2}\vec{x}^T A \vec{x} = \vec{x}^T A \vec{x}$$

Note that $\vec{x}^T A \vec{x}$ is a 1×1 matrix, so that $\vec{x}^T A \vec{x} = (\vec{x}^T A \vec{x})^T = \vec{x}^T A^T \vec{x}$.

24. $q(\vec{e}_i) = \vec{e}_i \cdot A\vec{e}_i = \vec{e}_i \cdot (i\text{th column of } A) = a_{ii}$

25. $q(\vec{v}) = \vec{v} \cdot A\vec{v} = \vec{v} \cdot (\lambda \vec{v}) = \lambda(\vec{v} \cdot \vec{v}) \underset{\uparrow}{=} \lambda$

$\vec{v}$ is a unit vector

26. False; If $A = \begin{bmatrix} 0 & 1 \\ 1 & 0 \end{bmatrix}$ then $q\left(\begin{bmatrix} 1 \\ 0 \end{bmatrix}\right) = \begin{bmatrix} 1 \\ 0 \end{bmatrix} \cdot \begin{bmatrix} 0 & 1 \\ 1 & 0 \end{bmatrix}\begin{bmatrix} 1 \\ 0 \end{bmatrix} = \begin{bmatrix} 1 \\ 0 \end{bmatrix} \cdot \begin{bmatrix} 0 \\ 1 \end{bmatrix} = 0$.

27. *True*; if $a_{ii} \neq b_{ii}$ for some i, then $p(\vec{e}_i) = a_{ii} \neq b_{ii} = p(\vec{e}_i)$. If $a_{ii} = b_{ii}$ for all i but $a_{ij} \neq b_{ij}$ for some $i \neq j$, then $q(\vec{e}_i + \vec{e}_j) = a_{ii} + a_{jj} + 2a_{ij} \neq b_{ii} + b_{jj} + 2b_{ij} = p(\vec{e}_i + \vec{e}_j)$.

28. *False*; consider $A = \begin{bmatrix} 1 & 2 \\ 2 & 1 \end{bmatrix}$ with $\det(A) < 0$, so that A is indefinite.

29. *False*; consider $A = \begin{bmatrix} 2 & -1 \\ -1 & 2 \end{bmatrix}$, which is positive definite.

30. *True*; since q is continuous and takes positive as well as negative values, q must take the value 0 at some $\vec{x} \neq \vec{0}$, by the Intermediate Value Theorem. More precisely, let $\vec{v}$ and $\vec{w}$ be orthogonal unit eigenvectors with $q(\vec{v}) > 0$ and $q(\vec{w}) < 0$. Then $f(t) = q[(\cos t)\vec{v} + (\sin t)\vec{w}]$ is continuous on $\left[0, \frac{\pi}{2}\right]$, so that there will be a t_0 such that $f(t_0) = 0$. Let $\vec{x} = (\cos t_0)\vec{v} + (\sin t_0)\vec{w}$, a unit vector.

31. Let $\vec{v}_1, \ldots, \vec{v}_n$ be an orthonormal eigenbasis for A with $A\vec{v}_i = \lambda_i \vec{v}_i$. We know that $q(\vec{v}_i) = \lambda_i$ (see Exercise 25), so that $q(\vec{v}_1) = \lambda_1$ and $q(\vec{v}_n) = \lambda_n$ are in the image.
We claim that all numbers between λ_n and λ_1 are in the image as well. To see this, apply the Intermediate Value Theorem to the continuous function $f(t) = q\big((\cos t)\vec{v}_n + (\sin t)\vec{v}_1\big)$ on $\left[0, \frac{\pi}{2}\right]$ (note that $f(0) = q(\vec{v}_n) = \lambda_n$ and $f\left(\frac{\pi}{2}\right) = q(\vec{v}_1) = \lambda_1$).

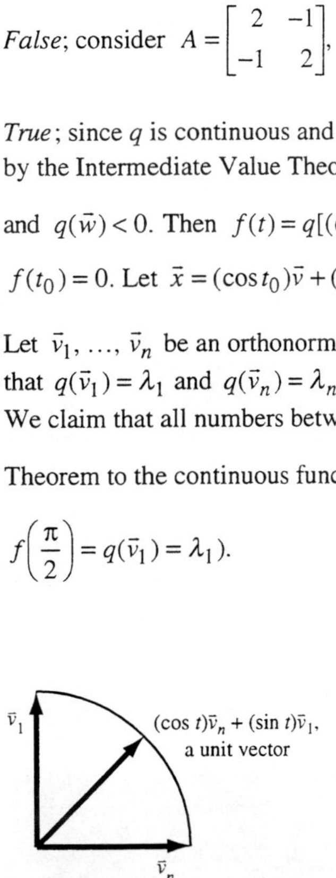

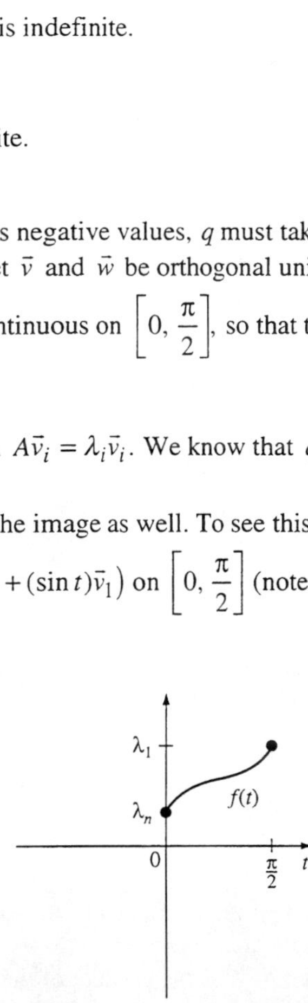

The Intermediate Value Theorem tells us that for any c between λ_n and λ_1, there is a t_0 such that $f(t_0) = q\big((\cos t_0)\vec{v}_n + (\sin t_0)\vec{v}_1\big) = c$. Note that $(\cos t_0)\vec{v}_n + (\sin t_0)\vec{v}_1$ is a unit vector.
Now we will show that, conversely, $q(\vec{v})$ is on $[\lambda_n, \lambda_1]$ for all unit vectors $\vec{v}$. Write $\vec{v} = c_1\vec{v}_1 + \cdots + c_n\vec{v}_n$ and note that $\|\vec{v}\|^2 = c_1^2 + \cdots + c_n^2 = 1$. Then $q(\vec{v}) = \lambda_1 c_1^2 + \lambda_2 c_2^2 + \cdots + \lambda_n c_n^2 \leq \lambda_1 c_1^2 + \lambda_1 c_2^2 + \cdots + \lambda_1 c_n^2 = \lambda_1$.
Likewise, $q(\vec{v}) \geq \lambda_n$. We have shown that the image of S^{n-1} under q is the closed interval $[\lambda_n, \lambda_1]$.

32. The hint almost gives it away. Since D is a diagonal matrix with positive diagonal entries, we can write $D = D_1^2$, where D_1 is diagonal with positive diagonal entries (the square roots of the entries of D). Now $A = SDS^T = SD_1D_1S^T = SD_1(SD_1)^T = BB^T$ where $B = SD_1$. The columns of B are scalar multiples of the corresponding columns of S, so that they are orthogonal.

33. From Example 1 we have $S = \frac{1}{\sqrt{5}}\begin{bmatrix} 2 & 1 \\ -1 & 2 \end{bmatrix}$ and $D = \begin{bmatrix} 9 & 0 \\ 0 & 4 \end{bmatrix}$, so that $D_1 = \begin{bmatrix} 3 & 0 \\ 0 & 2 \end{bmatrix}$ and $B = SD_1 = \frac{1}{\sqrt{5}}\begin{bmatrix} 6 & 2 \\ -3 & 4 \end{bmatrix}$.

34. Define D_1 as in Exercise 32. Then $A = SDS^{-1} = SD_1D_1S^{-1} = (SD_1S^{-1})(SD_1S^{-1}) = B^2$, where $B = SD_1S^{-1}$. B is positive definite, since $S^{-1}BS = D_1$ is diagonal with positive diagonal entries.

35. $S = \frac{1}{\sqrt{5}}\begin{bmatrix} 2 & 1 \\ -1 & 2 \end{bmatrix}$ and $D_1 = \begin{bmatrix} 3 & 0 \\ 0 & 2 \end{bmatrix}$ (see Exercise 33), so that $B = SD_1S^{-1} = \begin{bmatrix} 2.8 & -0.4 \\ -0.4 & 2.2 \end{bmatrix}$.

36. Recall that $a = q(\vec{e}_1) > 0$ and $\det A = ac - b^2 = \lambda_1\lambda_2 > 0$.

$\begin{bmatrix} a & b \\ b & c \end{bmatrix} = \begin{bmatrix} x & 0 \\ y & z \end{bmatrix}\begin{bmatrix} x & y \\ 0 & z \end{bmatrix} = \begin{bmatrix} x^2 & xy \\ xy & y^2+z^2 \end{bmatrix}$ means that $\left| \begin{aligned} x^2 &= a \\ xy &= b \\ y^2+z^2 &= c \end{aligned} \right|$.

It is required that x and z be positive. This system has the unique solution

$$x = \sqrt{a}$$
$$y = \frac{b}{x} = \frac{b}{\sqrt{a}}$$
$$z = \sqrt{c - y^2} = \sqrt{c - \frac{b^2}{a}} = \sqrt{\frac{ac-b^2}{a}}$$

37. Use the formulas for x, y, z derived in Exercise 36.

$$x = \sqrt{a} = \sqrt{8} = 2\sqrt{2}$$
$$y = \frac{b}{\sqrt{a}} = -\frac{2}{2\sqrt{2}} = -\frac{1}{\sqrt{2}}$$
$$z = \sqrt{\frac{ac-b^2}{a}} = \sqrt{\frac{36}{8}} = \frac{3}{\sqrt{2}}, \text{ so that}$$
$$L = \begin{bmatrix} 2\sqrt{2} & 0 \\ -\frac{1}{\sqrt{2}} & \frac{3}{\sqrt{2}} \end{bmatrix}.$$

38.
- (i) implies (ii): See the hint at the end of the exercise.
- (ii) implies (iii): $\det A^{(m)}$ is the product of the (positive) eigenvalues.
- (iii) implies (iv):

  $$A = \begin{bmatrix} A^{(n-1)} & \vec{v} \\ \vec{v}^T & k \end{bmatrix} = \begin{bmatrix} B & 0 \\ \vec{x}^T & 1 \end{bmatrix}\begin{bmatrix} B^T & \vec{x} \\ 0 & t \end{bmatrix} = \begin{bmatrix} BB^T & B\vec{x} \\ \vec{x}^TB^T & \vec{x}^T\vec{x}+t \end{bmatrix}$$

  The system $\left| \begin{aligned} B\vec{x} &= \vec{v} \\ \vec{x}^T\vec{x}+t &= k \end{aligned} \right|$ has the unique solution

  $$\vec{x} = B^{-1}\vec{v}$$
  $$t = k - \vec{x}^T\vec{x} = k - \|B^{-1}\vec{v}\|^2.$$

  Note that t is positive since $0 < \det(A^{(n)}) = \det(A) = \det\begin{bmatrix} B & 0 \\ \vec{x}^T & 1 \end{bmatrix}\det\begin{bmatrix} B^T & \vec{x} \\ 0 & t \end{bmatrix} = (\det B)^2 \cdot t$.

- (iv) $\Rightarrow$ (i)

  $\vec{x}^T A\vec{x} = \vec{x}^T LL^T \vec{x} = (L^T \vec{x})^T (L^T \vec{x}) = \left\| L^T \vec{x} \right\|^2 > 0$ if $\vec{x} \neq \vec{0}$, since L is invertible.

39. Solve the system $\begin{bmatrix} 4 & -4 & 8 \\ -4 & 13 & 1 \\ 8 & 1 & 26 \end{bmatrix} = \begin{bmatrix} x & 0 & 0 \\ y & w & 0 \\ z & t & s \end{bmatrix} \begin{bmatrix} x & y & z \\ 0 & w & t \\ 0 & 0 & s \end{bmatrix}$

$$\left.\begin{array}{l} x^2 = 4, \text{ so } x = 2 \\ 2y = -4, \text{ so } y = -2 \\ 2z = 8, \text{ so } z = 4 \\ 4 + w^2 = 13, \text{ so } w = 3 \\ -8 + 3t = 1, \text{ so } t = 3 \\ 16 + 9 + s^2 = 26, \text{ so } s = 1 \end{array}\right\} L = \begin{bmatrix} 2 & 0 & 0 \\ -2 & 3 & 0 \\ 4 & 3 & 1 \end{bmatrix}$$

40. If $A = QR$, then $A^T A = (QR)^T QR = R^T Q^T QR = R^T R = LL^T$, $L = R^T$.

41. $\frac{\partial q}{\partial x_1} = 2ax_1 + bx_2$ and $\frac{\partial q}{\partial x_2} = bx_1 + 2cx_2$, so that $q_{11} = \frac{\partial^2 q}{\partial x_1^2} = 2a$, $q_{22} = \frac{\partial^2 q}{\partial x_2^2} = 2c$, and $q_{12} = \frac{\partial^2 q}{\partial x_1 \partial x_2} = b$, and

$D = \det\begin{bmatrix} q_{11} & q_{12} \\ q_{12} & q_{22} \end{bmatrix} = \det\begin{bmatrix} 2a & b \\ b & 2c \end{bmatrix} = 4ac - b^2 > 0.$

The matrix $A = \begin{bmatrix} a & \frac{b}{2} \\ \frac{b}{2} & c \end{bmatrix}$ of q is positive definite, since $a > 0$ and $\det(A) = \frac{1}{4}D > 0$. This means by definition, that q has a minimum at $\vec{0}$, since $q(\vec{x}) > 0 = q(\vec{0})$ for all $\vec{x} \neq \vec{0}$.

42. The eigenvalues of B are $p - q$ and $nq + p - q = p + (n - 1)q$, so that B is positive definite if $p - q > 0$ and $p + (n - 1)q > 0$.

43. If $\vec{v}_1, \ldots, \vec{v}_n$ is such a basis consisting of unit vectors, and we let $A = [\vec{v}_1 \ \cdots \ \vec{v}_n]$, then

$A^T A = \begin{bmatrix} 1 & \cos\alpha & \cdots & \cos\alpha \\ \cos\alpha & \vdots & \cdots & \cos\alpha \\ \vdots & \vdots & \ddots & \vdots \\ \cos\alpha & \cos\alpha & \cdots & 1 \end{bmatrix}$ is positive definite, so that $1 - \cos\alpha > 0$ and $1 + (n - 1)\cos\alpha > 0$ or

$1 > \cos\alpha > \frac{1}{1-n}$ or $0 < \alpha < \arccos\left(\frac{1}{1-n}\right)$.

Conversely, if α is in this range, then the matrix $\begin{bmatrix} 1 & \cos\alpha & \cdots & \cos\alpha \\ \cos\alpha & \vdots & & \cos\alpha \\ \vdots & \vdots & \ddots & \vdots \\ \cos\alpha & \cos\alpha & \cdots & 1 \end{bmatrix}$ is positive definite, so that it has a Cholesky factorization LL^T. The columns of L^T give us a basis with the desired property.

7.5

1. $\sigma_1 = 2$, $\sigma_2 = 1$

2. The image of the unit circle is the unit circle, since the transformation defined by A preserves length. Thus $\sigma_1 = \sigma_2 = 1$ by Fact 7.5.3.

3. $A^TA = I_n$; the eigenvalues of A^TA are all 1, so that the singular values of A are all 1.

4. $A^TA = \begin{bmatrix} 1 & 1 \\ 1 & 2 \end{bmatrix}$, with eigenvalues $\lambda_{1,2} = \frac{3 \pm \sqrt{5}}{2}$. The singular values of A are $\sigma_1 = \sqrt{\frac{3+\sqrt{5}}{2}} = \frac{1+\sqrt{5}}{2} \approx 1.62$ and $\sigma_2 = \sqrt{\frac{3-\sqrt{5}}{2}} = \frac{-1+\sqrt{5}}{2} \approx 0.62$.

5. $A^TA = \begin{bmatrix} p^2+q^2 & 0 \\ 0 & p^2+q^2 \end{bmatrix}$, with e-values $\lambda_1 = \lambda_2 = p^2+q^2$. Singular values of A are $\sigma_1 = \sigma_2 = \sqrt{p^2+q^2}$.

A represents a rotation-dilation, with dilation factor $\sqrt{p^2+q^2}$, so that the image of the unit circle is a circle with radius $\sqrt{p^2+q^2}$.

6. The eigenvalues of A^TA are $\lambda_1 = 25$ and $\lambda_2 = 0$, so that the singular values of A are $\sigma_1 = 5$ and $\sigma_2 = 0$ (these are also the eigenvalues of A; compare with Exercise 24).

$E_5 = \text{span}\begin{bmatrix} 1 \\ 2 \end{bmatrix}$, so that $\vec{v}_1 = \frac{1}{\sqrt{5}}\begin{bmatrix} 1 \\ 2 \end{bmatrix}$ works. The image of the unit circle is the line segment connecting the tips of $A\vec{v}_1 = 5\vec{v}_1$ and $A(-\vec{v}_1) = -5\vec{v}_1$.

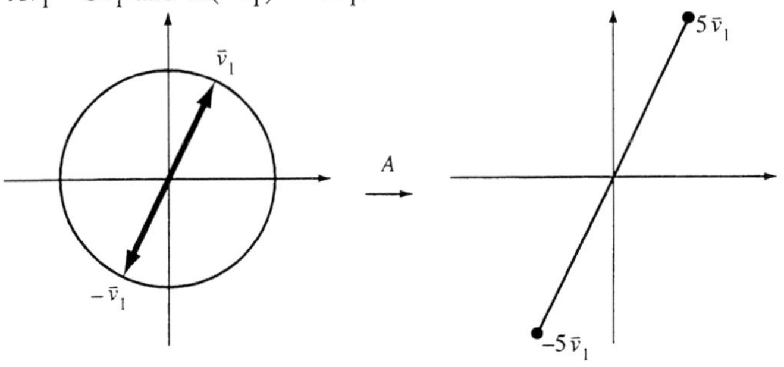

7. $A^TA = \begin{bmatrix} 1 & 0 \\ 0 & 4 \end{bmatrix}$

$\lambda_1 = 4,\ \lambda_2 = 1;\ \sigma_1 = 2,\ \sigma_2 = 1$

eigenvectors of A^TA: $\vec{v}_1 = \begin{bmatrix} 0 \\ 1 \end{bmatrix}$, $\vec{v}_2 = \begin{bmatrix} 1 \\ 0 \end{bmatrix}$, $\vec{u}_1 = \frac{1}{\sigma_1}(A\vec{v}_1) = \begin{bmatrix} 0 \\ -1 \end{bmatrix}$, $\vec{u}_2 = \frac{1}{\sigma_2}(A\vec{v}_2) = \begin{bmatrix} 1 \\ 0 \end{bmatrix}$, so that

$U = \begin{bmatrix} 0 & 1 \\ -1 & 0 \end{bmatrix}$, $\Sigma = \begin{bmatrix} 2 & 0 \\ 0 & 1 \end{bmatrix}$, $V = \begin{bmatrix} 0 & 1 \\ 1 & 0 \end{bmatrix}$.

8. $A^TA = \begin{bmatrix} p^2+q^2 & 0 \\ 0 & p^2+q^2 \end{bmatrix}$; $\lambda_1 = \lambda_2 = p^2+q^2$; $\sigma_1 = \sigma_2 = \sqrt{p^2+q^2}$

eigenvectors of A^TA: $\vec{v}_1 = \begin{bmatrix} 1 \\ 0 \end{bmatrix}$, $\vec{v}_2 = \begin{bmatrix} 0 \\ 1 \end{bmatrix}$, $\vec{u}_1 = \frac{1}{\sigma_1}A\vec{v}_1 = \frac{1}{\sqrt{p^2+q^2}}\begin{bmatrix} p \\ q \end{bmatrix}$, $\vec{u}_2 = \frac{1}{\sigma_2}A\vec{v}_2 = \frac{1}{\sqrt{p^2+q^2}}\begin{bmatrix} -q \\ p \end{bmatrix}$, so that

$U = \frac{1}{\sqrt{p^2+q^2}}\begin{bmatrix} p & -q \\ q & p \end{bmatrix}$, $\Sigma = \left(\sqrt{p^2+q^2}\right)I_2$, $V = I_2$.

9. $A^TA = \begin{bmatrix} 5 & 10 \\ 10 & 20 \end{bmatrix}$ (See Exercise 6)

$\lambda_1 = 25,\ \lambda_2 = 0;\quad \sigma_1 = 5,\ \sigma_2 = 0$

eigenvectors of A^TA: $\vec{v}_1 = \frac{1}{\sqrt{5}}\begin{bmatrix} 1 \\ 2 \end{bmatrix}$, $\vec{v}_2 = \frac{1}{\sqrt{5}}\begin{bmatrix} -2 \\ 1 \end{bmatrix}$, $\vec{u}_1 = \frac{1}{\sigma_1}A\vec{v}_1 = \frac{1}{\sqrt{5}}\begin{bmatrix} 1 \\ 2 \end{bmatrix}$, $\vec{u}_2$ = a unit vector orthogonal to $\vec{u}_1 = \frac{1}{\sqrt{5}}\begin{bmatrix} -2 \\ 1 \end{bmatrix}$ so that $U = V = \frac{1}{\sqrt{5}}\begin{bmatrix} 1 & -2 \\ 2 & 1 \end{bmatrix}$, $\Sigma = \begin{bmatrix} 5 & 0 \\ 0 & 0 \end{bmatrix}$.

10. In Example 4 we found $\begin{bmatrix} 6 & 2 \\ -7 & 6 \end{bmatrix} = \frac{1}{\sqrt{5}}\begin{bmatrix} 1 & 2 \\ -2 & 1 \end{bmatrix}\begin{bmatrix} 10 & 0 \\ 0 & 5 \end{bmatrix}\left(\frac{1}{\sqrt{5}}\begin{bmatrix} 2 & -1 \\ 1 & 2 \end{bmatrix}\right)$; now take the transpose of both sides:

$$\begin{bmatrix} 6 & -7 \\ 2 & 6 \end{bmatrix} = \underset{\uparrow\ U}{\frac{1}{\sqrt{5}}\begin{bmatrix} 2 & 1 \\ -1 & 2 \end{bmatrix}}\underset{\uparrow\ \Sigma}{\begin{bmatrix} 10 & 0 \\ 0 & 5 \end{bmatrix}}\underset{\uparrow\ V^T}{\left(\frac{1}{\sqrt{5}}\begin{bmatrix} 1 & -2 \\ 2 & 1 \end{bmatrix}\right)}.$$

11. $A^TA = \begin{bmatrix} 1 & 0 \\ 0 & 4 \end{bmatrix}$; $\lambda_1 = 4,\ \lambda_2 = 1;\ \sigma_1 = 2,\ \sigma_2 = 1$

eigenvectors of A^TA: $\vec{v}_1 = \begin{bmatrix} 0 \\ 1 \end{bmatrix}$, $\vec{v}_2 = \begin{bmatrix} 1 \\ 0 \end{bmatrix}$, $\vec{u}_1 = \frac{1}{\sigma_1}A\vec{v}_1 = \begin{bmatrix} 0 \\ 1 \\ 0 \end{bmatrix}$, $\vec{u}_2 = \frac{1}{\sigma_2}A\vec{v}_2 = \begin{bmatrix} 1 \\ 0 \\ 0 \end{bmatrix}$, $\vec{u}_3 = \begin{bmatrix} 0 \\ 0 \\ 1 \end{bmatrix}$,

$U = \begin{bmatrix} 0 & 1 & 0 \\ 1 & 0 & 0 \\ 0 & 0 & 1 \end{bmatrix}$, $\Sigma = \begin{bmatrix} 2 & 0 \\ 0 & 1 \end{bmatrix}$, $V = \begin{bmatrix} 0 & 1 \\ 1 & 0 \end{bmatrix}$.

12. In Example 5 we see that $\begin{bmatrix} 0 & 1 & 1 \\ 1 & 1 & 0 \end{bmatrix} = \frac{1}{\sqrt{2}}\begin{bmatrix} 1 & -1 \\ 1 & 1 \end{bmatrix}\begin{bmatrix} \sqrt{3} & 0 & 0 \\ 0 & 1 & 0 \end{bmatrix}\begin{bmatrix} \frac{1}{\sqrt{6}} & \frac{2}{\sqrt{6}} & \frac{1}{\sqrt{6}} \\ \frac{1}{\sqrt{2}} & 0 & -\frac{1}{\sqrt{2}} \\ \frac{1}{\sqrt{3}} & -\frac{1}{\sqrt{3}} & \frac{1}{\sqrt{3}} \end{bmatrix}$.

Now take the transpose of both sides.

$$\begin{bmatrix} 0 & 1 \\ 1 & 1 \\ 1 & 0 \end{bmatrix} = \underset{\uparrow\ U}{\begin{bmatrix} \frac{1}{\sqrt{6}} & \frac{1}{\sqrt{2}} & \frac{1}{\sqrt{3}} \\ \frac{2}{\sqrt{6}} & 0 & -\frac{1}{\sqrt{3}} \\ \frac{1}{\sqrt{6}} & -\frac{1}{\sqrt{2}} & \frac{1}{\sqrt{3}} \end{bmatrix}}\underset{\uparrow\ \Sigma}{\begin{bmatrix} \sqrt{3} & 0 \\ 0 & 1 \\ 0 & 0 \end{bmatrix}}\underset{\uparrow\ V^T}{\left(\frac{1}{\sqrt{2}}\begin{bmatrix} 1 & 1 \\ -1 & 1 \end{bmatrix}\right)}$$

13. $A^TA = \begin{bmatrix} 37 & 16 \\ 16 & 13 \end{bmatrix}$; $\lambda_1 = 45,\ \lambda_2 = 5;\ \sigma_1 = 3\sqrt{5},\ \sigma_2 = \sqrt{5}$

eigenvectors of A^TA: $\vec{v}_1 = \frac{1}{\sqrt{5}}\begin{bmatrix} 2 \\ 1 \end{bmatrix}$, $\vec{v}_2 = \frac{1}{\sqrt{5}}\begin{bmatrix} -1 \\ 2 \end{bmatrix}$, $\vec{u}_1 = \frac{1}{\sigma_1}A\vec{v}_1 = \begin{bmatrix} 1 \\ 0 \end{bmatrix}$, $\vec{u}_2 = \frac{1}{\sigma_2}A\vec{v}_2 = \begin{bmatrix} 0 \\ 1 \end{bmatrix}$, so that

$U = \begin{bmatrix} 1 & 0 \\ 0 & 1 \end{bmatrix}$, $\Sigma = \begin{bmatrix} 3\sqrt{5} & 0 \\ 0 & \sqrt{5} \end{bmatrix}$, $V = \frac{1}{\sqrt{5}}\begin{bmatrix} 2 & -1 \\ 1 & 2 \end{bmatrix}$.

14. $A^TA = \begin{bmatrix} 4 & 6 \\ 6 & 13 \end{bmatrix}$; $\lambda_1 = 16$, $\lambda_2 = 1$; $\sigma_1 = 4$, $\sigma_2 = 1$

eigenvectors of A^TA: $\vec{v}_1 = \frac{1}{\sqrt{5}}\begin{bmatrix} 1 \\ 2 \end{bmatrix}$, $\vec{v}_2 = \frac{1}{\sqrt{5}}\begin{bmatrix} -2 \\ 1 \end{bmatrix}$, $\vec{u}_1 = \frac{1}{\sigma_1}A\vec{v}_1 = \frac{1}{\sqrt{5}}\begin{bmatrix} 2 \\ 1 \end{bmatrix}$, $\vec{u}_2 = \frac{1}{\sigma_2}A\vec{v}_2 = \frac{1}{\sqrt{5}}\begin{bmatrix} -1 \\ 2 \end{bmatrix}$, so that

$U = \frac{1}{\sqrt{5}}\begin{bmatrix} 2 & -1 \\ 1 & 2 \end{bmatrix}$, $\Sigma = \begin{bmatrix} 4 & 0 \\ 0 & 1 \end{bmatrix}$, $V = \frac{1}{\sqrt{5}}\begin{bmatrix} 1 & -2 \\ 2 & 1 \end{bmatrix}$.

15. If $A\vec{v}_1 = \sigma_1\vec{u}_1$ and $A\vec{v}_2 = \sigma_2\vec{u}_2$, then $A^{-1}\vec{u}_1 = \frac{1}{\sigma_1}\vec{v}_1$ and $A^{-1}\vec{u}_2 = \frac{1}{\sigma_2}\vec{v}_2$, so that the singular values of A^{-1} are the reciprocals of the singular values of A. See Exercise 16 for a more detailed explanation.

16. If $A = U\Sigma V^T$ then $A^{-1} = V\Sigma^{-1}U^T$ and $(A^{-1})^TA^{-1} = U(\Sigma^{-1})^2U^{-1}$. Thus $(A^{-1})^TA^{-1}$ is similar to $(\Sigma^{-1})^2$, so that the eigenvalues of $(A^{-1})^TA^{-1}$ are the squares of the reciprocals of the singular values of A. It follows that the singular values of A^{-1} are the reciprocals of those of A.

17. We need to check that $A\left(\frac{\vec{b}\cdot\vec{u}_1}{\sigma_1}\vec{v}_1 + \cdots + \frac{\vec{b}\cdot\vec{u}_n}{\sigma_n}\vec{v}_n\right) = \text{proj}_{\text{im}A}\vec{b}$ (see page 222 of the text).

But $A\left(\frac{\vec{b}\cdot\vec{u}_1}{\sigma_1}\vec{v}_1 + \cdots + \frac{\vec{b}\cdot\vec{u}_n}{\sigma_n}\vec{v}_n\right) = \frac{\vec{b}\cdot\vec{u}_1}{\sigma_1}A\vec{v}_1 + \cdots + \frac{\vec{b}\cdot\vec{u}_n}{\sigma_n}A\vec{v}_n = (b\cdot\vec{u}_1)\vec{u}_1 + \cdots + (\vec{b}\cdot\vec{u}_n)\vec{u}_n = \text{proj}_{\text{im}A}\vec{b}$, since $\vec{v}_1, \ldots, \vec{v}_n$ is an orthonormal basis of im(A) (see Fact 4.1.6).

18. $\vec{b} = \begin{bmatrix} 1 \\ 2 \\ 3 \\ 4 \end{bmatrix}$, $\vec{u}_1 = \frac{1}{2}\begin{bmatrix} 1 \\ 1 \\ 1 \\ 1 \end{bmatrix}$, $\vec{u}_2 = \frac{1}{2}\begin{bmatrix} 1 \\ 1 \\ -1 \\ -1 \end{bmatrix}$, $\vec{v}_1 = \frac{1}{5}\begin{bmatrix} 3 \\ -4 \end{bmatrix}$, $\vec{v}_2 = \frac{1}{5}\begin{bmatrix} 4 \\ 3 \end{bmatrix}$, $\sigma_1 = 2$, $\sigma_2 = 1$, so that

$\vec{x}^* = \frac{\vec{b}\cdot\vec{u}_1}{\sigma_1}\vec{v}_1 + \frac{\vec{b}\cdot\vec{u}_2}{\sigma_2}\vec{v}_2 = \begin{bmatrix} -0.1 \\ -3.2 \end{bmatrix}$.

19. $\vec{x} = c_1\vec{v}_1 + \cdots + c_n\vec{v}_n$ is a least-squares solution if $A\vec{x} = c_1A\vec{v}_1 + \cdots + c_nA\vec{v}_n = c_1\sigma_1\vec{u}_1 + \cdots + c_r\sigma_r\vec{u}_r = \text{proj}_{\text{im}A}\vec{b}$. But $\text{proj}_{\text{im}A}\vec{b} = (\vec{b}\cdot\vec{u}_1)\vec{u}_1 + \cdots + (\vec{b}\cdot\vec{u}_r)\vec{u}_r$, since $\vec{u}_1, \ldots, \vec{u}_r$ is an orthonormal basis of im(A).

Comparing the coefficients of $\vec{u}_i$ above we find that it is required that $c_i\sigma_i = \vec{b}\cdot\vec{u}_i$ or $c_i = \frac{\vec{b}\cdot\vec{u}_i}{\sigma_i}$, for $i = 1, \ldots, r$, while no condition is imposed on $c_{r+1}, \ldots, c_n$. The least-squares solutions are of the form

$\vec{x}^* = \frac{b\cdot\vec{u}_1}{\sigma_1}\vec{v}_1 + \cdots + \frac{\vec{b}\cdot\vec{u}_r}{\sigma_r}\vec{v}_r + c_{r+1}\vec{v}_{r+1} + \cdots c_n\vec{v}_n$, where $c_{r+1}, \ldots, c_n$ are arbitrary (see Exercise 17 for a special case).

20. a. $A = U\Sigma V^T = UV^TV\Sigma V^T = QS$, where $Q = UV^T$ and $S = V\Sigma V^T$. Note that Q is orthogonal, being the product of orthogonal matrices; S is symmetric as $S^T = (V^T)^T\Sigma^TV^T = V\Sigma V^T = S$; and S is similar to Σ, so that the eigenvalues of S are the (positive) diagonal entries of Σ.

b. Yes, write $A = U\Sigma V^T = U\Sigma U^TUV^T = S_1Q_1$ where $S_1 = U\Sigma U^T$ and $Q_1 = UV^T$.

21. $A = \underbrace{\frac{1}{\sqrt{5}}\begin{bmatrix} 1 & 2 \\ -2 & 1 \end{bmatrix}}_{U} \underbrace{\begin{bmatrix} 10 & 0 \\ 0 & 5 \end{bmatrix}}_{\Sigma} \underbrace{\left(\frac{1}{\sqrt{5}}\begin{bmatrix} 2 & -1 \\ 1 & 2 \end{bmatrix}\right)}_{V^T} = \underbrace{\frac{1}{\sqrt{5}}\begin{bmatrix} 1 & 2 \\ -2 & 1 \end{bmatrix}}_{U} \underbrace{\left(\frac{1}{\sqrt{5}}\begin{bmatrix} 2 & -1 \\ 1 & 2 \end{bmatrix}\right)}_{V^T} \underbrace{\left(\frac{1}{\sqrt{5}}\begin{bmatrix} 2 & 1 \\ -1 & 2 \end{bmatrix}\right)}_{V} \underbrace{\begin{bmatrix} 10 & 0 \\ 0 & 5 \end{bmatrix}}_{\Sigma} \underbrace{\left(\frac{1}{\sqrt{5}}\begin{bmatrix} 2 & -1 \\ 1 & 2 \end{bmatrix}\right)}_{V^T}$

$= \underbrace{\frac{1}{5}\begin{bmatrix} 4 & 3 \\ -3 & 4 \end{bmatrix}}_{Q} \underbrace{\begin{bmatrix} 9 & -2 \\ -2 & 6 \end{bmatrix}}_{S}$

22. a. Note that the transformation S preserves length.

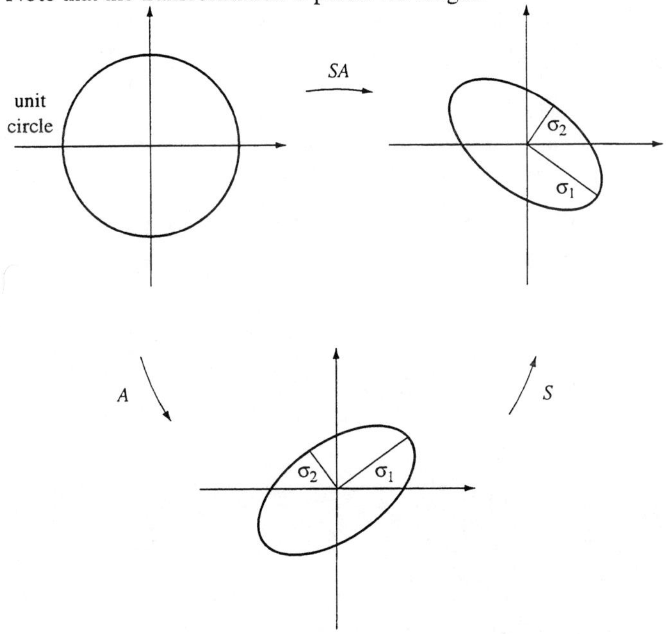

b. $(SA)^T SA = A^T S^T SA = A^T A$ since S is orthogonal. Thus, SA and A have the same singular values.

23. $AA^T U = U\Sigma V^T V \Sigma^T U^T U = U\Sigma\Sigma^T$, since $V^T V = I_n$ and $U^T U = I_m$, so that

$$AA^T \vec{u}_i = \begin{cases} \sigma_i^2 \vec{u}_i & \text{for } i = 1, \ldots, r \\ \vec{0} & \text{for } i = r+1, \ldots, m \end{cases}.$$

The *nonzero* eigenvalues of $A^T A$ and AA^T are the same.

24. The eigenvalues of $A^T A = A^2$ are the squares of the eigenvalues of A, so that the singular values of A are the absolute values of the eigenvalues of A.

25.

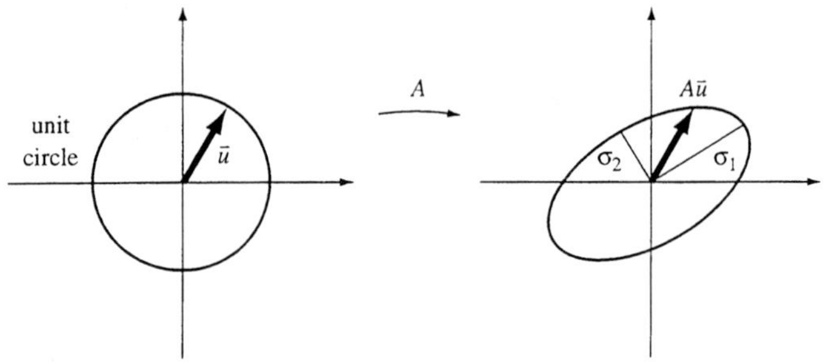

Algebraically: Write $\vec{u} = c_1\vec{v}_1 + c_2\vec{v}_2$ and note that $\|\vec{u}\|^2 = c_1^2 + c_2^2 = 1$.

Then $A\vec{u} = c_1\sigma_1\vec{u}_1 + c_2\sigma_2\vec{u}_2$, so that $\|A\vec{u}\|^2 = c_1^2\sigma_1^2 + c_2^2\sigma_2^2 \ge c_1^2\sigma_2^2 + c_2^2\sigma_2^2 = \sigma_2^2$ and $\|A\vec{u}\| \ge \sigma_2$. Likewise $\|A\vec{u}\| \le \sigma_1$.

26. Write $\vec{v} = c_1\vec{v}_1 + \cdots + c_n\vec{v}_n$ and note that $\|\vec{v}\|^2 = c_1^2 + \cdots + c_n^2$. Then $A\vec{v} = c_1\sigma_1\vec{u}_1 + \cdots + c_r\sigma_r\vec{u}_r$ and $\|A\vec{v}\|^2 = c_1^2\sigma_1^2 + c_2^2\sigma_2^2 + \cdots + c_r^2\sigma_r^2 \le c_1^2\sigma_1^2 + c_2^2\sigma_1^2 + \cdots + c_r^2\sigma_1^2 \le \sigma_1^2\|v\|^2$ so that $\|A\vec{v}\| \le \sigma_1\|\vec{v}\|$. Likewise, $\|A\vec{v}\| \ge \sigma_n\|\vec{v}\|$.

27. Let $\vec{v}$ be a unit eigenvector with eigenvalue λ and use Exercise 26.

28. If $\lambda_1, \ldots, \lambda_n$ are the eigenvalues of A^TA, then $(\det A)^2 = \det(A^TA) = \lambda_1 \cdots \lambda_n = \sigma_1^2 \cdots \sigma_n^2$, so that $|\det A| = \sigma_1 \cdots \sigma_n$. For a 2 × 2 matrix:

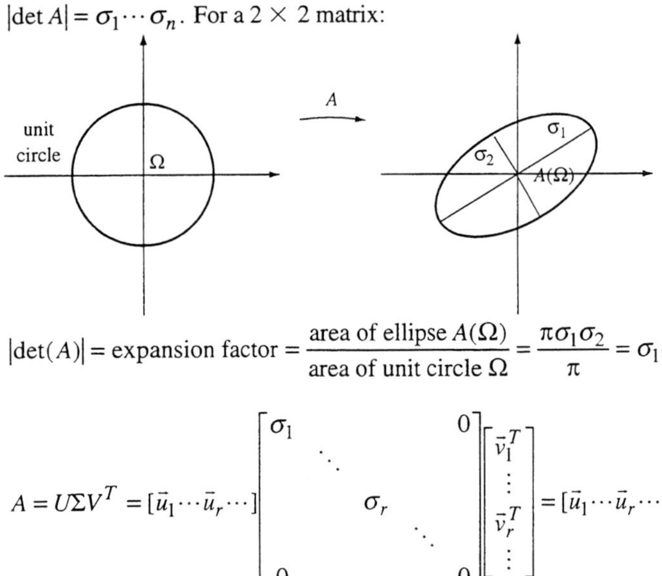

$$|\det(A)| = \text{expansion factor} = \frac{\text{area of ellipse } A(\Omega)}{\text{area of unit circle } \Omega} = \frac{\pi\sigma_1\sigma_2}{\pi} = \sigma_1\sigma_2.$$

29. $$A = U\Sigma V^T = [\vec{u}_1 \cdots \vec{u}_r \cdots]\begin{bmatrix} \sigma_1 & & & & 0 \\ & \ddots & & & \\ & & \sigma_r & & \\ & & & \ddots & \\ 0 & & & & 0 \end{bmatrix}\begin{bmatrix} \vec{v}_1^T \\ \vdots \\ \vec{v}_r^T \\ \vdots \end{bmatrix} = [\vec{u}_1 \cdots \vec{u}_r \cdots]\begin{bmatrix} \sigma_1\vec{v}_1^T \\ \vdots \\ \sigma_r\vec{v}_r^T \\ 0 \end{bmatrix} = \sigma_1\vec{u}_1\vec{v}_1^T + \cdots + \sigma_r\vec{u}_r\vec{v}_r^T$$

30. $$\begin{bmatrix} 6 & 2 \\ -7 & 6 \end{bmatrix} = \frac{1}{\sqrt{5}}\begin{bmatrix} 1 & 2 \\ -2 & 1 \end{bmatrix}\begin{bmatrix} 10 & 0 \\ 0 & 5 \end{bmatrix}\frac{1}{\sqrt{5}}\begin{bmatrix} 2 & -1 \\ 1 & 2 \end{bmatrix} = 10\frac{1}{\sqrt{5}}\begin{bmatrix} 1 \\ -2 \end{bmatrix}\left(\frac{1}{\sqrt{5}}[2 \ \ -1]\right) + 5\frac{1}{\sqrt{5}}\begin{bmatrix} 2 \\ 1 \end{bmatrix}\left(\frac{1}{\sqrt{5}}[1 \ \ 2]\right)$$
$$= \begin{bmatrix} 4 & -2 \\ -8 & 4 \end{bmatrix} + \begin{bmatrix} 2 & 4 \\ 1 & 2 \end{bmatrix}$$

31. The formula $A = \sigma_1\vec{u}_1\vec{v}_1^T + \cdots + \sigma_r\vec{u}_r\vec{v}_r^T$ gives such a representation.

32. $(SAR)^T SAR = R^T A^T S^T SAR = R^T A^T AR$ is similar to $A^T A$, so that the matrices $A^T A$ and $(SAR)^T SAR$ have the same eigenvalues. Thus A and SAR have the same singular values.

33. False; consider $A = \begin{bmatrix} 0 & 1 \\ 2 & 0 \end{bmatrix}$, with $\sigma_1 = 2$ and $\sigma_1 = 1$, but $A^2 = \begin{bmatrix} 2 & 0 \\ 0 & 2 \end{bmatrix}$ has singular values 2, 2.

34. $A = U\Sigma U^T$ means that $U^T AU = U^{-1}AU = \Sigma$, i.e., A is orthogonally diagonalizable, with eigenvalues ≥ 0. This is the case if and only if A is *symmetric* and *positive semidefinite*.

35. We will freely use the diagram on page 434 (with $r = n$).
We have $A^T A\vec{v}_i = A^T(\sigma_i \vec{u}_i) = \sigma_i^2 \vec{v}_i$ and therefore $(A^T A)^{-1}\vec{v}_i = \frac{1}{\sigma_i^2}\vec{v}_i$ for $i = 1, \ldots, n$.

Then $(A^T A)^{-1} A^T \vec{u}_i = (A^T A)^{-1}(\sigma_i \vec{v}_i) = \frac{1}{\sigma_i}\vec{v}_i$ for $i = 1, \ldots, n$ and $(A^T A)^{-1} A^T \vec{u}_i = \vec{0}$ for $i = n + 1, \ldots, m$ since $\vec{u}_i$ is in $\ker(A^T)$ in this case. Note that $(A^T A)^{-1} A^T \vec{u}_i$ is the least-squares solution of the equation $A\vec{x} = \vec{u}_i$; for $i = 1, \ldots, n$ this is the exact solution since $\vec{u}_i$ is in $\text{im}(A)$.

36. We will freely use the diagram on page 434. By construction of the $\vec{v}_i$ as eigenvectors of $A^T A$ we have $A^T A\vec{v}_i = \lambda_i \vec{v}_i = \sigma_i^2 \vec{v}_i$, or $(A^T A)^{-1}\vec{v}_i = \frac{1}{\sigma_i^2}\vec{v}_i$. Then

$A(A^T A)^{-1} A^T \vec{u}_i = A(A^T A)^{-1}(\sigma_i \vec{v}_i) = A\left(\frac{1}{\sigma_i}\vec{v}_i\right) = \frac{1}{\sigma_i} A\vec{v}_i = \vec{u}_i$ for $i = 1, \ldots, n$ and $A(A^T A)^{-1} A^T \vec{u}_i = \vec{0}$ for $i = n + 1, \ldots, m$ since $\vec{u}_i$ is in $\ker(A^T)$ in this case. The fact that $A(A^T A)^{-1} A^T \vec{u}_i = \begin{cases} \vec{u}_i & \text{if } i = 1, \ldots, n \\ \vec{0} & \text{if } i = n+1, \ldots, m \end{cases}$ means that the matrix $A(A^T A)^{-1} A^T$ represents the orthogonal projection onto $\text{im}(A) = \text{span}(\vec{u}_1, \ldots, \vec{u}_n)$.

37. *Yes*; since $A^T A$ is diagonalizable and has only 1 as an eigenvalue, we must have $A^T A = I_n$.

38. *False*; $A = \begin{bmatrix} 1 & 0 \\ 0 & 2 \end{bmatrix}$ (with $\sigma_1 = 2$ and $\sigma_2 = 1$) and $B = \begin{bmatrix} 1 & 1 \\ 0 & 2 \end{bmatrix}$ are similar, but the larger singular value of B is more than $\left\| \begin{bmatrix} 1 \\ 2 \end{bmatrix} \right\| = \sqrt{5}$.

Chapter 8

8.1

1. $x(t) = 7e^{5t}$, by Fact 8.1.1.

2. $x(t) = -e \cdot e^{-0.71t} = -e^{1-0.71t}$, by Fact 8.1.1.

3. $P(t) = 7e^{0.03t}$, by Fact 8.1.1.

4. This is just an antiderivative problem: $y(t) = 0.8\frac{t^2}{2} + C = 0.4t^2 + C$, and $C = -0.8$, so that $y(t) = 0.4t^2 - 0.8$.

5. $y(t) = -0.8e^{0.8t}$, by Fact 8.1.1.

6. $x\,dx = dt$
   $\frac{x^2}{2} = t + C$, and $\frac{1}{2} = 0 + C$, so that
   $\frac{x^2}{2} = t + \frac{1}{2}$
   $x^2 = 2t + 1$
   $x(t) = \sqrt{2t+1}$

7. $x^{-2}dx = dt$
   $-x^{-1} = t + C$
   $-\frac{1}{x} = t + C$, and $-1 = 0 + C$, so that
   $-\frac{1}{x} = t - 1$
   $x(t) = \frac{1}{1-t}$; note that $\lim\limits_{x \to 1^-} x(t) = \infty$.

8. $x^{-1/2}dx = dt$
   $2x^{1/2} = t + C$, and $2\sqrt{4} = 0 + C$, so that $2x^{1/2} = t + 4$.
   $x(t) = \left(\frac{t}{2} + 2\right)^2$ for $t \geq -4$.

9. $x^{-k}dx = dt$
   $\frac{1}{1-k}x^{1-k} = t + C$, and $\frac{1}{1-k} = C$, so that
   $\frac{1}{1-k}x^{1-k} = t + \frac{1}{1-k}$
   $x^{1-k} = (1-k)t + 1$
   $x(t) = ((1-k)t + 1)^{1/1-k}$.

10. $\cos x\,dx = dt$
    $\sin x = t + C$, and $C = 0$.
    $x(t) = \arcsin(t)$ for $|t| < 1$.

11. $\frac{dx}{1+x^2} = dt$
arctan$(x) = t + C$ and $C = 0$.
$x(t) = \tan(t)$ for $|t| < \frac{\pi}{2}$.

12. We want $e^{kt} = 3^t$ or $e^k = 3$ or $k = \ln(3)$: $\frac{dx}{dt} = \ln(3)x$.

13. **a.** The debt in millions is $0.45(1.06)^{212} \approx 104245$, or about 100 billion dollars.

b. The debt in millions is $0.45e^{0.06\cdot 212} \approx 150466$, or about 150 billion dollars.

14. **a.** $x(t) = e^{-\frac{t}{8270}}$, by Fact 8.1.1.
If T is the half-life, then $e^{-\frac{T}{8270}} = \frac{1}{2}$ or $-\frac{T}{8270} = \ln\left(\frac{1}{2}\right)$ or $T = -8270\ln\left(\frac{1}{2}\right) \approx 5732$.
The half-life is about 5732 years.

b. We want to find t such that $e^{-\frac{t}{8270}} = 1 - 0.47 = 0.53$ or $-\frac{t}{8270} = \ln(0.53)$ or $t = -8270\ln(0.53) \approx 5250$.
The Iceman died about 5250 years before A.D. 1991, or about 3250 B.C. The Austrian expert was wrong.

15. If $P(t) = P_0 \cdot e^{\frac{k}{100}t}$, then the doubling time T satisfies the equation $P(T) = P_0 \cdot e^{\frac{k}{100}T} = 2P_0$ or $e^{\frac{k}{100}T} = 2$ or $\frac{k}{100}T = \ln(2)$ or $T = \frac{100}{k}\cdot\ln(2) \approx \frac{69}{k}$ since $\ln(2) \approx 0.69$.

16.

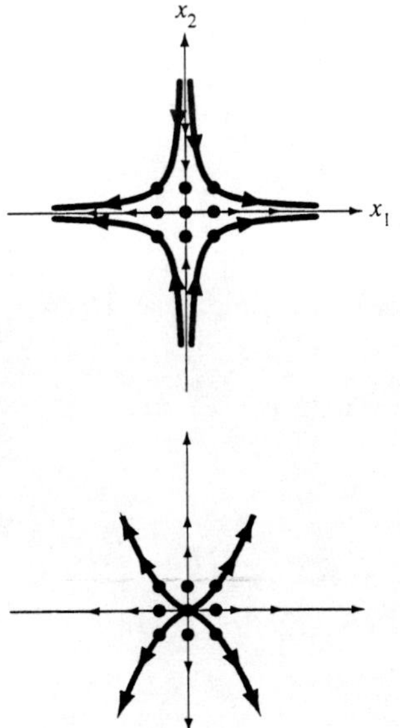

17.

18.

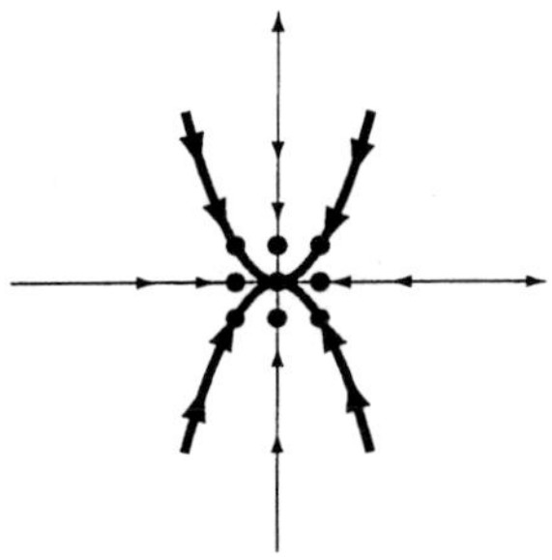

19.

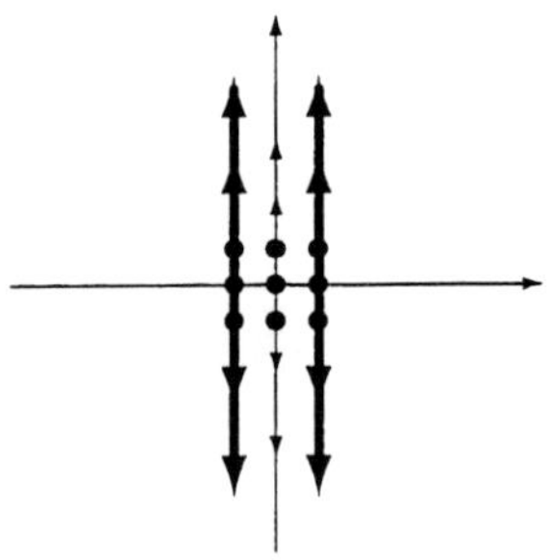

20. $A\vec{x} = \begin{bmatrix} 0 & -1 \\ 1 & 0 \end{bmatrix}\begin{bmatrix} x_1 \\ x_2 \end{bmatrix} = \begin{bmatrix} -x_2 \\ x_1 \end{bmatrix}$

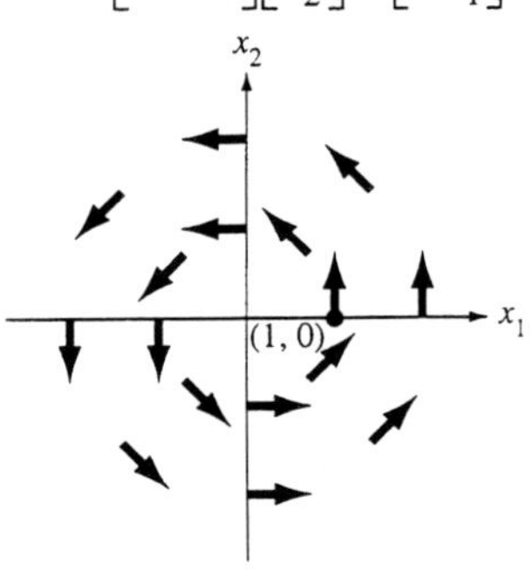

It appears that the trajectories will be circles. If we start at $\begin{bmatrix} 1 \\ 0 \end{bmatrix}$ we will trace out the unit circle $\vec{x}(t) = \begin{bmatrix} \cos(t) \\ \sin(t) \end{bmatrix}$.

We can verify that $\frac{d\vec{x}}{dt} = \begin{bmatrix} -\sin(t) \\ \cos(t) \end{bmatrix}$ equals $\begin{bmatrix} 0 & -1 \\ 1 & 0 \end{bmatrix}\vec{x}|t| = \begin{bmatrix} -\sin(t) \\ \cos(t) \end{bmatrix}$, as claimed.

21. $A\vec{x} = \begin{bmatrix} 0 & 1 \\ 0 & 0 \end{bmatrix}\begin{bmatrix} x_1 \\ x_2 \end{bmatrix} = \begin{bmatrix} x_2 \\ 0 \end{bmatrix}$

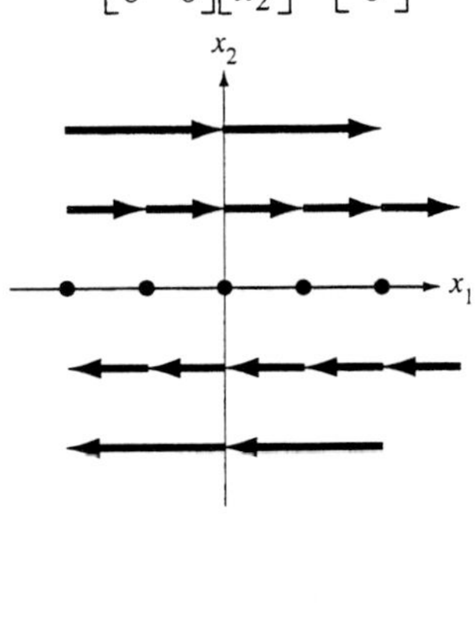

The trajectories will be horizontal lines. If we start at $\begin{bmatrix} p \\ q \end{bmatrix}$, then the horizontal velocity will be q, so that $\vec{x}(t) = \begin{bmatrix} x_1(t) \\ x_2(t) \end{bmatrix} = \begin{bmatrix} p+qt \\ q \end{bmatrix}$. We can verify that $\frac{d\vec{x}}{dt} = \begin{bmatrix} q \\ 0 \end{bmatrix}$ equals $\begin{bmatrix} 0 & 1 \\ 0 & 0 \end{bmatrix}\vec{x}(t) = \begin{bmatrix} q \\ 0 \end{bmatrix}$, as claimed.

22. We are told that $\frac{d\vec{x}_1}{dt} = A\vec{x}_1$ and $\frac{d\vec{x}_2}{dt} = A\vec{x}_2$. Let $\vec{x}(t) = \vec{x}_1(t) + \vec{x}_2(t)$. Then $\frac{d\vec{x}}{dt} = \frac{d\vec{x}_1}{dt} + \frac{d\vec{x}_2}{dt} = A\vec{x}_1 + A\vec{x}_2 = A(\vec{x}_1 + \vec{x}_2) = A\vec{x}$, as claimed.

23. We are told that $\frac{dx_1}{dt} = A\vec{x}_1$. Let $\vec{x}(t) = k\vec{x}_1(t)$. Then $\frac{d\vec{x}}{dt} = \frac{d}{dt}(k\vec{x}_1) = k\frac{d\vec{x}_1}{dt} = kA\vec{x}_1 = A(k\vec{x}_1) = A\vec{x}$, as claimed.

24. We are told that $\frac{d\vec{x}}{dt} = A\vec{x}$. Let $\vec{c}(t) = e^{kt}\vec{x}(t)$.

Then $\frac{d\vec{c}}{dt} = \frac{d}{dt}(e^{kt}\vec{x}) = \left(\frac{d}{dt}e^{kt}\right)\vec{x} + e^{kt}\frac{d\vec{x}}{dt} = ke^{kt}\vec{x} + e^{kt}A\vec{x} = (A + kI_n)(e^{kt}\vec{x}) = (A + kI_n)\vec{c}$, as claimed.

25. We are told that $\frac{d\vec{x}}{dt} = A\vec{x}$. Let $\vec{c}(t) = \vec{x}(kt)$. Using the chain rule we find that

$\frac{d\vec{c}}{dt} = \frac{d}{dt}(\vec{x}(kt)) = k\left.\frac{d\vec{x}}{dt}\right|_{kt} = kA(\vec{x}(kt)) = kA\vec{c}(t)$, as claimed.

To get the vector field $kA\vec{c}$ we scale the vectors of the field $A\vec{x}$ by k.

26. $\lambda_1 = 3$, $\lambda_2 = -2$; $\vec{v}_1 = \begin{bmatrix} 1 \\ 1 \end{bmatrix}$, $\vec{v}_2 = \begin{bmatrix} -2 \\ 3 \end{bmatrix}$, $c_1 = 5$, $c_2 = -1$, so that $\vec{x}(t) = 5e^{3t}\begin{bmatrix} 1 \\ 1 \end{bmatrix} - e^{-2t}\begin{bmatrix} -2 \\ 3 \end{bmatrix}$.

27. Use Fact 8.1.2.

The eigenvalues of $A = \begin{bmatrix} -4 & 3 \\ 2 & -3 \end{bmatrix}$ are $\lambda_1 = -6$ and $\lambda_2 = -1$, with associated eigenvectors $\vec{v}_1 = \begin{bmatrix} -3 \\ 2 \end{bmatrix}$ and $\vec{v}_2 = \begin{bmatrix} 1 \\ 1 \end{bmatrix}$. The coordinates of $\vec{x}(0) = \begin{bmatrix} 1 \\ 0 \end{bmatrix}$ with respect to $\vec{v}_1$ and $\vec{v}_2$ are $c_1 = -\frac{1}{5}$ and $c_2 = \frac{2}{5}$.

By Fact 8.1.2 the solution is $\vec{x}(t) = -\frac{1}{5}e^{-6t}\begin{bmatrix} -3 \\ 2 \end{bmatrix} + \frac{2}{5}e^{-t}\begin{bmatrix} 1 \\ 1 \end{bmatrix}$.

28. $\lambda_1 = 2$, $\lambda_2 = 10$; $\vec{v}_1 = \begin{bmatrix} -3 \\ 2 \end{bmatrix}$, $\vec{v}_2 = \begin{bmatrix} 1 \\ 2 \end{bmatrix}$; $c_1 = -\frac{1}{8}$, $c_2 = \frac{5}{8}$, so that $\vec{x}(t) = -\frac{1}{8}e^{2t}\begin{bmatrix} -3 \\ 2 \end{bmatrix} + \frac{5}{8}e^{10t}\begin{bmatrix} 1 \\ 2 \end{bmatrix}$.

29. $\lambda_1 = 0$, $\lambda_2 = 5$; $\vec{v}_1 = \begin{bmatrix} -2 \\ 1 \end{bmatrix}$, $\vec{v}_2 = \begin{bmatrix} 1 \\ 2 \end{bmatrix}$; $c_1 = -2$, $c_2 = 1$, so that $\vec{x}(t) = -2\begin{bmatrix} -2 \\ 1 \end{bmatrix} + e^{5t}\begin{bmatrix} 1 \\ 2 \end{bmatrix} = \begin{bmatrix} 4 \\ -2 \end{bmatrix} + e^{5t}\begin{bmatrix} 1 \\ 2 \end{bmatrix}$.

30. $\lambda_1 = 0$, $\lambda_2 = 5$, $\vec{v}_1 = \begin{bmatrix} -2 \\ 1 \end{bmatrix}$, $\vec{v}_2 = \begin{bmatrix} 1 \\ 2 \end{bmatrix}$; $c_1 = -1$, $c_2 = 0$, so that $\vec{x}(t) = -\begin{bmatrix} -2 \\ 1 \end{bmatrix} = \begin{bmatrix} 2 \\ -1 \end{bmatrix}$.

31. $\lambda_1 = 1,\ \lambda_2 = 6,\ \lambda_3 = 0;\ \vec{v}_1 = \begin{bmatrix} 1 \\ -2 \\ 1 \end{bmatrix}$. Since $\vec{x}(0) = \vec{v}_1$ we need not find $\vec{v}_2$ and $\vec{v}_3$.

$c_1 = 1,\ c_2 = c_3 = 0$, so that $\vec{x}(t) = e^t \begin{bmatrix} 1 \\ -2 \\ 1 \end{bmatrix}$.

In Exercises 32 to 35, find the eigenvalues and eigenspaces. Then determine the direction of the flow along the eigenspaces (outward if $\lambda > 0$ and inward if $\lambda < 0$). Use Figure 11 of Section 8.1 as a guide to sketch the other trajectories.

32.

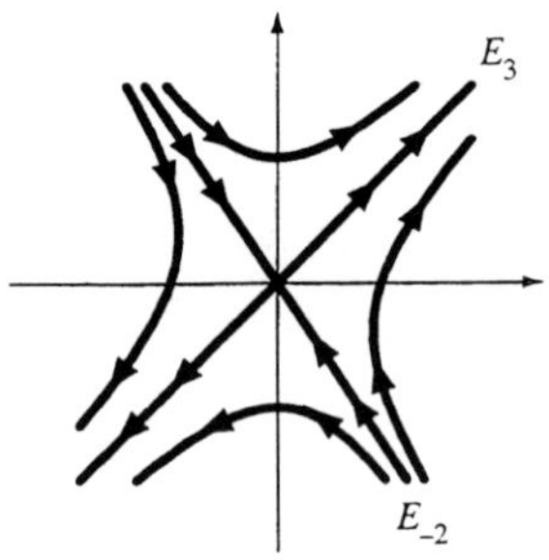

See Exercise 26.

33.

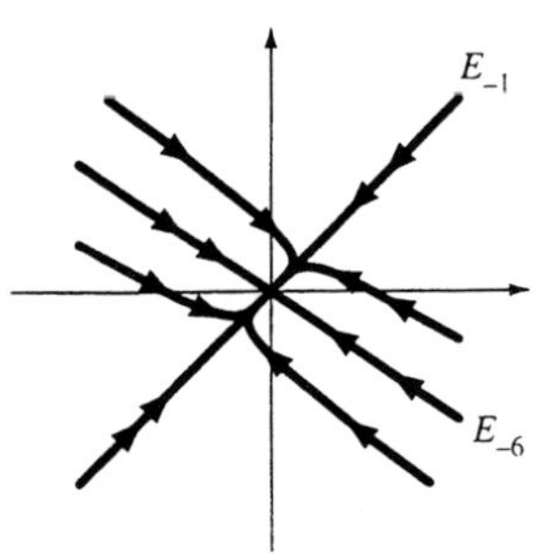

See Exercise 27.

34.

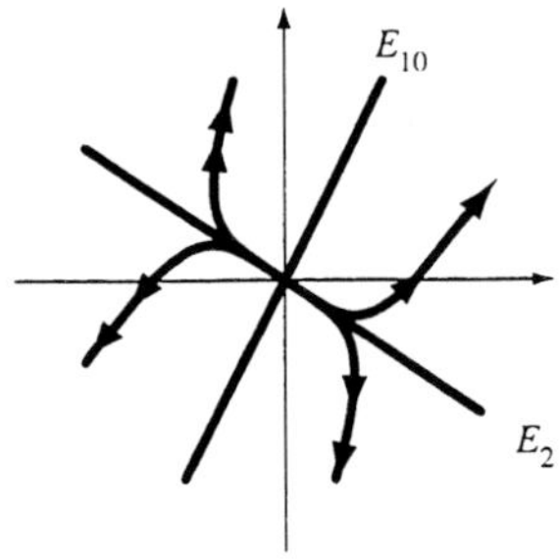

See Exercise 28.

35. 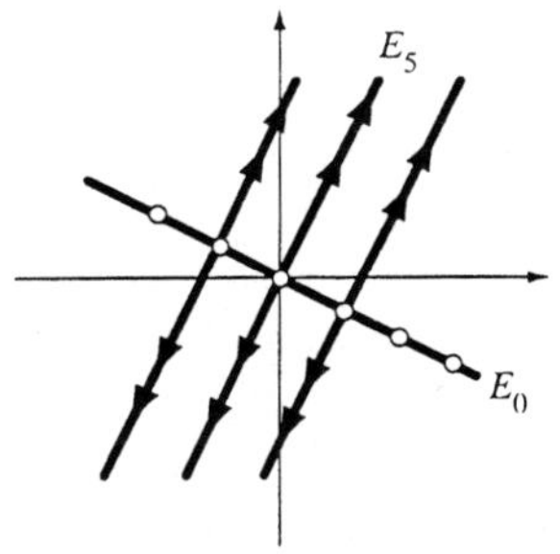

See Exercises 29 and 30.

In Exercises 36 to 39, find the eigenvalues and eigenspaces (the eigenvalues will always be positive). Then determine the direction of the flow along the eigenspaces (outward if $\lambda > 1$ and inward if $1 > \lambda > 0$). Use Figure 11 of Section 6.1 as a guide to sketch the other trajectories.

36. 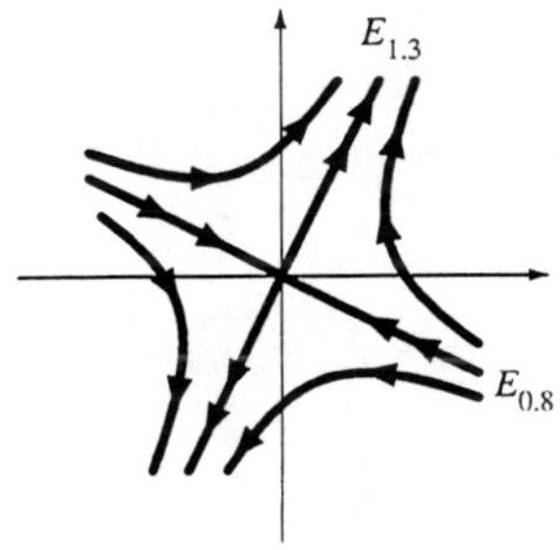

37. 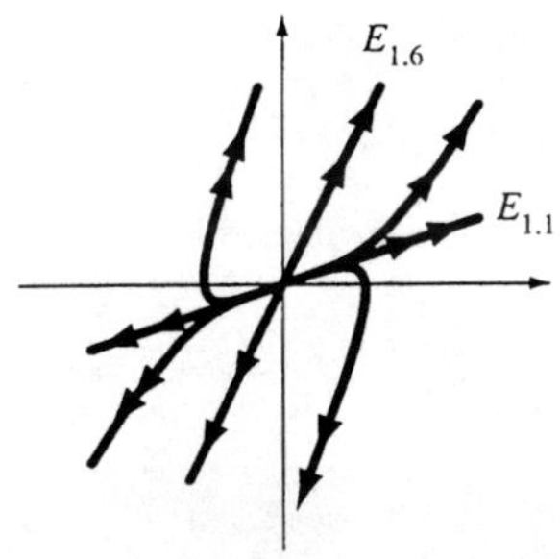

38. 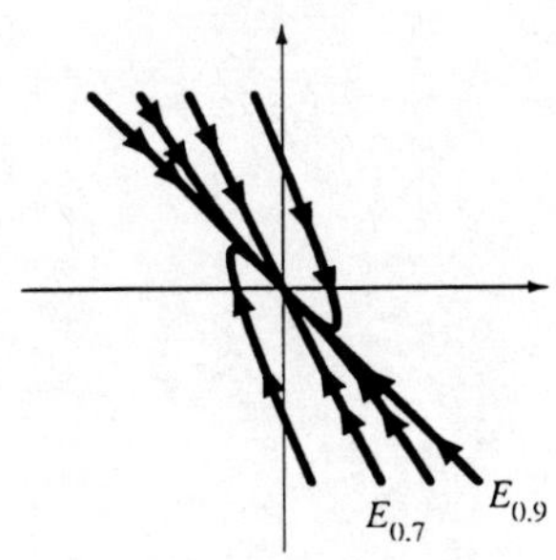

39.

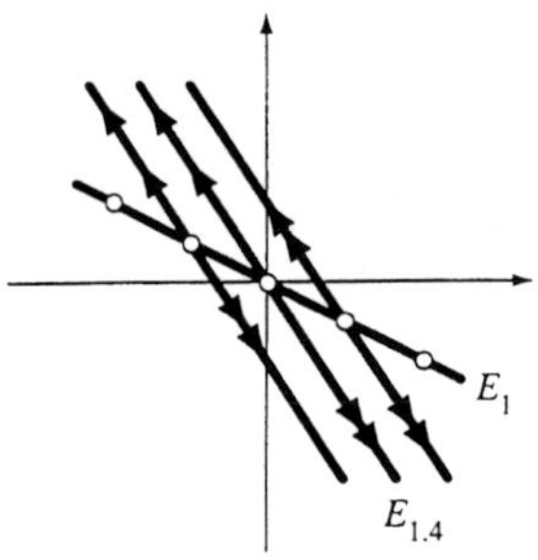

40. $\vec{x}(t) = e^{2t}\begin{bmatrix} 2 \\ 3 \end{bmatrix} + e^{3t}\begin{bmatrix} 3 \\ 4 \end{bmatrix}$

We want a 2×2 matrix A with eigenvalues $\lambda_1 = 2$ and $\lambda_2 = 3$ and associated eigenvectors $\vec{v}_1 = \begin{bmatrix} 2 \\ 3 \end{bmatrix}$ and $\vec{v}_2 = \begin{bmatrix} 3 \\ 4 \end{bmatrix}$; that is $A\begin{bmatrix} 2 & 3 \\ 3 & 4 \end{bmatrix} = \begin{bmatrix} 4 & 9 \\ 6 & 12 \end{bmatrix}$ or $A = \begin{bmatrix} 4 & 9 \\ 6 & 12 \end{bmatrix}\begin{bmatrix} 2 & 3 \\ 3 & 4 \end{bmatrix}^{-1} = \begin{bmatrix} 4 & 9 \\ 6 & 12 \end{bmatrix}\begin{bmatrix} -4 & 3 \\ 3 & -2 \end{bmatrix} = \begin{bmatrix} 11 & -6 \\ 12 & -6 \end{bmatrix}$.

41. The trajectories are of the form $\vec{x}|t| = c_1 e^{\lambda_1 t}\vec{v}_1 + c_2 e^{\lambda_2 t}\vec{v}_2 = c_1\vec{v}_1 + c_2 e^{\lambda_2 t}\vec{v}_2$.

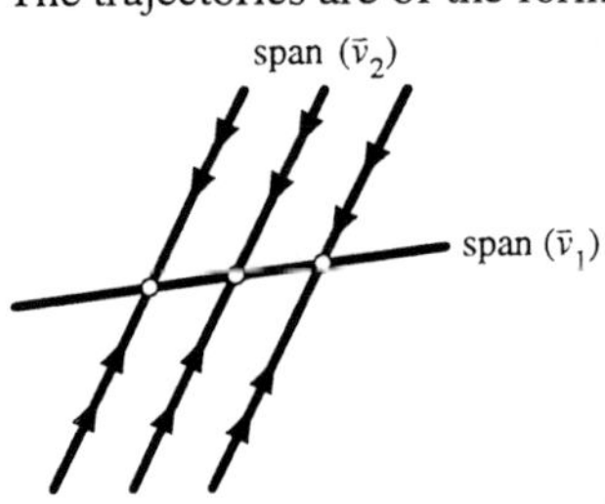

42. a. The term $0.8x$ in the second equation indicates that species y is helped by x, while species x is hindered by y (consider the term $-1.2y$ in the first equation). Thus y preys on x.

b.

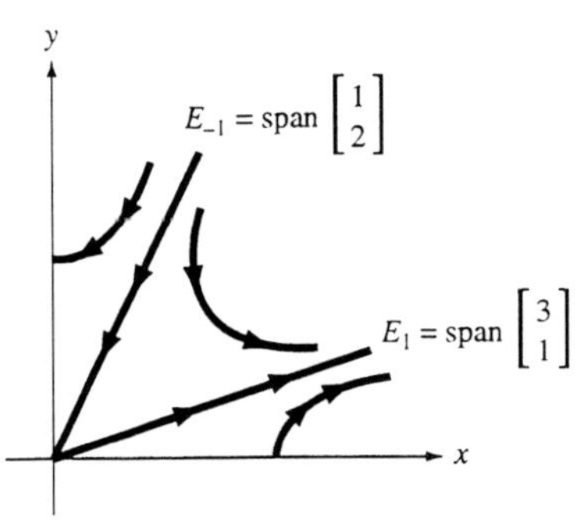

c. If $\dfrac{y(0)}{x(0)} < 2$ then both species will prosper, and $\lim\limits_{t \to \infty} \dfrac{y(t)}{x(t)} = \dfrac{1}{3}$.

If $\dfrac{y(0)}{x(0)} \geq 2$ then both species will die out.

43. a. These two species are *competing* as each is hindered by the other (consider the terms $-y$ and $-2x$).

b.

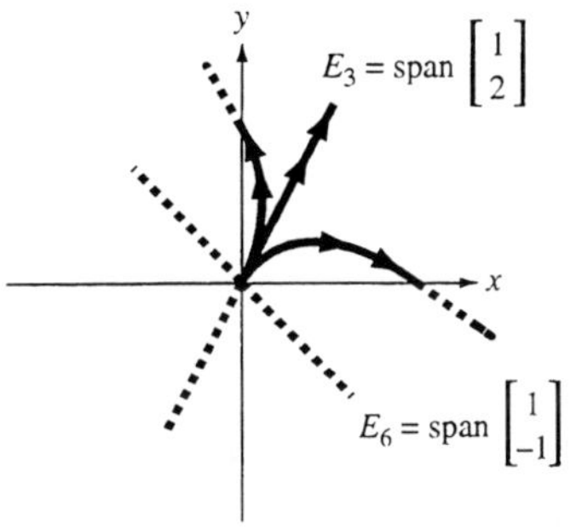

Although only the first quadrant is relevant for our model, it is useful to consider the phase portrait in the other quadrants as well.

c. If $\frac{y(0)}{x(0)} > 2$ then species y wins (x will die out); if $\frac{y(0)}{x(0)} < 2$ then x wins. If $\frac{y(0)}{x(0)} = 2$ then both will prosper and $\frac{y(t)}{x(t)} = 2$ for all t.

44. a. The two species are in symbiosis: Each is helped by the other (consider the terms $4y$ and $2x$).

b.

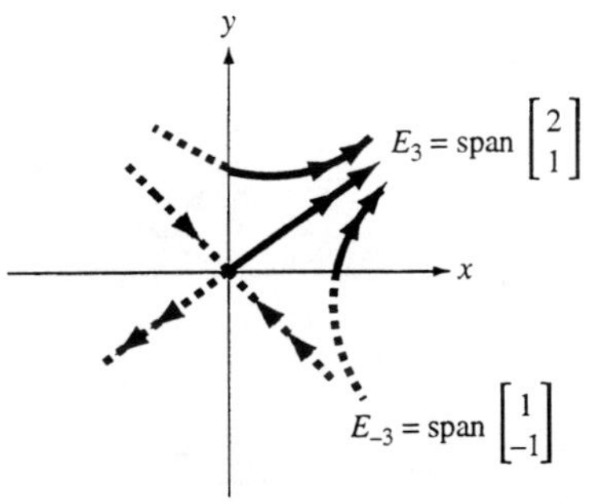

c. Both populations will prosper and $\lim_{t\to\infty} \frac{y(t)}{x(t)} = \frac{1}{2}$, regardless of the initial populations.

45. a. Species y has the more vicious fighters, since they kill members of species x at a rate of 4 per time unit, while the fighters of species x only kill at a rate of 1.

b.

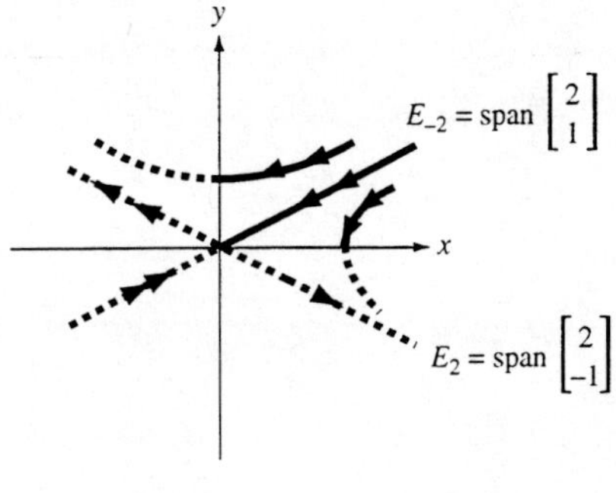

c. If $\frac{y(0)}{x(0)} < \frac{1}{2}$ then x wins; if $\frac{y(0)}{x(0)} > \frac{1}{2}$ then y wins; if $\frac{y(0)}{x(0)} = \frac{1}{2}$ nobody will survive the battle.

46. *Phase portrait*:

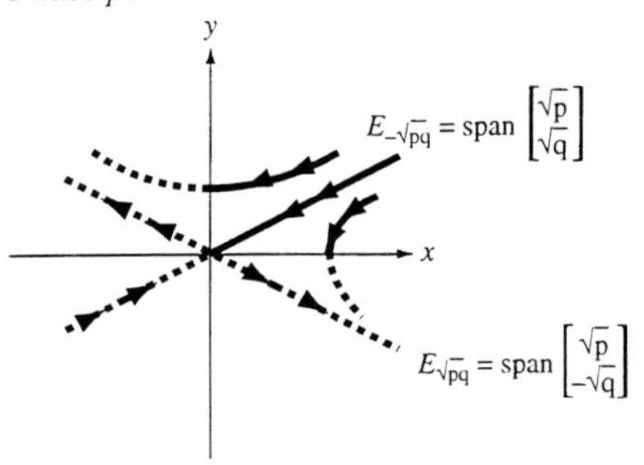

if $\frac{y(0)}{x(0)} < \frac{\sqrt{q}}{\sqrt{p}}$ then x wins; if $\frac{y(0)}{x(0)} > \frac{\sqrt{q}}{\sqrt{p}}$ then y wins; if $\frac{y(0)}{x(0)} = \frac{\sqrt{q}}{\sqrt{p}}$ then nobody will survive the battle.

47. a. The two species are in symbiosis: Each is helped by the other (consider the terms kx and ky).

b. $\lambda_{1,2} = \frac{-5 \pm \sqrt{9+4k^2}}{2}$

Both eigenvalues are negative if $\sqrt{9+4k^2} < 5$ or $9+4k^2 < 25$ or $4k^2 < 16$ or $k < 2$ (recall that k is positive).

If $k = 2$ then the eigenvalues are -5 and 0.

If $k > 2$ then there is a positive and a negative eigenvalue.

c.

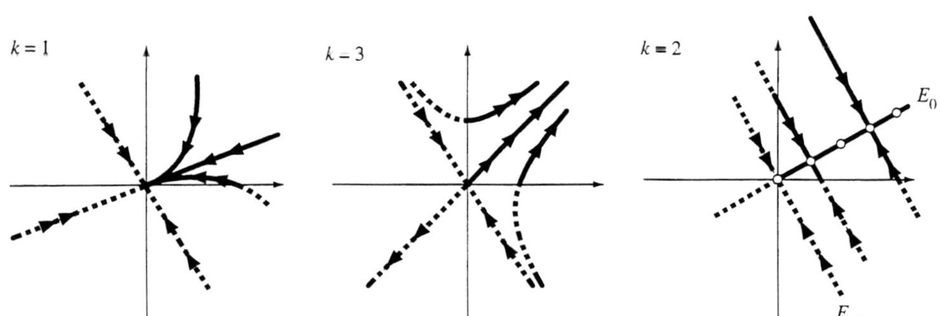

both species die out — both species prosper — system approaches an equilibrium state

48. a. Symbiosis

b. $\lambda_{1,2} = \frac{-5 \pm \sqrt{9+4k}}{2}$

Both eigenvalues are negative if $\sqrt{9+4k} < 5$ or $9+4k < 25$ or $4k < 16$ or $k < 4$.

If $k = 4$ then the eigenvalues are -5 and 0.

If $k > 4$ then there is a positive and a negative eigenvalue.

c. $k = 1$: See corresponding figure in Exercise 47.

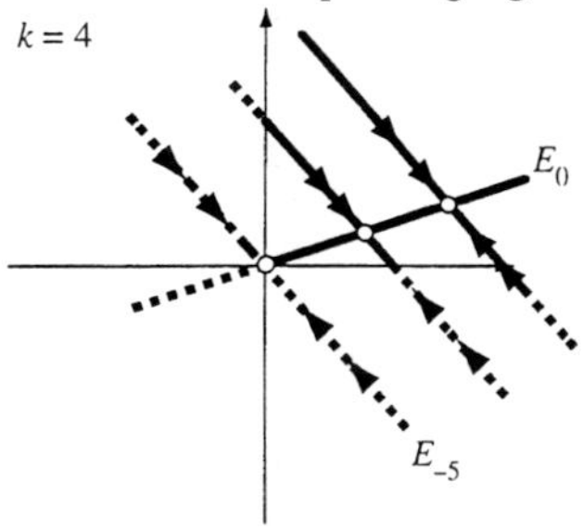

system approaches an equilibrium state

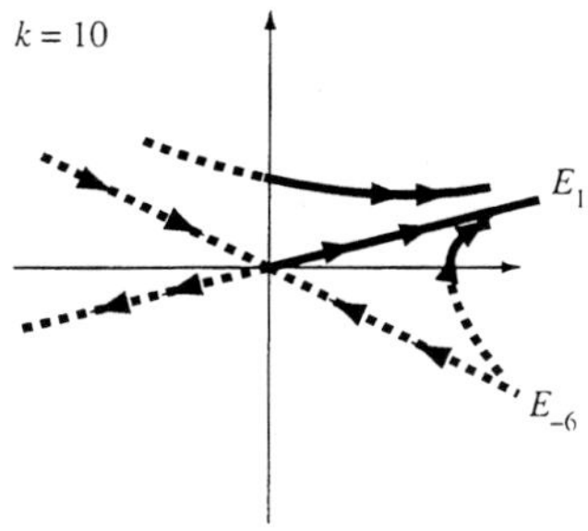

both species prosper

49. $A = \begin{bmatrix} -1 & -0.2 \\ 0.6 & -0.2 \end{bmatrix}$, $\lambda_1 = -0.4$, $\lambda_2 = -0.8$

$E_{-0.4} = \text{span}\begin{bmatrix} -1 \\ 3 \end{bmatrix}$, $E_{-0.8} = \text{span}\begin{bmatrix} 1 \\ -1 \end{bmatrix}$

$\begin{bmatrix} g(0) \\ h(0) \end{bmatrix} = 15\begin{bmatrix} -1 \\ 3 \end{bmatrix} + 45\begin{bmatrix} 1 \\ -1 \end{bmatrix}$, so that $c_1 = 15$, $c_2 = 45$.

$\begin{bmatrix} g(t) \\ h(t) \end{bmatrix} = 15e^{-0.4t}\begin{bmatrix} -1 \\ 3 \end{bmatrix} + 45e^{-0.8t}\begin{bmatrix} 1 \\ -1 \end{bmatrix}$, so that

$g(t) = -15e^{-0.4t} + 45e^{-0.8t}$

$h(t) = 45e^{-0.4t} - 45e^{-0.8t}$.

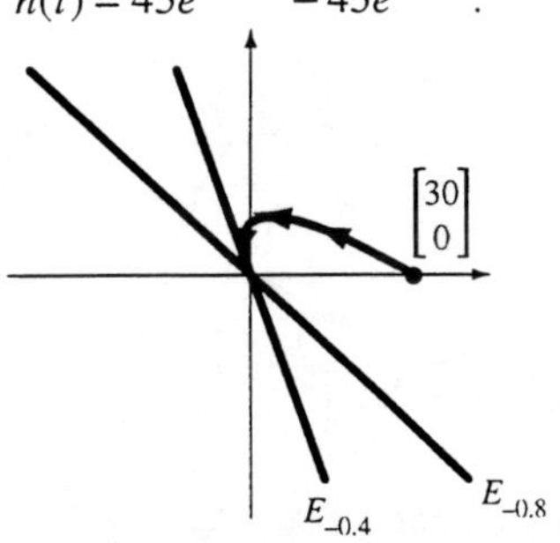

50. We want both eigenvalues λ_1 and λ_2 to be negative, so that $\text{tr}(A) = \lambda_1 + \lambda_2 < 0$ and $\det(A) = \lambda_1\lambda_2 > 0$.

Conversely, if $\text{tr}(A) < 0$ and $\det(A) > 0$, then the two eigenvalues $\lambda_{1,2} = \dfrac{\text{tr}(A) \pm \sqrt{(\text{tr}\,A)^2 - 4\det(A)}}{2}$ are both negative.

Answer: $\text{tr}(A) < 0$ and $\det(A) > 0$.

51. ith component of $\dfrac{d}{dt}(S\vec{x}) = \dfrac{d}{dt}\left(s_{i1}x_1(t) + s_{i2}x_2(t) + \cdots + s_{in}x_n(t)\right)$

$= s_{i1}\dfrac{dx_1}{dt} + s_{i2}\dfrac{dx_2}{dt} + \cdots + s_{in}\dfrac{dx_n}{dt}$

$=$ ith component of $S\dfrac{d\vec{x}}{dt}$

52. The solutions of $\frac{d\vec{x}}{dt} = \begin{bmatrix} 0 & 1 \\ 0 & 0 \end{bmatrix}\vec{x}$ are of the form $\begin{bmatrix} p+qt \\ q \end{bmatrix}$, where $\vec{x}(0) = \begin{bmatrix} p \\ q \end{bmatrix}$, by Exercise 21.

Since $\begin{bmatrix} \lambda & 1 \\ 0 & \lambda \end{bmatrix} = \lambda I_2 + \begin{bmatrix} 0 & 1 \\ 0 & 0 \end{bmatrix}$, the solutions of the given system are of the form $\vec{x}(t) = e^{\lambda t}\begin{bmatrix} p+qt \\ q \end{bmatrix}$, by Exercise 24. The zero state is a stable equilibrium solution if and only if $\lambda < 0$. The case $\lambda = 0$ is discussed in Exercise 21.

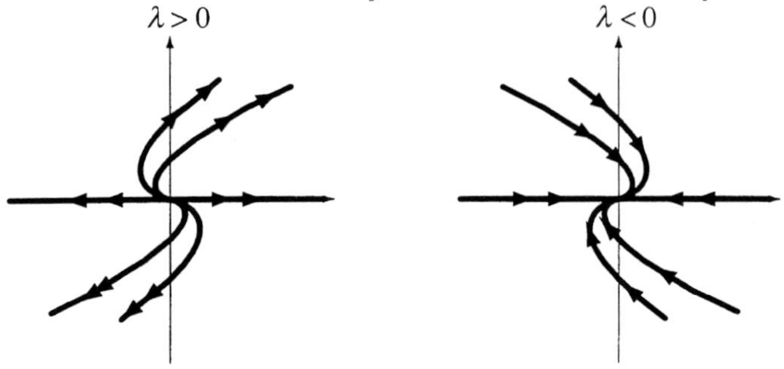

53. For the initial value $\begin{bmatrix} 1 \\ 0 \end{bmatrix}$, the system $\frac{d\vec{x}}{dt} = \begin{bmatrix} 0 & -1 \\ 1 & 0 \end{bmatrix}\vec{x}$ has the solution $\vec{x}(t) = \begin{bmatrix} \cos(t) \\ \sin(t) \end{bmatrix}$, by Exercise 20; the system $\frac{d\vec{x}}{dt} = \begin{bmatrix} 0 & -q \\ q & 0 \end{bmatrix}$ has the solution $\vec{x}(t) = \begin{bmatrix} \cos(qt) \\ \sin(qt) \end{bmatrix}$, by Exercise 25; and the system $\frac{d\vec{x}}{dt} = \begin{bmatrix} p & -q \\ q & p \end{bmatrix}\vec{x}$ has the solution $\vec{x}(t) = e^{pt}\begin{bmatrix} \cos(qt) \\ \sin(qt) \end{bmatrix}$, by Exercise 24 $\left(\text{write } \begin{bmatrix} p & -q \\ q & p \end{bmatrix} = pI_2 + \begin{bmatrix} 0 & -q \\ q & 0 \end{bmatrix}\right)$.

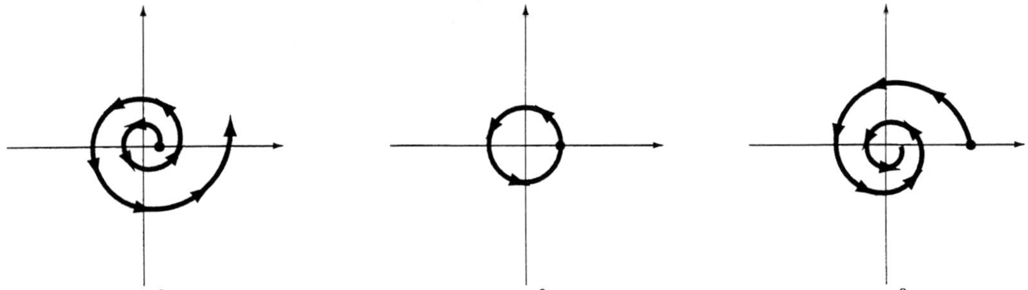

54. $A = \begin{bmatrix} 0 & 1 \\ -2 & -3 \end{bmatrix}$, $\lambda_1 = -1$, $\lambda_2 = -2$; $\vec{v}_1 = \begin{bmatrix} 1 \\ -1 \end{bmatrix}$ and $\vec{v}_2 = \begin{bmatrix} 1 \\ -2 \end{bmatrix}$.

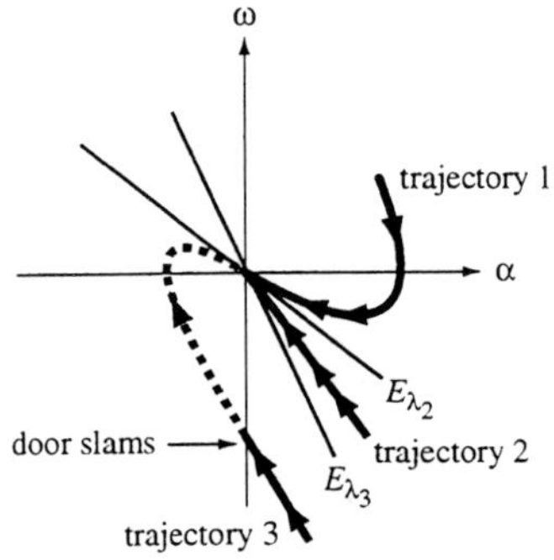

In the case of trajectory 3 the door will slam: Initially the door is opened just a little (α is small) and given a strong push to close it (ω is large negative). More generally, the door will slam if the point $\begin{bmatrix}\alpha(0)\\ \omega(0)\end{bmatrix}$ representing the initial state is located below the line $E_{-2} = \text{span}\begin{bmatrix}1\\ -2\end{bmatrix}$, that is, if $\dfrac{\omega(0)}{\alpha(0)} < -2$.

55. $A = \begin{bmatrix}0 & 1\\ -p & -q\end{bmatrix}$, $\lambda_{1,2} = \frac{1}{2}\left(-q \pm \sqrt{q^2 - 4p}\right)$; note that both eigenvalues are negative.

$E_{\lambda_1} = \text{span}\begin{bmatrix}1\\ \lambda_1\end{bmatrix}$ and $E_{\lambda_2} = \begin{bmatrix}1\\ \lambda_2\end{bmatrix}$.

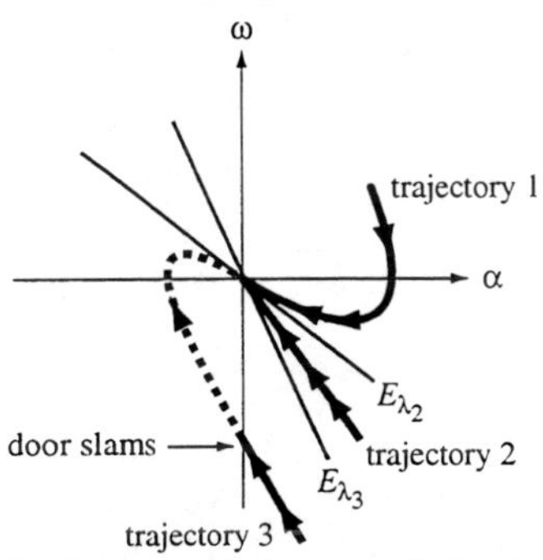

In the case of trajectory 3 the door will slam: Initially the door is opened just a little (α is small) and given a strong push to close it (ω is large negative). More generally, the door will slam if the point $\begin{bmatrix}\alpha(0)\\ \omega(0)\end{bmatrix}$ representing the initial state is located below the line $E_{\lambda_2} = \text{span}\begin{bmatrix}1\\ \lambda_2\end{bmatrix}$, that is, if $\dfrac{\omega(0)}{\alpha(0)} < \lambda_2$.

8.2

1. By Euler's formula (Fact 8.2.2), $e^{2\pi i} = \cos(2\pi) + i\sin(2\pi) = 1$.

2. By Euler's formula (Fact 8.2.2), $e^{\frac{1}{2}\pi i} = \cos\left(\frac{\pi}{2}\right) + i\sin\left(\frac{\pi}{2}\right) = i$.

3. 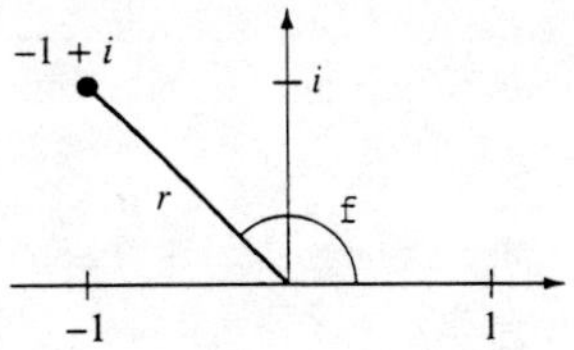

$r = \sqrt{(-1)^2 + 1^2} = \sqrt{2}$

$\phi = \dfrac{3\pi}{4}$, so that $z = \sqrt{2}e^{\frac{3}{4}\pi i}$.

4. $e^{3it} = \cos(3t) + i\sin(3t)$

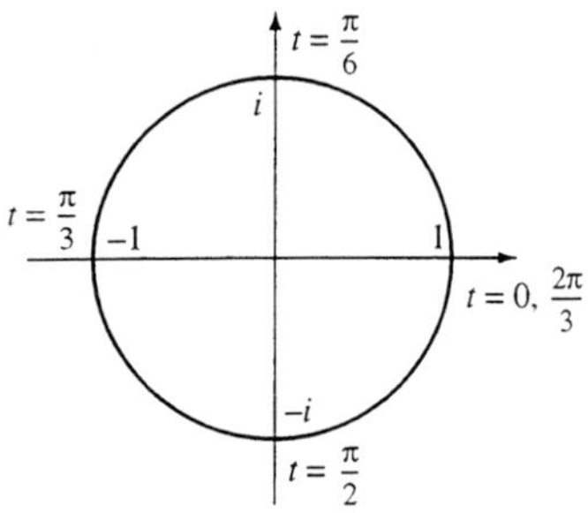

Period T given by $3T = 2\pi$ or $T = \frac{2\pi}{3}$.

5. $e^{-0.1t-2it} = e^{-0.1t}e^{-2it} = e^{-0.1t}(\cos(2t) - i\sin(2t))$
spirals inward, in clockwise direction

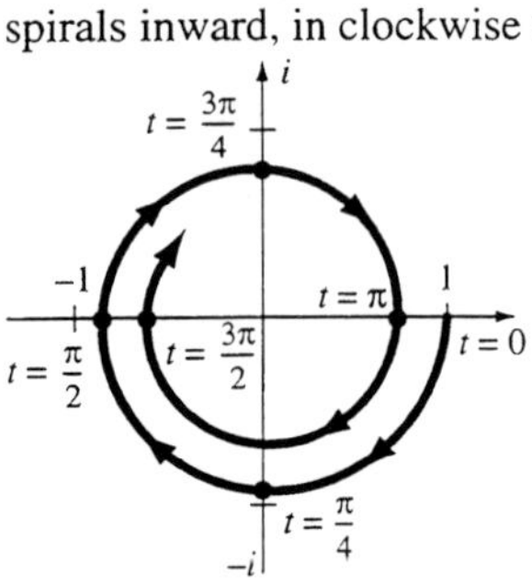

6. $\lambda_{1,2} = \pm i$; $E_i = \ker\begin{bmatrix} i-3 & 2 \\ -5 & i+3 \end{bmatrix} = \operatorname{span}\begin{bmatrix} 3+i \\ 5 \end{bmatrix}$ and $E_{-i} = \operatorname{span}\begin{bmatrix} 3-i \\ 5 \end{bmatrix}$.

General solution:

$\vec{x}(t) = c_1 e^{it}\begin{bmatrix} 3+i \\ 5 \end{bmatrix} + c_2 e^{-it}\begin{bmatrix} 3-i \\ 5 \end{bmatrix}$

If $c_1 = c_2 = 1$ then $\vec{x}(t) = (\cos(t) + i\sin(t))\begin{bmatrix} 3+i \\ 5 \end{bmatrix} + (\cos(t) - i\sin(t))\begin{bmatrix} 3-i \\ 5 \end{bmatrix} = \begin{bmatrix} 6\cos(t) - 2\sin(t) \\ 10\cos(t) \end{bmatrix}$.

7. $\det(A) = -2$, so that the zero state is not stable, by Fact 8.2.5.

8. Recall that all eigenvalues of a symmetric matrix are real. The zero state is stable if (and only if) all eigenvalues are negative (by Fact 8.2.4); this is the case if (and only if) the matrix is negative definite (by Fact 7.4.2).

9. The eigenvalues are conjugate complex, $\lambda_{1,2} = p \pm iq$, and $\operatorname{tr}(A) = 2p < 0$, so that p is negative. By Fact 8.2.4, the zero state is stable.

10. If $A = \begin{bmatrix} a & b \\ b & c \end{bmatrix}$ then $q(x_1, x_2) = ax_1^2 + 2bx_1x_2 + cx_2^2$, and the system takes the form $\left| \begin{array}{l} \frac{dx_1}{dt} = 2ax_1 + 2bx_2 \\ \frac{dx_2}{dt} = 2bx_1 + 2cx_2 \end{array} \right|$.

a. The matrix of the system is $\begin{bmatrix} 2a & 2b \\ 2b & 2c \end{bmatrix}$, which is $2A$, so that $\frac{d\vec{x}}{dt} = \operatorname{grad}(q) = 2A\vec{x}$.

b. By Fact 7.4.7, the level curves are ellipses. From multivariable calculus we know that grad(q) is perpendicular to the level curve, pointing inwards, so that all trajectories will approach the origin: The zero state is stable.

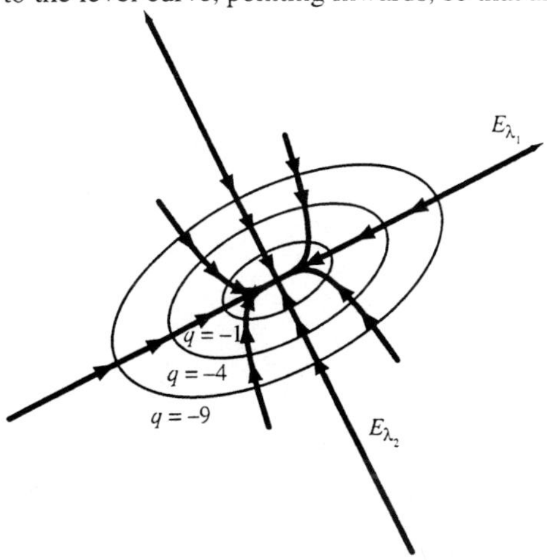

c. By Fact 7.4.7, the level curves are hyperbolas.

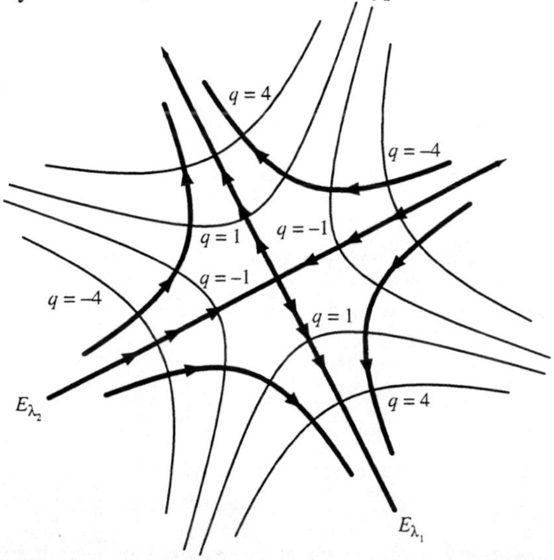

$\lambda_1 > 0 > \lambda_2$

d. The zero state is a stable equilibrium solution of the system $\frac{d\vec{x}}{dt} = \text{grad}(q) = 2A\vec{x}$ if (and only if) the eigenvalues of $2A$ (and A) are negative. This means that the quadratic form $q(\vec{x}) = \vec{x} \cdot A\vec{x}$ is negative definite.

11. a. $q(\vec{x}) = 2a_{i1}x_ix_1 + 2a_{i2}x_ix_2 + \cdots + a_{ii}x_i^2 + \cdots + 2a_{in}x_ix_n +$ terms not involving x_i, so that $\frac{\partial q}{\partial x_i} = 2a_{i1}x_1 + 2a_{i2}x_2 + \cdots + 2a_{ii}x_i + \cdots + 2a_{in}x_n$ and $\frac{d\vec{x}}{dt} = \text{grad}(q) = 2A\vec{x}$.

The matrix of the system is $B = 2A$.

d. The zero state is a stable equilibrium solution of the system $\frac{d\vec{x}}{dt} = \text{grad}(q) = 2A\vec{x}$ if (and only if) all the eigenvalues of $2A$ are negative. This means that the quadratic form $q(\vec{x}) = \vec{x} \cdot A\vec{x}$ is negative definite.

12. Yes, the zero state is stable. To see this, use technology to determine that the real parts of the eigenvalues are all negative. Or draw a rough sketch of the characteristic polynomial to see that there is one real eigenvalue λ_1 between –1 and –2; the two other eigenvalues must be $\lambda_{2,3} = p \pm iq$ with $p < 0$, since
$\lambda_1 + \lambda_2 + \lambda_3 = \lambda_1 + 2p = \text{tr}(A) = -2.$

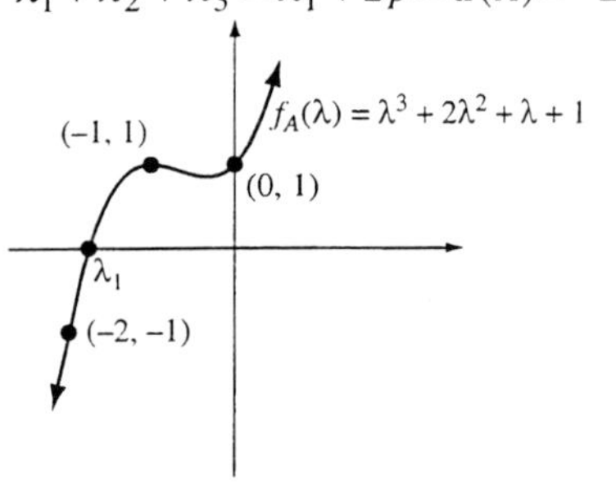

13. Recall that the zero state is stable if (and only if) the real parts of all eigenvalues are negative. Now the eigenvalues of A^{-1} are the reciprocals of those of A; the real parts have the same sign
$\left(\text{if } \lambda = p + iq, \text{ then } \frac{1}{\lambda} = \frac{1}{p+iq} = \frac{p-iq}{p^2+q^2}\right).$

14. a. For $i > 1$, $\frac{dx_i}{dt} = -k_i x_i + x_{i-1}$. This means that in the absence of quantity $x_{i-1}(t)$, the quantity $x_i(t)$ will decay exponentially, but the presence of x_{i-1} helps x_i to grow.

For $i = 1$, the beginning of the loop, $\frac{dx_1}{dt} = -k_1 x_1 - bx_n$, so that the presence of x_n contributes to the decrease of x_1.

b. If $n = 2$ then the matrix of the system is $A = \begin{bmatrix} -k_1 & -b \\ 1 & -k_2 \end{bmatrix}$ with $\text{tr}(A) = -k_1 - k_2 < 0$ and $\det(A) = k_1k_2 + b > 0$, so that the zero state is stable, by Fact 8.2.5.

c. No; consider the case $k_1 = k_2 = k_3 = 1$ for simplicity; then the matrix of the system is $A = \begin{bmatrix} -1 & 0 & b \\ 1 & -1 & 0 \\ 0 & 1 & -1 \end{bmatrix}$ and $f_A(\lambda) = (\lambda + 1)^3 - b$. If b exceeds 1, then the matrix A will have a positive eigenvalue, so that the zero state is not stable.

15. The eigenvalues are $\lambda_1 = \text{tr}(A) > 0$ and $\lambda_2 = 0$.

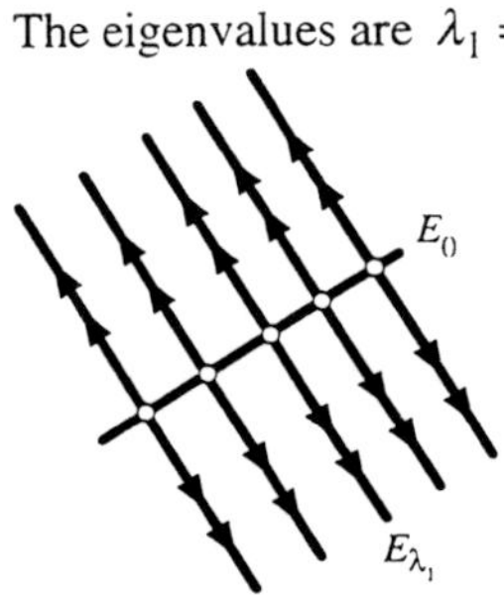

16. If $A = \begin{bmatrix} 0 & 1 \\ a & b \end{bmatrix}$ then $\text{tr}(A) = b$ and $\det(A) = -a$. By Fact 8.2.5, the zero state is stable if a and b are both negative.

17. If $A=\begin{bmatrix}-1 & k\\ k & -1\end{bmatrix}$ then $\operatorname{tr}(A)=-2$ and $\det(A)=1-k^2$. By Fact 8.2.5, the zero state is stable if $\det(A)=1-k^2>0$, that is, if $|k|<1$.

18. If $\lambda_1, \lambda_2, \lambda_3$ are real and negative, then $\operatorname{tr}(A)=\lambda_1+\lambda_2+\lambda_3<0$ and $\det(A)=\lambda_1\lambda_2\lambda_3<0$.
If λ_1 is real and negative and $\lambda_{2,3}=p\pm iq$, where p is negative, then $\operatorname{tr}(A)=\lambda_1+2p<0$ and $\det(A)=\lambda_1(p^2+q^2)<0$. Either way, both trace and determinant are negative.

19. False, consider $A=\begin{bmatrix}1 & 0 & 0\\ 0 & 2 & 0\\ 0 & 0 & -4\end{bmatrix}$.

20. Use Fact 8.2.6, with $p=0$, $q=\pi$; $a=1$, $b=0$.

$$\vec{x}(t)=[\vec{w}\ \ \vec{v}]\begin{bmatrix}\cos(\pi t) & -\sin(\pi t)\\ \sin(\pi t) & \cos(\pi t)\end{bmatrix}\begin{bmatrix}1\\0\end{bmatrix}=(\cos(\pi t))\vec{w}+(\sin(\pi t))\vec{v}.$$

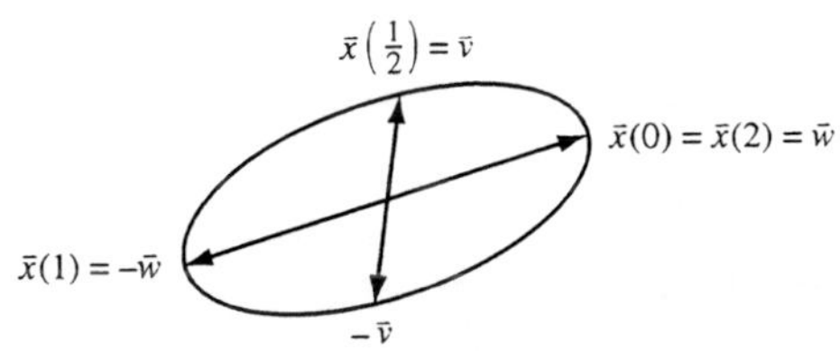

21. a. $\left|\begin{array}{l}\dfrac{db}{dt}=0.05b+s\\ \dfrac{ds}{dt}=\quad\quad 0.07s\end{array}\right|$ and $\begin{bmatrix}b(0)\\ s(0)\end{bmatrix}=\begin{bmatrix}1000\\1000\end{bmatrix}$

b. $\lambda_1=0.07$, $\lambda_2=0.05$; $\vec{v}_1=\begin{bmatrix}50\\1\end{bmatrix}$, $\vec{v}_2=\begin{bmatrix}1\\0\end{bmatrix}$; $\vec{x}(0)=1000\vec{v}_1-49000\vec{v}_2$; so that

$b(t)=50000e^{0.07t}-49000e^{0.05t}$ and $s(t)=1000e^{0.07t}$.

22. a. $\lambda_1=3$, $\lambda_2=0.5$; $E_3=\operatorname{span}\begin{bmatrix}1\\-1\end{bmatrix}$, $E_{0.5}=\operatorname{span}\begin{bmatrix}0\\1\end{bmatrix}$
System is discrete so choose VII.

b. $\lambda_{1,2}=-\frac{1}{2}\pm i$, $r>1$, so that trajectory spirals outwards. Choose II.

c. $\lambda_1=3$, $\lambda_2=0.5$, $E_3=\begin{bmatrix}1\\-1\end{bmatrix}$, $E_{0.5}=\begin{bmatrix}0\\1\end{bmatrix}$.
System is continuous, so choose I.

d. $\lambda_{1,2}=-\frac{1}{2}\pm i$; real part is negative so that trajectories spiral inwards in the counterclockwise direction $\left(\text{if } \vec{x}=\begin{bmatrix}1\\0\end{bmatrix} \text{ then } \dfrac{d\vec{x}}{dt}=\begin{bmatrix}-1.5\\2\end{bmatrix}\right)$. Choose IV.

e. $\lambda_1 = 1$, $\lambda_2 = -2$; $E_1 = \text{span}\begin{bmatrix}0\\1\end{bmatrix}$, $E_{-2} = \text{span}\begin{bmatrix}1\\-1\end{bmatrix}$.

System is continuous so choose V.

23. $\lambda_{1,2} = \pm 3i$, $E_{3i} = \text{span}\left(\begin{bmatrix}1\\0\end{bmatrix} + i\begin{bmatrix}0\\-1\end{bmatrix}\right)$, so that $p = 0$, $q = 3$, $\vec{w} = \begin{bmatrix}0\\-1\end{bmatrix}$, $\vec{v} = \begin{bmatrix}1\\0\end{bmatrix}$.

Now use Fact 8.2.6:

$$\vec{x}(t) = e^{0t}\begin{bmatrix}0 & 1\\-1 & 0\end{bmatrix}\begin{bmatrix}\cos(3t) & -\sin(3t)\\ \sin(3t) & \cos(3t)\end{bmatrix}\begin{bmatrix}a\\b\end{bmatrix} = \begin{bmatrix}\sin(3t) & \cos(3t)\\ -\cos(3t) & \sin(3t)\end{bmatrix}\begin{bmatrix}a\\b\end{bmatrix}$$

There are other ways to write these solutions; for an example see page A-21 of the text.

24. $\lambda_{1,2} = \pm 6i$, $E_{6i} = \text{span}\left(\begin{bmatrix}2\\0\end{bmatrix} + i\begin{bmatrix}0\\3\end{bmatrix}\right)$, so that

$$\vec{x}(t) = \begin{bmatrix}0 & 2\\3 & 0\end{bmatrix}\begin{bmatrix}\cos(6t) & -\sin(6t)\\ \sin(6t) & \cos(6t)\end{bmatrix}\begin{bmatrix}a\\b\end{bmatrix} = \begin{bmatrix}2\sin(6t) & 2\cos(6t)\\ 3\cos(6t) & -3\sin(6t)\end{bmatrix}\begin{bmatrix}a\\b\end{bmatrix}.$$

25. $\lambda_{1,2} = 2 \pm 4i$, $E_{2+4i} = \text{span}\left(\begin{bmatrix}1\\0\end{bmatrix} + i\begin{bmatrix}0\\1\end{bmatrix}\right)$, so that

$$\vec{x}(t) = e^{2t}\begin{bmatrix}0 & 1\\1 & 0\end{bmatrix}\begin{bmatrix}\cos(4t) & -\sin(4t)\\ \sin(4t) & \cos(4t)\end{bmatrix}\begin{bmatrix}a\\b\end{bmatrix} = e^{2t}\begin{bmatrix}\sin(4t) & \cos(4t)\\ \cos(4t) & -\sin(4t)\end{bmatrix}\begin{bmatrix}a\\b\end{bmatrix}.$$

See page A-22 of the text for another representation of the answer.

26. $\lambda_{1,2} = -2 \pm 3i$, $E_{-2+3i} = \text{span}\left(\begin{bmatrix}5\\3\end{bmatrix} + i\begin{bmatrix}0\\1\end{bmatrix}\right)$, so that

$$\vec{x}(t) = e^{-2t}\begin{bmatrix}0 & 5\\1 & 3\end{bmatrix}\begin{bmatrix}\cos(3t) & -\sin(3t)\\ \sin(3t) & \cos(3t)\end{bmatrix}\begin{bmatrix}a\\b\end{bmatrix} = e^{-2t}\begin{bmatrix}5\sin(3t) & 5\cos(3t)\\ \cos(3t)+3\sin(3t) & -\sin(3t)+3\cos(3t)\end{bmatrix}\begin{bmatrix}a\\b\end{bmatrix}.$$

27. $\lambda_{1,2} = -1 \pm 2i$, $E_{-1+2i} = \text{span}\left(\begin{bmatrix}1\\0\end{bmatrix} + i\begin{bmatrix}0\\-1\end{bmatrix}\right)$, so that $p = -1$, $q = 2$, $\vec{w} = \begin{bmatrix}0\\-1\end{bmatrix}$, $\vec{v} = \begin{bmatrix}1\\0\end{bmatrix}$.

Now $\begin{bmatrix}1\\-1\end{bmatrix} = \vec{x}(0) = \vec{w} + \vec{v}$, so that $a = 1$ and $b = 1$.

Then $\vec{x}(t) = e^{-t}\begin{bmatrix}0 & 1\\-1 & 0\end{bmatrix}\begin{bmatrix}\cos(2t) & -\sin(2t)\\ \sin(2t) & \cos(2t)\end{bmatrix}\begin{bmatrix}1\\1\end{bmatrix} = e^{-t}\begin{bmatrix}\sin(2t)+\cos(2t)\\ \sin(2t)-\cos(2t)\end{bmatrix}$.

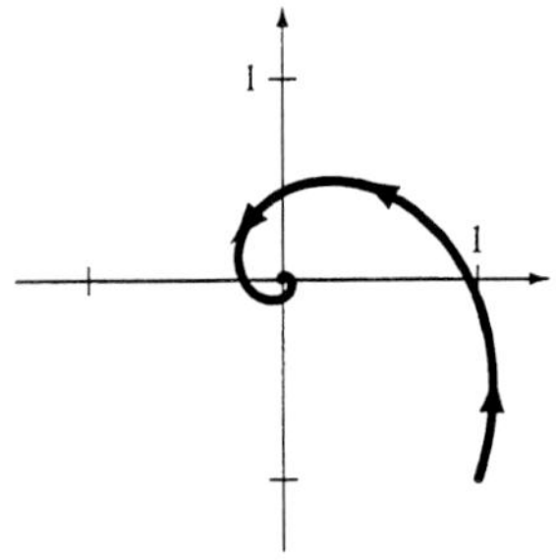

28. $\lambda_{1,2} = \pm 2i$, $E_{2i} = \text{span}\left(\begin{bmatrix}1\\0\end{bmatrix} + i\begin{bmatrix}0\\2\end{bmatrix}\right)$. $\vec{x}(0) = 0\begin{bmatrix}0\\2\end{bmatrix} + 1\begin{bmatrix}1\\0\end{bmatrix}$, so that $a = 0$ and $b = 1$.

$$\vec{x}(t) = \begin{bmatrix}0 & 1\\2 & 0\end{bmatrix}\begin{bmatrix}\cos(2t) & -\sin(2t)\\ \sin(2t) & \cos(2t)\end{bmatrix}\begin{bmatrix}0\\1\end{bmatrix} = \begin{bmatrix}\cos(2t)\\-2\sin(2t)\end{bmatrix}$$

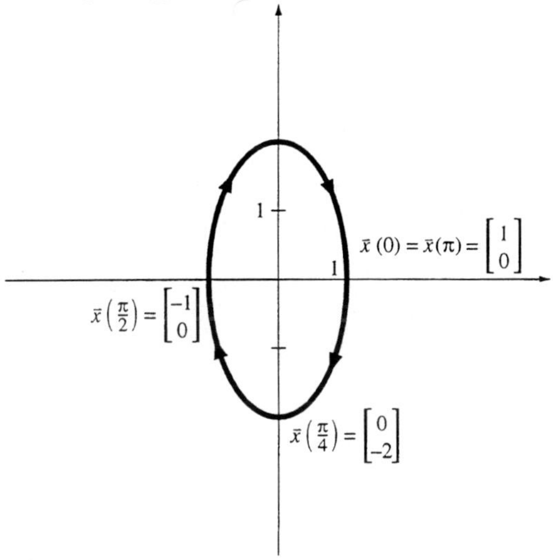

29. $\lambda_{1,2} = \pm i$, $E_i = \text{span}\left(\begin{bmatrix}1\\1\end{bmatrix} + i\begin{bmatrix}0\\1\end{bmatrix}\right)$

$a = 1$, $b = 0$, so that $\vec{x}(t) = \begin{bmatrix}0 & 1\\1 & 1\end{bmatrix}\begin{bmatrix}\cos(t) & -\cos(t)\\ \sin(t) & \cos(t)\end{bmatrix}\begin{bmatrix}1\\0\end{bmatrix} = \begin{bmatrix}\sin(t)\\ \sin(t) + \cos(t)\end{bmatrix} = \cos(t)\begin{bmatrix}0\\1\end{bmatrix} + \sin(t)\begin{bmatrix}1\\1\end{bmatrix}$.

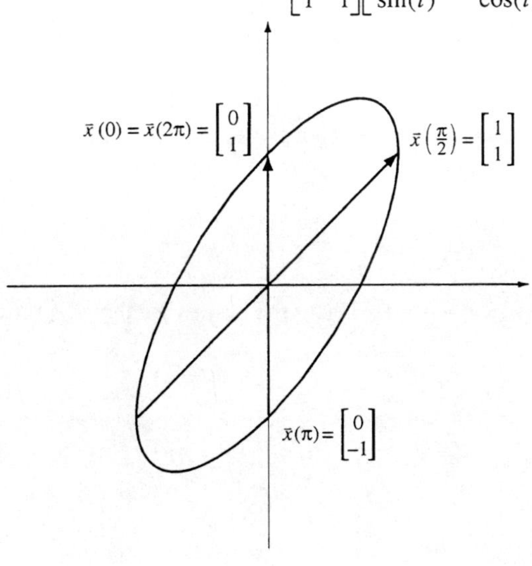

30. $\lambda_{1,2} = 1 \pm 2i$, $E_{1+2i} = \text{span}\left(\begin{bmatrix} 3 \\ -2 \end{bmatrix} + i\begin{bmatrix} 1 \\ 0 \end{bmatrix}\right)$

$a = 1$, $b = 0$, so that $\vec{x}(t) = e^t \begin{bmatrix} 1 & 3 \\ 0 & -2 \end{bmatrix}\begin{bmatrix} \cos(2t) & -\sin(2t) \\ \sin(2t) & \cos(2t) \end{bmatrix}\begin{bmatrix} 1 \\ 0 \end{bmatrix} = e^t \begin{bmatrix} \cos(2t) + 3\sin(2t) \\ -2\sin(2t) \end{bmatrix}$.

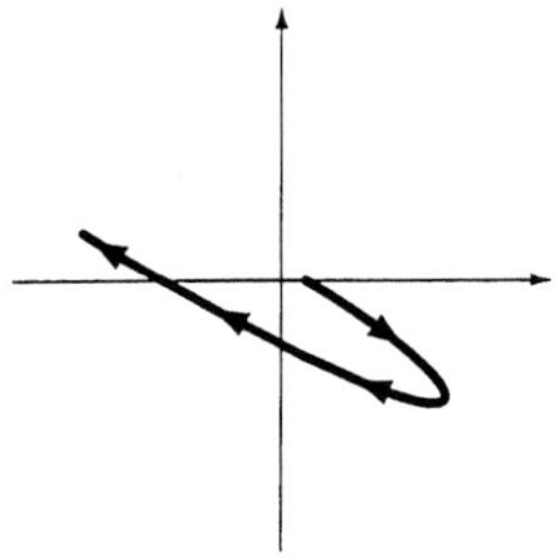

31. $A = \begin{bmatrix} 0 & 1 \\ -b & -c \end{bmatrix}$ and $f_A(\lambda) = \lambda^2 + c\lambda + b$, with eigenvalues $\lambda_{1,2} = \dfrac{-c \pm \sqrt{c^2 - 4b}}{2}$.

a. If $c = 0$ then $\lambda_{1,2} = \pm i\sqrt{b}$. The trajectories are ellipses.

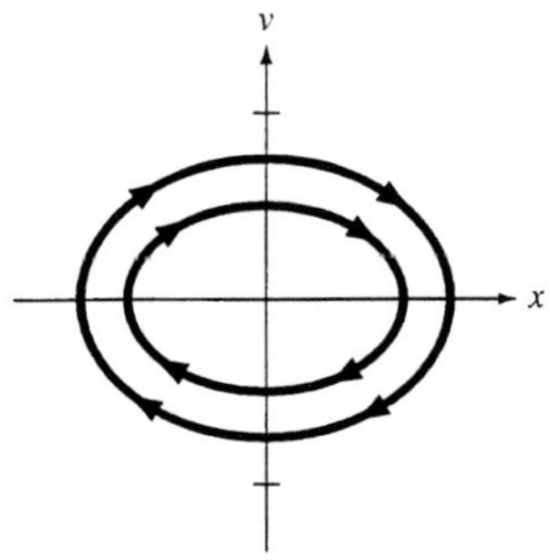

The block *oscillates harmonically*, with period $\dfrac{2\pi}{\sqrt{b}}$. The zero state is not asymptotically stable.

b. $\lambda_{1,2} = \dfrac{-c \pm i\sqrt{4b - c^2}}{2}$

The trajectories spiral inwards, since $\text{Re}(\lambda_1) = \text{Re}(\lambda_2) = -\dfrac{c}{2} < 0$. This is the case of a *damped oscillation*. The zero state is asymptotically stable.

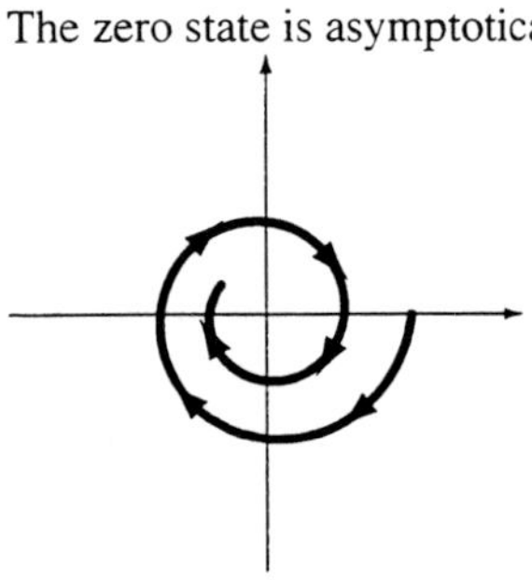

c. This case is discussed in Exercise 8.1.55. The zero state is stable here.

32. If $z = f + ig$ and $w = p + iq$ then $zw = (fp - gq) + i(fq + gp)$, so that
$(zw)' = (f'p + fp' - g'q - gq') + i(f'q + fq' + g'p + gp')$.
Also $z'w = (f' + ig')(p + iq) = (f'p - g'q) + i(f'q + g'p)$ and $zw' = (f + ig)(p' + iq') = (fp' - gq') + i(gp' + fq')$.
We can see that $(zw)' = z'w + zw'$, as claimed.

33. a. $\dfrac{1}{z(t)}$ is differentiable when $z(t) \neq 0$, since both the real and the imaginary parts are differentiable $\left(\text{if } z = p + iq \text{ then } \dfrac{1}{z} = \dfrac{p - iq}{p^2 + q^2}\right)$. To find $\left(\dfrac{1}{z}\right)'$, apply the product rule to the equation $z\left(\dfrac{1}{z}\right) = 1$:

$$z'\left(\frac{1}{z}\right) + z\left(\frac{1}{z}\right)' = 0, \text{ so that } \left(\frac{1}{z}\right)' = -\frac{z'}{z^2}.$$

b. $\left(\dfrac{z}{w}\right)' = \left(z\dfrac{1}{w}\right)' = z'\dfrac{1}{w} + z\left(\dfrac{1}{w}\right)' = \dfrac{z'}{w} - \dfrac{zw'}{w'} = \dfrac{z'w - zw'}{w^2}$

34. a. $\left(\dfrac{z_1}{z_2}\right)' = \dfrac{z_1'z_2 - z_1z_2'}{z_2^2} = \dfrac{\lambda z_1z_2 - \lambda z_1z_2}{z_2^2} = 0$, so that $\dfrac{z_1(t)}{z_2(t)} = k$, a constant.
Now $z_1(t) = kz_2(t)$; substituting $t = 0$ gives $1 = z_1(0) = kz_2(0) = k$, so that $z_1(t) = z_2(t)$, as claimed.

b. Let $z_2(t) = e^{pt}(\cos(qt) + i\sin(q + t))$ be the solution constructed in the text (page 461). Since $z_2(t) \neq 0$ for all t, this is the only solution, by part a.

35. If $\vec{x}(t) = \vec{x}_0 + tA\vec{x}_0$ then $\dfrac{d\vec{x}}{dt} = A\vec{x}_0$ equals $A\vec{x} = A\vec{x}_0 + t(A^2\vec{x}_0) = A\vec{x}_0$ and $\vec{x}(0) = \vec{x}_0$, as claimed.
For any $\vec{x}$, $A\vec{x}$ is in ker(A) since $A(A\vec{x}) = A^2\vec{x} = \vec{0}$.

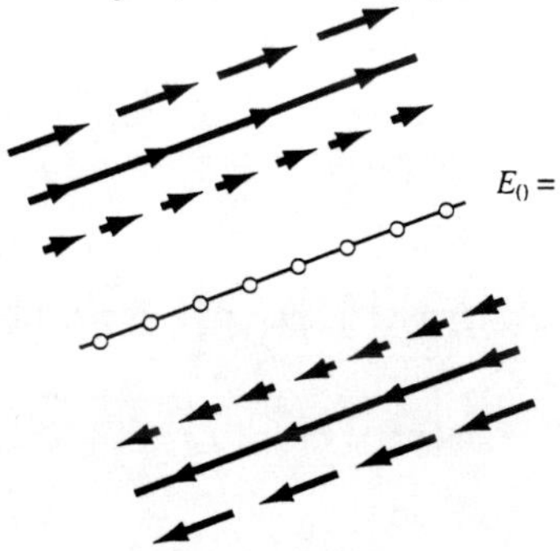

36. By Exercise 36, the solution is $\vec{x}(t) = \begin{bmatrix} 0 \\ 1 \end{bmatrix} + t\begin{bmatrix} 1 & 1 \\ -1 & -1 \end{bmatrix}\begin{bmatrix} 0 \\ 1 \end{bmatrix} = \begin{bmatrix} t \\ 1 - t \end{bmatrix}$.

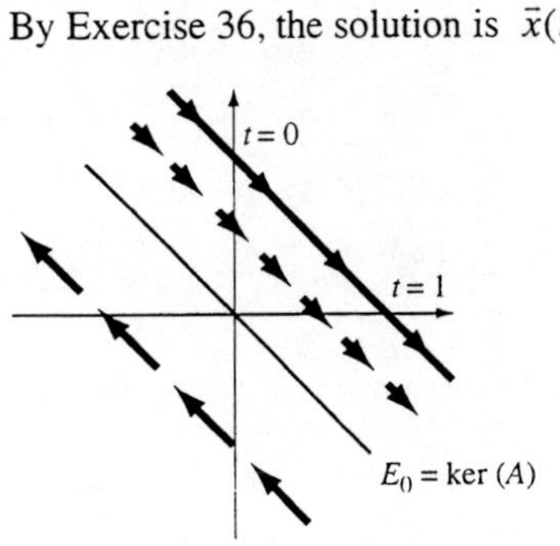

37. Since A has eigenvalues λ_0, λ_0, the matrix $(A-\lambda_0 I_2)$ has the eigenvalues 0, 0, and we can use Exercise 35 to solve the system $\frac{d\vec{c}}{dt}=(A-\lambda_0 I_2)\vec{c}$. The solutions are $\vec{c}(t)=\vec{x}_0+t(A-\lambda_0 I_2)\vec{x}_0$.

Applying Exercise 8.1.24, with $k=-\lambda_0$, we find that $\vec{c}(t)=e^{-\lambda_0 t}\vec{x}(t)$ or

$\vec{x}(t)=e^{\lambda_0 t}\vec{c}(t)=e^{\lambda_0 t}\left(\vec{x}_0+t(A-\lambda_0 I_2)\vec{x}_0\right)$.

The zero state is stable if (and only if) λ_0 is negative.

38. The eigenvalues are 3, 3. By Exercise 37, $\vec{x}(t)=e^{3t}\left(\begin{bmatrix}0\\1\end{bmatrix}+t\begin{bmatrix}-2&4\\-1&2\end{bmatrix}\begin{bmatrix}0\\1\end{bmatrix}\right)=e^{3t}\begin{bmatrix}4t\\1+2t\end{bmatrix}$.

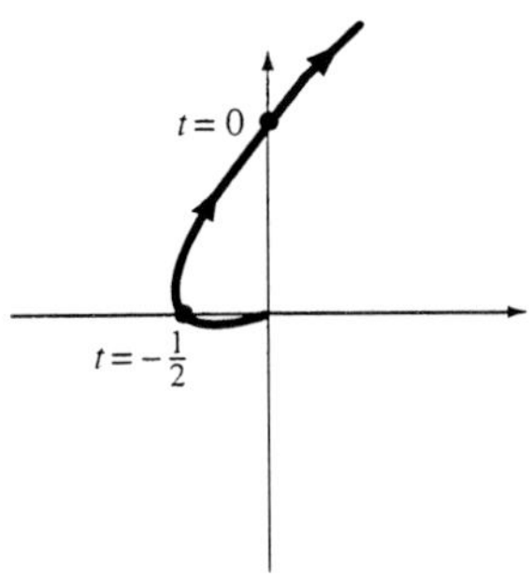

39. Let $A=\begin{bmatrix}\lambda&1&0\\0&\lambda&1\\0&0&\lambda\end{bmatrix}$. We first solve the system $\frac{d\vec{c}}{dt}=(A-\lambda I_3)\vec{c}=\begin{bmatrix}0&1&0\\0&0&1\\0&0&0\end{bmatrix}\vec{c}$, or

$\frac{dc_1}{dt}=c_2(t),\ \frac{dc_2}{dt}=c_3(t),\ \frac{dc_3}{dt}=0.$

$c_3(t)=k_3$, a constant, so that $\frac{dc_2}{dt}=k_3$ and $c_2(t)=k_3t+k_2$. Likewise $c_1(t)=\frac{k_3}{2}t^2+k_2t+k_1$.

Applying Exercise 8.1.24, with $k=-\lambda$, we find that $\vec{c}(t)=e^{-\lambda t}\vec{x}(t)$ or $\vec{x}(t)=e^{\lambda t}\vec{c}(t)=e^{\lambda t}\begin{bmatrix}k_1+k_2t+\frac{k_3}{2}t^2\\k_2+k_3t\\k_3\end{bmatrix}$

where k_1, k_2, k_3 are arbitrary constants. The zero state is stable if (and only if) the real part of λ is negative.

40. **a.** $B(t)=1000(1+0.05i)^t=1000(r(\cos\phi+i\sin\phi))^t=1000r^t(\cos(\phi t)+i\sin(\phi t))$, where $r=\sqrt{1+0.05^2}>1$ and $\phi=\arctan(0.05)\approx 0.05$.

b. $B(t) = 1000e^{0.05i} = 1000(\cos(0.05t) + i\sin(0.05t))$

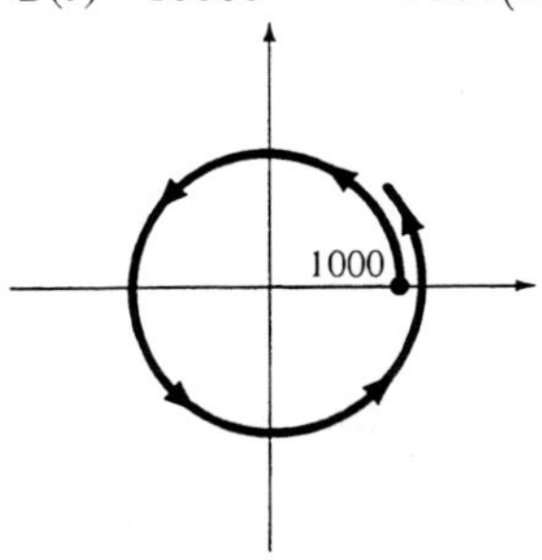

trajectory to part a
Slowly spiral outwards,

with $B(126) \approx 1170 + 14i$

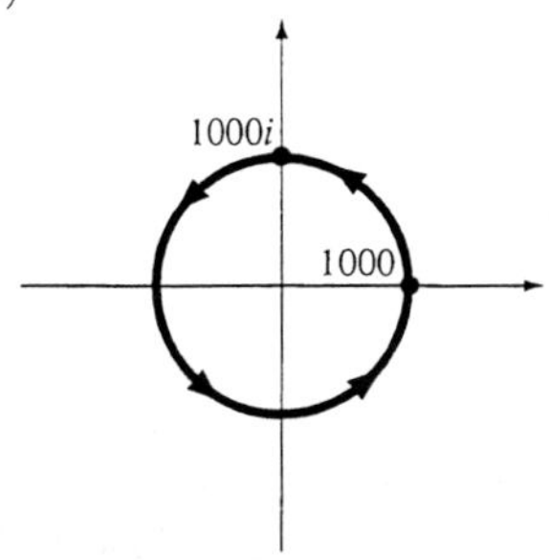

trajectory to part b
circle

period $= \dfrac{2\pi}{0.05} \approx 126$ (years)

c. We would choose an account with annual compounding, since the modulus of the balance grows in this case. In the case of continuous compounding the modulus of the balance remains unchanged.

Chapter 9

9.1

1. Not a subspace since it does not contain the neutral element, that is, the function $f(t) = 0$, for all t.

2. This subset V is a subspace of P_3:
   - The neutral element $f(t) = 0$ (for all t) is in V.
   - If f and g are in V (so that $f(2) = g(2) = 0$), then $(f + g)(2) = f(2) + g(2) = 0 + 0 = 0$, so that $f + g$ is in V.
   - If f is in V (so that $f(2) = 0$), and k is any constant, then $(kf)(2) = kf(2) = 0$, so that kf is in V.

   A polynomial $f(t) = a + bt + ct^2 + dt^3$ is in V if $f(2) = a + 2b + 4c + 8d = 0$, or $a = -2b - 4c - 8d$. The general element of V is of the form $f(t) = (-2b - 4c - 8d) + bt + ct^2 + dt^3 = b(t-2) + c(t^2 - 4) + d(t^3 - 8)$, so that $t - 2,\ t^2 - 4,\ t^3 - 8$ is a basis of V.

3. This subset V is a subspace of P_3:
   - The neutral element $f(t) = 0$ (for all t) is in V since $f'(1) = f(2) = 0$.
   - If f and g are in V (so that $f'(1) = f(2)$ and $g'(1) = g(2)$), then $(f+g)'(1) = (f'+g')(1) = f'(1) + g'(1) = f(2) + g(2) = (f+g)(2)$, so that $f + g$ is in V.
   - If f is in V (so that $f'(1) = f(2)$) and k is any constant, then $(kf)'(1) = (kf')(1) = kf'(1) = kf(2) = (kf)(2)$, so that kf is in V.

   If $f(t) = a + bt + ct^2 + dt^3$ then $f'(t) = b + 2ct + 3dt^2$, and f is in V if $f'(1) = b + 2c + 3d = a + 2b + 4c + 8d = f(2)$, or $a + b + 2c + 5d = 0$. The general element of V is of the form $f(t) = (-b - 2c - 5d) + bt + ct^2 + dt^3 = b(t-1) + c(t^2 - 2) + d(t^3 - 5)$, so that $t - 1,\ t^2 - 2,\ t^3 - 5$ is a basis of V.

4. This subset V is a subspace of P_3:
   - The neutral element $f(t) = 0$ (for all t) is in V since $\int_0^1 0\,dt = 0$.
   - If f and g are in V $\left(\text{so that } \int_0^1 f = \int_0^1 g = 0\right)$, then $\int_0^1 (f+g) = \int_0^1 f + \int_0^1 g = 0$, so that $f + g$ in V.
   - If f is in V $\left(\text{so that } \int_0^1 f = 0\right)$ and k is any constant, then $\int_0^1 kf = k\int_0^1 f = 0$, so that kf is in V.

   If $f(t) = a + bt + ct^2 + dt^3$ then $\int_0^1 f(t)dt = \left[at + \frac{b}{2}t^2 + \frac{c}{3}t^3 + \frac{d}{4}t^4\right]_0^1 = a + \frac{b}{2} + \frac{c}{3} + \frac{d}{4} = 0$ if $a = -\frac{b}{2} - \frac{c}{3} - \frac{d}{4}$.

   The general element of V is $f(t) = \left(-\frac{b}{2} - \frac{c}{3} - \frac{d}{4}\right) + bt + ct^2 + dt^3 = b\left(t - \frac{1}{2}\right) + c\left(t^2 - \frac{1}{3}\right) + d\left(t^3 - \frac{1}{4}\right)$, so that $t - \frac{1}{2},\ t^2 - \frac{1}{3},\ t^3 - \frac{1}{4}$ is a basis of V.

5. If $p|t| = a + bt + ct^2 + dt^3$ then $p(-t) = a - bt + ct^2 - dt^3$ and $-p(-t) = -a + bt - ct^2 + dt^3$.
   Comparing coefficients we see that $p(t) = -p(-t)$ for all t if (and only if) $a = c = 0$.
   The general element of the subset is of the form $p(t) = bt + dt^3$.
   These polynomials form a subspace of P_3, with basis $t,\ t_3$.

6. Not a subspace, since I_3 and $-I_3$ are invertible, but their sum is not.

7. The set V of all symmetric 3×3 matrices is a subspace of $\mathbb{R}^{3\times3}$:
- The zero matrix is symmetric.
- If A and B are symmetric (that is, $A^T = A$ and $B^T = B$), then $(A+B)^T = A^T + B^T = A + B$, so that $A + B$ is symmetric as well.
- If A is symmetric (that is, $A^T = A$), and k is any constant, then $(kA)^T = kA^T = kA$, so that kA is symmetric as well.

8. The set V of all skew-symmetric 3×3 matrices is a subspace of $\mathbb{R}^{3\times3}$:
- The zero matrix is skew-symmetric.
- If A and B are skew-symmetric (that is, $A^T = -A$ and $B^T = -B$), then $(A+B)^T = A^T + B^T = -A - B = -(A+B)$, so that $(A + B)$ is skew-symmetric as well.
- If A is skew-symmetric (that is, $A^T = -A$) and k is any constant, then $(kA)^T = kA^T = k(-A) = -(kA)$, so that kA is skew-symmetric as well.

9. Yes, the zero matrix is diagonal, the sum of two diagonal matrices is diagonal, and so is any scalar multiple of a diagonal matrix.

10. Yes: Let $\vec{v} = \begin{bmatrix} 1 \\ 2 \\ 3 \end{bmatrix}$ and let V be the set of all 3×3 matrices for which $\vec{v}$ is an eigenvector. Then
- the zero matrix is in V.
- if A and B are in V (that is, $A\vec{v} = \lambda\vec{v}$ and $B\vec{v} = \mu\vec{v}$, for some λ and μ), then $(A+B)\vec{v} = A\vec{v} + B\vec{v} = \lambda\vec{v} + \mu\vec{v} = (\lambda+\mu)\vec{v}$, so that $A + B$ is in V.
- If A is in V (that is, $A\vec{v} = \lambda\vec{v}$ for some λ), then $(kA)\vec{v} = k(A\vec{v}) = k(\lambda\vec{v}) = (k\lambda)\vec{v}$, so that kA is in V, for any scalar k.

11. Not a subspace: I_3 is in rref, but $2I_3$ is not.

12. Yes, the set W of all arithmetic sequences is a subspace of V. Use the fact that a sequence $(x_0, x_1, x_2, \ldots)$ is arithmetic if $x_n = x_0 + kn$ for some constant k.
- The sequence $(0, 0, 0, \ldots)$ is an arithmetic sequence, with $k = 0$.
- If (x_n) and (y_n) are arithmetic sequences (with $x_n = x_0 + p_n$ and $y_n = y_0 + q_n$), then $x_n + y_n = x_0 + y_0 + (p+q)n$, so that $(x_n + y_n)$ is an arithmetic sequence as well.
- If (x_n) is an arithmetic sequence (with $x_n = x_0 + pn$) and k is an arbitrary constant, then $kx_n = kx_0 = (kp)n$, so that (kx_n) is an arithmetic sequence as well.

13. Not a subspace: $(1, 2, 4, 8, \ldots)$ and $(1, 1, 1, 1, \ldots)$ are both geometric sequences, but their sum $(2, 3, 5, 9, \ldots)$ is not.

14. Yes
- $(0, 0, 0, \ldots, 0, \ldots)$ converges to 0.
- If $\lim_{n\to\infty} x_n = 0$ and $\lim_{n\to\infty} y_n = 0$, then $\lim_{n\to\infty} (x_n + y_n) = \lim_{n\to\infty} x_n + \lim_{n\to\infty} y_n = 0$.
- If $\lim_{n\to\infty} x_n = 0$ and k is any constant, then $\lim_{n\to\infty} (kx_n) = k \lim_{n\to\infty} x_n = 0$.

15. The W of all square-summable sequences is a subspace of V:

• The sequence (0, 0, 0, ...) is in W.

• Suppose (x_n) and (y_n) are in W, with $\sum_{i=1}^{\infty} x_i^2 = L$ and $\sum_{i=1}^{\infty} y_i^2 = M$. We need to show that $\sum_{i=1}^{\infty}(x_i + y_i)^2$ converges, that is, there is a K such that $\sum_{i=1}^{n}(x_i + y_i)^2 \le K$ for all n.

The *triangle inequality* (Exercise 4.1.12) tells us that $\left\| \begin{bmatrix} x_1 + y_1 \\ x_2 + y_2 \\ \vdots \\ x_n + y_n \end{bmatrix} \right\| \le \left\| \begin{bmatrix} x_1 \\ x_2 \\ \vdots \\ x_n \end{bmatrix} \right\| + \left\| \begin{bmatrix} y_1 \\ y_2 \\ \vdots \\ y_n \end{bmatrix} \right\| \le \sqrt{L} + \sqrt{M}$.

Therefore, $\sum_{i=1}^{n}(x_i + y_i)^2 = \left\| \begin{bmatrix} x_1 + y_1 \\ x_2 + y_2 \\ \vdots \\ x_n + y_n \end{bmatrix} \right\|^2 \le \left(\sqrt{L} + \sqrt{M}\right)^2$, so that $K = \left(\sqrt{L} + \sqrt{M}\right)^2$ does the job.

• If (x_n) is in W $\left(\text{so that } \sum_{i=1}^{\infty} x_i^2 \text{ converges}\right)$, then (kx_n) is in W as well, for any constant k, since $\sum_{i=1}^{\infty}(kx_i)^2 = k^2 \sum_{i=1}^{\infty} x_i^2$ will converge.

16. $\begin{bmatrix} a & b \\ c & d \\ e & f \end{bmatrix} = a\begin{bmatrix} 1 & 0 \\ 0 & 0 \\ 0 & 0 \end{bmatrix} + b\begin{bmatrix} 0 & 1 \\ 0 & 0 \\ 0 & 0 \end{bmatrix} + c\begin{bmatrix} 0 & 0 \\ 1 & 0 \\ 0 & 0 \end{bmatrix} + d\begin{bmatrix} 0 & 0 \\ 0 & 1 \\ 0 & 0 \end{bmatrix} + e\begin{bmatrix} 0 & 0 \\ 0 & 0 \\ 1 & 0 \end{bmatrix} + f\begin{bmatrix} 0 & 0 \\ 0 & 0 \\ 0 & 1 \end{bmatrix}$

The matrices $\begin{bmatrix} 1 & 0 \\ 0 & 0 \\ 0 & 0 \end{bmatrix}, \begin{bmatrix} 0 & 1 \\ 0 & 0 \\ 0 & 0 \end{bmatrix}, \begin{bmatrix} 0 & 0 \\ 1 & 0 \\ 0 & 0 \end{bmatrix}, \begin{bmatrix} 0 & 0 \\ 0 & 1 \\ 0 & 0 \end{bmatrix}, \begin{bmatrix} 0 & 0 \\ 0 & 0 \\ 1 & 0 \end{bmatrix}, \begin{bmatrix} 0 & 0 \\ 0 & 0 \\ 0 & 1 \end{bmatrix}$ form a basis of $\mathbb{R}^{3\times 2}$, so that $\dim(\mathbb{R}^{3\times 2}) = 6$.

17. Let E_{ij} be the $m \times n$ matrix with a 1 as its ijth entry, and zeros everywhere else. Any A in $\mathbb{R}^{m\times n}$ can be written as the sum of all $a_{ij}E_{ij}$, and the E_{ij} are linearly independent, so that they form a basis of $\mathbb{R}^{m\times n}$. Thus $\dim(\mathbb{R}^{m\times n}) = mn$.

18. Any f in P_n can be written as a linear combination of $1, t, t^2, \ldots, t^n$, by definition of P_n. Also, $1, t, \ldots, t^n$ are linearly independent; to see this consider a relation $c_0 + c_1 t + \cdots + c_n t^n = 0$; since the polynomial $c_0 + c_1 t + \cdots + c_n t^n$ has more than n zeros, we must have $c_0 = c_1 = \cdots = c_n = 0$, as claimed. Thus, $\dim(P_n) = n + 1$.

19. $\begin{bmatrix} a + bi \\ c + di \end{bmatrix} = a\begin{bmatrix} 1 \\ 0 \end{bmatrix} + b\begin{bmatrix} i \\ 0 \end{bmatrix} + c\begin{bmatrix} 0 \\ 1 \end{bmatrix} + d\begin{bmatrix} 0 \\ i \end{bmatrix}$

The vectors $\begin{bmatrix} 1 \\ 0 \end{bmatrix}, \begin{bmatrix} i \\ 0 \end{bmatrix}, \begin{bmatrix} 0 \\ 1 \end{bmatrix}, \begin{bmatrix} 0 \\ i \end{bmatrix}$ form a basis of $\mathbb{C}^2$ as a *real* linear space, so that $\dim(\mathbb{C}^2) = 4$.

20. $\begin{bmatrix} a & b \\ c & -a \end{bmatrix} = a\begin{bmatrix} 1 & 0 \\ 0 & -1 \end{bmatrix} + b\begin{bmatrix} 0 & 1 \\ 0 & 0 \end{bmatrix} + c\begin{bmatrix} 0 & 0 \\ 1 & 0 \end{bmatrix}$

The matrices $\begin{bmatrix} 1 & 0 \\ 0 & -1 \end{bmatrix}, \begin{bmatrix} 0 & 1 \\ 0 & 0 \end{bmatrix}, \begin{bmatrix} 0 & 0 \\ 1 & 0 \end{bmatrix}$ form a basis of this space; its dimension is the three.

21. $\begin{bmatrix} a & b \\ b & c \end{bmatrix} = a\begin{bmatrix} 1 & 0 \\ 0 & 0 \end{bmatrix} + b\begin{bmatrix} 0 & 1 \\ 1 & 0 \end{bmatrix} + c\begin{bmatrix} 0 & 0 \\ 0 & 1 \end{bmatrix}$

The matrices $\begin{bmatrix} 1 & 0 \\ 0 & 0 \end{bmatrix}, \begin{bmatrix} 0 & 1 \\ 1 & 0 \end{bmatrix}, \begin{bmatrix} 0 & 0 \\ 0 & 1 \end{bmatrix}$ form a basis of this space; its dimension is three.

22. Define E_{ij} as in Exercise 17.

A symmetric $n \times n$ matrix A can be written as

$$A = \begin{bmatrix} a_{11} & a_{12} & \cdots & \cdots & \cdots & \cdots & a_{1n} \\ a_{12} & a_{22} & & & & & \vdots \\ \vdots & & \ddots & & & & \vdots \\ \vdots & & & a_{ii} & \cdots & a_{ij} & \vdots \\ \vdots & & & \vdots & \ddots & \vdots & \vdots \\ \vdots & & & a_{ij} & \cdots & a_{jj} & \vdots \\ a_{1n} & \cdots & \cdots & \cdots & \cdots & \cdots & a_{nn} \end{bmatrix} = \sum_{i=1}^{n} a_{ii}E_{ii} + \sum_{i<j} a_{ij}(E_{ij} + E_{ji})$$

This shows that the matrices E_{ii} ($i = 1, \ldots, n$) and $E_{ij} + E_{ji}$ (for $i<j$) form a basis of the space of all symmetric $n \times n$ matrices (they are clearly linearly independent).

The number of matrices in this basis equals the number of entries in an $n \times n$ matrix on and above the diagonal, which is $1+2+\cdots+n = \dfrac{n(n+1)}{2} = \dfrac{n^2+n}{2}$.

23. $\begin{bmatrix} 0 & a \\ -a & 0 \end{bmatrix} = a\begin{bmatrix} 0 & 1 \\ -1 & 0 \end{bmatrix}$

The matrix $\begin{bmatrix} 0 & 1 \\ -1 & 0 \end{bmatrix}$ forms a basis of the space of all skew-symmetric 2×2 matrices; this space is one-dimensional.

24. Use Exercises 8 and 23 as a guide. Define E_{ij} as in Exercise 17.

A skew-symmetric $n \times n$ matrix can be written as

$$A = \begin{bmatrix} 0 & a_{12} & \cdots & \cdots & \cdots & \cdots & a_{1n} \\ -a_{12} & 0 & & & & & \vdots \\ \vdots & & \ddots & & & & \vdots \\ \vdots & & & 0 & \cdots & a_{ij} & \vdots \\ \vdots & & & \vdots & \ddots & \vdots & \vdots \\ \vdots & & & -a_{ij} & \cdots & 0 & \vdots \\ -a_{1n} & \cdots & \cdots & \cdots & \cdots & \cdots & 0 \end{bmatrix} = \sum_{i<j} a_{ij}(E_{ij} - E_{ji}).$$

This shows that the matrices $E_{ij} - E_{ji}$ (for $i<j$) form a basis of the space of all skew-symmetric $n \times n$ matrices (they are clearly linearly independent). The number of matrices in this basis equals the number of entries in an $n \times n$ matrix above the diagonal, which is (count column by column) $1+2+\cdots+(n-1) = \dfrac{(n-1)n}{2} = \dfrac{n^2-n}{2}$.

25. A polynomial $f(t)=a+bt+ct^2$ is in this subspace if $f(1)=a+b+c=0$, or $a=-b-c$.
The polynomials in the subspace are of the form $f(t)=(-b-c)+bt+ct^2=b(t-1)+c(t^2-1)$, so that $t-1$, t^2-1 is a basis of the subspace, whose dimension is 2.

26. Denote the subspace by W. A polynomial $f(t)=a+bt+ct^2+dt^3$ is in W if $f(1)=a+b+c+d=0$ and $\int_{-1}^{1} f(t)\,dt=\left[at+\frac{b}{2}t^2+\frac{c}{3}t^3+\frac{d}{4}t^4\right]_{-1}^{1}=2a+\frac{2}{3}c=0.$

The system $\left|\begin{matrix} a+b+c+d=0 \\ 2a\;\;+\frac{2}{3}c\;\;\;\;=0\end{matrix}\right|$ reduces to $\left|\begin{matrix} a\;\;+\frac{1}{3}c\;\;\;\;=0 \\ b+\frac{2}{3}c+d=0\end{matrix}\right|$, with general solution $\begin{bmatrix} -\frac{1}{3}c \\ -\frac{2}{3}c-d \\ c \\ d \end{bmatrix}$.

The polynomials in W are of the form $f(t)=-\frac{1}{3}c-\left(\frac{2}{3}c+d\right)t+ct^2+dt^3=c\left(t^2-\frac{2}{3}t-\frac{1}{3}\right)+d(t^3-1)$, so that $t^2-\frac{2}{3}t-\frac{1}{3}$, t^3-1 is a basis of W, and $\dim(W)=2$.

27. $\begin{bmatrix} a & b \\ c & d \end{bmatrix}$ is in the subspace if $\begin{bmatrix} a & b \\ c & d \end{bmatrix}\begin{bmatrix} 1 & 0 \\ 0 & 2 \end{bmatrix}=\begin{bmatrix} a & 2b \\ c & 2d \end{bmatrix}$ equals $\begin{bmatrix} 1 & 0 \\ 0 & 2 \end{bmatrix}\begin{bmatrix} a & b \\ c & d \end{bmatrix}=\begin{bmatrix} a & b \\ 2c & 2d \end{bmatrix}$, which is the case if $b=c=0$. The matrices in the subspace are of the form $\begin{bmatrix} a & 0 \\ 0 & d \end{bmatrix}=a\begin{bmatrix} 1 & 0 \\ 0 & 0 \end{bmatrix}+d\begin{bmatrix} 0 & 0 \\ 0 & 1 \end{bmatrix}$, so that $\begin{bmatrix} 1 & 0 \\ 0 & 0 \end{bmatrix}, \begin{bmatrix} 0 & 0 \\ 0 & 1 \end{bmatrix}$ is a basis, and the dimension is 2.

28. $\begin{bmatrix} a & b \\ c & d \end{bmatrix}$ is in the subspace if $\begin{bmatrix} a & b \\ c & d \end{bmatrix}\begin{bmatrix} 1 & 1 \\ 0 & 1 \end{bmatrix}=\begin{bmatrix} a & a+b \\ c & c+d \end{bmatrix}$ equals $\begin{bmatrix} 1 & 1 \\ 0 & 1 \end{bmatrix}\begin{bmatrix} a & b \\ c & d \end{bmatrix}=\begin{bmatrix} a+c & b+d \\ c & d \end{bmatrix}$ which is the case if $c=0$ and $a=d$.

The matrices in the subspace are of the form $\begin{bmatrix} a & b \\ 0 & a \end{bmatrix}=a\begin{bmatrix} 1 & 0 \\ 0 & 1 \end{bmatrix}+b\begin{bmatrix} 0 & 1 \\ 0 & 0 \end{bmatrix}$, so that $\begin{bmatrix} 1 & 0 \\ 0 & 1 \end{bmatrix}, \begin{bmatrix} 0 & 1 \\ 0 & 0 \end{bmatrix}$ is a basis, and the dimension is 2.

29. The arithmetic sequences are of the form $(a, a+k, a+2k, a+3k, \ldots)=a(1,1,1,1,\ldots)+k(0,1,2,3,\ldots)$, so that the sequences $(1,1,1,\ldots)$ (all 1's) and $(0,1,2,3,\ldots)$ (the nth entry is n) form a basis of this space, whose dimension is 2.

30. We show that the set of all even functions is a subspace of $F(\mathbb{R},\mathbb{R})$:
- If $f(t)=0$ for all t, then $f(-t)=f(t)=0$ for all t.
- If f and g are even (that is, $f(-t)=f(t)$ and $g(-t)=g(t)$ for all t), then $(f+g)(-t)=f(-t)+g(-t)=f(t)+g(t)=(f+g)(t)$, so that $f+g$ is even as well.
- If f is even and k is any constant, then $(kf)(-t)=kf(-t)=kf(t)=(kf)(t)$, so that kf is even as well.

An analogous proof shows that the odd functions form a subspace of $F(\mathbb{R},\mathbb{R})$.

31. If $f(t)=a+bt+ct^2+dt^3+et^4$, then $f(-t)=a-bt+ct^2-dt^3+et^4$ and $-f(-t)=-a+bt-ct^2+dt^3-et^4$.

a. f is even if $f(-t)=f(t)$ for all t. Comparing coefficients we find that $b=d=0$, so that $f(t)$ is of the form $f(t)=a+ct^2+et^4$, with basis $1, t^2, t^4$. The dimension is 3.

b. f is odd if $f(-t)=-f(t)$, which is the case if $a=c=e=0$. The odd polynomials are of the form $f(t)=bt+dt^3$, with basis t, t^3 and dimension 2.

32. We show that $L(\mathbb{R}^n, \mathbb{R}^m)$ is a subspace of $F(\mathbb{R}^n, \mathbb{R}^m)$:

- The zero transformation $T(\vec{x}) = \vec{0}$ (for all $\vec{x}$) is linear, represented by the zero matrix.
- If T and S are linear transformations from $\mathbb{R}^n$ to $\mathbb{R}^m$ (with $T(\vec{x}) = A\vec{x}$ and $S(\vec{x}) = B\vec{x}$ for some $m \times n$ matrices A and B), then $(T+S)(\vec{x}) = T(\vec{x}) + S(\vec{x}) = A\vec{x} + B\vec{x} = (A+B)\vec{x}$, so that $T + S$ is linear as well, given by the matrix $A + B$.
- If T is a linear transformation from $\mathbb{R}^n$ to $\mathbb{R}^m$ (with $T(\vec{x}) = A\vec{x}$) and k is any constant, then $(kT)(\vec{x}) = kT(\vec{x}) = kA\vec{x} = (kA)\vec{x}$, so that kT is linear as well, with matrix kA.

33. T is linear:

$T(f+g) = \int_{-2}^{3}(f+g) = \int_{-2}^{3} f + \int_{-2}^{3} g = T(f) + T(g)$

$T(kf) = \int_{-2}^{2} kf = k\int_{-2}^{2} f = kT(f)$

T is not an isomorphism since $\dim(P_3) \neq \dim(\mathbb{R})$ (see Fact 9.1.8d).

34. T is not linear since $T(2 \cdot I_2) = 3I_2$ is not equal to $2 \cdot T(I_2) = 2(2I_2) = 4I_2$.

35. T is linear: $T(A+B) = \text{tr}(A+B) = \sum_{i=1}^{3}(a_{ii} + b_{ii}) = \sum_{i=1}^{3} a_{ii} + \sum_{i=1}^{3} b_{ii} = \text{tr}(A) + \text{tr}(B) = T(A) + T(B)$

$T(kA) = \text{tr}(kA) = \sum_{i=1}^{3}(ka_{ii}) = k\sum_{i=1}^{3} a_{ii} = k\text{tr}(A) = kT(A)$

T is not an isomorphism since $\dim(\mathbb{R}^{3\times 3}) \neq \dim(\mathbb{R})$.

36. T is not linear since $T(2I_2) = 4$ does not equal $2T(I_2) = 2$.

37. T is linear:

$T(z+w) = (3+4i)(z+w) = (3+4i)z + (3+4i)w = T(z) + T(w)$

$T(kz) = (3+4i)kz = k(3+4i)z = kT(z)$

T is an isomorphism, since we can solve the equation $w = (3+4i)z$ for z: $z = \frac{1}{3+4i}w$.

38. T is linear:

$T(A+B) = S^{-1}(A+B)S = S^{-1}(AS+BS) = S^{-1}AS + S^{-1}BS = T(A) + T(B)$

$T(kA) = S^{-1}(kA)S^{-1} = kS^{-1}AS = kT(A)$

T is an isomorphism since we can solve the equation $B = S^{-1}AS$ for A: $A = SBS^{-1}$

39. Let $S = \begin{bmatrix} 1 & 2 \\ 2 & 4 \end{bmatrix}$. T is linear:

$T(A+B) = S(A+B) = SA + SB = T(A) + T(B)$

$T(kA) = S(kA) = kSA = kT(A)$

T is not an isomorphism since $S = \begin{bmatrix} 1 & 2 \\ 2 & 4 \end{bmatrix}$ is not invertible (note that I_2 is not in im(T); see Fact 9.1.8b).

40. T is linear:

$T(f+g) = (f+g)'' + 4(f+g)' = f'' + 4f' + g'' + 4g' = T(f) + T(g)$

$T(kf) = (kf)'' + 4(kf)' = kf'' + 4kf' = kT(f)$

T is not an isomorphism since $T(t) = 0$, so that $\ker(T) \neq \{0\}$ (see Fact 9.1.8b). (Alternatively, we could point out that t^6 is not in im(T), so that $\text{im}(T) \neq P^6$.)

41. T is linear:

$$T(f+g)=\begin{bmatrix}(f+g)(7)\\(f+g)(11)\end{bmatrix}=\begin{bmatrix}f(7)+g(7)\\f(11)+g(11)\end{bmatrix}=\begin{bmatrix}f(7)\\f(11)\end{bmatrix}\times\begin{bmatrix}g(7)\\g(11)\end{bmatrix}=T(f)+T(g)$$

$$T(kf)=\begin{bmatrix}(kf)(7)\\(kf)(11)\end{bmatrix}=\begin{bmatrix}kf(7)\\kf(11)\end{bmatrix}=k\begin{bmatrix}f(7)\\f(11)\end{bmatrix}=kT(f)$$

T is not an isomorphism since $\dim(P_2)\neq\dim(\mathbb{R}^2)$.

42. No, since I_2 is positive semidefinite, but $(-1)I_2=-I_2$ is not.

43. Yes, the functions with period 3 do form a subspace of $F(\mathbb{R},\mathbb{R})$:
- If $f(t)=0$ for all t, then $f(t+3)=f(t)=0$ for all t.
- If f and g both have period 3 (that is, $f(t+3)=f(t)$ and $g(t+3)=g(t)$ for all t), then $(f+g)(t+3)=f(t+3)+g(t+3)=f(t)+g(t)=(f+g)(t)$, so that $f+g$ has period 3 as well.
- If f has period 3 and k is any constant, then $(kf)(t+3)=kf(t+3)=kf(t)=(kf)(t)$, so that kf has period 3 as well.

44. T is linear:
- We need to show that $T(A+B)=T(A)+T(B)$ for any two $m\times n$ matrices A and B. This means that $(T(A+B))(\vec{v})=(T(A)+T(B))(\vec{v})$ for all $\vec{v}$ in $\mathbb{R}^n$:
  $(T(A+B))(\vec{v})=(A+B)\vec{v}=A\vec{v}+B\vec{v}=(T(A))(\vec{v})+(T(B))(\vec{v})=(T(A)+T(B))\vec{v}$
- Likewise, $(T(kA))(\vec{v})=(kA)\vec{v}=kA\vec{v}=kT(A)(\vec{v})=(kT(A))(\vec{v})$, so that $T(kA)=kT(A)$.

An $m\times n$ matrix A is in the kernel of T if $T(A)$ is the zero transformation from $\mathbb{R}^n$ to $\mathbb{R}^m$, that is, $T(A)(\vec{v})=A\vec{v}=\vec{0}$ for all $\vec{v}$ in $\mathbb{R}^n$. This is the case only for the zero matrix A, so that $\ker(T)=\{0\}$.

By definition 2.1.1, a transformation S from $\mathbb{R}^n$ to $\mathbb{R}^m$ is linear if there is an $m\times n$ matrix A such that $S(\vec{v})=A\vec{v}=(T(A))\vec{v}$ for all $\vec{v}$ in $\mathbb{R}^n$, that is, $S=T(A)$. This shows that $\operatorname{im}(T)=L(\mathbb{R}^n,\mathbb{R}^m)$.

Since $\ker(T)=\{0\}$ and $\operatorname{im}(T)=L(\mathbb{R}^n,\mathbb{R}^m)$, T defines an isomorphism from $\mathbb{R}^{m\times n}$ to $L(\mathbb{R}^n,\mathbb{R}^m)$, so that $\dim(L(\mathbb{R}^n,\mathbb{R}^m))=\dim(\mathbb{R}^{m\times n})=mn$, by Fact 9.1.8.

45.

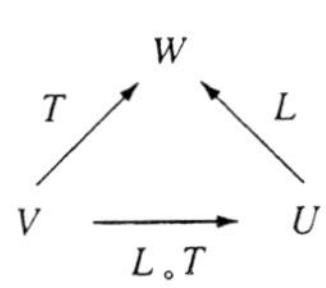

If T and L are linear, then
$(L\circ T)(f+g)=L(T(f+g))=L(T(f)+T(g))=L(T(f))+L(T(g))=(L\circ T)(f)+(L\circ T)(g)$, and
$(L\circ T)(kf)=L(T(kf))=L(kT(f))=kL(T(f))=k(L\circ T)(f)$, so that $L\circ T$ is linear as well.

If T and L are isomorphisms, then $L\circ T$ is an isomorphism as well, since the composite of invertible functions is invertible.

46. T is linear:

$T(\vec{x}(t)+\vec{y}(t))=\vec{x}(0)+\vec{y}(0)=T(\vec{x}(t))+T(\vec{y}(t))$

$T(k\vec{x}(t))=k\vec{x}(0)+kT(x(t))$

T is invertible (and therefore an isomorphism) by Fact 8.1.2: For every $\vec{x}_0$ in $\mathbb{R}^n$ there is a unique $\vec{x}(t)$ in W such that $T(x(t))=\vec{x}(0)=\vec{x}_0$, namely, $\vec{x}(t)=c_1e^{\lambda_1 t}\vec{v}_1+\cdots+c_ne^{\lambda_n t}\vec{v}_n$, where the c_i are the coordinates of $\vec{x}_0$ with respect to $\vec{v}_1,\ldots,\vec{v}_n$.

47. We need to show that $(T(kf))(t) = (k(Tf))(t)$ for all t.
$(T(kf))(t) = (kf)(t-1) = kf(t-1)$ and $(k(Tf))(t) = k(Tf)(t) = kf(t-1)$: the results agree.

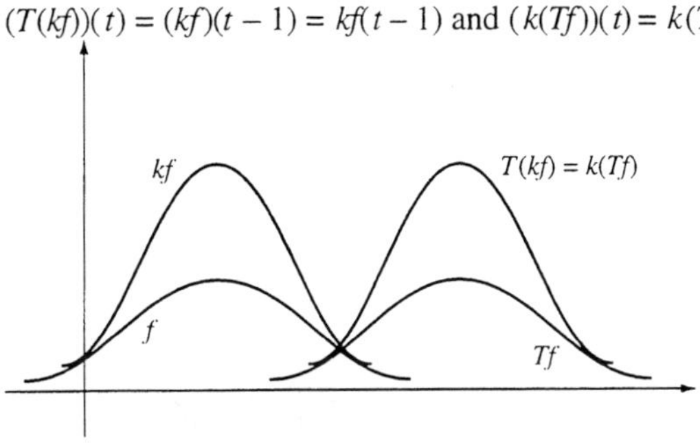

48. To see that $0 + 0 = 0$, apply the formula $f + 0 = f$ to $f = 0$ (see Definition 9.1.1.3).
Then $0 = k0 - k0 = k(0+0) - k0 = k0 + k0 - k0 = k0$.

49. $0_w = T(0_v) - T(0_v) = T(0_v + 0_v) - T(0_v) = T(0_v) + T(0_v) - T(0_v) = T(0_v)$
We have used the equation $0_v + 0_v = 0_v$ derived in Exercise 48.

50. We need to solve the system $\dfrac{d\vec{x}}{dt} = A\vec{x}$, where $A = \begin{bmatrix} 0 & 1 \\ -1 & 0 \end{bmatrix}$.

The eigenvalues of A are $\lambda_{1,2} = \pm i$, and $E_i = \text{span}\begin{bmatrix} 1 \\ i \end{bmatrix}$, so that $\vec{v} = \begin{bmatrix} 1 \\ 0 \end{bmatrix}$ and $\vec{w} = \begin{bmatrix} 0 \\ 1 \end{bmatrix}$.

Fact 8.2.6 tells us that $\vec{x}(t) = \begin{bmatrix} 0 & 1 \\ 1 & 0 \end{bmatrix}\begin{bmatrix} \cos t & -\sin t \\ \sin t & \cos t \end{bmatrix}\begin{bmatrix} a \\ b \end{bmatrix} = \begin{bmatrix} a\sin(t) + b\cos(t) \\ a\cos(t) - b\sin(t) \end{bmatrix}$, where a and b are arbitrary constants. The first component is $x(t) = a\sin(t) + b\cos(t)$, where a and b are arbitrary.

51. Introduce the auxiliary function $y = \dfrac{dx}{dt}$. Then $\left| \begin{aligned} \frac{dx}{dt} &= y \\ \frac{dy}{dt} &= -2x - 3y \end{aligned} \right|$.

We need to solve the system $\dfrac{d\vec{x}}{dt} = A\vec{x}$, where $A = \begin{bmatrix} 0 & 1 \\ -2 & -3 \end{bmatrix}$. The eigenvalues of A are $\lambda_1 = -1$ and $\lambda_2 = -2$,

with $E_{-1} = \text{span}\begin{bmatrix} 1 \\ -1 \end{bmatrix}$ and $E_{-2} = \text{span}\begin{bmatrix} 1 \\ -2 \end{bmatrix}$.

Fact 8.1.2 tells us that $\vec{x}(t) = c_1 e^{-t}\begin{bmatrix} 1 \\ -1 \end{bmatrix} + c_2 e^{-2} + \begin{bmatrix} 1 \\ -2 \end{bmatrix}$, where c_1 and c_2 are arbitrary constants. The first component is $x(t) = c_1 e^{-t} + c_2 e^{-2t}$, where c_1, c_2 are arbitrary.

52. Introduce the auxiliary function $y = \dfrac{dx}{dt}$. Then $\left| \begin{aligned} \frac{dx}{dt} &= y \\ \frac{dy}{dt} &= -13x + 4y \end{aligned} \right|$.

We need to solve the system $\dfrac{d\vec{x}}{dt} = A\vec{x}$, where $A = \begin{bmatrix} 0 & 1 \\ -13 & 4 \end{bmatrix}$. The eigenvalues of A are $\lambda_{1,2} = 2 \pm 3i$, and

$E_{2+3i} = \text{span}\begin{bmatrix} 1 \\ 2+3i \end{bmatrix}$, so that $\vec{v} = \begin{bmatrix} 1 \\ 2 \end{bmatrix}$ and $\vec{w} = \begin{bmatrix} 0 \\ 3 \end{bmatrix}$.

Fact 8.2.6 tells us that $\vec{x}(t)=e^{2t}\begin{bmatrix}0 & 1\\ 3 & 2\end{bmatrix}\begin{bmatrix}\cos(3t) & -\sin(3t)\\ \sin(3t) & \cos(3t)\end{bmatrix}\begin{bmatrix}a\\ b\end{bmatrix}$, where a and b are arbitrary. The first component is $x(t)=e^{2t}(a\sin(3t)+b\cos(3t))$, where a, b are arbitrary.

53. If $A^T=\lambda A$ then $L(A)=A+A^T=A+\lambda A=(1+\lambda)A$, and vice versa.
Thus L has the same "eigenmatrices" as the transformation considered in Example 26, but the corresponding eigenvalues are 1 higher:

$E_2=\text{symmetric matrices}=\text{span}\left(\begin{bmatrix}1 & 0\\ 0 & 0\end{bmatrix},\begin{bmatrix}0 & 0\\ 0 & 1\end{bmatrix},\begin{bmatrix}0 & 1\\ 1 & 0\end{bmatrix}\right)$

$E_0=\text{skew-symmetric matrices}=\text{span}\begin{bmatrix}0 & 1\\ -1 & 0\end{bmatrix}$

$\begin{bmatrix}1 & 0\\ 0 & 0\end{bmatrix},\begin{bmatrix}0 & 0\\ 0 & 1\end{bmatrix},\begin{bmatrix}0 & 1\\ 1 & 0\end{bmatrix},\begin{bmatrix}0 & 1\\ -1 & 0\end{bmatrix}$ is an eigenbasis.

54. $T(f)=f+f'=\lambda f$ means that $f'=(\lambda-1)f$, so that $f(t)=ce^{(\lambda-1)t}$, where c is a constant (by Fact 8.1.1).
Hence the eigenvalues are all real λ, and E_λ is one-dimensional, spanned by $e^{(\lambda-1)t}$.

55. If $T(z)=\lambda z$, then $z=T(T(z))=\lambda^2 z$, so that $\lambda^2=1$ and $\lambda=\pm 1$.
$E_1=\text{real numbers}=\text{span}(1)$
$E_{-1}=\text{imaginary numbers}=\text{span}(i)$
$1, i$ is an eigenbasis.

56. $(x_0, x_1, x_2, \ldots)$ is an "eigen sequence" with eigenvalue λ if
$T(x_0, x_1, x_2, \ldots)=(x_1, x_2, x_3, \ldots)=\lambda(x_0, x_1, x_2, \ldots)$, that is, if $x_{i+1}=\lambda x_i$ for all i.
This is just the definition of a *geometric sequence* with multiplication factor λ. Thus, all reals λ are eigenvalues, and E_λ is one-dimensional, spanned by the geometric sequence $(1, \lambda, \lambda^2, \ldots)$.

57. T reflects the graph of f in the y-axis. Note that $T(Tf)=f$.
If $T(f)=\lambda f$, then $f=T(T(f))=\lambda^2 f$, so that $\lambda^2=1$ and $\lambda=\pm 1$.
By definition $E_1=\{\text{even functions}\}$, and $E_{-1}=\{\text{odd functions}\}$ (see Exercise 30).

58. If $f(t)=a+bt+ct^2$ then $(Tf)(t)=t(b+2ct)=bt+2ct^2$.

It is required that $(bt+2ct^2)=\lambda(a+bt+ct^2)$, that is $\left|\begin{matrix}0=\lambda a\\ b=\lambda b\\ 2c=\lambda c\end{matrix}\right|$, where a, b, c are not all zero.

The solutions are
- $\lambda=0$, with $b=c=0$ and a arbitrary
- $\lambda=1$, with $a=c=0$ and b arbitrary
- $\lambda=2$, with $b=a=0$ and c arbitrary

Thus, $E_0=\text{span}(1)$, $E_1=\text{span}(t)$, $E_2=\text{span}(t^2)$, and $1, t, t^2$ is an eigenbasis.

59. Let $S=\begin{bmatrix}1&2\\4&3\end{bmatrix}$. Note that the eigenvalues of S are $\lambda_1=-1$, $\lambda_2=5$, with $E_{-1}=\text{span}\begin{bmatrix}1\\-1\end{bmatrix}$ and $E_5=\begin{bmatrix}1\\2\end{bmatrix}$.

The condition $T(A)=SA=\lambda A$ means that both columns of A are in E_λ, the eigenspace of S associated with λ. Therefore, the eigenvalues of T are those of S, $\lambda_1=-1$ and $\lambda_2=5$. The eigenspace associated with -1 is spanned by $\begin{bmatrix}1&0\\-1&0\end{bmatrix}$ and $\begin{bmatrix}0&1\\0&-1\end{bmatrix}$, and the eigenspace associated with 5 is spanned by $\begin{bmatrix}1&0\\2&0\end{bmatrix}$ and $\begin{bmatrix}0&1\\0&2\end{bmatrix}$.

$\begin{bmatrix}1&0\\-1&0\end{bmatrix},\begin{bmatrix}0&1\\0&-1\end{bmatrix},\begin{bmatrix}1&0\\2&0\end{bmatrix},\begin{bmatrix}0&1\\0&2\end{bmatrix}$ is an eigenbasis.

60. It is required that $T(A)=S^{-1}AS=\lambda A$, or $AS=\lambda SA$. If we write $A=\begin{bmatrix}a&b\\c&d\end{bmatrix}$ it is required that

$$\begin{bmatrix}a&2b\\c&2d\end{bmatrix}=\begin{bmatrix}\lambda a&\lambda b\\2\lambda c&2\lambda d\end{bmatrix}.$$

We have the solutions

- $\lambda=1$, $b=c=0$, a, d arbitrary
- $\lambda=2$, $a=c=d=0$, b arbitrary
- $\lambda=\frac{1}{2}$, $a=b=d=0$, c arbitrary.

This means that

$$E_1=\{\text{diagonal matrices}\}=\text{span}\left(\begin{bmatrix}1&0\\0&0\end{bmatrix},\begin{bmatrix}0&0\\0&1\end{bmatrix}\right)$$

$$E_2=\text{span}\begin{bmatrix}0&1\\0&0\end{bmatrix}$$

$$E_{1/2}=\text{span}\begin{bmatrix}0&0\\1&0\end{bmatrix}$$

$\begin{bmatrix}1&0\\0&0\end{bmatrix},\begin{bmatrix}0&0\\0&1\end{bmatrix},\begin{bmatrix}0&1\\0&0\end{bmatrix},\begin{bmatrix}0&0\\1&0\end{bmatrix}$ is an eigenbasis.

61. Using Exercise 60 as a guide, we observe that the basis of $\mathbb{R}^{n\times n}$ consisting of all E_{ij} is an eigenbasis (see Exercise 17). We need to show that $S^{-1}E_{ij}S=\lambda E_{ij}$ for some λ, or $E_{ij}S=\lambda SE_{ij}$.

Since S is diagonal, we have $E_{ij}S=s_{jj}E_{ij}$ and $SE_{ij}=s_{ii}E_{ij}$, so that $E_{ij}S=\lambda SE_{ij}$ for $\lambda=\frac{s_{jj}}{s_{ii}}$, as claimed.

62. a. Check all the conditions in Definition 9.1.1.

b. $T(x\oplus y)=T(xy)=\ln(xy)=\ln(x)+\ln(y)=T(x)+T(y)$

$T(k\odot x)=T(x^k)=\ln(x^k)=k\ln(x)=kT(x)$

T is an isomorphism since $\ker(T)=\{1\}$ (note that 1 is the neutral element in $\mathbb{R}^+$) and $\ln(T)=\mathbb{R}$; use Fact 9.1.8b.

63. Yes; let T be an invertible function from $\mathbb{R}^2$ to $\mathbb{R}$ (see page 87). On $\mathbb{R}^2$ we can define the "exotic" operations $\vec{v} \oplus \vec{w} = T^{-1}(T(\vec{v})+T(\vec{w}))$ and $k \odot \vec{v} = T^{-1}(kT(\vec{v}))$ (check the conditions of Definition 9.1.1 hold).

Then T is a linear transformation from $\mathbb{R}^2$ (with these exotic operations) to $\mathbb{R}$ (with the usual operations): $T(\vec{v} \oplus \vec{w}) = T(T^{-1}(T(\vec{v})+T(\vec{w}))) = T(\vec{v})+T(\vec{w})$, and $T(k \odot v) = kT(\vec{v})$.

Now T is an isomorphism, so that the dimension of $\mathbb{R}^2$ (with our exotic operations) equals the dimension of the "usual" $\mathbb{R}$, which is 1.

64. If a real linear space V has more than one element, then it is infinite; to see this, note that the scalar multiples kf of a nonzero f in V are all distinct (think about it!). Since Timbuktu has more than one citizen, but their number is finite, we cannot make the set X into a real linear space.

9.2

1. Let B be the standard basis of P_2: $1, t, t^2$. Then the coordinates of the given polynomials with respect to $\mathcal{B}$ are $[f]_{\mathcal{B}} = \begin{bmatrix}7\\3\\1\end{bmatrix}$, $[g]_{\mathcal{B}} = \begin{bmatrix}9\\9\\4\end{bmatrix}$, $[h]_{\mathcal{B}} = \begin{bmatrix}3\\2\\1\end{bmatrix}$. Finding $\text{rref}\begin{bmatrix}7 & 9 & 3\\3 & 9 & 2\\1 & 4 & 1\end{bmatrix} = I_3$, we conclude that $[f]_{\mathcal{B}}, [g]_{\mathcal{B}}, [h]_{\mathcal{B}}$ are linearly independent, hence so are f, g, h since the coordinate transformation is an isomorphism.

2. Let $\mathcal{B}$ be the basis $\begin{bmatrix}1 & 0\\0 & 0\end{bmatrix}, \begin{bmatrix}0 & 1\\0 & 0\end{bmatrix}, \begin{bmatrix}0 & 0\\1 & 0\end{bmatrix}, \begin{bmatrix}0 & 0\\0 & 1\end{bmatrix}$ of $\mathbb{R}^{2\times 2}$. Then the coordinates of the given matrices with respect to $\mathcal{B}$ are $\begin{bmatrix}11\\11\end{bmatrix}_{\mathcal{B}} = \begin{bmatrix}1\\1\\1\\1\end{bmatrix}, \begin{bmatrix}12\\34\end{bmatrix}_{\mathcal{B}} = \begin{bmatrix}1\\2\\3\\4\end{bmatrix}, \begin{bmatrix}2 & 3\\5 & 7\end{bmatrix}_{\mathcal{B}} = \begin{bmatrix}2\\3\\5\\7\end{bmatrix}, \begin{bmatrix}1 & 4\\6 & 8\end{bmatrix}_{\mathcal{B}} = \begin{bmatrix}1\\4\\6\\8\end{bmatrix}$. Finding

$\text{rref}\begin{bmatrix}1 & 1 & 2 & 1\\1 & 2 & 3 & 4\\1 & 3 & 5 & 6\\1 & 4 & 7 & 8\end{bmatrix} = \begin{bmatrix}1 & 0 & 0 & -1\\0 & 1 & 0 & 4\\0 & 0 & 1 & -1\\0 & 0 & 0 & 0\end{bmatrix} \neq I_4$, we conclude that the four vectors $\begin{bmatrix}1\\1\\1\\1\end{bmatrix}, \begin{bmatrix}1\\2\\3\\4\end{bmatrix}, \begin{bmatrix}2\\3\\5\\7\end{bmatrix}, \begin{bmatrix}1\\4\\6\\8\end{bmatrix}$ are linearly dependent, and so are the four given matrices. In fact, $\begin{bmatrix}1 & 4\\6 & 8\end{bmatrix} = -\begin{bmatrix}1 & 1\\1 & 1\end{bmatrix} + 4\begin{bmatrix}1 & 2\\3 & 4\end{bmatrix} - \begin{bmatrix}2 & 3\\5 & 7\end{bmatrix}$.

3. We proceed as in Exercise 1. Since $\text{rref}\begin{bmatrix}1 & 1 & 1 & 1\\2 & 7 & 8 & 8\\9 & 0 & 1 & 4\\1 & 7 & 5 & 8\end{bmatrix} = I_4$, the four given polynomials do form a basis of P_3.

4. Consider the coordinate vectors of the 3 given polynomials with respect to the standard basis of P_2: $1, t, t^2$.

$\begin{bmatrix}1\\1\\0\end{bmatrix}, \begin{bmatrix}0\\1\\1\end{bmatrix}, \begin{bmatrix}2k\\2+k\\1\end{bmatrix}$

Since $\det\begin{bmatrix}1 & 0 & 2k\\1 & 1 & 2+k\\0 & 1 & 1\end{bmatrix} = k-1$, these three vectors form a basis of $\mathbb{R}^3$ unless $k = 1$. Therefore, the three polynomials $f(t), tf(t), g(t)$ form a basis of P_2 *unless* $k = 1$.

5. Let $\mathcal{B}$ be the standard basis of $\mathbb{R}^2$, so that $[\vec{x}]_{\mathcal{B}} = \vec{x}$.

First column of $M = [T(1)]_{\mathcal{B}} = \begin{bmatrix} 1 \\ 0 \end{bmatrix}_{\mathcal{B}} = \begin{bmatrix} 1 \\ 0 \end{bmatrix}$

Second column of $M = [T(t)]_{\mathcal{B}} = \begin{bmatrix} 1 \\ 1 \end{bmatrix}_{\mathcal{B}} = \begin{bmatrix} 1 \\ 1 \end{bmatrix}$

Third column of $M = [T(t^2)]_{\mathcal{B}} = \begin{bmatrix} 1 \\ 2 \end{bmatrix}_{\mathcal{B}} = \begin{bmatrix} 1 \\ 2 \end{bmatrix}$

so $M = \begin{bmatrix} 1 & 1 & 1 \\ 0 & 1 & 2 \end{bmatrix}$.

6. Let $\mathcal{B}$ be the standard basis of P_2: $1, t, t^2$.

First column of $M = [T(1)]_{\mathcal{B}} = [1]_{\mathcal{B}} = \begin{bmatrix} 1 \\ 0 \\ 0 \end{bmatrix}$

Second column of $M = [T(t)]_{\mathcal{B}} = [t+1]_{\mathcal{B}} = \begin{bmatrix} 1 \\ 1 \\ 0 \end{bmatrix}$

Third column of $M = [T(t^2)]_{\mathcal{B}} = [(t+1)^2]_{\mathcal{B}} = \begin{bmatrix} 1 \\ 2 \\ 1 \end{bmatrix}$

so $M = \begin{bmatrix} 1 & 1 & 1 \\ 0 & 1 & 2 \\ 0 & 0 & 1 \end{bmatrix}$.

7. Let $\mathcal{B}$ be the standard basis of $\mathbb{R}^2$, so that $[\vec{x}]_{\mathcal{B}} = \vec{x}$.

First column of $M = T(1) = \begin{bmatrix} 1 \\ 2 \end{bmatrix}$

Second column of $M = T(t) = \begin{bmatrix} 1 \\ 0 \end{bmatrix}$

Third column of $M = T(t^2) = \begin{bmatrix} 1 \\ \frac{2}{3} \end{bmatrix}$

Fourth column of $M = T(t^3) = \begin{bmatrix} 1 \\ 0 \end{bmatrix}$

so $M = \begin{bmatrix} 1 & 1 & 1 & 1 \\ 2 & 0 & \frac{2}{3} & 0 \end{bmatrix}$.

8. Let $\mathcal{B}$ be the standard basis of $\mathbb{R}^3$, so that $[\vec{x}]_{\mathcal{B}} = \vec{x}$.

First column of $M = T(1) = \begin{bmatrix} 1 \\ 0 \\ 1 \end{bmatrix}$

Second column of $M = T(t) = \begin{bmatrix} 0 \\ 1 \\ 2 \end{bmatrix}$

Third column of $M = T(t^2) = \begin{bmatrix} 0 \\ 2 \\ 4 \end{bmatrix}$

so $M = \begin{bmatrix} 1 & 0 & 0 \\ 0 & 1 & 2 \\ 1 & 2 & 4 \end{bmatrix}$

9. Let $\mathcal{B} = 1$ be the standard basis of $\mathbb{R}$, so that $[t]_{\mathcal{B}} = t$.

First column of $M = T\begin{bmatrix} 1 & 0 \\ 0 & 0 \end{bmatrix} = 1$

Second column of $M = T\begin{bmatrix} 0 & 1 \\ 0 & 0 \end{bmatrix} = 0$

Third column of $M = T\begin{bmatrix} 0 & 0 \\ 1 & 0 \end{bmatrix} = 0$

Fourth column of $M = T\begin{bmatrix} 0 & 0 \\ 0 & 1 \end{bmatrix} = 1$

so $M = [1 \quad 0 \quad 0 \quad 1]$.

10. Let $\mathcal{B}$ be the standard basis of $\mathbb{R}^{2\times 2}$.

First column of $M = \left[F\begin{bmatrix} 1 & 0 \\ 0 & 0 \end{bmatrix} \right]_{\mathcal{B}} = \begin{bmatrix} 1 & 0 \\ 0 & 0 \end{bmatrix}_{\mathcal{B}} = \begin{bmatrix} 1 \\ 0 \\ 0 \\ 0 \end{bmatrix}$

Second column of $M = \left[F\begin{bmatrix} 0 & 1 \\ 0 & 0 \end{bmatrix} \right]_{\mathcal{B}} = \begin{bmatrix} 0 & \frac{1}{2} \\ \frac{1}{2} & 0 \end{bmatrix}_{\mathcal{B}} = \begin{bmatrix} 0 \\ \frac{1}{2} \\ \frac{1}{2} \\ 0 \end{bmatrix}$

Third column of $M = \left[F\begin{bmatrix} 0 & 0 \\ 1 & 0 \end{bmatrix} \right]_{\mathcal{B}} = \begin{bmatrix} 0 & \frac{1}{2} \\ \frac{1}{2} & 0 \end{bmatrix}_{\mathcal{B}} = \begin{bmatrix} 0 \\ \frac{1}{2} \\ \frac{1}{2} \\ 0 \end{bmatrix}$

Fourth column of $M = \left[F\begin{bmatrix} 0 & 0 \\ 0 & 1 \end{bmatrix} \right]_{\mathcal{B}} = \begin{bmatrix} 0 & 0 \\ 0 & 1 \end{bmatrix}_{\mathcal{B}} = \begin{bmatrix} 0 \\ 0 \\ 0 \\ 1 \end{bmatrix}$ so $M = \begin{bmatrix} 1 & 0 & 0 & 0 \\ 0 & \frac{1}{2} & \frac{1}{2} & 0 \\ 0 & \frac{1}{2} & \frac{1}{2} & 0 \\ 0 & 0 & 0 & 1 \end{bmatrix}$.

11. $M = \begin{bmatrix} 1 & 1 & 1 \\ 0 & 1 & 2 \end{bmatrix}$ and $\text{rref}(M) = \begin{bmatrix} 1 & 0 & -1 \\ 0 & 1 & 2 \end{bmatrix}$.

$\ker(M) = \text{span}\begin{bmatrix} 1 \\ -2 \\ 1 \end{bmatrix}$ and $\text{im}(M) = \mathbb{R}^2$.

$\ker(T) = \text{span}(1 - 2t + t^2)$ and $\text{im}(T) = \mathbb{R}^2 = \text{span}(\vec{e}_1, \vec{e}_2)$.

12. $M = \begin{bmatrix} 1 & 1 & 1 \\ 0 & 1 & 2 \\ 0 & 0 & 1 \end{bmatrix}$ is invertible, so that T is an isomorphism.

The only eigenvalue of M is 1, with $E_1 = \text{span}(\vec{e}_1)$. Thus the only eigenvalue of T is 1, with $E_1 = \text{span}(1)$. T is not diagonalizable.

13. $M = \begin{bmatrix} 1 & 1 & 1 & 1 \\ 2 & 0 & \frac{2}{3} & 0 \end{bmatrix}$ and $\text{rref}(M) = \begin{bmatrix} 1 & 0 & \frac{1}{3} & 0 \\ 0 & 1 & \frac{2}{3} & 1 \end{bmatrix}$.

$\ker(M) = \text{span}\left(\begin{bmatrix} -1 \\ -2 \\ 3 \\ 0 \end{bmatrix}, \begin{bmatrix} 0 \\ -1 \\ 0 \\ 1 \end{bmatrix}\right)$, so that $\ker(T) = \text{span}(-1 - 2t + 3t^2, -t + t^3)$.

14. $M = \begin{bmatrix} 1 & 0 & 0 \\ 0 & 1 & 2 \\ 1 & 2 & 4 \end{bmatrix}$ and $\text{rref}(M) = \begin{bmatrix} 1 & 0 & 0 \\ 0 & 1 & 2 \\ 0 & 0 & 0 \end{bmatrix}$.

$\text{im}(M) = \text{span}\left(\begin{bmatrix} 1 \\ 0 \\ 1 \end{bmatrix}, \begin{bmatrix} 0 \\ 1 \\ 2 \end{bmatrix}\right)$ and $\ker(M) = \text{span}\begin{bmatrix} 0 \\ -2 \\ 1 \end{bmatrix}$.

$\text{im}(T) = \text{span}\left(\begin{bmatrix} 1 \\ 0 \\ 1 \end{bmatrix}, \begin{bmatrix} 0 \\ 1 \\ 2 \end{bmatrix}\right)$ and $\ker(T) = \text{span}(-2t + t^2)$.

T is not an isomorphism, since M is not invertible.

15. $M = \begin{bmatrix} 1 & 0 & 0 & 0 \\ 0 & \frac{1}{2} & \frac{1}{2} & 0 \\ 0 & \frac{1}{2} & \frac{1}{2} & 0 \\ 0 & 0 & 0 & 1 \end{bmatrix}$ and $\text{rref}(M) = \begin{bmatrix} 1 & 0 & 0 & 0 \\ 0 & 1 & 1 & 0 \\ 0 & 0 & 0 & 1 \\ 0 & 0 & 0 & 0 \end{bmatrix}$.

$\text{im}(M) = \text{span}\left(\begin{bmatrix} 1 \\ 0 \\ 0 \\ 0 \end{bmatrix}, \begin{bmatrix} 0 \\ \frac{1}{2} \\ \frac{1}{2} \\ 0 \end{bmatrix}, \begin{bmatrix} 0 \\ 0 \\ 0 \\ 1 \end{bmatrix}\right)$ and $\ker(M) = \text{span}\begin{bmatrix} 0 \\ -1 \\ 1 \\ 0 \end{bmatrix}$.

$\text{im}(F) = \text{span}\left(\begin{bmatrix} 1 & 0 \\ 0 & 0 \end{bmatrix}, \begin{bmatrix} 0 & \frac{1}{2} \\ \frac{1}{2} & 0 \end{bmatrix}, \begin{bmatrix} 0 & 0 \\ 0 & 1 \end{bmatrix}\right) =$ symmetric matrices

$\ker(F) = \text{span}\begin{bmatrix} 0 & -1 \\ 1 & 0 \end{bmatrix} = \text{skew-symmetric } 2\times 2 \text{ matrices}$

Eigenspaces:

$E_0 = \ker(F) = \text{skew-symmetric matrices}$

$E_1 = \text{im}(F) = \text{symmetric matrices}$

F is diagonalizable, with eigenbasis $\begin{bmatrix} 1 & 0 \\ 0 & 0 \end{bmatrix}, \begin{bmatrix} 0 & \frac{1}{2} \\ \frac{1}{2} & 0 \end{bmatrix}, \begin{bmatrix} 0 & 0 \\ 0 & 1 \end{bmatrix}, \begin{bmatrix} 0 & -1 \\ 1 & 0 \end{bmatrix}$.

16. a. $[T(x_1^2)]_{\mathcal{B}} = [2x_1x_2]_{\mathcal{B}} = \begin{bmatrix} 0 \\ 2 \\ 0 \end{bmatrix}$

$[T(x_1x_2)]_{\mathcal{B}} = [x_2^2 - x_1x_2]_{\mathcal{B}} = \begin{bmatrix} 0 \\ -1 \\ 1 \end{bmatrix}$

$[T(x_2^2)]_{\mathcal{B}} = [-2x_2^2]_{\mathcal{B}} = \begin{bmatrix} 0 \\ 0 \\ -2 \end{bmatrix}$

so $M = \begin{bmatrix} 0 & 0 & 0 \\ 2 & -1 & 0 \\ 0 & 1 & -2 \end{bmatrix}$

b. $\text{rref}(M) = \begin{bmatrix} 1 & 0 & -1 \\ 0 & 1 & -2 \\ 0 & 0 & 0 \end{bmatrix}$

$\text{im}(M) = \text{span}\left(\begin{bmatrix} 0 \\ 2 \\ 0 \end{bmatrix}, \begin{bmatrix} 0 \\ -1 \\ 1 \end{bmatrix}\right), \ \ker(M) = \text{span}\begin{bmatrix} 1 \\ 2 \\ 1 \end{bmatrix}$

$\text{im}(T) = \text{span}(2x_1x_2,\, x_2^2 - x_1x_2),\ \ker(T) = \text{span}(x_1^2 + 2x_1x_2 + x_2^2)$

c. Since M is lower triangular, the eigenvalues of M and T are the diagonal entries of $M \cdot 0, -1$, and -2.

$E_0 = \ker(T) = \text{span}(x_1^2 + 2x_1x_2 + x_2^2)$

$E_{-1} = \text{span}(x_1x_2 + x_2^2)$

$E_{-2} = \text{span}(x_2^2)$

17. a. $[T(1)]_{\mathcal{B}} = \begin{bmatrix} 0 & 3 \\ 0 & -1 \end{bmatrix}_{\mathcal{B}} = \begin{bmatrix} 0 \\ 3 \\ 0 \\ -1 \end{bmatrix}$

$[T(t)]_{\mathcal{B}} = \begin{bmatrix} 1 & 3 \\ 0 & -5 \end{bmatrix}_{\mathcal{B}} = \begin{bmatrix} 1 \\ 3 \\ 0 \\ -5 \end{bmatrix}$

$$[T(t^2)]_{\mathcal{B}} = \begin{bmatrix} 6 & 3 \\ 0 & -25 \end{bmatrix}_{\mathcal{B}} = \begin{bmatrix} 6 \\ 3 \\ 0 \\ -25 \end{bmatrix}$$

$$\text{so } M = \begin{bmatrix} 0 & 1 & 6 \\ 3 & 3 & 3 \\ 0 & 0 & 0 \\ -1 & -5 & -25 \end{bmatrix}.$$

b. $$\text{rref}(M) = \begin{bmatrix} 1 & 0 & -5 \\ 0 & 1 & 6 \\ 0 & 0 & 0 \\ 0 & 0 & 0 \end{bmatrix}$$

$$\text{im}(M) = \text{span}\left(\begin{bmatrix} 0 \\ 3 \\ 0 \\ -1 \end{bmatrix}, \begin{bmatrix} 1 \\ 3 \\ 0 \\ -5 \end{bmatrix} \right) \text{ and } \ker(M) = \text{span} \begin{bmatrix} 5 \\ -6 \\ 1 \end{bmatrix}$$

$$\text{im}(T) = \text{span}\left(\begin{bmatrix} 0 & 3 \\ 0 & -1 \end{bmatrix}, \begin{bmatrix} 1 & 3 \\ 0 & -5 \end{bmatrix} \right) \text{ and } \ker(T) = \text{span}(5 - 6t + t^2)$$

18. a. $$[T(1)]_{\mathcal{B}} = [1]_{\mathcal{B}} = \begin{bmatrix} 1 \\ 0 \\ 0 \end{bmatrix}$$

$$[T(t)]_{\mathcal{B}} = [t + a]_{\mathcal{B}} = \begin{bmatrix} a \\ 1 \\ 0 \end{bmatrix}$$

$$[T(t^2)]_{\mathcal{B}} = [t^2 + 2at + 2b]_{\mathcal{B}} = \begin{bmatrix} 2b \\ 2a \\ 1 \end{bmatrix}$$

$$\text{so } M = \begin{bmatrix} 1 & a & 2b \\ 0 & 1 & 2a \\ 0 & 0 & 1 \end{bmatrix}.$$

b. M and T are invertible, so that the equation $T(f) = f + af' + bf'' = g$ has a unique solution f in P_2 for any g in P_2.

19. $T(f) = t \cdot f$ is linear and $\ker(T) = \{0\}$, but T is not an isomorphism since the constant function 1 is not in the image of T.

20. Yes; suppose $\dim(V) = \dim(W) = n$, and let $\mathcal{B}$ and $\mathcal{C}$ be bases of V and W, respectively. Then the coordinate transformations $T_{\mathcal{B}}$ and $T_{\mathcal{C}}$ define isomorphisms from V to $\mathbb{R}^n$ and from W to $\mathbb{R}^n$, respectively.

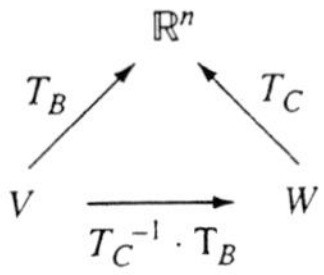

$T_{\mathcal{C}}^{-1} \circ T_{\mathcal{B}}$ is an isomorphism from V to W (see Exercise 9.1.45).

21. a. $[T(\cos t)]_{\mathcal{B}} = [(b-1)\cos(t) - a\sin(t)]_{\mathcal{B}} = \begin{bmatrix} b-1 \\ -a \end{bmatrix}$

$[T(\sin t)]_{\mathcal{B}} = [a\cos(t) + (b-1)\sin(t)]_{\mathcal{B}} = \begin{bmatrix} a \\ b-1 \end{bmatrix}$

so $M = \begin{bmatrix} b-1 & a \\ -a & b-1 \end{bmatrix}$, a rotation-dilation matrix.

b. The equation $T(f) = \cos(t)$ corresponds to $M\vec{x} = \begin{bmatrix} 1 \\ 0 \end{bmatrix}$. If M is invertible, the unique solution is

$\vec{x} = M^{-1}\begin{bmatrix} 1 \\ 0 \end{bmatrix} = \dfrac{1}{(b-1)^2 + a^2}\begin{bmatrix} b-1 & -a \\ a & b-1 \end{bmatrix}\begin{bmatrix} 1 \\ 0 \end{bmatrix} = \dfrac{1}{(b-1)^2 + a^2}\begin{bmatrix} b-1 \\ a \end{bmatrix}$, or

$f(t) = \dfrac{1}{(b-1)^2 + a^2}((b-1)\cos t + a\sin(t))$.

Note that M is invertible unless $b = 1$ and $a = 0$, in which case there is no solution.

22. a. $[T(\cos t)]_{\mathcal{B}} = [0]_{\mathcal{B}} = \vec{0}$

$[T(\sin t)]_{\mathcal{B}} = [0]_{\mathcal{B}} = \vec{0}$

$[T(t\cos(t))]_{\mathcal{B}} = [-2\sin(t)]_{\mathcal{B}} = \begin{bmatrix} 0 \\ -2 \\ 0 \\ 0 \end{bmatrix}$

$[T(t\sin(t))]_{\mathcal{B}} = [2\cos(t)]_{\mathcal{B}} = \begin{bmatrix} 2 \\ 0 \\ 0 \\ 0 \end{bmatrix}$

so $M = \begin{bmatrix} 0 & 0 & 0 & 2 \\ 0 & 0 & -2 & 0 \\ 0 & 0 & 0 & 0 \\ 0 & 0 & 0 & 0 \end{bmatrix}$.

b. The equation $T(f) = \cos(t)$ corresponds to $M\vec{x} = \begin{bmatrix}1\\0\\0\\0\end{bmatrix}$, with solutions $\vec{x} = \begin{bmatrix}p\\q\\0\\\frac{1}{2}\end{bmatrix}$, where p and q are arbitrary.

Thus $f(t) = p\cos(t) + q\sin(t) + \frac{1}{2}t\sin(t)$. Before we graph $f(t)$ for $p = q = 0$; note that $p\cos(t) + q\sin(t)$ is just a sinusoidal function.

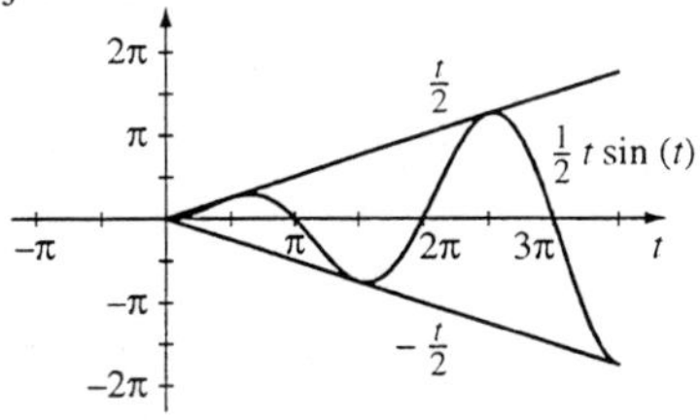

23. a. If f is in the kernel of T, then $T(f) = \begin{bmatrix}f(a_0)\\f(a_1)\\\vdots\\f(a_n)\end{bmatrix} = \begin{bmatrix}0\\0\\\vdots\\0\end{bmatrix}$, so that $f(a_i) = 0$ for $i = 0, 1, \ldots, n$. Since f (in P_n) has at least $n + 1$ zeros, it must be the zero polynomial. Thus $\ker(T) = \{0\}$.

b. Yes, by Fact 9.2.7, since $\ker(T) = \{0\}$ and $\dim P_n = \dim \mathbb{R}^{n+1} = n+1$.

c. Since T is an isomorphism from P_n to $\mathbb{R}^{n+1}$, there is a unique polynomial f in P_n such that

$T(f) = \begin{bmatrix}f(a_0)\\f(a_1)\\\vdots\\f(a_n)\end{bmatrix} = \begin{bmatrix}b_0\\b_1\\\vdots\\b_n\end{bmatrix}$, or $f(a_i) = b_i$ for $i = 0, 1, \ldots, n$. In other words: There is a unique polynomial f in P_n whose graph goes through the points $(a_0, b_0), (a_1, b_1), \ldots, (a_n, b_n)$.

24. a.
- The sequence $(0, 0, 0, \ldots)$ is in W.
- If the sequences (x_n) and (y_n) are in W (that is, $x_{n+2} = x_{n+1} + 6x_n$ and $y_{n+2} = y_{n+1} + 6y_n$ for all n), then $x_{n+2} + y_{n+2} = x_{n+1} + y_{n+1} + 6(x_n + y_n)$ so that the sequence $(x_n + y_n)$ is in W as well.
- If the sequence (x_n) is in W and k is any constant, then $kx_{n+2} = kx_{n+1} + 6kx_n$, so that the sequence (kx_n) is in W as well.

b. A sequence in W is determined by its first two components (a and b, say), which we can choose freely. All the later components can then be expressed in terms of a and b, since $x_{n+2} = x_{n+1} + 6x_n$:
$(a, b, b + 6a, 6a + 7b, 42a + 13b, \ldots) = a(1, 0, 6, 6, 42, \ldots) + b(0, 1, 1, 7, 13, \ldots)$.
The two sequences $(1, 0, 6, 6, 42, \ldots)$ and $(0, 1, 1, 7, 13, \ldots)$ form a basis of W, so that $\dim(W) = 2$.

c. It is required that $c^{n+2} = c^{n+1} + 6c^n$ for all $n \ge 0$
or $c^2 = c + 6$
or $c^2 - c - 6 = 0$
or $(c - 3)(c + 2) = 0$
The solutions are $c_1 = 3$ and $c_2 = -2$: The geometric sequences $(1, 3, 9, 27, 81, \ldots)$ and $(1, -2, 4, -8, 16, \ldots)$ are in W.

d. Yes, the two sequences we found in part c do the job, since $\dim(W) = 2$.

e. $(x_0, x_1, x_2, x_3, x_4, \ldots) = (0, 1, 1, 7, 13, \ldots)$
We are looking for constants p and q such that
$(x_0, x_1, x_2, x_3, \ldots) = (0, 1, 1, 7, \ldots) = p(1, 3, 9, 27, \ldots) + q(1, -2, 4, -8, \ldots)$.
To find p and q, it suffices to consider the first two components: $\begin{vmatrix} 0 = p + q \\ 1 = 3p - 2q \end{vmatrix}$, so that $p = \frac{1}{5}$ and $q = -\frac{1}{5}$.

Thus $x_n = \frac{1}{5}3^n - \frac{1}{5}(-2)^n$.

25. Suppose $\mathcal{B}$ consists of $f_1, \ldots, f_n$. Suppose $f = c_1f_1 + c_2f_2 + \cdots + c_nf_n$ and $g = d_1f_1 + d_2f_2 + \cdots + d_nf_n$, so that $f + g = (c_1 + d_1)f_1 + (c_2 + d_2)f_2 + \cdots + (c_n + d_n)f_n$.

Then $[f+g]_{\mathcal{B}} = \begin{bmatrix} c_1 + d_1 \\ c_2 + d_2 \\ \vdots \\ c_n + d_n \end{bmatrix} = \begin{bmatrix} c_1 \\ c_2 \\ \vdots \\ c_n \end{bmatrix} + \begin{bmatrix} d_1 \\ d_2 \\ \vdots \\ d_n \end{bmatrix} = [f]_{\mathcal{B}} + [g]_{\mathcal{B}}$, as claimed.

26. Note that $\int_0^1 x^n dx = \frac{1}{n+1}$, so that $f(x) = x^n - \frac{1}{n+1}$ is in ker(I) for all $n \geq 0$. This allows us to generate as many linearly independent functions in ker(I) as we want, showing that ker(I) is infinite-dimensional.

27. L is an isomorphism by Fact 9.2.7, since $\dim(\mathbb{R}^n) = \dim(\text{im}(A)) = \text{rank}(A) = n$ and $\ker(L) = \ker(A) = \{\vec{0}\}$, by Fact 3.3.5.
If $\vec{y}$ is in im(A), then we can use Fact 4.4.7 to solve the equation $L(\vec{x}) = A\vec{x} = \vec{y}$ for $\vec{x}$: $\vec{x} = (A^TA)^{-1}A^T\vec{y}$ (note that the least-squares solution is the exact solution when $\vec{y}$ is in im(A)).

28. We can use Fact 9.2.7 to show that L is an isomorphism: $\dim(\text{im}(A^T)) = \text{rank}(A^T)$ equals $\dim(\text{im}(A)) = \text{rank}(A)$, and the kernel of L consists of all $\vec{x}$ in $\text{im}(A^T)$ such that $A\vec{x} = \vec{0}$, that is, $\ker(L) = \text{im}(A^T) \cap \ker(A) = \{\vec{0}\}$, by Fact 4.4.1. In general, $F(\vec{y}) = A^T\vec{y}$ fails to be the inverse of L, as the example $A = \begin{bmatrix} 2 & 0 \\ 0 & 2 \end{bmatrix}$ shows.

29. a. Let A be the matrix of T with respect to some basis $\mathcal{B}$ of V. The matrix A has at least one complex eigenvalue λ, by the Fundamental Theorem of Algebra, and λ will be an eigenvalue of T as well.

b. Suppose V is in the eigenspace of A associated with the eigenvalue λ. We have to show that $B\vec{x}$ is in V for all $\vec{x}$ in V, that is, $A(B\vec{x}) = \lambda(B\vec{x})$. Now $AB\vec{x} = BA\vec{x} = B\lambda\vec{x} = \lambda B\vec{x}$, as claimed. By part a, the transformation $T(\vec{x}) = B\vec{x}$ from V to V has an eigenvector $\vec{v}$, and $\vec{v}$ is an eigenvector for A as well since $\vec{v}$ is in V, so that $A\vec{v} = \lambda\vec{v}$.

30. Define T from $\mathbb{C}$ to C by $T(p+iq) = \begin{bmatrix} p & -q \\ q & p \end{bmatrix}$. It is easy to check that T is an isomorphism, with inverse

$T^{-1}\begin{bmatrix} p & -q \\ q & p \end{bmatrix} = p + iq$.

If $z = p + iq$ and $w = a + ib$, then $T(zw) = T(pa - qb + i(pb + qa)) = \begin{bmatrix} pa - qb & -pb - qa \\ pb + qa & pa - qb \end{bmatrix}$ and

$T(z)T(w) = \begin{bmatrix} p & -q \\ q & p \end{bmatrix}\begin{bmatrix} a & -b \\ b & a \end{bmatrix} = \begin{bmatrix} pa - qb & -pb - qa \\ qa + pb & -qb + pa \end{bmatrix}$; the results agree, as claimed.

31. As in Exercise 30, define T from $\mathbb{C}$ to C by $T(p+iq)=\begin{bmatrix} p & -q \\ q & p \end{bmatrix}$. T is an isomorphism such that $T(zw)=T(z)T(w)$.

Now define F from $\mathbb{H}$ to H by $F\begin{bmatrix} w & -\bar{z} \\ z & \bar{w} \end{bmatrix}=\begin{bmatrix} L(w) & -L(\bar{z}) \\ L(z) & L(\bar{w}) \end{bmatrix}$. It is straightforward to check that F is an isomorphism of linear spaces.

Also, $F\left(\begin{bmatrix} w_1 & -\bar{z}_1 \\ z_1 & \bar{w}_1 \end{bmatrix}\begin{bmatrix} w_2 & -\bar{z}_2 \\ z_2 & \bar{w}_2 \end{bmatrix}\right) F\begin{bmatrix} w_1w_2-\bar{z}_1z_2 & * \\ z_1w_2+\bar{w}_1z_2 & * \end{bmatrix}=\begin{bmatrix} L(w_1)L(w_2)-L(\bar{z}_1)L(z_2) & * \\ L(z_1)L(w_2)+L(\bar{w}_1)L(z_2) & * \end{bmatrix}$ and

$F\begin{bmatrix} w_1 & -\bar{z}_1 \\ z_1 & \bar{w}_1 \end{bmatrix}F\begin{bmatrix} w_2 & -\bar{z}_2 \\ z_2 & \bar{w}_2 \end{bmatrix}=\begin{bmatrix} L(w_1) & -L(\bar{z}_1) \\ L(z_1) & L(\bar{w}_1) \end{bmatrix}\begin{bmatrix} L(w_2) & -L(\bar{z}_2) \\ L(z_2) & L(\bar{w}_2) \end{bmatrix}=\begin{bmatrix} L(w_1)L(w_2)-L(\bar{z}_1)L(z_2) & * \\ L(z_1)L(w_2)+L(\bar{w}_1)L(z_2) & * \end{bmatrix}$.

The results agree (we know that the second columns will "work out," since the results are in H).

32. As the hint suggests, we find the kernel of $M=\begin{bmatrix} f_1(a_1) & f_2(a_1) & \cdots & f_n(a_1) \\ f_1(a_2) & f_2(a_2) & \cdots & f_n(a_2) \\ \vdots & \vdots & & \vdots \\ f_1(a_n) & f_2(a_n) & \cdots & f_n(a_n) \end{bmatrix}$.

If $M\begin{bmatrix} c_1 \\ c_2 \\ \vdots \\ c_n \end{bmatrix}=\begin{bmatrix} f_1(a_1) & f_2(a_1) & \cdots & f_n(a_1) \\ f_1(a_2) & f_2(a_2) & \cdots & f_n(a_2) \\ \vdots & \vdots & & \vdots \\ f_1(a_n) & f_2(a_n) & \cdots & f_n(a_n) \end{bmatrix}\begin{bmatrix} c_1 \\ c_2 \\ \vdots \\ c_n \end{bmatrix}=\begin{bmatrix} c_1f_1(a_1)+\cdots+c_nf_n(a_1) \\ c_1f_1(a_2)+\cdots+c_nf_n(a_n) \\ \vdots \\ c_1f_1(a_n)+\cdots+c_nf_n(a_n) \end{bmatrix}=\begin{bmatrix} 0 \\ 0 \\ \vdots \\ 0 \end{bmatrix}$, then the polynomial $c_1f_1+\cdots+c_nf_n$ in P_{n-1} has at least n zeros, namely, $a_1, a_2, \ldots, a_n$. It follows that $c_1f_1+c_2f_2+\cdots+c_nf_n=0$ and therefore $c_1=c_2=\cdots=c_n=0$ since the f_i are linearly independent. We have shown that $\ker(M)=\{\vec{0}\}$, so that M is invertible.

33. We need to show that there are constants $w_1, \ldots, w_n$ such that

$$w_1f_1(a_1)+w_2f_1(a_2)+\cdots+w_nf_1(a_n)=\int_{-1}^{1} f_1$$

$$w_1f_2(a_1)+w_2f_2(a_2)+\cdots+w_nf_2(a_n)=\int_{-1}^{1} f_2$$

$$\vdots \qquad \vdots$$

$$w_1f_n(a_1)+w_2f_n(a_2)+\cdots+w_nf_n(a_n)=\int_{-1}^{1} f_n.$$

The coefficient matrix of this system is invertible, since it is the transpose of the invertible matrix M discussed in Exercise 32; therefore, the system has a unique solution $w_1, \ldots, w_n$. Now if f is any polynomial in P_{n-1} and

$f=\sum_{j=1}^{n} c_jF_j$, then $\int_{-1}^{1} f=\sum_{j=1}^{n} c_j\int_{i=1}^{n} f_j=\sum_{j=1}^{n} c_j\sum_{i=1}^{n} w_if_j(a_i)=\sum_{i=1}^{n} w_i\sum_{j=1}^{n} c_jf_j(a_i)=\sum_{i=1}^{n} w_if(a_i)$, as claimed.

34. If we work with the basis $f_1(t)=1$, $f_2(t)=t$, and $f_3(t)=t^2$ of P_2, then we have to solve the system

$\begin{vmatrix} w_1+w_2+w_3=2 \\ -w_1+w_2 \quad =0 \\ w_1+w_2 \quad =\frac{2}{3} \end{vmatrix}$ (see Exercise 33). The solution is $w_1=w_3=\frac{1}{3}$ and $w_2=\frac{4}{3}$, so that

$\int_{-1}^{1} f=\frac{1}{3}f(-1)+\frac{4}{3}f(0)+\frac{1}{3}f(1)$ for f in P_2. This is what you get when you apply Simpson's Rule (with two subintervals) to f; note that Simpson's Rule gives the exact value of the integral in this case.

35. Let M be the matrix of T with respect to some basis $\mathcal{B}$ of V, and define $\det(T) = \det(M)$.
We need to check that this value is independent of the choice of the basis. Indeed, if N is the matrix of T with respect to some other basis $\mathcal{C}$, then $N = S^{-1}MS$, where S is the matrix of the transformation $T_{\mathcal{B}} \circ T_{\mathcal{C}}^{-1}$. Thus M and N are similar and $\det(N) = \det(M)$.

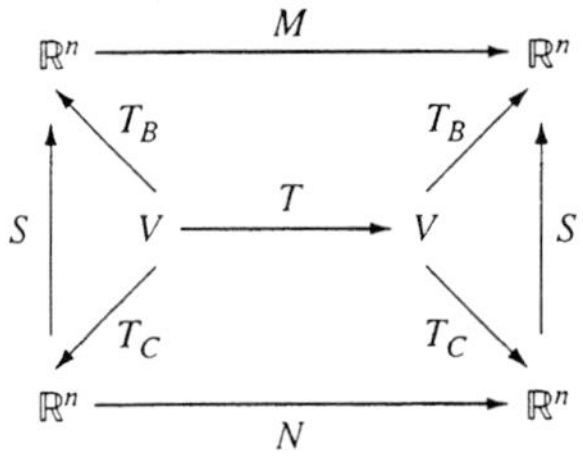

9.3

1. Since f is nonzero, there is a c on $[a, b]$ such that $f(c) = d \neq 0$. By continuity, there is an $\varepsilon > 0$ such that $|f(x)| > \frac{|d|}{2}$ for all x on the open interval $(c - \varepsilon, c + \varepsilon)$ where f is defined (see any good Calculus text).

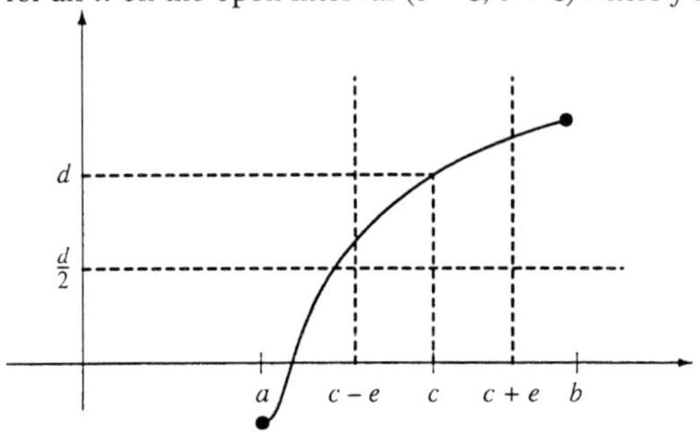

We assume that $c - \varepsilon \geq a$ and $c + \varepsilon \leq b$ and leave the other cases to the reader. Then
$\langle f, f\rangle = \int_a^b (f(t))^2\, dt \geq \int_{c-\varepsilon}^{c+\varepsilon} \left(\frac{d}{2}\right)^2 dt = \frac{d^2\varepsilon}{2} > 0$, as claimed.

2. $\langle f, g+h\rangle = \langle g+h, f\rangle = \langle g, f\rangle + \langle h, f\rangle = \langle f, g\rangle + \langle f, h\rangle$

↑	↑	↑
Definition 9.3.1a	Definition 9.3.1b	Definition 9.3.1a

3. Note that $\langle \vec{x}, \vec{y}\rangle = (S\vec{x})^T s\vec{y} = S\vec{x} \cdot S\vec{y}$.

a. We will check the four parts of Definition 9.3.1.

α. $\langle \vec{x}, \vec{y}\rangle = S\vec{x} \cdot S\vec{y} = S\vec{y} \cdot S\vec{x} = \langle \vec{y}, \vec{x}\rangle$

β. $\langle \vec{x}+\vec{y}, \vec{z}\rangle = S(\vec{x}+\vec{y}) \cdot S\vec{z} = (S\vec{x}+S\vec{y}) \cdot S\vec{z} = (S\vec{x} \cdot S\vec{z}) + (S\vec{y} \cdot S\vec{z}) = \langle \vec{x}, \vec{z}\rangle + \langle \vec{y}, \vec{z}\rangle$

γ. $\langle c\vec{x}, \vec{y}\rangle = S(c\vec{x}) \cdot S\vec{y} = c(S\vec{x}) \cdot S\vec{y} = c\langle \vec{x}, \vec{y}\rangle$

δ. If $\vec{x} \neq 0$, then $\langle \vec{x}, \vec{x}\rangle = S\vec{x} \cdot S\vec{x} = \|S\vec{x}\|^2$ is positive if $S\vec{x} \neq \vec{0}$, that is, if $\vec{x}$ is not in the kernel of S. It is required that $S\vec{x} \neq \vec{0}$ whenever $\vec{x} \neq \vec{0}$, that is, $\ker(S) = \{\vec{0}\}$, that is, S must be invertible.

Answer: S must be invertible.

b. It is required that $\langle \vec{x}, \vec{y}\rangle = (S\vec{x})^T S\vec{y} = \vec{x}^T S^T S\vec{y}$ equal $\vec{x} \circ \vec{y} = \vec{x}^T \vec{y}$ for all $\vec{x}$ and $\vec{y}$. This is the case if and only if $S^T S = I_n$, that is, S *is orthogonal*.

4. **a.** For column vectors $\vec{v}$, $\vec{w}$, $\langle \vec{v}, \vec{w} \rangle = \text{trace}(\vec{v}^T \vec{w}) = \text{trace}(\vec{v} \cdot \vec{w}) = \vec{v} \cdot \vec{w}$, the dot product.

b. For row vectors $\vec{v}$, $\vec{w}$, the ijth entry of $\vec{v}^T \vec{w}$ is $v_i w_j$, so that $\langle \vec{v}, \vec{w} \rangle = \text{trace}(\vec{v}^T \vec{w}) = \sum_{i=1}^{n} v_i w_i = \vec{v} \cdot \vec{w}$, again the dot product.

5. **a.** $\langle\langle A, B \rangle\rangle = \text{tr}(AB^T) = \text{tr}((AB^T)^T) = \text{tr}(BA^T) = \langle\langle B, A \rangle\rangle$

In the second step we have used the fact that $\text{tr}(M) = \text{tr}(M^T)$, for any square matrix M.

b. $\langle\langle A+B, C \rangle\rangle = \text{tr}((A+B)C^T) = \text{tr}(AC^T + BC^T) = \text{tr}(AC^T) + \text{tr}(BC^T) = \langle\langle A, C \rangle\rangle + \langle\langle B, C \rangle\rangle$

c. $\langle\langle cA, B \rangle\rangle = \text{tr}(cAB^T) = c\text{tr}(AB^T) = c\langle\langle A, B \rangle\rangle$

d. $\langle\langle A, A \rangle\rangle = \text{tr}(AA^T) = \sum_{i=1}^{n} \|\vec{v}_i\|^2 > 0$ if $A \neq 0$, where $\vec{v}_i$ is the ith row of A.

We have shown that $\langle\langle ., . \rangle\rangle$ does indeed define an inner product.

6. **a.** The iith entry of PQ is $\sum_{k=1}^{n} p_{ik} q_{ki}$, so that $\text{tr}(PQ) = \sum_{i=1}^{m} \sum_{k=1}^{n} p_{ik} q_{ki}$.

Likewise, $\text{tr}(QP) = \sum_{i=1}^{n} \sum_{k=1}^{m} q_{ik} p_{ki}$.

Reversing the roles of i and k and the order of summation we see that $\text{tr}(PQ) = \text{tr}(QP)$, as claimed.

b. Using part a and the fact that $\text{tr}(M) = \text{tr}(M^T)$ for any square matrix M, we find that $\langle A, B \rangle = \text{tr}(A^T B) = \text{tr}(BA^T) = \text{tr}((BA^T)^T) = \text{tr}(AB^T) = \langle\langle A, B \rangle\rangle$

7. Axioms a, b, and c hold for any choice of k (check this!). Also, it is required that $\langle\langle v, v \rangle\rangle = k\langle v, v \rangle$ be positive for nonzero v. Since $\langle v, v \rangle$ is positive, this is the case if (and only if) *k is positive*.

8. By parts b and c of Definition 9.3.1, we have $T(u+v) = \langle u+v, w \rangle = \langle u, w \rangle + \langle v, w \rangle = T(u) + T(v)$ and $T(cv) = \langle cv, w \rangle = c\langle v, w \rangle = cT(v)$, so that T is linear. If $w = 0$, then $\text{im}(T) = \{0\}$ and $\ker(T) = V$. If $w \neq 0$, then $\text{im}(T) = \mathbb{R}$ and $\ker(T)$ consists of all v perpendicular to w.

9. *True*; if f is even and g is odd, then fg is odd, so that $\langle f, g \rangle = \int_{-1}^{1} fg = 0$.

10. A function $g(t) = a + bt + ct^2$ is orthogonal to $f(t) = t$ if

$\langle f, g \rangle = \int_{-1}^{1} (at + bt^2 + ct^3)dt = \left[\frac{a}{2}t^2 + \frac{b}{3}t^3 + \frac{c}{4}t^4\right]_{-1}^{1} = \frac{2}{3}b = 0$, that is, if $b = 0$.

Thus, the functions 1 and t^2 form a basis of the space of all functions in P_2 orthogonal to $f(t) = t$.

To find an *orthonormal* basis $g_1(t)$, $g_2(t)$, we apply Gram-Schmidt. Now $\|1\| = \frac{1}{2}\int_{-1}^{1} 1\, dt = 1$, so that we can let

$g_1(t) = 1$. Then $g_2|t| = \dfrac{t^2 - \langle 1, t^2\rangle 1}{\left\|t^2 - \langle 1, t^2\rangle 1\right\|} = \dfrac{t^2 - \frac{1}{3}}{\left\|t^2 - \frac{1}{3}\right\|} = \dfrac{\sqrt{5}}{2}(3t^2 - 1)$

Answer: $g_1|t| = 1$, $g_2|t| = \dfrac{\sqrt{5}}{2}(3t^2 - 1)$

11. $\langle f, g\rangle = \langle \cos(t), \cos(t+\delta)\rangle = \langle \cos(t), \cos(t)\cos(\delta) - \sin(t)\sin(\delta)\rangle$
$= \cos(\delta)\langle \cos(t), \cos(t)\rangle - \sin(\delta)\langle \cos(t), \sin(t)\rangle = \cos(\delta)$, by Fact 9.3.4.
Also, $\langle g, g\rangle = 1$ (left to reader).
Thus, $\angle(f, g) = \arccos\left(\dfrac{\langle f, g\rangle}{\|f\|\|g\|}\right) = \arccos(\cos\delta) = \delta$.

12. By Fact 9.3.5, $a_0 = \left\langle |t|, \dfrac{1}{\sqrt{2}}\right\rangle = \dfrac{1}{\sqrt{2}\pi}\displaystyle\int_{-\pi}^{\pi} |t|dt = \dfrac{\pi}{\sqrt{2}}$, $b_k = \langle |t|, \sin(kt)\rangle = \dfrac{1}{\pi}\displaystyle\int_{-\pi}^{\pi} |t|\sin(kt)dt = 0$, since the integrand is an odd function.

$$c_k = \langle |t|, \cos(kt)\rangle = \frac{1}{\pi}\int_{-\pi}^{\pi} |t|\cos(kt)dt = \frac{2}{\pi}\int_0^{\pi} t\cos(kt)dt = \frac{2}{\pi}\left\{\left[\frac{1}{k}t\sin(kt)\right]_0^{\pi} - \int_0^{\pi}\frac{1}{k}\sin(kt)dt\right\}$$

$$= \frac{2}{\pi}\frac{1}{k^2}[\cos(kt)]_0^{\pi} = \frac{2}{\pi k^2}(\cos(k\pi) - 1) = \begin{cases} 0 & \text{if } k \text{ is even} \\ -\dfrac{4}{k^2\pi} & \text{if } k \text{ is odd} \end{cases}$$

Summary:
$a_0 = 0$
$b_k = 0$
$$c_k = \begin{cases} 0 & \text{if } k \text{ is even} \\ -\dfrac{4}{k^2\pi} & \text{if } k \text{ is odd} \end{cases}$$

13. The sequence $(a_0, b_1, c_1, b_2, c_2, \ldots)$ is "square-summable" by Fact 9.3.6, so that it is in ℓ_2. Also, $\|(a_0, b_1, c_1, b_2, c_2, \ldots)\|^2 = a_0^2 + b_1^2 + c_1^2 + b_2^2 + c_2^2 + \cdots = \|f\|^2$, by Fact 9.3.6, so that the two norms are equal.

14. **a.** This is not an inner product since there are nonzero polynomials $f(t)$ in P_2 with $f(1) = f(2) = 0$, so that $\langle f, f\rangle = (f(1))^2 + (f(2))^2 = 0$. (For example, let $f(t) = (t-1)(t-2)$.)

b. This is an inner product. We leave it to the reader to check axioms a to c. As for d: A nonzero polynomial f in P_2 has at most two zeros, so that $f(1) \neq 0$ *or* $f(2) \neq 0$ *or* $f(3) \neq 0$, and $\langle f, f\rangle = (f(1))^2 + (f(2))^2 + (f(3))^2 > 0$.

15. We can write $\langle \vec{x}, \vec{y}\rangle = \vec{x}^T A\vec{y}$, where $A = \begin{bmatrix} a & b \\ c & d \end{bmatrix}$. First it is required that $\langle \vec{x}, \vec{y}\rangle = \vec{x}^T A\vec{y}$ equal $\langle \vec{y}, \vec{x}\rangle = \vec{y}^T A\vec{x} = (\vec{y}^T A\vec{x})^T = \vec{x}^T A^T\vec{y}$ for all $\vec{x}$ and $\vec{y}$; this is the case if (and only if) A *is symmetric*.
We leave it to the reader to check that axioms b and c hold for any choice of A.
Finally, we want that $\langle \vec{x}, \vec{x}\rangle = \vec{x}^T A\vec{x} > 0$ for all nonzero $\vec{x}$; in other words, A *must be positive definite*.

Summary: The matrix $A = \begin{bmatrix} a & b \\ c & d \end{bmatrix}$ must be symmetric and positive definite. This means that $b = c$, $a > 0$, and $ad - bc = ad - b^2 > 0$.

16. a. We start with the standard basis 1, t and use the Gram-Schmidt process to construct an *orthonormal* basis $g_1(t)$, $g_2(t)$.

$\|1\| = \sqrt{\int_0^1 1\,dt} = 1$, so that we can let $g_1(t) = 1$. Then $g_2(t) = \dfrac{t - \langle 1, t\rangle 1}{\|t - \langle 1, t\rangle 1\|} = \dfrac{t - \frac{1}{2}}{\|t - \frac{1}{2}\|} = \sqrt{3}(2t-1)$.

Summary: $g_1(t) = 1$ and $g_2(t) = \sqrt{3}(2t-1)$ is an orthonormal basis.

b. We are looking for $\text{proj}_{P_1}(t^2) = \left\langle g_1(t), t^2\right\rangle g_1(t) + \left\langle g_2(t), t^2\right\rangle g_2(t)$, by Fact 9.3.3.

We find that $\left\langle g_1(t), t^2\right\rangle = \int_0^1 t^2 dt = \frac{1}{3}$ and $\left\langle g_2(t), t^2\right\rangle = \sqrt{3}\int_0^1 (2t^3 - t^2)dt = \frac{\sqrt{3}}{6}$, so that

$\text{proj}_{P_1} t^2 = \frac{1}{3} + \frac{1}{2}(2t-1) = t - \frac{1}{6}$.

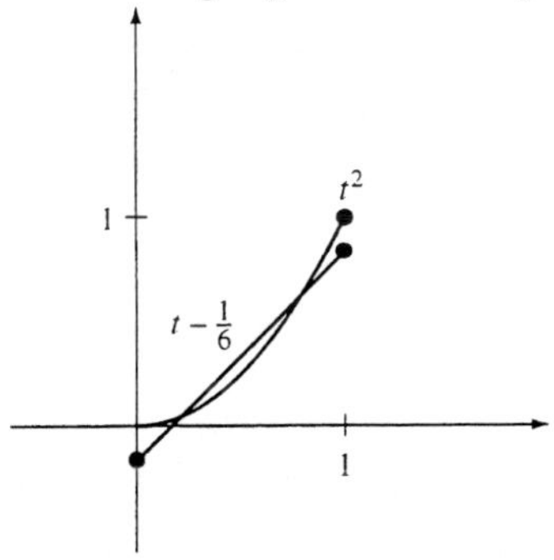

17. We leave it to the reader to check that the first three axioms are satisfied for any such T. As for axiom d: It is required that $\langle v, v\rangle = T(v)\cdot T(v) = \|T(v)\|^2$ be positive for any nonzero v, that is, $T(v) \neq \vec{0}$. This means that the *kernel of T must be* $\{0\}$.

18. Let the orthonormal basis be $f_1, \ldots, f_n$ and $f = c_1 f_1 + \cdots + c_n f_n$; then $[f]_{\mathcal{B}} = \begin{bmatrix} c_1 \\ c_2 \\ \vdots \\ c_n \end{bmatrix}$. The Pythagorean Theorem tells us that $\|f\|^2 = c_1^2 + \cdots + c_n^2 = \|[f]_{\mathcal{B}}\|^2$, so that $\|f\| = \|[f]_{\mathcal{B}}\|$.

19. Arguing exactly as in Exercise 15, we see that A must be symmetric and positive definite.

20. a. We want all vectors $\vec{v}$ such that $\vec{v}^T \begin{bmatrix} 5 & 2 \\ 2 & 8 \end{bmatrix}\begin{bmatrix} 1 \\ 0 \end{bmatrix} = \vec{v}^T \begin{bmatrix} 5 \\ 2 \end{bmatrix} = \vec{v} \cdot \begin{bmatrix} 5 \\ 2 \end{bmatrix} = 0$. These are the vectors on $\text{span}\begin{bmatrix} -2 \\ 5 \end{bmatrix}$.

b. We want all vectors $\vec{v}$ such that $\|\vec{v}\|^2 = \langle \vec{v}, \vec{v} \rangle = \vec{v}^T \begin{bmatrix} 5 & 2 \\ 2 & 8 \end{bmatrix} \vec{v} = 1$. Using the methods of Section 7.4, we find that these vectors form the *ellipse* sketched below.

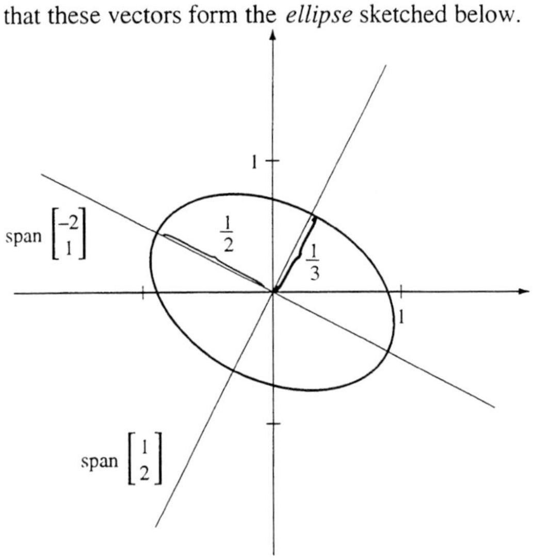

21. We can write $q(\vec{x}) = \vec{x}^T A \vec{x}$ for some symmetric matrix A. Then

$$\langle \vec{v}, \vec{w} \rangle = q(\vec{v} + \vec{w}) - q(\vec{v}) - q(\vec{w}) = (\vec{v} + \vec{w})^T A(\vec{v} + \vec{w}) - \vec{v}^T A\vec{v} - \vec{w}^T A\vec{w} = \vec{v}^T A\vec{w} + \vec{w}^T A\vec{v}$$

$$= \vec{v}^T A\vec{w} + (\vec{w}^T A\vec{v})^T \quad \left(\text{since } \vec{w}^T A\vec{v} \text{ is } |x|\right)$$

$$= \vec{v}^T A\vec{w} + \vec{v}^T A\vec{w} \text{ (since } A \text{ is symmetric)}$$

$$= \vec{v}^T (2A)\vec{w}$$

By Exercise 19, this is an inner product.

22. Apply the Cauchy-Schwarz inequality to $f(t)$ and $g(t) = 1$; note that $\|g\| = 1$:

$|\langle f, g \rangle| \le \|f\|\|g\| = \|f\|$ or $\langle f, g \rangle^2 \le \|f\|^2$ or $\left(\int_0^1 f(t)dt \right)^2 \le \int_0^1 (f(t))^2 \, dt$, as claimed.

23. We start with the standard basis $1, t$ of P_1 and use the Gram-Schmidt process to construct an *orthonormal basis* $g_1(t), g_2(t)$.

$\|1\| = \sqrt{\frac{1}{2}(1 \cdot 1) + 1 \cdot 1} = 1$, so that we can let $g_1|t| = 1$. Then $g_2|t| = \dfrac{t - \langle 1, t \rangle 1}{\|t - (1, t)1\|} = \dfrac{t - \frac{1}{2}}{\|t - \frac{1}{2}\|} = 2t - 1$.

Summary: $g_1|t| = 1$ and $g_2|t| = 2t - 1$ is an orthonormal basis.

24. a. $\langle f, g+h \rangle = \langle f, g \rangle + \langle f, h \rangle = 0 + 8 = 8$

b. $\|g+h\| = \sqrt{\langle g+h, g+h \rangle} = \sqrt{\langle g, g \rangle + 2\langle g, h \rangle + \langle h, h \rangle} = \sqrt{1+6+50} = \sqrt{57}$

c. Since $\langle f, g \rangle = 0$, $\|g\| = 1$, and $\|f\| = 2$, $\dfrac{f}{2}, g$ is an orthonormal basis of span(f, g).

Now $\text{proj}_E h = \left\langle \dfrac{f}{2}, h \right\rangle \dfrac{f}{2} + \langle g, h \rangle g = \dfrac{1}{4} \langle f, h \rangle f + \langle g, h \rangle g = 2f + 3g$.

d. From part c we know that $\frac{1}{2}f,\ g$ are orthonormal, so we apply Algorithm 4.2.1 to obtain the third polynomial in an orthonormal basis of span(f, g, h):

$$\frac{h-\text{proj}_E h}{\|h-\text{proj}_E h\|}=\frac{h-2f-3g}{\|h-2f-3g\|}=\frac{h-2f-3g}{5}=-\frac{2}{5}f-\frac{3}{5}g+\frac{1}{5}h$$

Orthonormal basis: $\frac{1}{2}f,\ g,\ -\frac{2}{5}f-\frac{3}{5}g+\frac{1}{5}h$

25. Using the inner product defined in Example 2, we find that

$\|\vec{x}\|=\sqrt{\langle\vec{x},\vec{x}\rangle}=\sqrt{1+\frac{1}{4}+\frac{1}{9}+\cdots+\frac{1}{n^2}+\cdots}=\sqrt{\frac{\pi^2}{6}}=\frac{\pi}{\sqrt{6}}$ (see the text right after Fact 9.3.6).

26. $a_0=\frac{1}{\sqrt{2\pi}}\int_{-\pi}^{\pi}f(t)dt=0$

$$b_k=\frac{1}{\pi}\int_{-\pi}^{\pi}f(t)\sin(kt)dt=\frac{1}{\pi}\left\{-\int_{-\pi}^{0}\sin(kt)dt+\int_{0}^{\pi}\sin(kt)dt\right\}=\frac{2}{\pi}\int_{0}^{\pi}\sin(kt)dt=-\frac{2}{k\pi}[\cos(kt)]_0^{\pi}=\begin{cases}0 & \text{if } k \text{ is even}\\ \frac{4}{\pi k} & \text{if } k \text{ is odd}\end{cases}$$

$c_k=\frac{1}{\pi}\int_{-\pi}^{\pi}f(t)\cos(kt)dt=0$, since the integrand is odd.

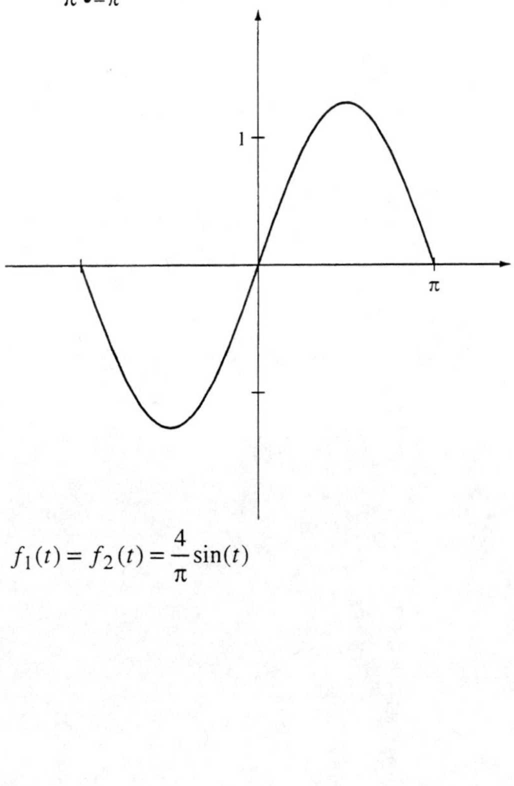

$f_1(t)=f_2(t)=\frac{4}{\pi}\sin(t)$

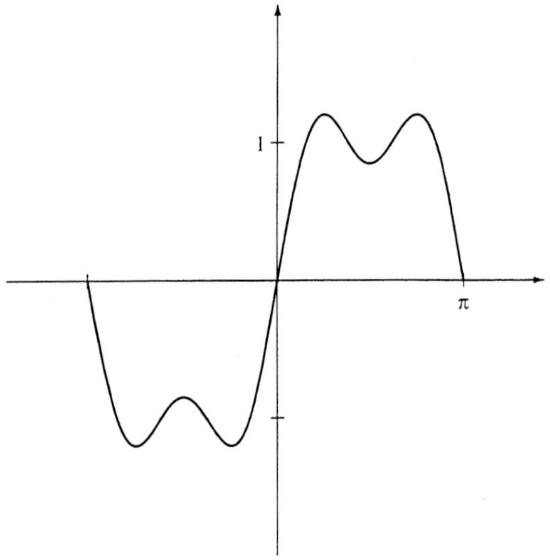

$$f_3(t) = f_4(t) = \frac{4}{\pi}\sin(t) + \frac{4}{3\pi}\sin(3t)$$

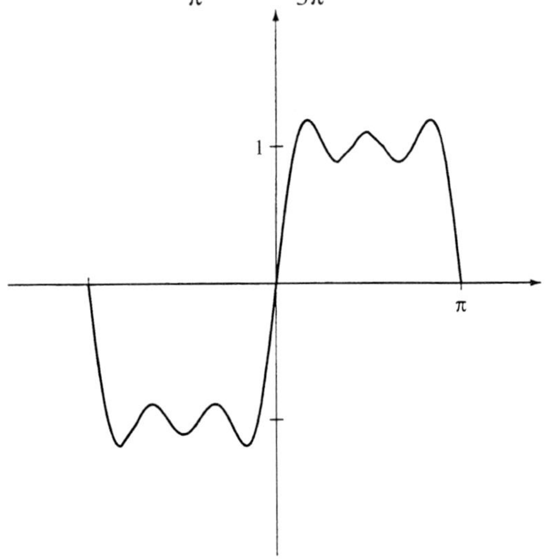

$$f_5(t) = f_6(t) = \frac{4}{\pi}\sin(t) + \frac{4}{3\pi}\sin(3t) + \frac{4}{5\pi}\sin(5t)$$

27. $a_0 = \frac{1}{\sqrt{2}\pi}\int_{-\pi}^{\pi} f(t)dt = \frac{1}{\sqrt{2}}$

$$b_k = \frac{1}{\pi}\int_{-\pi}^{\pi} f(t)\sin(kt)dt = \frac{1}{\pi}\int_0^{\pi} \sin(kt)dt = -\frac{1}{k\pi}[\cos(kt)]_0^{\pi} = \begin{cases} 0 & \text{if } k \text{ is even} \\ \frac{2}{k\pi} & \text{if } k \text{ is odd} \end{cases}$$

$$c_k = \frac{1}{\pi}\int_{-\pi}^{\pi} f(t)\cos(kt)dt = \frac{1}{\pi}\int_0^{\pi} \cos(kt)dt = \frac{1}{k\pi}[\sin(kt)]_0^{\pi} = 0$$

$f_1(t) = f_2(t) = \frac{1}{2} + \frac{2}{\pi}\sin(t)$

$f_3(t) = f_4(t) = \frac{1}{2} + \frac{2}{\pi}\sin(t) + \frac{2}{3\pi}\sin(3t)$

$\vdots$

28. $\|f\|^2 = \langle f, f\rangle = \frac{1}{\pi}\int_{-\pi}^{\pi}(f(t))^2\,dt = \frac{1}{\pi}\int_{-\pi}^{\pi}1\,dt = 2$

Now Fact 9.3.6 tells us that $\frac{16}{\pi^2} + \frac{16}{9\pi^2} + \frac{16}{25\pi^2} + \cdots = \frac{16}{\pi^2}\left(\sum_{k\text{ odd}}\frac{1}{k^2}\right) = 2$, or

$\sum_{k\text{ odd}}\frac{1}{k^2} = 1 + \frac{1}{q} + \frac{1}{25} + \frac{1}{49} + \cdots = \frac{\pi^2}{8}$.

29. $\|f\|^2 = \langle f, f\rangle = \frac{1}{\pi}\int_{-\pi}^{\pi}(f(t))^2\,dt = \frac{1}{\pi}\int_{0}^{\pi}1\,dt = 1$

Now Fact 9.3.6 tells us that $\frac{1}{2} + \frac{4}{\pi^2} + \frac{4}{9\pi^2} + \frac{4}{25\pi^2} + \cdots = 1$, or $\frac{1}{2} + \frac{4}{\pi^2}\left(\sum_{k\text{ odd}}\frac{1}{k^2}\right) = 1$ or

$\sum_{k\text{ odd}}\frac{1}{k^2} = 1 + \frac{1}{9} + \frac{1}{25} + \frac{1}{49} + \cdots = \frac{\frac{1}{2}}{\frac{4}{\pi^2}} = \frac{\pi^2}{8}$.

30. There is an invertible linear transformation $T(\vec{x}) = A\vec{x}$ from $\mathbb{R}^2$ to $\mathbb{R}^2$ that transforms E into the unit circle. (If E is parametrized by $\cos(t)\vec{w}_1 + \sin(t)\vec{w}_2$, let $A = [\vec{w}_1\ \ \vec{w}_2]^{-1}$. Compare with Exercises 2.2.50 and 2.2.51.) Now $\vec{x}$ is on E if $\|T(\vec{x})\| = 1$, or $T(\vec{x})\cdot T(\vec{x}) = 1$. This means that the inner product $\langle\vec{x}, \vec{y}\rangle = T(\vec{x})\cdot T(\vec{y})$ does the job (see Exercise 17). (There are other, very different, approaches to this problem.)

31. An orthonormal basis of P_2 of the desired form is $f_0(t) = \frac{1}{\sqrt{2}}$, $f_1(t) = \sqrt{\frac{3}{2}}t$, $f_2(t) = \frac{1}{2}\sqrt{\frac{5}{2}}(3t^2 - 1)$ (compare with Exercise 10), and the zeros of $f_2(t)$ are $a_{1,2} = \pm\frac{1}{\sqrt{3}}$.

Next we find the weights w_1, w_2 such that $\int_{-1}^{1}f(t)dt = \sum_{i=1}^{2}w_i f(a_i)$ for all f in P_1. We need to make sure that the equation holds for 1 and t.

$$\left|\begin{array}{l} 2 = w_1 + w_2 \\ 0 = \frac{1}{\sqrt{3}}w_1 - \frac{1}{\sqrt{3}}w_2 \end{array}\right|, \text{ with solution } w_1 = w_2 = 1.$$

It follows that the equation $\int_{-1}^{1}f(t)dt = f(a_1) + f(a_2) = f\left(\frac{1}{\sqrt{3}}\right) + f\left(-\frac{1}{\sqrt{3}}\right)$ holds for all polynomials f in P_1. We can check that it holds for t^2 and t^3 as well, that is, it holds in fact for all cubic polynomials.

$\int_{-1}^{1}t^2dt = \frac{2}{3}$ equals $\left(\frac{1}{\sqrt{3}}\right)^2 + \left(-\frac{1}{\sqrt{3}}\right)^2 = \frac{2}{3}$, and $\int_{-1}^{1}t^3dt = 0$ equals $\left(\frac{1}{\sqrt{3}}\right)^3 + \left(-\frac{1}{\sqrt{3}}\right)^3 = 0$.

9.4

1. The characteristic polynomial of this differential equation is $\lambda - 5$, so that $\lambda_1 = 5$. By Fact 9.4.8 the general solution is $f(t) = Ce^{5t}$, where C is an arbitrary constant.

2. The solutions of $\frac{dx}{dt} + 3x = 0$ are of the form $x(t) = Ce^{-3t}$, where C is an arbitrary constant, and the differential equation $\frac{dx}{dt} + 3x = 7$ has the particular solution $x_p(t) = \frac{7}{3}$, so that the general solution is $x(t) = Ce^{-3t} + \frac{7}{3}$ (where C is a constant). Alternatively, we could use Fact 9.4.13.

3. Use Fact 9.1.13, where $a = -2$ and $g(t) = e^{3t}$:
$$f(t) = e^{-2t}\int e^{2t}e^{3t}dt = e^{-2t}\int e^{5t}dt = e^{-2t}\left(\frac{1}{5}e^{5t} + C\right) = \frac{1}{5}e^{3t} + Ce^{-2t},$$ where C is a constant.

4. We can look for a sinusoidal solution $x_p(t) = P\cos(3t) + Q\sin(3t)$, as in Example 7. P and Q need to be chosen in such a way that $-3P\sin(3t) + 3Q\cos(3t) - 2P\cos(3t) - 2Q\sin(3t) = \cos(3t)$ or $\left|\begin{array}{l} -2P + 3Q = 1 \\ -3P - 2Q = 0 \end{array}\right|$ with solution $P = -\frac{2}{13}$ and $Q = \frac{3}{13}$. Since the general solution of $\frac{dx}{dt} - 2x = 0$ is $x(t) = Ce^{2t}$, the general solution of $\frac{dx}{dt} - 2x = \cos(3t)$ is $x(t) = Ce^{2t} - \frac{2}{13}\cos(3t) + \frac{3}{13}\sin(3t)$, where C is an arbitrary constant.

5. Using Fact 9.4.13, $f(t) = e^t\int e^{-t}t\,dt = e^t(-te^{-t} - e^{-t} + C) = Ce^t - t - 1$, where C is an arbitrary constant.

6. Using Fact 9.4.13, $f(t) = e^{2t}\int e^{-2t}e^{2t}dt = e^{2t}\int dt = e^{2t}(t + C)$, where C is an arbitrary constant.

7. By Definition 9.4.6, $p_T(\lambda) = \lambda^2 + \lambda - 12 = (\lambda + 4)(\lambda - 3)$.
Since $p_T(\lambda)$ has distinct roots $\lambda_1 = -4$ and $\lambda_2 = 3$, the solutions of the differential equation are of the form $f(t) = c_1e^{-4t} + c_2e^{3t}$, where c_1 and c_2 are arbitrary constants (by Fact 9.4.8).

8. $p_T(\lambda) = \lambda^2 + 3\lambda - 10 = (\lambda + 5)(\lambda - 2) = 0$
$x(t) = c_1e^{-5t} + c_2e^{2t}$, where c_1, c_2 are arbitrary constants.

9. $p_T(\lambda) = \lambda^2 - 9 = (\lambda - 3)(\lambda + 3) = 0$
$f(t) = c_1e^{3t} + c_2e^{-3t}$, where c_1, c_2 are arbitrary constants.

10. $p_T(\lambda) = \lambda^2 + 1 = 0$ has roots $\lambda_{1,2} = \pm i$. By Fact 9.4.9, $f(t) = c_1\cos(t) + c_2\sin(t)$, where c_1, c_2 are arbitrary constants.

11. $p_T(\lambda) = \lambda^2 - 2\lambda + 2 = 0$ has roots $\lambda_{1,2} = 1 \pm i$. By Fact 9.4.9, $x(t) = e^t(c_1\cos(t) + c_2\sin(t))$, where c_1, c_2 are arbitrary constants.

12. $p_T(\lambda) = \lambda^2 - 4\lambda + 13 = 0$ has roots $\lambda_{1,2} = 2 \pm 3i$. By Fact 9.4.9, $f(t) = e^{2t}(c_1\cos(3t) + c_2\sin(3t))$, where c_1, c_2 are arbitrary constants.

13. $p_T(\lambda) = \lambda^2 + 2\lambda + 1 = (\lambda+1)^2 = 0$ has roots $\lambda_{1,2} = -1$. Following the method of example 10, we find $f(t) = e^{-t}(c_1 t + c_2)$, where c_1, c_2 are arbitrary constants.

14. $p_T(\lambda) = \lambda^2 + 3\lambda = \lambda(\lambda+3) = 0$ has roots $\lambda_1 = 0$, $\lambda_2 = -3$.
$f(t) = c_1 e^{0t} + c_2 e^{-3t} = c_1 + c_2 e^{-3t}$, where c_1, c_2 are arbitrary constants.

15. By integrating twice we find $f(t) = c_1 + c_2 t$, where c_1, c_2 are arbitrary constants.

16. By Fact 9.4.10, the differential equation has a particular solution of the form $f_p(t) = P\cos(t) + Q\sin(t)$. Plugging f_p into the equation we find

$(-P\cos(t) - Q\sin(t)) + 4(-P\sin(t) + Q\cos(t)) + 13(P\cos(t) + Q\sin(t)) = \cos(t)$ or $\begin{vmatrix} 12P + 4Q = 1 \\ -4P + 12Q = 0 \end{vmatrix}$, so

$$P = \frac{3}{40}$$
$$Q = \frac{1}{40}.$$

Therefore, $f_p(t) = \frac{3}{40}\cos(t) + \frac{1}{40}\sin(t)$.

Next we find a basis of the solution space of $f''(t) + 4f'(t) + 13f(t) = 0$. $p_T(\lambda) = \lambda^2 + 4\lambda + 13 = 0$ has roots $-2 \pm 3i$. By Fact 9.4.9, $f_1(t) = e^{-2t}\cos(3t)$ and $f_2(t) = e^{-2t}\sin(3t)$ is a basis of the solution space.

By Fact 9.4.4, the solutions of the original differential equation are of the form $f(t) = c_1 f_1(t) + c_2 f_2(t) + f_p(t) = c_1 e^{-2t}\cos(3t) + c_2 e^{-2t}\sin(3t) + \frac{3}{40}\cos(t) + \frac{1}{40}\sin(t)$, where c_1, c_2 are arbitrary constants.

17. By Fact 9.4.10, the differential equation has a particular solution of the form $f_p(t) = P\cos(t) + Q\sin(t)$. Plugging f_p into the equation we find $(-P\cos(t) - Q\sin(t)) + 2(-P\sin(t) + Q\cos(t)) + P\cos(t) + Q\sin(t) = \sin(t)$ or

$\begin{vmatrix} 2Q = 0 \\ -2P = 1 \end{vmatrix}$, so $P = -\frac{1}{2}$, $Q = 0$.

Therefore, $f_p(t) = -\frac{1}{2}\cos(t)$.

Next we find a basis of the solution space of $f''(t) + 2f'(t) + f(t) = 0$. In Exercise 13 we see that $f_1(t) = e^{-t}$, $f_2(t) = te^{-t}$ is such a basis.

By Fact 9.4.4, the solutions of the original differential equation are of the form $f(t) = c_1 f_1(t) + c_2 f_2(t) + f_p(t) = c_1 e^{-t} + c_2 te^{-t} - \frac{1}{2}\cos(t)$, where c_1, c_2 are arbitrary constants.

18. We follow the approach outlined in Exercises 16 and 17.

- Particular solution $f_p = \frac{1}{10}\cos(t) + \frac{3}{10}\sin(t)$
- Solutions of $f''(t) + 3f'(t) + 2f(t) = 0$ are $f_1(t) = e^{-t}$ and $f_2(t) = e^{-2t}$.
- The solutions of the original differential equation are of the form $f(t) = c_1e^{-t} + c_2e^{-2t} + \frac{1}{10}\cos(t) + \frac{3}{10}\sin(t)$, where c_1 and c_2 are arbitrary constants.

19. We follow the approach outlined in Exercises 16 and 17.

- Particular solution $x_p(t) = \cos(t)$
- Solutions of $\frac{d^2x}{dt^2} + 2x = 0$ are $x_1(t) = \cos\left(\sqrt{2}t\right)$ and $x_2(t) = \sin\left(\sqrt{2}t\right)$.
- The solutions of the original differential equation are of the form $x(t) = c_1\cos\left(\sqrt{2}t\right) + c_2\sin\left(\sqrt{2}t\right) + \cos(t)$, where c_1 and c_2 are arbitrary constants.

20. $p_T(\lambda) = \lambda^3 - 3\lambda^2 + 2\lambda = \lambda(\lambda - 1)(\lambda - 2) = 0$ has roots $\lambda_1 = 0$, $\lambda_2 = 1$, $\lambda_3 = 2$.
By Fact 9.4.8, the general solution is $f(t) = c_1 + c_2e^t + c_3e^{2t}$, where c_1, c_2, c_3 are arbitrary constants.

21. $p_T(\lambda) = \lambda^3 + 2\lambda^2 - \lambda - 2 = (\lambda - 1)(\lambda + 1)(\lambda + 2) = 0$ has roots $\lambda_1 = 1$, $\lambda_2 = -1$, $\lambda_3 = -2$.
By Fact 9.4.8, the general solution is $f(t) = c_1e^t + c_2e^{-t} + c_3e^{-2t}$, where c_1, c_2, c_3 are arbitrary constants.

22. $p_T(\lambda) = \lambda^3 - \lambda^2 - 4\lambda + 4 = (\lambda - 1)(\lambda - 2)(\lambda + 2) = 0$ has roots $\lambda_1 = 1$, $\lambda_2 = 2$, $\lambda_3 = -2$.
By Fact 9.4.8, the general solution is $f(t) = c_1e^t + c_2e^{2t} + c_3e^{-2t}$, where c_1, c_2, c_3 are arbitrary constants.

23. General solution $f(t) = Ce^{5t}$
Plug in: $3 = f(0) = Ce^0 = C$, so that $f(t) = 3e^{5t}$.

24. General solution $x(t) = Ce^{-3t} + \frac{7}{3}$ (Exercise 2)
Plug in: $0 = x(0) = C + \frac{7}{3}$, so that $C = -\frac{7}{3}$ and $x(t) = -\frac{7}{3}e^{-3t} + \frac{7}{3}$.

25. General solution $f(t) = Ce^{-2t}$
Plug in: $1 = f(1) = Ce^{-2}$, so that $C = e^2$ and $f(t) = e^2e^{-2t} = e^{2-2t}$.

26. General solution $f(t) = r_1e^{3t} + c_2e^{-3t}$ (Exercise 9), with $f'(t) = 3c_1e^{3t} - 3c_2e^{-3t}$
Plug in: $0 = f(0) = c_1 + c_2$ and $1 = f'(0) = 3c_1 - 3c_2$, so that $c_1 = \frac{1}{6}$, $c_2 = -\frac{1}{6}$, and $f(t) = \frac{1}{6}e^{3t} - \frac{1}{6}e^{-3t}$.

27. General solution $f(t) = c_1\cos(3t) + c_2\sin(3t)$ (Fact 9.4.9)
Plug in: $0 = f(0) = c_1$ and $1 = f\left(\frac{\pi}{2}\right) = -c_2$, so that $c_1 = 0$, $c_2 = -1$, and $f(t) = -\sin(3t)$.

28. General solution $f(t) = c_1e^{-4t} + c_2e^{3t}$, with $f'(t) = -4c_1e^{-4t} + 3c_2e^{3t}$
Plug in: $0 = f(0) = c_1 + c_2$ and $0 = f'(0) = -4c_1 + 3c_2$, so that $c_1 = c_2 = 0$ and $f(t) = 0$.

29. General solution $f(t) = c_1\cos(2t) + c_2\sin(2t) + \frac{1}{3}\sin(t)$, so that $f'(t) = -2c_1\sin(2t) + 2c_2\cos(2t) + \frac{1}{3}\cos(t)$ (use the approach outlined in Exercises 16 and 17)

Plug in: $0 = f(0) = c_1$ and $0 = f'(0) = 2c_2 + \frac{1}{3}$, so that $c_1 = 0$, $c_2 = -\frac{1}{6}$, and $f(t) = -\frac{1}{6}\sin(2t) + \frac{1}{3}\sin(t)$.

30. **a.** k is a positive constant that depends on the rate of cooling of the coffee (it varies with the material of the cup, for example).
A is the room temperature.

b. $T'(t) + kT(t) = kA$
Constant particular solution: $T_p(t) = A$

General solution of $T'(t) + kT(t) = 0$ is $T(t) = Ce^{-kt}$.

General solution of the original differential equation: $T(t) = Ce^{-kt} + A$

Plug in: $T_0 = T(0) = C + A$, so that $C = T_0 - A$ and $T(t) = (T_0 - A)e^{-kt} + A$.

31. $\frac{dv}{dt} + \frac{k}{m}v = g$

constant particular solution: $v_p = \frac{mg}{k}$

General solution of $\frac{dv}{dt} + \frac{k}{m}v = 0$ is $v(t) = Ce^{-\frac{k}{m}t}$.

General solution of the original differential equation: $v(t) = Ce^{-\frac{k}{m}t} + \frac{mg}{k}$

Plug in: $0 = v(0) = C + \frac{mg}{k}$, so that $C = -\frac{mg}{k}$ and $v(t) = \frac{mg}{k}\left(1 - e^{-\frac{k}{m}t}\right)$

$\lim_{t\to\infty} v(t) = \frac{mg}{k}$ (the "terminal velocity")

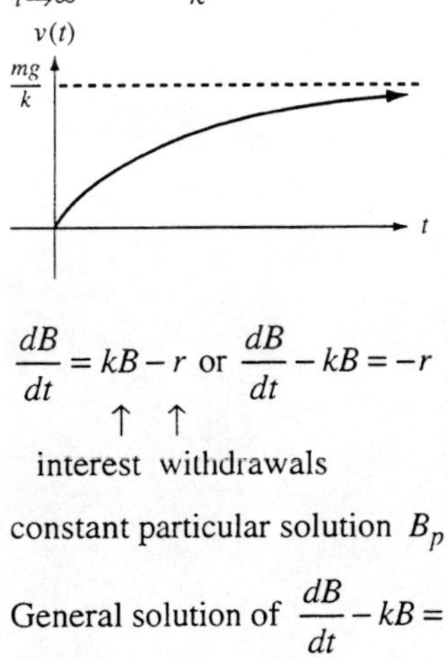

32. $\frac{dB}{dt} = kB - r$ or $\frac{dB}{dt} - kB = -r$

↑ ↑
interest withdrawals

constant particular solution $B_p = \frac{r}{k}$

General solution of $\frac{dB}{dt} - kB = 0$ is $B(t) = Ce^{kt}$

General solution of the original differential equation: $B(t) = Ce^{kt} + \frac{r}{k}$

Plug in: $B_0 = B(0) = C + \frac{r}{k}$, so that $C = B_0 - \frac{r}{k}$ and $B(t) = \left(B_0 - \frac{r}{k}\right)e^{kt} + \frac{r}{k}$

if $B_0 > \frac{r}{k}$ then interest will exceed withdrawals and balance will grow.

if $B_0 < \frac{r}{k}$ then withdrawals will exceed interest and account will be depleted.

if $B_0 = \frac{r}{k}$ then the balance will remain the same.

The graphs below show the three possible scenarios.

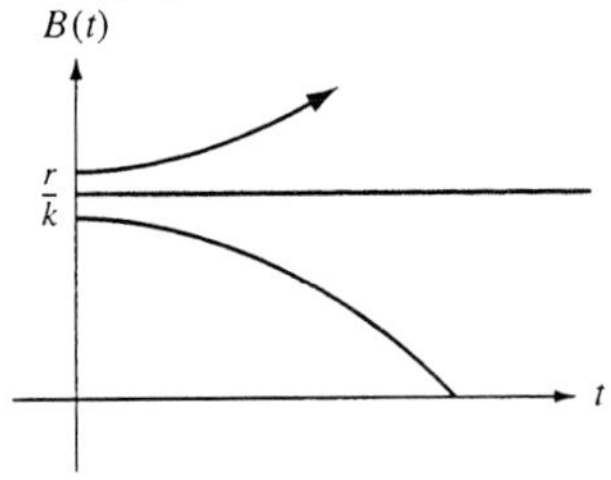

33. By Fact 9.4.9 $x(t) = c_1 \cos\left(\sqrt{\frac{g}{L}}t\right) + c_2 \sin\left(\sqrt{\frac{g}{L}}t\right)$, with period $P = \frac{2\pi}{\sqrt{\frac{g}{L}}} = 2\pi\frac{\sqrt{L}}{\sqrt{g}}$. It is required that

$2 = P = 2\pi\frac{\sqrt{L}}{\sqrt{g}}$ or $L = \frac{g}{\pi^2} \approx 0.994$ (meters).

34. a. We will take downward forces as positive.
Let g = acceleration due to gravity,
ρ = density of block
a = length of edge of block
Then (weight of block) = (mass of block) $\cdot g$ = (density of block)(volume of block)$g = \rho a^3 g$
buoyancy = (weight of displaced water) = (mass of displaced water) $\cdot g$
= (density of water)(volume of displaced water)$g = 1a^2 x(t) g = a^2 g x(t)$

b. Newton's Second Law of Motion tells us that

$m\frac{d^2x}{dt^2} = F = \text{weight} - \text{buoyancy} = \rho a^3 g - a^2 g x(t)$, where $m = \rho a^3$ is the mass of the block.

$$\rho a^3 \frac{d^2x}{dt^2} = \rho a^3 g - a^2 g x(t)$$

$$\frac{d^2x}{dt^2} = g - \frac{g}{\rho a}x(t)$$

$$\frac{d^2x}{dt^2} + \frac{g}{\rho a}x = g$$

constant solution $x_p = \rho a$

general solution (use Fact 9.4.9): $x(t) = c_1 \cos\left(\sqrt{\frac{g}{\rho a}}t\right) + c_2 \sin\left(\sqrt{\frac{g}{\rho a}}t\right) + \rho a$

Now $c_2 = 0$ since block is at rest at $t = 0$.
Plug in: $a = x(0) = c_1 + \rho a$, so that $c_1 = a - \rho a$ and

$x(t) = (a - \rho a)\cos\left(\sqrt{\frac{g}{\rho a}}t\right) + \rho a \approx 2\cos(11t) + 8$ (measured in centimeters)

c. The period is $P = \frac{2\pi}{\sqrt{\frac{g}{\rho a}}} = \frac{2\pi\sqrt{\rho a}}{\sqrt{g}}$. Thus the period increases as ρ or a increases (denser wood or larger block), or as g decreases (on the moon). The period is independent of the initial state.

35. a. $p_T(\lambda) = \lambda^2 + 3\lambda + 2 = (\lambda+1)(\lambda+2) = 0$ with roots $\lambda_1 = -1$ and $\lambda_2 = -2$, so $x(t) = c_1 e^{-t} + c_2 e^{-2t}$.

b. $x''(t) = -c_1 e^{-t} - 2c_2 e^{-2t}$

Plug in: $1 = x(0) = c_1 + c_2$ and $0 = x'(0) = -c_1 - 2c_2$, so that $c_1 = 2$, $c_2 = -1$ and $x(t) = 2e^{-t} - e^{-2t}$.

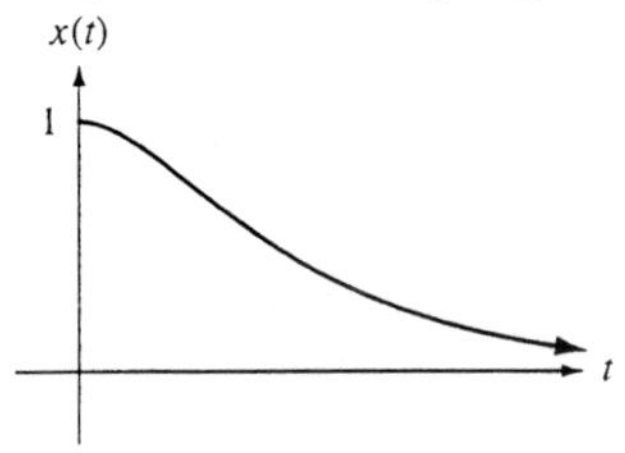

c. Plug in: $1 = x(0) = c_1 + c_2$ and $-3 = x'(0) = -c_1 - 2c_2$, so that $c_1 = -1$, $c_2 = 2$, and $x(t) = -e^{-t} + 2e^{-2t}$.

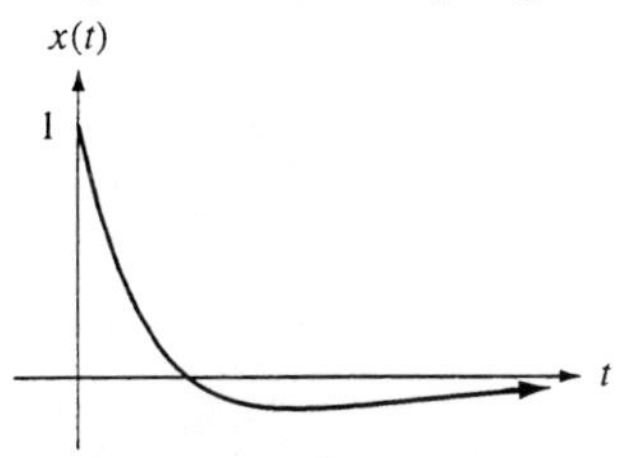

d. The oscillator in part (b) never reaches the equilibrium, while the oscillator in part (c) goes through the equilibrium once.

36. $f_T(\lambda) = \lambda^2 + 2\lambda + 101 = 0$ has roots $\lambda_{1,2} = -1 \pm 20i$.

By Fact 9.4.9, $x(t) = e^{-t}(c_1 \cos(20t) + c_2 \sin(20t))$.

Any nonzero solution goes through the equilibrium infinitely many times.

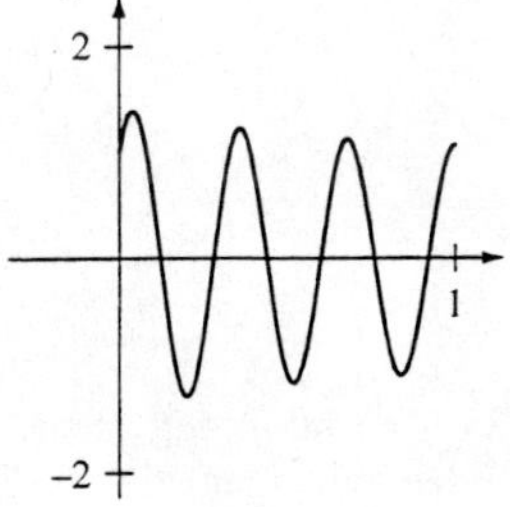

37. $f_T(\lambda) = \lambda^2 + 6\lambda + 9 = (\lambda + 3)^2$ has roots $\lambda_{1,2} = -3$.

Following the method of Example 10, we find the general solution $x(t) = e^{-3t}(c_1 + c_2 t)$ with $x'(t) = e^{-3t}(c_2 - 3c_1 - 3c_2 t)$.

Plug in: $0 = x(0) = c_1$, and $1 = x'(0) = c_2 - 3c_1$, so that $c_1 = 0$, $c_2 = 1$, and $x(t) = te^{-3t}$.

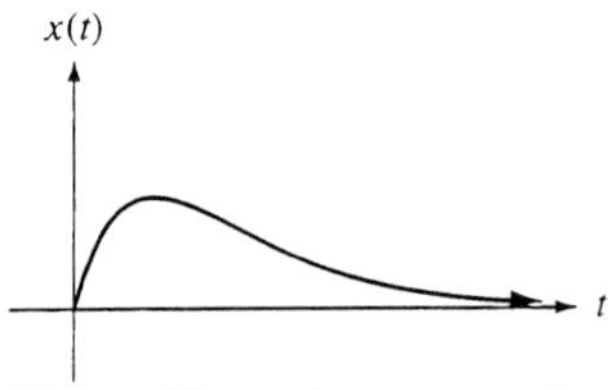

The oscillator does not go through the equilibrium at $t > 0$.

38. a. $(D - \lambda)(p(t)e^{\lambda t}) = [p(t)e^{\lambda t}]' - \lambda p(t)e^{\lambda t} = p'(t)e^{\lambda t} + \lambda p(t)e^{\lambda t} - \lambda p(t)e^{\lambda t} = p'(t)e^{\lambda t}$, as claimed.

b. Applying the result from part (a) m times we find $(D - \lambda)^m (p(t)e^{\lambda t}) = p^{(m)}(t)e^{\lambda t} = 0$, since $p^{(m)}(t) = 0$ for a polynomial of degree less than m.

c. By Fact 9.4.3, we are looking for m linearly independent functions. By part (b), the functions $e^{\lambda t}, te^{\lambda t}, t^2 e^{\lambda t}, \ldots, t^{m-1}e^{\lambda t}$ do the job.

d. Note that the kernel of $(D - \lambda_i)^{m_i}$ is contained in the kernel of $(D - \lambda_1)^{m_1} \cdots (D - \lambda_r)^{m_r}$, for any $1 \le i \le r$. Therefore, we have the following basis:

$e^{\lambda_1 t}, te^{\lambda_1 t}, \ldots, t^{m_1 - 1}e^{\lambda_1 t}$,

$e^{\lambda_2 t}, te^{\lambda_2 t}, \ldots, t^{m_2 - 1}e^{\lambda_2 t}$,

$\ldots$

$e^{\lambda_r t}, te^{\lambda_r t}, \ldots, t^{m_1 - 1}e^{\lambda_r t}$.

39. $f_T(\lambda) = \lambda^3 + 3\lambda^2 + 3\lambda + 1 = (\lambda + 1)^3 = 0$ has roots $\lambda_{1,2,3} = -1$. In other words, we can write the differential equation as $(D + 1)^3 f = 0$.

By Exercise 38, part (c), the general solution is $f(t) = e^{-t}(c_1 + c_2 t + c_3 t^2)$.

40. $f_T(\lambda) = \lambda^3 + \lambda^2 - \lambda - 1 = (\lambda + 1)^2(\lambda - 1) = 0$ has roots $\lambda_{1,2} = -1$, $\lambda_3 = 1$.

In other words, we can write the differential equation as $(D + 1)^2(D - 1) = 0$.

By Exercise 38, part (d), the general solution is $x(t) = e^{-t}(c_1 + c_2 t) + c_3 e^t$.

41. We are looking for functions x such that $T(x) = \lambda x$, or $T(x) - \lambda x = 0$. Now $T(x) - \lambda x$ is an nth-order linear differential operator, so that its kernel is n-dimensional, by Fact 9.4.3. Thus λ is indeed an eigenvalue of T, with an n-dimensional eigenspace.

42. a. We need to solve the second-order differential equation $Tx = D^2x = \dfrac{d^2x}{dt^2} = \lambda x$. This differential equation has a two-dimensional solution space E_λ for any λ, so that all λ are eigenvalues of T.

if $\lambda > 0$ then $E_\lambda = \text{span}\left(e^{\sqrt{\lambda}t}, e^{-\sqrt{\lambda}t}\right)$

if $\lambda = 0$ then $E_\lambda = \text{span}(1, t)$

if $\lambda < 0$ then $E_\lambda = \text{span}\left(\sin\left(\sqrt{-\lambda}t\right), \cos\left(\sqrt{-\lambda}t\right)\right)$

b. Among the eigenfunctions of T we found in part (a), we seek those of period 1. In the case $\lambda < 0$ the period is $P = \dfrac{2\pi}{\sqrt{-\lambda}}$, and $P = 1$ if $\lambda = -4\pi^2$. Thus $\lambda = -4\pi^2$ is an eigenvalue of L, with $E_\lambda = \text{span}(\sin(2\pi t), \cos(2\pi t))$.
In the case $\lambda > 0$ there are no periodic solutions. In the case $\lambda = 0$ we have the constant solutions, so that $\lambda = 0$ is an eigenvalue with $E_0 = \text{span}(1)$.
Summary: There are two eigenvalues:
$\lambda = -4\pi^2$ with $E_\lambda = \text{span}(\cos 2\pi, \sin 2\pi)$
$\lambda = 0$ with $E_\lambda = \text{span}(1)$

43. a. Using the approach of Exercises 16 and 17 we find $x(t) = c_1e^{-2t} + c_2e^{-3t} + \dfrac{1}{10}\cos t + \dfrac{1}{10}\sin t$.

b. For large t, $x(t) \approx \dfrac{1}{10}\cos t + \dfrac{1}{10}\sin t$.

44. a. Using the approach of Exercises 16 and 17 we find $x(t) = e^{-2t}(c_1 \cos t + c_2 \sin t) - \dfrac{1}{40}\cos(3t) + \dfrac{3}{40}\sin(3t)$.

b. For large t, $x(t) \approx -\dfrac{1}{40}\cos(3t) + \dfrac{3}{40}\sin(3t)$.

45. We can write the system as $\left|\begin{array}{l|l} \dfrac{dx_1}{dt} = x_1 + 2x_2 & x_1(0) = 1 \\ \dfrac{dx_2}{dt} = x^2 & x_2(0) = -1 \end{array}\right|$.

The solution of the second equation, with the given initial value, is $x_2(t) = -e^t$.

Now the first equation takes the form $\dfrac{dx_1}{dt} - x_1 = -2e^t$.

Using Example 9 (with $a = 1$ and $c = -2$) we find $x_1(t) = e^t(-2t + C)$.

plug in: $1 = x_1(0) = C$, so that $x_1(t) = e^t(1 - 2t)$ and $\vec{x}(t) = e^t\begin{bmatrix} 1-2t \\ -1 \end{bmatrix}$.

46. We can write the system as $\left|\begin{array}{l} \frac{dx_1}{dt} = 2x_1 + 3x_2 + x_3 \\ \frac{dx_2}{dt} = \quad x_2 + 2x_3 \\ \frac{dx_3}{dt} = \quad x_3 \end{array}\right| \begin{array}{l} x_3(0) = 2 \\ x_1(0) = 1 \\ x_3(0) = -1 \end{array}$.

We solve for x_2 and x_3 as in Exercise 45:

$x_2(t) = e^t(1-2t)$

$x_3(t) = -e^t$

Now the first equation takes the form $\frac{dx_1}{dt} - 2x_1 = 3e^t(1-2t) - e^t = e^t(2-6t),\ x_1(0) = 2.$

We use Fact 9.4.13 to solve this differential equation:

$x_1(t) = e^{2t}\int e^{-2t}e^t(2-6t)dt = e^{2t}\int(2e^{-t} - 6te^{-t})dt = e^{2t}[-2e^{-t} + 6te^{-t} + 6e^{-t} + C]$

plug in: $2 = x_1(0) = (-2+6+c)$, so that $c = -2$ and $x_1(t) = e^{2t}(4e^{-t} + 6te^{-t} - 2) = 4e^t + 6te^t - 2e^{2t}$.

$$\vec{x}(t) = \begin{bmatrix} 4e^t + 6te^t - 2e^{2t} \\ e^t - 2te^t \\ -e^t \end{bmatrix}$$

47. a. We start with a *preliminary remark* that will be useful below: If $f(t) = p(t)e^{\lambda t}$, where $p(t)$ is a polynomial, then $f(t)$ has an antiderivative of the form $q(t)e^{\lambda t}$, where $q(t)$ is another polynomial. We leave this remark as a calculus exercise.

The function $x_n(t)$ satisfies the differential equation $\frac{dx_n}{dt} = a_{nn}x_n$, so that $x_n = C\cdot e^{annt}$, which is of the desired form.

Now we will show that x_k is of the desired form, assuming that $x_{k+1}, \ldots, x_n$ have this form. x_k satisfies the differential equation $\frac{dx_k}{dt} = a_{kk}x_k + a_{k,\,k+1}x_{k+1} + \cdots + a_{kn}x_n$ or $\frac{dx_k}{dt} - a_{kk}x_k = a_{k,\,k+1}x_{k+1} + \cdots + a_{kn}x_n$.

Note that by assumption of the function on the right-hand side has the form $p_1(t)e^{\lambda_1 t} + \cdots + p_m(t)e^{\lambda mt}$. If we set $a_{kk} = a$ for simplicity, we can write $\frac{dx_k}{dt} - ax_k = p_1(t)e^{\lambda_1 t} + \cdots + p_m(t)e^{\lambda mt}$.

By Fact 9.4.13, the solution is

$x_k(t) = e^{at}\int e^{-at}(p_1(t)e^{\lambda_1 t} + \cdots + p_m(t)e^{\lambda_m t})dt = e^{at}\int(p_1(t)e^{(\lambda_1 - a)t} + \cdots + p_m(t)e^{(\lambda_m - a)t})dt$

$= e^{at}(q_1(t)e^{(\lambda_1 - a)} + \cdots + q_m(t)e^{(\lambda m - a)t} + C) = q_1(t)e^{\lambda_1 t} + \cdots + q_m(t)e^{\lambda_m t} + Ce^{at}$ as claimed (note that a is one of the λ_i). The constant C is determined by $x_k(0)$. Note that we used the *preliminary remark* in the second to last step.

b. It is shown in introductory calculus classes that $\lim_{t\to\infty}(t^m e^{\lambda t}) = 0$ if and only if λ is negative (here m is a fixed positive integer). In light of part (a), this proves the claim.

48. There is an invertible S such that $S^{-1}AS = B$ is upper triangular. Then the system $\frac{d\vec{x}}{dt} = A\vec{x} = SBS^{-1}\vec{x}$ can be written as $\frac{d}{dt}(S^{-1}\vec{x}) = B(S^{-1}\vec{x})$ or $\frac{d\vec{u}}{dt} = B\vec{u}$, where $\vec{u} = S^{-1}\vec{x}$. Note that B has the m distinct diagonal entries $\lambda_1, \ldots, \lambda_m$.

a. By Exercise 47, the system $\frac{d\vec{u}}{dt} = B\vec{u}$ has a unique solution $\vec{u}(t)$. Then the system $\frac{d\vec{x}}{dt} = A\vec{x}$ has the unique solution $\vec{x}(t) = S\vec{u}(t)$.

b. It suffices to note that $\lim_{t\to\infty} \vec{x}(t) = \vec{0}$ if and only if $\lim_{t\to\infty} \vec{u}(t) = \vec{0}$, where $\vec{u} = S^{-1}\vec{x}$.